ALLE ZEIT WACH
SJ
1842

Hans Eschenauer · Juhani Koski
Andrzej Osyczka (Editors)

Multicriteria Design Optimization

Procedures and Applications

With 171 Figures

Springer-Verlag Berlin Heidelberg NewYork
London Paris Tokyo Hong Kong Barcelona

Prof. Dr.-Ing. Hans Eschenauer
University of Siegen
Research Laboratory for Structural Optimization
at the Institute of Mechanics and Control Engineering
D-5900 Siegen
Germany

Assoc. Prof. Dr. Eng. Juhani Koski
Dept. of Mechanical Engineering
Tampere University of Technology
P.O. Box 527
SF-33101 Tampere
Finland

Assoc. Prof. Dr. hab. inż. Andrzej Osyczka
Technical University of Cracow
Institute of Machine Technology
PL-31-155 Cracow
Poland

ISBN 3-540-50604-7 Springer-Verlag Berlin Heidelberg NewYork
ISBN 0-387-50604-7 Springer-Verlag NewYork Berlin Heidelberg

Library of Congress Cataloging-in-Publication Data
Multicriteria design optimization : procedures and applications
[edited by] Hans Eschenauer, Juhani Koski, Andrzej Osyczka.
Includes bibliographical references and indexes.
ISBN 3-540-50604-7. -- ISBN 0-387-50604-7
1. Engineering design--Mathematical models. 2. Mathematical optimization.
I. Eschenauer, Hans. II. Koski, Juhani. III. Osyczka, Andrzej. TA174.M85 1990
620'.00425--dc20 90-9931

Offsetprinting: Mercedes-Druck, Berlin; Bookbinding: Lüderitz & Bauer, Berlin
2161/3020-543210 – Printed on acid-free paper

To Gerda

To Anu

To Laura

PREFACE

The modern era of design optimization began about twenty years ago with the recognition of the usefulness of mathematical programming techniques. Methods based on mathematical programming were first adapted to single-criterion optimum design problems. Now, more attention is given to multicriteria modelling, as in many engineering applications often several conflicting criteria have to be considered by the designer.

Even though multicriteria optimization goes back as far as V. Pareto's study in 1898, a greater interest in such fields as optimization theory, operations research, and control theory was not aroused until the late 1960s. Since that time numerous studies on this topic have been published. Most of them deal with the theory of decision making from a general point of view whereas only a relatively small number of publications can be found in the field of engineering design. Thus, the aim of this book is to fill this gap and to provide the designer with a new tool for solving the optimization problems in which several conflicting and noncommensurable criteria are to be satisfied. In order to get a representative survey of the current works, the editors asked for contributions from some leading researchers so that a broad range of applications could be gathered in a coherent volume.

In order to introduce the subject to the readers, Chapter 1 outlines the background of multicriteria optimization, broadly describes the relevant mathematical procedures, and also shows some real-life examples which motivate the designer to apply multicriteria techniques.

The first part of this volume (Chapters 2-4) deals with multicriteria optimization procedures. Chapter 2 presents the optimization procedure SAPOP which provides the designer with a general tool for solving structural optimum design problems. Most activities in multicriteria design optimization concentrate on the application of interactive procedures. Chapter 3 outlines these procedures irrespective of their role in the design process and also describes two software packages which facilitate the interactive processes for optimum design. Knowledge-based systems recently aroused great interest. Their use in multicriteria design optimization is described in Chapter 4.

The second part of this volume is devoted to the application of multicriteria techniques to different design problems which are divided into subject groups. The first group deals with mechanisms and dynamic systems. Here, Chapters 5.1 and 5.2 are devoted to the problem of optimal balancing of robot arms using counterweights and spring mechanisms. For the optimum design of spring balancing mechanisms , a general method for dealing with computationally expensive objective functions has been proposed. Optimization of automotive drive train and multibody systems are discussed in Chapters 5.3 and 5.4. Chapter 5.5 shows a special method for finding a relationship between FEM-analysis and optimization procedures using regression models. The second subject group explores aircraft and space technology problems. In Chapter 6.1 multicriteria optimal layouts of aircraft and spacecraft structures are discussed whereas Chapter 6.2 presents poblems of spacecraft structures with emphasis on mass and stiffness. Multicriteria optimization of machine tool systems is the subject of the third group. In Chapter 7.1 design problems of machine tool structures are presented, and in Chapter 7.2 the optimum design of machine tool spindle systems using a decomposition method is discussed. The fourth subject group deals with metal forming and cast metal technology. In Chapter 8.1, a multicriteria optimal control approach is applied to die designs for symmetric strip drawing. Optimal layouts of heterogeneous thick-walled, chilled cast-iron rollers are presented in Chapter 8.2, and a metal forming process is optimized and simulated in Chapter 8.3. Problems of civil and architectural engineering are considered in two chapters. Chapter 9.1 presents the multicriteria optimization of concrete beams, trusses, and cable structures, and in Chapter 9.2 multicriteria optimization techniques are applied to architectural planning. Finally, the optimization of structures made of advanced materials is discussed. Chapters 10.1, 10.2, and 10.4 deal with fibre-reinforced plate and shell structures and ceramic components, respectively. In Chapter 10.3 multicriteria optimization and advanced materials in telescope design are presented.

The editors wish to express their appreciation to all authors for their contributions and their cooperation in revising the chapters. We are especially grateful to Ms Ursula Schmitz (Stud.Ass.) who has performed the type-setting of the book with great skill and efficiency. She has also assisted as a translation editor for all chapters and tried to meet the editors' requirements with much care and patience. In preparing and organizing the publishing process, she did a splendid job. We would also like to express our sincere thanks to Ms Birgit Hollstein and Mr Michael Wengenroth for supervising the work on the book in its final phase. Thanks are also due to Ms Petra Franke, Ms Regina Knepper and Ms Henrike Strohbücker who have done the drawing of figures.

The editors wish to express their special thanks and appreciation to Dr. R.D. Parbery (University of Newcastle/Australia) for carefully proofreading the typescript. On this occasion, Dr. Parbery would like to thank the German Research Community (DFG) for sponsoring his stay as a visiting professor at the University of Siegen.

The DFG is also owed a debt of gratitude for its sponsorship of Professor A. Osyczka's eight-month-stay at the University of Siegen where he did the main part of his work on the book.

Finally, we are indebted to Ms. E. Raufelder and Mr. A. von Hagen of Springer Publishing Company, Heidelberg for the excellent cooperation.

Hans A. Eschenauer   Juhani Koski   Andrzej Osyczka

Siegen, Tampere, Cracow   March 1990

CONTENTS

List of Contributors XXI

List of Symbols XXIII

1 Multicriteria Optimization– Fundamentals and Motivation . . 1

H.A. Eschenauer, J. Koski, A. Osyczka

1.1 Introduction 1

1.1.1 On the Historical Development of Optimization Techniques 1

1.1.2 Necessity of Applying Optimization Techniques to the Design Process 3

1.1.3 Multicriteria Optimization as a Strategy in the Design Process 4

1.2 Mathematical Fundamentals 6

1.2.1 General Definitions and Notations in Scalar Optimization 6

1.2.2 The Multicriteria Programming Problem 9

1.2.3 The Multicriteria Control Problem 19

1.3 Components and Plants with their Objectives 20

1.3.1 Optimum Design of Highly Accurate Parabolic Antennas 20

1.3.2 Optimal Layout of a Novel Solar Energy Collector 23

1.3.3 Shape Optimization of Satellite Tanks 24

1.3.4 Optimal Shape Design 25

1.3.5 Optimal Layout of Tube-Flange Structures 26

1.4 Conclusion 28

References 29

Part I Procedures

2 Optimization Procedure SAPOP–A General Tool for Multicriteria Structural Designs 35

M. Bremicker, H.A. Eschenauer, P.U. Post

2.1 Demands on an Optimization Procedure 35
2.2 Structure of the Optimization Procedure 36
2.2.1 Definitions 36
2.2.2 Three-Columns Concept 37
2.3 Basic Ideas of the Procedure SAPOP 40
2.3.1 Problem Formulation and Input Data 41
2.3.2 SAPOP Main Module 41
2.3.3 Evaluation of the Optimization Results 44
2.4 Optimization Modelling 44
2.4.1 Design Models 45
2.4.2 Evaluation Models 46
2.4.3 Sensitivity Analysis 47
2.5 Description of Two Optimization Algorithms 52
2.5.1 Sequential Linear Programming Method (SEQLI). . . 52
2.5.2 Hybrid Method of Quadratic Programming with Reduced Line-Search Technique (QPRLT). 55
2.6 Comparison with other Structural Optimization Software Systems 58
2.7 Application Example 58
2.8 Conclusion 65
References 66

3 Interactive Multicriteria Optimization in Design Process 71

H.A. Eschenauer, A. Osyczka, E. Schäfer

3.1 Introduction 71
3.2 Interactive Multicriteria Optimization Procedures 72
3.2.1 General Remarks on Classification. 72
3.2.2 Approach by Fandel 74
3.2.3 STEP-Method 78
3.2.4 Approach by Jahn 80
3.2.5 Approach by Geoffrion 83

3.3 Software Package DIALOG 86
3.3.1 Basic Structure 86
3.3.2 Interactive MO-Layout of a Conical Shell. 88
3.4 Software Package CAMOS 101
3.4.1 Optimization Algorithms Used in CAMOS 102
3.4.2 Multicriteria Strategy Approaches 104
3.4.3 Description of CAMOS 104
3.4.4 Interactive MO-Layouts of a Machine Tool Spindle 107
3.5 Conclusion 109
References 111

4 Knowledge Engineering and Multicriteria Optimization 115

M. Balachandran, J.S. Gero

4.1 Introduction 115
4.2 Knowledge Engineering 117
4.2.1 General Concept 117
4.2.2 Architecture of a Knowledge-Based System 118
4.3 Role of a Knowledge-Based Approach in Multicriteria Optimization 119
4.4 Knowledge-Based Optimization Formulation 120
4.4.1 Description and Representation of Optimization Problems 122
4.4.2 Mathematical Symbolic Manipulation 125
4.4.3 Mathematical Formulation of Optimization Problems 127
4.5 Knowledge-Based Optimization Control 133
4.5.1 Recognition of Optimization Formulation 133
4.5.2 Optimization Algorithm Selection 135
4.5.3 Knowledge-Based Control in Pareto-Optimal Set Generation 137
4.6 Illustration Example 140
4.6.1 System Implementation 140
4.6.2 Knowledge-Based Optimization Formulation and Recognition 141
4.6.3 Knowledge-Based Control in Pareto-Optimal Set Generation 142
4.7 Conclusion 145
References 146

Part II Applications

5 Mechanisms and Dynamic Systems . . . 151

5.1 Optimal Counterweight Balancing of Robot Arms Using Multicriteria Approach . . . 151

J. Koski, A. Osyczka

5.1.1 Introduction . . . 151
5.1.2 Kinetic Model of a Robot Arm . . . 152
5.1.3 Formulation of the Optimization Problem . . . 157
5.1.4 Solution Method . . . 160
5.1.5 Pareto-Optimal Designs . . . 162
5.1.6 Conclusion . . . 165
References . . . 166
Appendix . . . 167

5.2 Multicriteria Optimization of Computationally Expensive Functions and its Application to Robot Spring Balancing Mechanism Design . . . 168

A. Osyczka, J. Zajac

5.2.1 Introduction . . . 168
5.2.2 General Description of the Method . . . 169
5.2.3 Optimum Design of Spring Balancing Mechanisms of Industrial Robots . . . 174
5.2.4 Computer Aided Optimum Design of a Robot Spring Balancing Mechanism . . . 181
5.2.5 Conclusion . . . 182
References . . . 183

5.3 On the Optimal Synthesis of an Automotive Drive Train . . . 184

F. Pfeiffer

5.3.1 The Automotive Drive Train . . . 184
5.3.2 The Mechanical Model . . . 185
5.3.3 The Optimization Model . . . 187
5.3.4 The Solution Procedure . . . 188
5.3.5 Results . . . 190

5.3.6 Conclusion 191
References 192

5.4 Modelling of Multibody Systems by Means of Optimization Procedures 193

H.H. Müller-Slany

5.4.1 Adapted Models for Elasto-Mechanical Systems 193
5.4.2 Adaptation of the Model as Optimization Procedure . . 195
5.4.3 Example: Adapted Model for the Reflector of a Parabolic Antenna 199
5.4.4 Conclusion 203
References 204

5.5 Approximation Methods in Structural Optimization Using Experimental Designs for Multiple Responses 205

D.H. van Campen, R. Nagtegaal, A.J.G. Schoofs

5.5.1 Introduction 205
5.5.2 Experimental Designs in the Case of a Single Response Variable 207
5.5.3 Experimental Designs in the Case of Multiple Responses 213
5.5.4 Model Fitting and Testing 217
5.5.5 Computer Program for Experimental Design 218
5.5.6 Application 218
5.5.7 Conclusion 226
References 227

6 Aircraft and Space Technology 229

6.1 Multicriteria Optimal Layouts of Aircraft and Spacecraft Structures 229

G. Kneppe

6.1.1 Introduction 229
6.1.2 Optimization Procedure 230
6.1.3 Multicriteria Optimization of a Composite Fin 232

6.1.4 Heat Flux and Frequency Optimization of a Satellite Structure 238
6.1.5 Conclusion 242
References 243

6.2 Multicriteria Design of Spacecraft Structures with Special Emphasis on Mass and Stiffness 244

H. Baier

6.2.1 Introduction and Technical Background 244
6.2.2 Problem Statement and Discussion 245
6.2.3 Considerations on Optimality Criteria 248
6.2.4 Solution Strategies 250
6.2.5 Examples of Spacecraft Structural Optimization 252
6.2.6 Conclusion 258
References 259

7 Machine Tool Design 261

7.1. Application of Multicriteria Optimization Methods to Machine Tool Structural Design 261

M. Yoshimura

7.1.1 Evaluative Factors of Machine Tools 261
7.1.2 Competitive and Cooperative Relationships between Evaluative Factors 266
7.1.3 Design Optimization Procedures 271
7.1.4 Application Examples 274
7.1.5 Conclusion 279
References 280

7.2 Decomposition Method for Multicriteria Optimization and its Application to Machine Tool Spindle Design 282

J. Montusiewicz, A. Osyczka, J. Zamorski

7.2.1 Introduction 282
7.2.2 Four-Stage Multicriteria Optimization Strategy 283
7.2.3 Application of the Strategy to the Design of a Machine Tool Spindle System with Hydrostatic Bearings 291

7.2.4 Numerical Example 298
7.2.5 Conclusion 301
References 301

8 Metal Forming and Cast Metal Technology 303

8.1 Optimal Die Design for Symmetric Strip Drawing 303

W. Stadler

8.1.1. Introduction and Review .303
8.1.2 The Mathematical Model .304
8.1.3 The Ubiquitous Conical Die 308
8.1.4 Conclusion 317
References 317

8.2 Optimal Layouts of Chilled Cast-Iron High Quality Rollers 319

M. Bremicker, H. Eschenauer, H.-W. Wodtke

8.2.1 Introduction 319
8.2.2 Definition of the Optimization Problem 321
8.2.3 Structural Analysis 325
8.2.4 Realization of an Optimization Loop 331
8.2.5 Optimization Results 331
8.2.6 Conclusion 335
Acknowledgement 337
References 337

8.3 A Mechanical Model for the Optimization and Simulation of the Metal Forming Process in Roller Levelling of Sheets 339

L. Henrich, K. Schiffner

8.3.1 Introduction 339
8.3.2 Mechanical Model 340
8.3.3 Formulation of the Optimization Problem 344
8.3.4 Results and Discussion 346
References 352

9 Civil and Architectural Engineering 355

9.1 Multicriteria Optimization of Concrete Beams, Trusses, and Cable Structures 355

S. Jendo

9.1.1 Introduction 355
9.1.2 Optimization of a Prestressed Concrete I-Beam 357
9.1.3 Optimization of Isostatic Trusses 360
9.1.4 Optimization of Cable Structures 364
9.1.5 Further Applications of Muticriteria Optimization 371
References 372

9.2 Design Decision Making Using Pareto-Optimal Dynamic Programming 376

J.S. Gero, M.A. Rosenman

9.2.1 The Stage-State Model of Optimization 376
9.2.2 Dynamic Programming 378
9.2.3 The Computer System - SID 381
9.2.4 A Building Design Example Using SID 381
9.2.5 Conclusion 394
References 395

10 Structures Made of Advanced Materials 397

10.1 Fibre-Reinforced Plate and Shell Structures under Various Loads - Proposals for their Optimization 397

H.A. Eschenauer, W. Fuchs, P.U. Post

10.1.1 Introduction 397
10.1.2 Structural Modelling and Analysis 398
10.1.3 Objective Functions - Design Variables - Constraints 405
10.1.4 Examples of Application 407
10.1.5 Conclusion 414
References 415

10.2 Multicriteria Thermoelastic Design of Antisymmetric Angle-Ply Laminates 417

S. Adali, K.J. Duffy

10.2.1 Introduction 417
10.2.2 Thermoelastic Buckling and Bending Equations 418
10.2.3 Multicriteria Design Problem 420
10.2.4 Solution Method 421
10.2.5 Numerical Results 424
10.2.6 Conclusion 427
References 427

10.3 Multicriteria Optimization and Advanced Materials in Telescope Design 429

K.-H. Stenvers

10.3.1 Motivation and Problem Formulation 429
10.3.2 Program System for Structural Analysis and Optimization 430
10.3.3 Application to a Submillimeter-Telescope 433
10.3.4 Conclusion and Future Prospects 443
References 445

10.4 Multicriteria Design of Ceramic Components 447

J. Koski, R. Silvennoinen

10.4.1 Introduction 447
10.4.2 Structural Analysis and Reliability of Ceramic Components 448
10.4.3 Multicriteria Shape Optimization Problem 451
10.4.4 Pareto-Optima of Ceramic Cantilever 453
10.4.5 Design of a Ceramic Piston Crown 456
10.4.6 Conclusion 461
References 462

Author Index 465

Subject Index 469

LIST OF CONTRIBUTORS

S. Adali
Faculty of Engineering, Dept. of Mechanical Engineering, King George V Ave., Durban 4001, SOUTH AFRICA

H. Baier
Dornier-System GmbH, Dept. MEB, Postfach 1360, D 7990 Friedrichshafen 1, FRG

M. Balachandran
Dept. of Architectural Science, University of Sydney, NSW 2006, AUSTRALIA

M. Bremicker
University of Michigan, 2212 G.G. Brown Building, Ann Arbor, Mi 48109, USA

D.H. van Campen
Eindhoven University of Technology, Building WHoog-01.144, P.O. Box 513, 5600 MB Eindhoven, NETHERLANDS

H.A. Eschenauer
Research Laboratory for Applied Structural Optimization at the Institute of Mechanics and Control Engineering, University of Siegen, Paul-Bonatz-Str., D 5900 Siegen, FRG

W. Fuchs
Lemmerz-Werke KGaA, Dept. TEF, Postfach 1120, D 5330 Königswinter 1, FRG

J.S. Gero
Dept. of Architectural Science, University of Sydney, NSW 2006, AUSTRALIA

L. Henrich
Institute of Mechanics and Control Engineering, University of Siegen, Paul-Bonatz-Str., D 5900 Siegen, FRG

S. Jendo
Polish Academy of Sciences, Institute of Fundamental Technological Research, 00-049 Warszawa, Swietokrzyska 21, POLAND

G. Kneppe
MBB Hubschrauber und Flugzeuge, CAE/Informationssysteme LWD 132, Postfach 801160, D 8000 München 80, FRG

J. Koski
Tampere University of Technology, P.O. Box 527, 33101 Tampere, FINLAND

J. Montusiewicz
Technical University of Lublin, 20-618 Lublin, ul. Nadbystrzycka 38, POLAND

H.H. Müller-Slany
Research Laboratory for Applied Structural Optimization, University of Siegen, Paul-Bonatz-Str., D 5900 Siegen, FRG
R. Nagtegaal
Eindhoven University of Technology, Building WHoog-01.144, P.O. Box 513, 5600 Eindhoven, NETHERLANDS
A. Osyczka
Technical University of Cracow, 31-155 Cracow, ul. Warszawska 24, POLAND
F. Pfeiffer
Institute of Mechanics, TU München, Postfach, D 8000 München 2, FRG
P.U. Post
Festo KG, Dept. EF-BE, Ruiterstr. 82, D 7300 Esslingen 1, FRG
M.A. Rosenman
Dept. of Architectural Science, University of Sydney, NSW 2006, AUSTRALIA
R. Silvennoinen
Tampere University of Technology, P.O. Box 527, 33101 Tampere, FINLAND
E. Schäfer
Research Laboratory for Applied Structural Optimization, University of Siegen, Paul-Bonatz-Str., D 5900 Siegen, FRG
K. Schiffner
Institute of Mechanics and Control Engineering, University of Siegen, Paul-Bonatz-Str., D 5900 Siegen, FRG
A.J.G. Schoofs
Eindhoven University of Technology, Building W-Hoog-01.144, P.O. Box 513, 5600 MB Eindhoven, NETHERLANDS
W. Stadler
San Francisco State University, Division of Engineering, 1600 Holloway Ave., San Francisco, California 94132, USA
K.H. Stenvers
Krupp Industrietechnik, GmbH, Dept. Systems Technology, Postfach 141960, D 4100 Duisburg 14, FRG
H.W. Wodtke
Research Laboratory for Applied Structural Optimization, University of Siegen, Paul-Bonatz-Str., D 5900 Siegen, FRG
M. Yoshimura
Dept. of Precision Engineering, Kyoto University, Sakyo-ku, Kyoto 606, JAPAN
J. Zamorski
Technical University of Cracow, 31-155 Cracow, ul. Warszawska 24, POLAND
J. Zajac
Technical University of Cracow, 31-155 Cracow, ul. Warszawska 24, POLAND

LIST OF SYMBOLS

Note: The following list is restricted to the most important subscripts, notations and letters in the book. Further terms are explained in the text.
Scalar quantities are printed in roman letters, vectors in boldface, tensors or matrices in capital letters and in boldface.

1. Subscripts and Notations

$*$	superscript asterisk indicates an optimal value
$\nabla_{\mathbf{x}}$	Nabla operator with respect to a vector **x**
Δ	difference
$a := b$	quantity a is defined by quantity b
$a := b,c$	quantity a is valid under the assumptions b,c
$\forall a : A$	all elements a have the attribute A
$a \overset{!}{=} b$	quantity a equals quantity b
$A \Rightarrow B$	A is a sufficient condition for B
$x \in M$	x is an element of set M
$x \notin M$	x is not an element of set M
$M \subset N$	set M is a subset of set N
$M \cup N$	union of set M and set N
$M \cap N$	intersection of set M and set N
$\lVert \mathbf{x} \rVert$	Euclidean norm of a vector **x**
$\lvert x \rvert$	absolute value of a scalar x
$(\tilde{\ })$	substitute problem
$(\tilde{\ })$	initial deformation state
$(\hat{\ })$	subspace
$(\hat{\ })$	total deformation state
$(\underset{\sim}{\ })$	stochastic value
$(\ )\vert$	covariant derivative

2. Latin Letters

A	1) coordination matrix
	2) matrix of inertia terms
$\mathbf{A_f}$	sensitivity matrix
A	concrete area

$\mathbf{a}$	vector of acceleration
$a_{f_{ij}}$	coefficients of the sensitivity matrix
$\mathbf{B}$	1) pay-off table matrix
	2) matrix of centripetal and Coriolis terms
$\mathbf{b}$	systematic departure (bias) from the true physical relationship
$\mathbf{C}_i$	i-th transfer matrix
C^{ijkl}	elasticity tensor
$\mathbf{c}$	1) vector of constant parameters
	2) vector of gravitational terms
c	moisture concentration
$\mathbf{D}$	damping matrix
$\mathbf{d}^b$, $\mathbf{d}^{nb}$	search directions for basic and nonbasic variables
E	1) expected value
	2) Young's modulus
$\mathbf{e}_i$	i-th unit vector
$\underset{\sim}{e}$	random error
$\mathbf{F}$	system matrix
F	1) control constraint set
	2) force
	3) nonempty set of admissible controls
f, $\mathbf{f}$, f_j	objective function, objective function vector, j-th coordinate function of the vector $\mathbf{f}$ ($j=1, \ldots, m$)
$f(\cdot)$	criterion map
$\bar{f}_j$	individual minimum of the j-th objective function
G, G_O	reduced gain, gain of an ideal parabolic antenna
$\mathbf{g}$, g_j	vector of inequality constraints, j-th inequality constraint ($j=1, \ldots, p$)
$g(\cdot)$	velocity function
$g(\mathbf{x}, \beta)$	interpolating function
$\mathbf{H}$	Hessian matrix
H	1) Hamiltonian functional
	2) unit step function
$\mathbf{h}$, h_i	vector of equality constraints, i-th equality constraint ($i=1, \ldots, q$)
$\mathbf{I}$	unity tensor, unity matrix
$\mathbf{J}$	Jacobian matrix
$\mathbf{K}$	stiffness matrix

L	Lagrangian function
M, $\mathbf{M}$	mass, mass matrix
$\mathbf{m}$	vector of torques
$\mathbf{P}$	pseudo-load matrix
P_f	probability of failure
$\mathbf{p}_A$	load vector for aeroelastics
$p[\mathbf{f}(\mathbf{x})]$	preference function, substitute objective function or quality criterion
$\mathbf{Q}$	1) aerodynamic influence coefficient matrix
	2) stiffness matrix reduced to plane stresses
$\mathbf{q}$	vector of degrees of freedom
$\mathbb{R}$	set of all real numbers
$\mathbb{R}^n$	n-dimensional Euclidean vector space
R	receptance
$\mathbf{r}$	1) load vector
	2) vector of random or uncontrollable variables
	3) excitation vector
$\mathbf{S}$	sensitivity matrix
S	1) convergence certainty
	2) prestressing steel area
S_i	system quality
S^2	variances
$\mathbf{s}_k$	search direction vector of the k-th step
$\mathbf{T}$	transformation matrix
T	total kinetic energy
t_i	layer thickness
U	1) control constraint
	2) utility function
$U_\varepsilon(\mathbf{x}^*)$	ε-neighbourhood of the point $\mathbf{x}^*$
$\mathbf{u}$	1) state variable vector
	2) control vector
V	1) total potential energy
	2) variance
	3) volume
$\mathbf{v}$	1) vector of parametric variables
	2) displacement vector
v_j	weighting coefficients (Ch. 3)
W	1) work of the applied forces, compliance
	2) weight

$\mathbf{w}$, w_j	vector of the weighting coefficients, j-th weighting coefficient or factor
$\mathbf{X}$	design matrix
X	domain of definition, feasible domain
X^*	complete Pareto-optimal solution set
$\mathbf{x}$, x_i	vector of $\mathbb{R}^n$, design variable vector, i-th design variable
$\mathbf{x}$	1) vector of input variables
	2) vector of coded or standardized input quantities
$\tilde{\mathbf{x}}$	compromise (suitable) solution
$\mathbf{x}_0$	starting vector, initial design vector
$\mathbf{x}^*$	Pareto-optimal vector, efficiency vector, functional-efficient vector
$\mathbf{x}^l$, $\mathbf{x}^u$	lower and upper limit of **x**, bounds
x_{n+j}	slack variables
Y	mapping of the set X into criterion space
$\hat{Y}^i$	subspace
∂Y^*	functional-efficient boundary of the set Y, minimal set
$\mathbf{y}$	1) analysis variables from linear transformation
	2) vector of output or response variables
$\underline{\mathbf{y}}$	point of the functional-efficient boundary, minimal point
$\bar{\mathbf{y}}$, $\bar{y}_j$	vector of the demand level, j-th demand level
Z_ν	set of feasible solutions
$\mathbf{z}$	state vector
z_j	relative deviation of the j-th objective function

3. Greek Letters

$\boldsymbol{\alpha}$, $\boldsymbol{\beta}$, $\boldsymbol{\lambda}$	vector of Lagrangian multipliers
α	1) step length
	2) level of significance
α^k	correction values
α_i	fibre angles
α_T	thermal expansion coefficient
$\alpha_{\alpha\beta}$	strains of the middle surface
$\boldsymbol{\beta}$	vector of unknown coefficients
β	scalar preference function and auxiliary design variable
$\beta_{\alpha\beta}$	curvature of the middle surface
β_T^{kl}	coefficients of the thermal stress tensor
β_C^{kl}	coefficients of the hygrothermal stress tensor

$\mathbf{\Gamma}$	functional space
Γ_i	quadratic penalty function
γ_{ij}	coefficients of the strain tensor
$\gamma_{\alpha 3}$	coefficients of the shear strain tensor
$\boldsymbol{\delta}$	direction of minimization step
ε	physical strain
ζ	damping ratio
η	1) aeroelastic efficiency
	2) surface efficiency
$\mathbf{\Theta}$	vector of physical constants
Θ	temperature distribution
$\varkappa_{ak}$	residual curvature
$\mathbf{\Lambda}$	adjoint variable matrix
λ	wavelength
λ^k	step length
ν	Poisson's ratio
$\boldsymbol{\xi}$	vector of input quantities
ρ	mass density
σ	standard deviation, root mean square value
σ_{mn}	physical stresses
τ^{ij}	stress tensor
$\mathbf{\Phi}$	modal matrix
Φ	heat flow (flux)
φ_F	fibre volume ratio
φ_i	eigenmode
Ψ_m	modal flexibility
Ψ_s	static compliance
ω_i	eigenfrequency, natural frequency

1 MULTICRITERIA OPTIMIZATION – FUNDAMENTALS AND MOTIVATION

H.A. Eschenauer, J. Koski, A. Osyczka

1.1 Introduction

1.1.1 On the Historical Development of Optimization Techniques

A well-known statement of the energy principles says that among all possible displacements the actual displacements make the total potential energy an absolute minimum. This means that the application of the principle of minimum potential energy leads to the fundamental equations of the boundary value problem in the theory of elasticity. The principles of mechanics go back to the 17th century. They allowed the formulation of classical problems in numerous fields of the natural and engineering sciences by means of the calculus of variation [1,2]. G.W. Leibniz (1646-1716) and L. Euler (1707-1783) found a suitable mathematical tool for finding the extreme values of given functions by introducing the infinitesimal calculus, with which it is possible to carry out an integrated treatment of energy principles in all fields of mechanics with application to dynamics of rigid bodies, general elasticity theory, analysis of load supporting structures (frames, trusses, plates, shells), the theory of buckling, the theory of vibrations, etc. Some very interesting examples from the field of classical mechanics are the "curve of the shortest falling time" ("brachistochrone") and the isoperimetric problem investigated by J. Bernoulli (1655-1705) and D. Bernoulli (1700-1782). Another important problem is that of the "smallest resistance of a body of revolution" solved by Sir I. Newton (1643-1727). With the principle of least action and the integral principles J.L. de Lagrange (1736-1813) and W.R. Hamilton (1805-1865) contributed to the perfection of the calculus of variation which still serves as the basis for various optimization methods. A further important application of minimal principles is to introduce special approach functions. Instead of solving the governing differential equations together with the boundary conditions, often a mathematically difficult task, the problem can be interpreted as finding functions which satisfy some or all boundary conditions and minimize the potential energy and the complementary energy, respectively. Useful approximation methods based on the variational principles of mechanics were devised by Lord Rayleigh (1842–1919), W. Ritz (1878–1909), B.G. Galerkin (1871–1945), and others.

In a first application on "Optimum Structural Design" variational methods have been treated by J.L. de Lagrange, T. Clausen (1801–1885), and B. de Saint-Venant (1797–1886). The investigations of finding out optimal designs of one-dimensional structures under various loadings should be mentioned here. Typical examples are the bar subjected to buckling loads or the cantilever beam under single load or dead weight, for which optimal cross-sectional shapes were found by means of the calculus of variation. For this purpose, optimality criteria are derived in terms of necessary conditions, e.g. Euler's equations in the case of unconstrained problems. If constraints are considered, Lagrange's multiplier method is applied. This corresponds to the solution of an isoperimetric problem [3].

Fig. 1/1 shows an interesting comparison between the natural shape of a tree branch and the optimal shape design of a cantilever beam under dead weight. W. Stadler treated this problem using structural control and multicriteria optimization techniques [7]. Nowadays, some

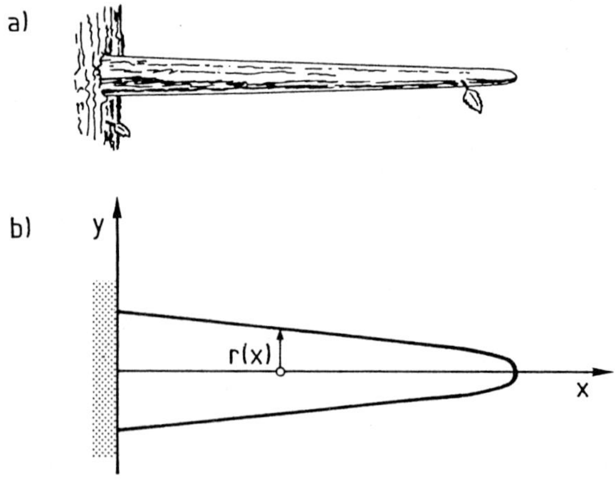

Fig. 1/1. A comparison of optimal shape designs [7]
a) A tree branch as a natural shape
b) A cantilever beam under dead weight

researchers in the field of applied structural optimization use variational principles for their investigations [4]. In developing optimization procedures together with mathematical programming methods, the "optimality criteria"- method can be effectively integrated into the problem system.

1.1.2 Necessity of Applying Optimization Techniques to the Design Process

The following aspects show the necessity of introducing optimization procedures into the practical design phase:

1) Increasing the quality and quantity of products and plants and at the same time reducing costs and thereby being competitive.
2) Fulfilling the permanently increasing specification demands as well as considering reliability and safety, observing severe pollution regulations and saving energy and raw materials.
3) Introducing inevitable rationalization measures in development and design offices (CAD, CAE) in order to save more time for the staff to work creatively.

According to VDI-guideline 2212 [37], the demand for "optimized solutions" can only be met if

- sufficient time is available for the development of alternative solutions,
- fast working software and hardware can be utilized,
- mathematical optimization methods are available and applicable.

Although there are numerous activities in establishing CAD-work stations, the two first points are only partially fulfilled. In addition, many branches of industry have reservations concerning structural optimization. This is due to the yet unrealized demand of practitioners for reliable, efficient and robust methods which are simple in application and furthermore proven to be problem-independent. It is a major concern of research in the field of design optimization to meet these practical demands.

The rapid development of efficient computer systems allows the integration of an optimization procedure into the process of engineering design [40]. This requires detailed mathematical-physical modelling for any structural analysis of a technical design problem and coupling it with a suitable optimization algorithm as well. Both are combined via the problem-programs of optimization modelling for tasks in engineering design (see Chapter 2).

For all these tasks of design optimization mentioned above, it is equally essential to note that the application of optimization theories in the design process depends upon the fundamental aspects of the technical problems.

1.1.3 Multicriteria Optimization as a Strategy in the Design Process

Nowadays, a multitude of new constructions and their corresponding designs require much closer attention because more than one main criterion corresponding to the aspects 1) and 2) in Section 1.1.2 is given respectively. Such optimization problems for *multiple* criteria are called either *Vector* or *Multicriteria Optimization Problems.* With reference to V. Pareto (1848-1923), the French-Italian economist and sociologist, who established an optimality concept in the field of economics based on a multitude of objectives, i.e. on the permanent conflict of interests and antagonisms in social life, this special field of optimization is also called *Pareto optimization* [5] .

The application of multicriteria optimization (MO) to problems in structural mechanics or technology in general took quite a long time. It was W. Stadler who referred to the scientific application of *Pareto's optimality concept* in the 1970's for the first time and who published several papers especially on natural shapes [6,7]. Since the end of the seventies, vector optimization has been more and more integrated into problems of optimal designs in the papers of a number of scientists (e.g. [8-16]).

Within the scope of a design process a designer and an analyst have to figure out which dimensions and shapes will fulfill certain main criteria of a structure in the best possible way. From common experience it is known, however, that all these demands can rarely be met simultaneously. Competing objectives require an estimation of the importance of the individual objectives by weighing them correspondingly. The application of vector optimization as a strategy in a design process is particularly suitable for decision making during single phases of development of new constructions or components. Fig. 1/2 shows the appropriate phases of such a process. In order to

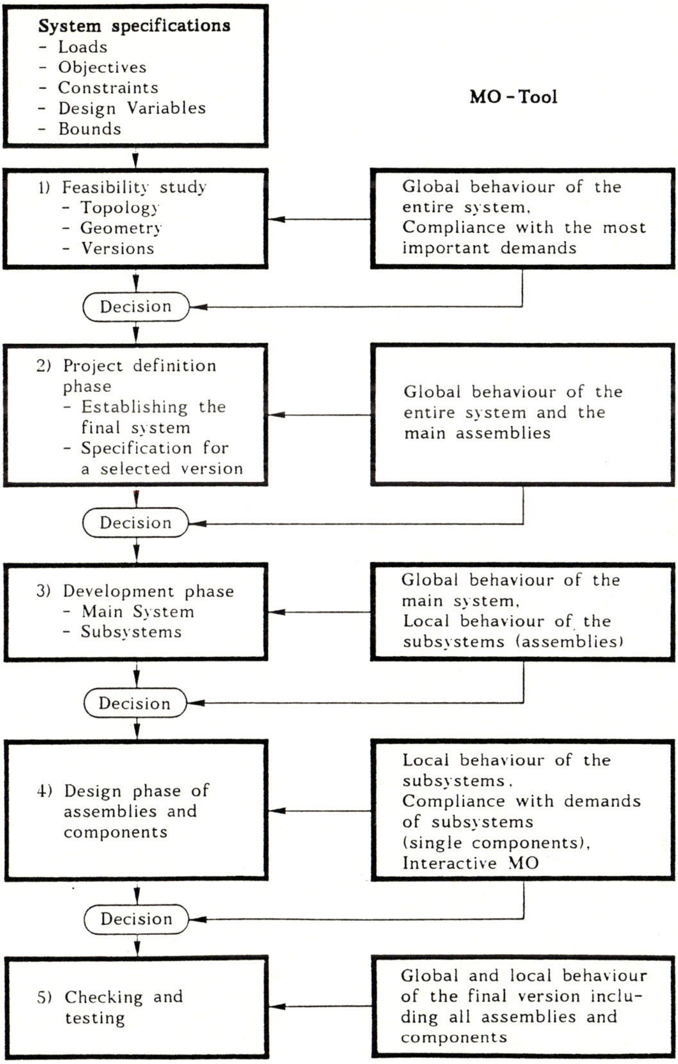

Fig. 1/2. Use of multicriteria techniques during the several phases of a design process

obtain statements concerning the global behaviour of different design versions, it is advantageous to find a number of Pareto-optimal or functional-efficient solutions (see Section 1.3). The particular local behaviour of a structural component or a group of structural components (assemblies) can efficiently be found via an interactive design process (see Chapter 3).

1.2 Mathematical Fundamentals

1.2.1 General Definitions and Notations in Scalar Optimization

The objective of design optimization is to select the values of the design variables x_i (i=1,...,n) under consideration of various constraints in such a way that an objective function $f=f(\mathbf{x})$ attains an extreme value. This can be expressed in the abbreviated form

$$\min_{\mathbf{x} \in \mathbb{R}^n} \{ f(\mathbf{x}) : \mathbf{h}(\mathbf{x}) = \mathbf{0},\ \mathbf{g}(\mathbf{x}) \leq \mathbf{0} \} \tag{1-1}$$

with $\mathbb{R}^n$ the set of real numbers, f an objective function, $\mathbf{x} \in \mathbb{R}^n$ a vector of n design variables, $\mathbf{g}$ a vector of p inequality constraints, $\mathbf{h}$ a vector of q equality constraints (e.g., system equations for the determination of stresses and deformations), and $X := \{ \mathbf{x} \in \mathbb{R}^n : \mathbf{h}(\mathbf{x}) = \mathbf{0},\ \mathbf{g}(\mathbf{x}) \leq \mathbf{0} \}$ the "feasible" domain where $\leq$ has to be interpreted for each individual component.

An additional problem in design optimization is that the objective function(s) and the constraints are generally nonlinear functions of the design variable vector $\mathbf{x} \in \mathbb{R}^n$. The continuity of the functionals as well as of their derivatives is usually assumed (Fig.1/3).

Before treating multicriteria optimization problems some relevant definitions and conditions from scalar optimization will be considered.

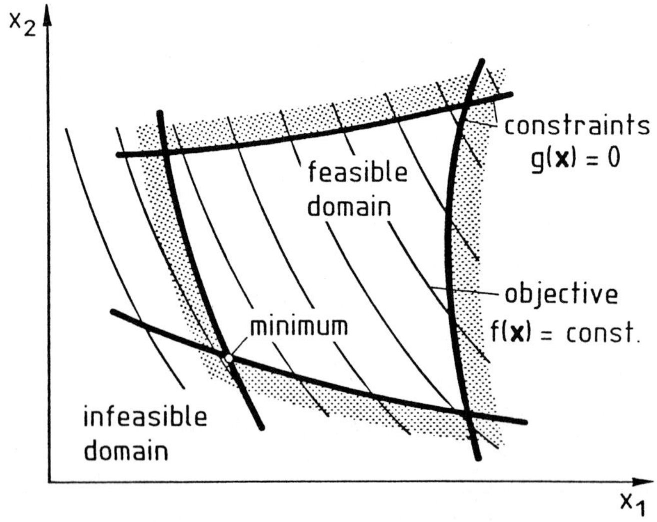

Fig. 1/3. Interpretation of a single criterion optimization problem

The subset $X \subset \mathbb{R}^n$ will be given as the domain of definition. $U_\varepsilon(\mathbf{x}^*)$ describes the ε-neighbourhood of the point $\mathbf{x}^*$; i.e. the number of all points $\mathbf{x}$ whose distance from $\mathbf{x}^*$ is smaller than $\varepsilon > 0$. The distance is given by the Euclidean metric.

Definition 1.1 *Global and local minima*

i. A point $\mathbf{x}^* \in X$ is a global minimum, if and only if

$$f(\mathbf{x}^*) \leq f(\mathbf{x}) \qquad \forall\, \mathbf{x} \in X\ . \tag{1-2a}$$

ii. A point $\mathbf{x}^* \in X$ is called a local minimum point of f on X, if and only if for some ε

$$f(\mathbf{x}^*) \leq f(\mathbf{x}) \qquad \forall\, \mathbf{x} \in X \cap U_\varepsilon(\mathbf{x}^*)\ . \tag{1-2b}$$

The value $f^* = f(\mathbf{x}^*)$ is accordingly called a local (relative) minimum.

Definition 1.2 *Convexity*

i. A subset X of $\mathbb{R}^n$ is convex if and only if

$$\mu \mathbf{x}_1 + (1-\mu)\mathbf{x}_2 \in X \tag{1-3a}$$

for each $\mathbf{x}_1, \mathbf{x}_2 \in X$ and for each real number $0 \leq \mu \leq 1$.

ii. A real-valued function f on the convex subset X is convex on X if and only if

$$f\,[\mu\mathbf{x}_1 + (1-\mu)\mathbf{x}_2] \leq \mu f(\mathbf{x}_1) + (1-\mu)\, f(\mathbf{x}_2) \tag{1-3b}$$

for each $\mathbf{x}_1, \mathbf{x}_2 \in X$ and for each $0 \leq \mu \leq 1$.

Theorem 1.1 *Optimality conditions for unconstrained minimum problems*

i. *Necessary Condition:*

$$\nabla f(\mathbf{x}^*) := \left(\frac{\partial f}{\partial x_1}, \frac{\partial f}{\partial x_2}, \ldots \frac{\partial f}{\partial x_n}\right)^T_{\mathbf{x}^*} = \mathbf{0} \tag{1-4a}$$

ii. *Sufficient Condition*: If the Hessian matrix

$$\mathbf{H}^* := \mathbf{H}(\mathbf{x}^*) = \left(\frac{\partial^2 f}{\partial x_i\, \partial x_j}\right)_{\mathbf{x}^*} \tag{1-4b}$$

is positive definite, $f(\mathbf{x})$ has a local minimum at $\mathbf{x}^*$.

Theorem 1.2 *Conditions for constrained minimum problems*

For the determination of optimality conditions, the Lagrangian function is now introduced [23]

$$L(\mathbf{x},\alpha,\beta) = f(\mathbf{x}) + \sum_{i=1}^{q} \alpha_i \, h_i(\mathbf{x}) + \sum_{j=1}^{p} \beta_j \, g_j(\mathbf{x}) \,, \tag{1-5}$$

where α_i, β_j denote Lagrange multipliers. Certain mathematical requirements yield the following conditions:

i. *Necessary conditions for a local minimum*

The Kuhn-Tucker conditions [24] are applied to test local optimality at a point $\mathbf{x}^*$

$$\nabla L(\mathbf{x}^*) = \nabla f(\mathbf{x}^*) + \sum_{i=1}^{q} \alpha_i^* \nabla h_i(\mathbf{x}^*) + \sum_{j=1}^{p} \beta_j^* \nabla g_j(\mathbf{x}^*) = \mathbf{0}$$

$$\begin{aligned} \text{and} \quad & h_i(\mathbf{x}^*) = 0 \,, && i = 1,\dots q \,, \\ & g_j(\mathbf{x}^*) \le 0 \,, && j = 1,\dots,p \,, \\ & \beta_j^* g_j(\mathbf{x}^*) = 0 \,, && \beta_j^* \ge 0 \qquad j = 1,\dots,p \,. \end{aligned} \tag{1-6}$$

ii. *Sufficient conditions*

For problems for which $f(\mathbf{x})$ is convex, the equality constraints are linear, and the inequality constraints are convex functions, i.e. for so-called convex problems, the Kuhn-Tucker conditions are also sufficient conditions (see [23]).

Fig. 1/4 shows a geometric interpretation in the presence of three inequality constraints. According to the constraints (1-6), the points A and B in Fig. 1/4 satisfy the following conditions:

1. At point A $\longrightarrow$ $\nabla f(\mathbf{x}^*) = \beta_1^* \nabla g_1(\mathbf{x}^*) + \beta_3^* \nabla g_3(\mathbf{x}^*) \,.$ (1-7a)

The gradient does *not* lie in the cone ($\beta_1 < 0$) set by the gradients of the constraint functions; A is not a minimum point because the function value can be reduced within the feasible domain.

2. At point B $\longrightarrow$ $\nabla f(\mathbf{x}^*) = \beta_2^* \nabla g_2(\mathbf{x}^*) + \beta_3^* \nabla g_3(\mathbf{x}^*) \,.$ (1-7b)

The considered point B is a local optimum because there is no direction within the feasible domain in which the function value can be reduced.

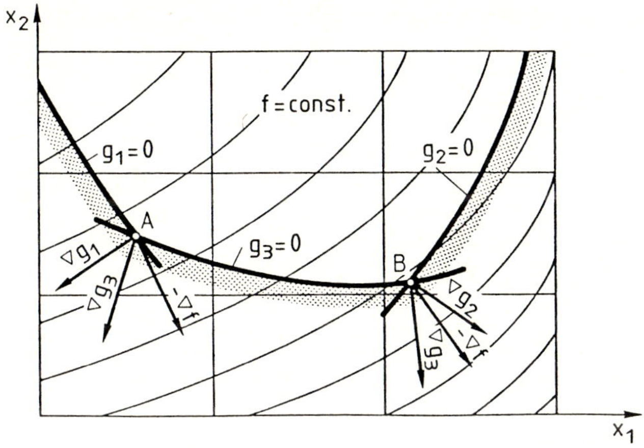

Fig. 1/4. Geometric interpretation of the Kuhn-Tucker conditions under consideration of three inequality constraints

1.2.2 The Multicriteria Programming Problem

In problems with multiple criteria one deals with a design variable vector $\mathbf{x}$ which fulfills all constraints and renders the m components of an objective function vector $\mathbf{f}(\mathbf{x})$ as small as possible. A completion of the problem (1-1) yields the vector optimization problem:

$$\min_{\mathbf{x} \in \mathbb{R}^n} \{\mathbf{f}(\mathbf{x}) \; : \; \mathbf{h}(\mathbf{x}) = \mathbf{0}, \; \mathbf{g}(\mathbf{x}) \leq \mathbf{0}\} \quad . \tag{1-8}$$

A characteristic feature of such optimization problems with multiple criteria is the appearance of an *objective conflict*, i.e. none of the feasible solutions allows the simultaneous minimization of all objectives, or the individual solutions of each single objective function differ. Consequently, the subject of multicriteria optimization deals with all kinds of conflicting problems.

Definition 1.3 *Convexity of MO*

A multicriteria optimization problem on $\mathbb{R}^m$ is convex if and only if

(a) the components of the objective function vector $\mathbf{f}(\mathbf{x})$ are convex,

(b) the components of the vector of the inequality constraints $\mathbf{g}(\mathbf{x})$ are convex, and

(c) the components of the vector of the equality constraints $\mathbf{h}$ are affine-linear functions of $\mathbf{x}$.

Definition 1.4 *Functional-efficiency or Pareto-optimality* ([6,25,30])

A vector $\mathbf{x}^* \in X$ is Pareto-optimal resp. p-efficient or functional-efficient for the problem (1-8), if and only if there is no vector $\mathbf{x} \in X$ with the characteristics

$$f_j(\mathbf{x}) \le f_j(\mathbf{x}^*) \qquad \text{for all } j \in \{1,\dots,m\}$$

and (1-9)

$$f_j(\mathbf{x}) < f_j(\mathbf{x}^*) \qquad \text{for at least one } j \in \{1,\dots,m\}\ .$$

For all non-Pareto-optimal vectors, the value of at least one objective function f_j can be reduced without increasing the functional values of the other components. Fig. 1/5 shows a mapping of the two-dimensional design space $X \subset \mathbb{R}^2$ into the criterion space $Y \subset \mathbb{R}^2$ where the Pareto-optimal solutions lie on the curved section AB. Solutions of nonlinear vector optimization problems can be found in different ways. By defining so-called substitute problems these are normally reduced to scalar optimization problems.

Definition 1.5 *Substitute problem and preference function*

The problem

$$\min_{\mathbf{x} \in X} p[\mathbf{f}(\mathbf{x})] \tag{1-10a}$$

is a substitute problem if there exists $\tilde{\mathbf{x}} \in X^*$ such that

$$p[\mathbf{f}(\tilde{\mathbf{x}})] = \min_{\mathbf{x} \in X} p[\mathbf{f}(\mathbf{x})]\ . \tag{1-10b}$$

The function p is called a preference function or a substitute objective function or a criterion of control effectiveness (the last term is mainly used in control engineering) [8,10-12,15,26].

It is obviously important to study whether the solutions $\tilde{\mathbf{x}}$ of the substitute problems are Pareto-optimal with respect to X and to the set of objective functions $f_1,\dots,f_m$, i.e. that a point $\mathbf{y} = \mathbf{f}(\tilde{\mathbf{x}})$ actually lies on the efficient boundary ∂Y^* [6,11].

A number of publications have dealt with various methods for transforming vector optimization problems into substitute problems [11-16, 32,38]. In the following these transformation rules will be denoted *"strategy"* when referring to the optimization procedure. Since the problem-dependence of the various strategies may be highly relevant, it is of interest to test their efficiency and thus their preference behaviour on typical structures [16].

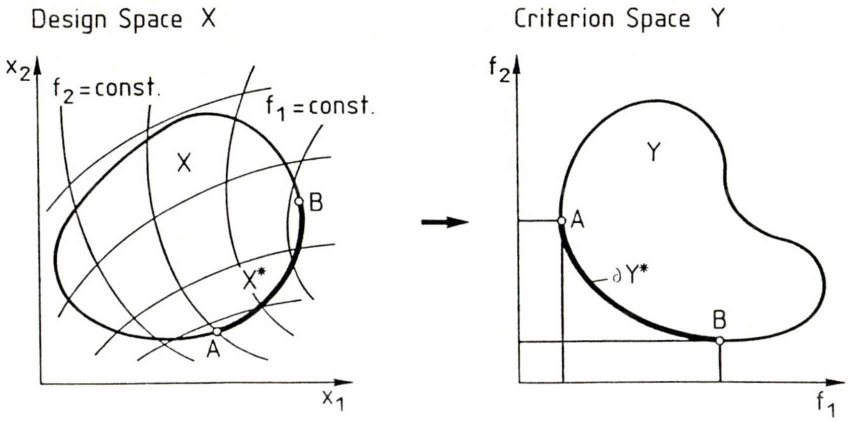

Fig. 1/5. Mapping of a feasible set into the criterion space

Some of the strategies used are described below:

a) Method of Objective Weighting

Objective weighting is obviously one of the most usual substitute models for vector optimization problems. It permits a preference formulation that is independent of the individual minima for positive weights; it also guarantees that all points will lie on the efficient boundary for convex problems. The preference function or utility function is here determined by the linear combination of the criteria $f_1,...,f_m$ together with the corresponding weighting factors $w_1,...,w_m$:

$$p[\mathbf{f}(\mathbf{x})] := \sum_{j=1}^{m} [w_j\, f_j\,(\mathbf{x})] = \mathbf{w}^T \mathbf{f}(\mathbf{x}), \qquad \mathbf{x} \in X. \tag{1-11}$$

It is usually assumed that

$$0 \le w_j \le 1, \qquad \sum_{j=1}^{m} w_j = 1 .$$

It is possible to generate Pareto-optima for the original problem (1-8) by varying the weights w_j in the preference function. In engineering and in economics this approach has been applied for quite some time [9,27, 28]. The deficiency of this stategy in structural optimization has been discussed for example in [39].

b) Method of Distance Functions

Distance functions are frequently applied and also lead to a scalarization of the vector optimization problem. A decision maker specifies a so-called demand level vector $\bar{\mathbf{y}} = (\bar{y}_1,\ldots,\bar{y}_m)^T$ with the objective function value to be achieved in the best possible way. In design optimization this corresponds to a set of assumed specification values or demands for the single objective functions. The respective substitute problem then reads

$$p[\mathbf{f}(\mathbf{x})] := \Big[\sum_{j=1}^{m} |f_j(\mathbf{x}) - \bar{y}_j|^r \Big]^{1/r}, \qquad 1 \le r \le \infty, \qquad \mathbf{x} \in X, \qquad (1\text{-}12)$$

where the variation of r meets various interpretations of the "distance" between the demand levels $\bar{\mathbf{y}}$ and the functional-efficient solutions. The following *distance functions* are most frequently used:

$$r = 1: \qquad p[\mathbf{f}(\mathbf{x})] = \sum_{j=1}^{m} \big| f_j(\mathbf{x}) - \bar{y}_j \big| , \qquad (1\text{-}13a)$$

$$r = 2: \qquad p[\mathbf{f}(\mathbf{x})] = \Big[\sum_{j=1}^{m} (f_j(\mathbf{x}) - \bar{y}_j)^2 \Big]^{1/2} \qquad \text{Euclidean metric}, \qquad (1\text{-}13b)$$

$$r \longrightarrow \infty: \qquad p[\mathbf{f}(\mathbf{x})] = \max_{j=1,\ldots,m} \big| f_j(\mathbf{x}) - \bar{y}_j \big| \qquad \text{Chebyshev metric}. \qquad (1\text{-}13c)$$

The choice of a demand level may cause problems. Therefore, Fig. 1/6 qualitatively gives the solutions of the substitute problem for various demand levels. It shows that the choice of $\bar{\mathbf{y}}_1$ yields a solution $\tilde{\mathbf{x}}$ of the substitute problem for which $\tilde{\mathbf{y}}_1 = \mathbf{f}(\tilde{\mathbf{x}}) \in \partial Y^*$ is efficient with respect to Y. The choice of $\bar{\mathbf{y}}_2$, however, yields an $\tilde{\mathbf{y}}_2 \in \partial Y^*$ not lying on the efficient boundary, and with the choice of the inner point $\bar{\mathbf{y}}_3$, the respective solution $\tilde{\mathbf{y}}_3$ is not an efficient point of the boundary of Y.

The use of distance functions is subject to the following disadvantages [11]:

1. The selection of "wrong" demand levels $\bar{\mathbf{y}}$ will lead to nonefficient solutions (Fig. 1/6).
2. The selection of "correct" or "valid" demand levels $\bar{\mathbf{y}}$ requires knowledge of the individual minima of the m objective functions $f_j(\mathbf{x})$, j=1,...,m which is not easy to achieve with nonconvex problems.

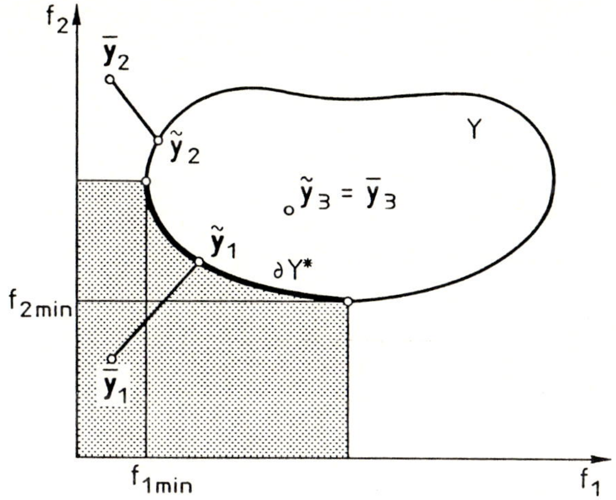

Fig. 1/6. Solution of the substitute problem for various demand levels

The methods of the distance functions can also be parametrized to generate Pareto-optima for the original problem (1-8). For example in [38] several possibilities for choosing the parameters and their relations to Pareto-optima have been considered in detail.

c) Method of Constraint Oriented Transformation (Trade-off Method)
Retransformation of the vector optimization problem into a scalar substitute problem may also be achieved by minimizing only one objective function with all others bounded [11,12]):

$$p[\mathbf{f}(\mathbf{x})] = f_1(\mathbf{x}) , \quad \mathbf{x} \in X$$

with (1-14)

$$f_j(\mathbf{x}) \leq \bar{y}_j , \quad j=2,\ldots,m .$$

Thus, f_1 is called the *main objective*, and $f_2,\ldots,f_m$ are called *secondary* or *side objectives*. The given problem can be interpreted in such a way that when minimizing f_1 the other components are not allowed to exceed the values $\bar{y}_2 ,\ldots, \bar{y}_m$. The dependence of the solution on the selection of these *constraint levels* for the two-dimensional case is shown in Fig. 1/7.

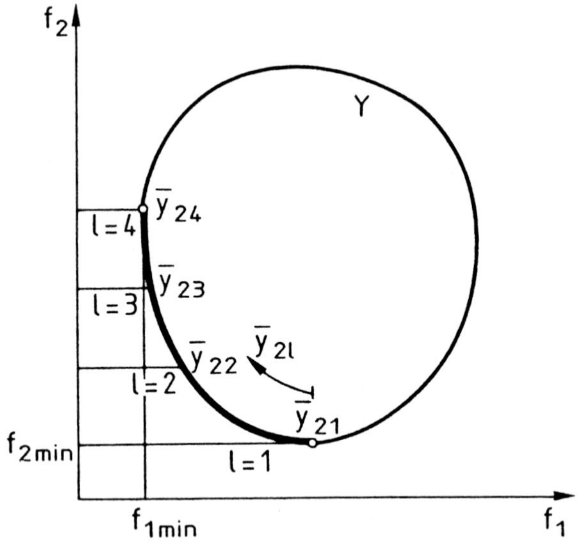

Fig. 1/7. Solution of a constraint-oriented transformation depending on the constraint level

The main objective function f_1 is generally one for which no a priori estimation of an upper limit $\bar{y}_1$ is available.

If the constraint levels are taken as equality constraints, and if other constraints are not considered, the problem corresponds to the minimization of the respective Lagrangian function [29]

$$L(\mathbf{x},\boldsymbol{\alpha}) := f_1(\mathbf{x}) + \sum_{j=2}^{m} \alpha_j [f_j(\mathbf{x}) - \bar{y}_j] , \qquad (1\text{-}15)$$

which is used in this case as a preference function. The necessary *optimality criteria* corresponding to the Kuhn-Tucker conditions (1-6) without inequality constraints are

$$\frac{\partial L}{\partial x_i} = \frac{\partial f_1}{\partial x_i} + \sum_{j=2}^{m} \alpha_j \frac{\partial f_j}{\partial x_i} \overset{!}{=} 0, \qquad i=1,\dots,n , \qquad (1\text{-}16a)$$

$$\frac{\partial L}{\partial \alpha_j} = f_j(\mathbf{x}) - \bar{y}_j \overset{!}{=} 0 , \qquad j=2,\dots,m . \qquad (1\text{-}16b)$$

They are the basis for calculating the optimal values for $x_1,\dots,x_n$ and those of the adequate Lagrange multipliers $\alpha_2,\dots,\alpha_m$. The introduction of the abbreviations

$$\mathbf{w}^T = (1, \alpha_2, \alpha_3, \ldots, \alpha_m) = (1, \boldsymbol{\alpha}^T) \quad , \tag{1-17a}$$

and

$$C = -\sum_{j=2}^{m} \alpha_j \bar{y}_j = -\boldsymbol{\alpha}^T \bar{\mathbf{y}} \tag{1-17b}$$

in equation (1-15) yields the following formula:

$$L(\mathbf{x}, \boldsymbol{\alpha}) = f_1(\mathbf{x}) + \sum_{j=2}^{m} \alpha_j f_j(\mathbf{x}) - \sum_{j=2}^{m} \alpha_j \bar{y}_j =$$

$$= \mathbf{w}^T \mathbf{f}(\mathbf{x}) - \boldsymbol{\alpha}^T \bar{\mathbf{y}} = \mathbf{w}^T \mathbf{f}(\mathbf{x}) + C \; . \tag{1-18}$$

Thus, the expression (1-18) corresponds locally to the substitute problem with objective weighting if one disregards the normalization of the weighting factors and the additive parameter C, which are irrelevant for solving the problem.

Finally, one can state that the considered "Trade-off"-formulation yields the Pareto-optimal set of solutions if one criterion is replaced by a sequence of inequality constraints. Therefore, this strategy is sometimes called "multi-constraint" or "bound formulation" (see [4, 41]).

d) *Method of Min-Max Formulation*

Besides the preference functions described above, the min-max formulation plays a very important role in solving substitute problems. It is based on the minimization of relative deviations of the single objective functions from the respective individual minimum [12, 32].

For the interpretation of a min-max formulation three given objective functions with the domain of definition $x_1 \le x \le x_2$ are considered (Fig. 1/8). If the extrema f_j are established separately for each objective function (criterion), the desired solution is that **x** which results in the smallest value of the maximum deviation of all objective functions. Therefore, the scalar substitute problem can be defined according to the min- max formulation as follows:

$$p[\mathbf{f}(\mathbf{x})] = \max_{j=1,\ldots,m} [z_j(\mathbf{x})] \tag{1-19a}$$

with

$$z_j(\mathbf{x}) = \frac{f_j(\mathbf{x}) - \bar{f}_j}{\bar{f}_j} , \quad \bar{f}_j > 0 \; , \quad j = 1, \ldots, m \; . \tag{1-19b}$$

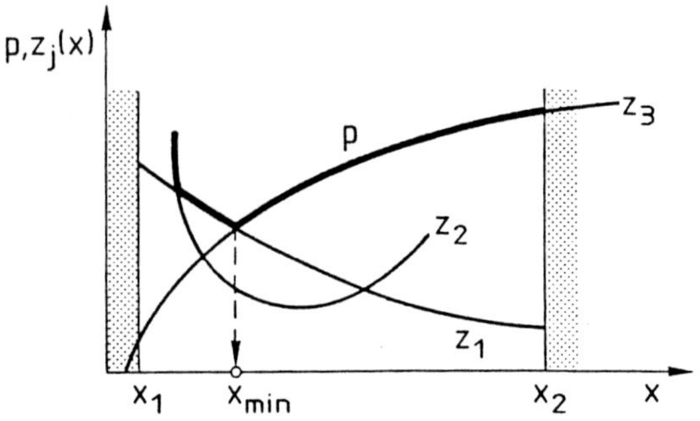

Fig. 1/8. Preference function of a min-max formulation in the one-dimensional case

In [4], a well-known modification of equation (1-18) is applied for practical computations. It consists of the minimization of a new variable x_{n+1} (comparable to a slack variable, e.g. see [31]) while simultaneously considering the additional constraints:

$$p[\mathbf{f}(\mathbf{x})] = x_{n+1} \wedge z_j(\mathbf{x}) - x_{n+1} \leq 0 \quad \mathbf{x} \in X \quad , \quad j=1,\ldots,m \ . \tag{1-20}$$

Equation (1-20) is especially appropriate for nonlinear optimization problems using effectively the inequality constraints in the optimization algorithms (e.g. method of sequential linearization) [21].

For the min-max formulation (1-20) a geometric interpretation can be given on the basis of the hypothesis that all inequality constraints $z_j(\mathbf{x}) - x_{n+1} \leq 0$ (j=1,...,m) are active within the min-max optimum $\tilde{\mathbf{x}}$, i.e. $z_j(\tilde{\mathbf{x}}) - x_{n+1} = 0$. Without going into detailed proof here, it can be stated that there will be a parameter graph within the hyperspace $\mathbb{R}^m$ from which one can conclude that the optimal solution point $f(\tilde{\mathbf{x}})$ must lie on a line within the criterion space. Therefore, it yields a first geometric position for $f(\tilde{\mathbf{x}})$. The second one results from minimizing the distance (r=2, Euclidean norm) between the reference point $\overline{\mathbf{f}}$ and any random point of the line in the criterion space. It can be shown that this precisely corresponds to the minimization of x_{n+1}. The min-max optimum can therefore be interpreted as the intersection of a line in the space with the functional-efficient solution set ∂Y^*. Fig. 1/9 shows this behaviour for two objective functions.

These investigations show that there are certain interdependencies between the min-max formulation and the method of distance functions.

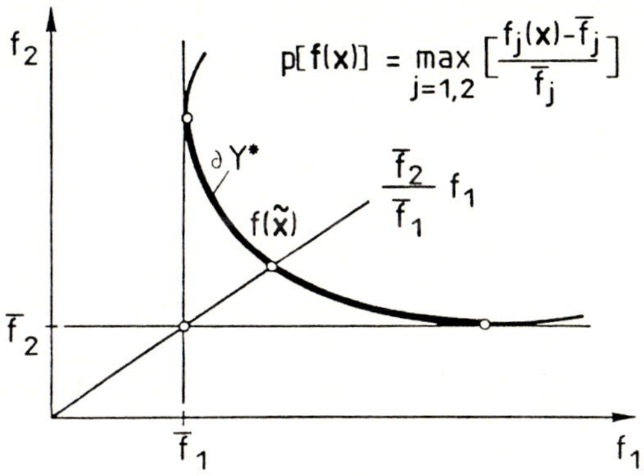

Fig. 1/9. Geometric interpretation of a min-max formulation for two criteria

Starting from the general distance formulation according to (1-13), the min-max formulation for $r \to \infty$ (Chebyshev metric) results in

$$p[\mathbf{f}(\mathbf{x})] = \max_{j=1,\dots,m} |f_j(\mathbf{x}) - \bar{y}_j| \,, \qquad \mathbf{x} \in X \tag{1-21}$$

with the components of the demand level vector $\bar{y}_j$. If the minima $\bar{f}_j$ of the individual objective function components are selected as components for the demand level vector, and if every objective function is related to the respective $\bar{f}_j$, then the distance function formulation is transformed into the min-max formulation in accordance with Eq. (1-19).

The min-max formulation described above yields the compromise solution $\tilde{\mathbf{x}}$ considering all objective functions with equal priority. But if the single objectives have to meet a special order or if the functional-efficient solution set X^* is of great importance for the decision maker, the min-max formulations can be modified or extended (see [16,21,34]):

- *Min-Max Formulation with Objective Weighting*

The introduction of dimensionless weighting factors $w_j \geq 0$ transforms the substitute problem (1-19) into

$$p[\mathbf{f}(\mathbf{x})] = \max_{j=1,\dots,m} [w_j z_j(\mathbf{x})] \,, \qquad \mathbf{x} \in X \,, \tag{1-22}$$

where $z_j(\mathbf{x})$ denotes the j-th relative deviation as in (1-19). The weighting factors describe the priority of the single objective functions. Thus, it is possible to select definite compromise solutions from

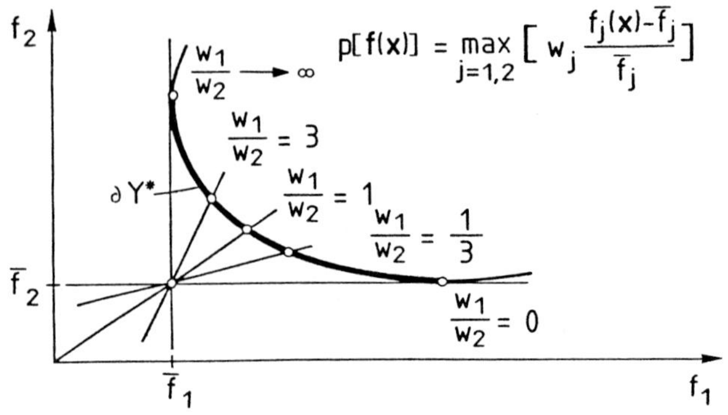

Fig. 1/10. Min-max solutions for two objective functions under consideration of different weighting-factor relations

random fields of functional-efficient sets. Moreover, the variation of w_j allows the establishment of the representative solution set. A similar modification also exists for Eq. (1-20)

$$p[\mathbf{f}(\mathbf{x})] = x_{n+1} \wedge \quad w_j z_j(\mathbf{x}) - x_{n+1} \leq 0 , \quad \mathbf{x} \in X \quad , \quad j=1,\dots,m . \tag{1-23}$$

Fig. 1/10 shows the geometric interpretation of Eq. (1-23) for the two-dimensional case. It is obvious that depending on the ratio w_1/w_2 of the two weighting factors one obtains different compromise solutions describing the whole functional-efficient boundary.

- *Min-Max Formulations Presuming a Demand-Level Vector*

If the definition of the relative deviations in (1-19b) is not based on the individual minima f_j but on the given components $\bar{y}_j$ of the demand level vector with the characteristics $\bar{y}_j = \bar{f}_j$, one can get analogous substitute problems to (1-22) and (1-23). However, the problem formulation does not guarantee that all inequality constraints become active at the solution point $\mathbf{x}$. In other words that they can be regarded as equality constraints. Only if all inequality constraints become active, the solution vector $\mathbf{x}$ lies on the intersection of the line in the space with the functional-efficient solution set ∂Y^*. The difference with respect to the previously mentioned formulation is illustrated in Fig. 1/11. If the line passing through the point $\mathbf{y}$ and defined by the relation w_1/w_2 intersects the functional-efficient boundary, the intersection point is also the compromise solution. If

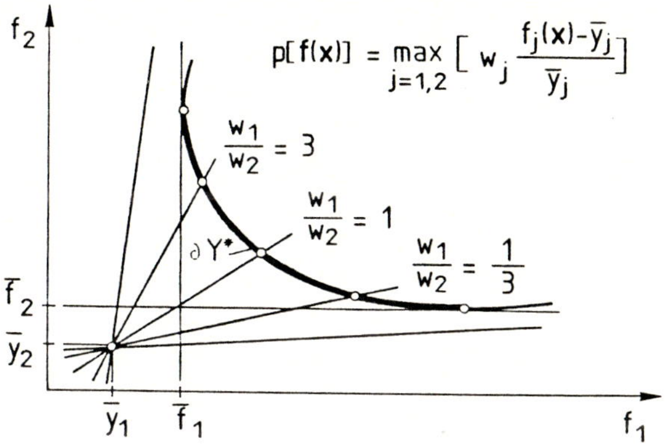

Fig. 1/11. Min-max solutions under consideration of a demand-level vector $\bar{\mathbf{y}}$

there is *no* intersection point, the point corresponding to f_1 or to f_2 is the solution depending on the ratio w_1/w_2.

The special selection of a demand-level vector $\bar{\mathbf{y}} = \mathbf{0}$ along with omitting the division by $\bar{y}_j$ within the relative deviation $z_j(\mathbf{x})$ yields a further modification of the min-max formulation a formulation

$$p[\mathbf{f}(\mathbf{x})] = \max_{j=1,\dots,m} [w_j f_j(\mathbf{x})], \qquad \mathbf{x} \in X \quad , \tag{1-24}$$

frequently applied in practice [4, 21] .

1.2.3 The Multicriteria Control Problem [35, 36]

As mentioned in the preceding section, some of the optimum design problems can be modelled using an optimal control approach [35,36]. Let the state $\mathbf{x} \in A \subset \mathbb{R}^n$ be controlled by means of a control vector $\mathbf{u}(\cdot) : [t_o,t_1] \longrightarrow U \subset \mathbb{R}^r$ in the state equation

$$\dot{\mathbf{x}} = g(\mathbf{x},\mathbf{u}) \tag{1-25}$$

with $\mathbf{x}(t_o) \in X$ corresponding to the initial set and $\mathbf{x}(t_1) \in X$ corresponding the terminal set and with $x_n = t$, the independent variable, so that $g^n(\mathbf{x},\mathbf{u}) = 1$. Furthermore,

$$g(\cdot): A \times U \longrightarrow B \text{ (open)} \subset \mathbb{R}^n \tag{1-26}$$

is the velocity function and U is the control constraint set, the set of all possible values of $\mathbf{u}(\cdot)$. It is usual to assume that $\mathbf{u}(\cdot)$ belongs to a nonempty set F of admissible controls. A criterion map $\mathbf{f}(\cdot): F \rightarrow \mathbb{R}^n$ is defined in terms of the component integrals

$$f_i(\mathbf{u}(\cdot)) = \int_{t_o}^{t_1} g_{oi}(\mathbf{x}(t), \mathbf{u}(t))\, dt , \tag{1-27}$$

where

$$g_{oi}(\cdot) : A \times U \rightarrow C_i \text{ (open)} \subset \mathbb{R}^n, \quad i=1,\dots,m . \tag{1-28}$$

The state space $\mathbb{R}^n$ is augmented with

$$\dot{\mathbf{y}} = \mathbf{g}_o(\mathbf{x},\mathbf{u}), \quad \mathbf{y}(t_o) = 0 , \tag{1-29}$$

where $\mathbf{y} \in \mathbb{R}^m$ is the criterion space and $\mathbf{g}_o=(g_{o1},\dots,g_{om})$. Let $\mathbf{u}(\cdot) \in F$, and $\mathbf{x}(\cdot)$ be a corresponding solution of the state equation (1-25), and let $\mathbf{s}(\cdot)$ be a solution of (1-29) corresponding to the pair $(\mathbf{x}(\cdot), \mathbf{u}(\cdot))$. The attainable criteria set is then defined by

$$Y = \{\mathbf{y} \in \mathbb{R}^m : \mathbf{y} = \mathbf{s}(t_1)\} . \tag{1.30}$$

Multicriteria control problems can be stated as finding an "optimal" control $\mathbf{u}^*(\cdot) \in F$ for $f(\mathbf{u}(\cdot))$ subject to $\mathbf{u}(\cdot) \in F$.

1.3 Components and Plants with their Objectives

In this chapter several examples illustrate the advantages of multi-criteria optimization techniques in decision making during the planning and the design process of complex components and plants.

1.3.1 Optimum Design of Highly Accurate Parabolic Antennas [17,18]

A practical application of the optimization strategies and procedures is to figure out the layout of the main components for highly accurate focusing parabolic antennas. Antennas can be defined as so-called wavetype transducers. As transmitting antennas they transform cable guided high-frequency energy into wave types convenient for an extension into free space, and as receiving antennas they retransform the energy taken from free space into cable guided waves. Moreover, one tries to achieve a transformation from one condition into the other with least possible losses in order to get optimal antenna gain. The transmission and reception of waves in the dm-, cm-, and mm-range (micro-wave range) are usually realized by means of parabolic reflectors based on the laws of geometrical optics.

The rays radiated from the focus of a paraboloid during transmission are reflected on its surface and leave the mirror as parallel, in-phase rays. This process is reversed for wave reception. The in-phase condition of the rays essentially depends on the existence of an accurate parabolic surface. As the ray reception is analogous to that of optical astronomy, radio astronomists usually call their parabolic antennas

Fig. 1/12. View of a 30-m-radio telescope for millimeter-wave range (Max-Planck Institute for Radio Astronomy, Bonn, FRG)

"radio telescopes" in contrast to the "mirror telescopes" in optical astronomy. Ideally, all incident rays should intersect in the focus assuming an ideal surface as exact as possible in any given position. It is obvious that due to this demand, the reflector and its supporting structure are the most important components of a movable parabolic antenna. In practice, however, such a highly accurate surface is hardly attainable. Fig. 1/12 shows the latest radio telescope for the mm-wave range (MRT) with 30 m aperture diameter developed, designed and manufactured by two German companies (Krupp Industrietechnik, Duisburg and MAN, Gustavsburg) and ordered by the Max-Planck Association, Munich (FRG).

The reflector consisting of single adjustable panels (Fig. 1/13) supported on a rear spatial framework is deformed by dead weight, by wind, and by temperature loads. Furthermore, there are manufacturing tolerances as well as measuring and adjusting faults during the positioning of the reflector surface. Due to these systematic and statistical differences, the phases of the individual rays will be different. Part of the energy will be diffused and radiated towards other directions. According

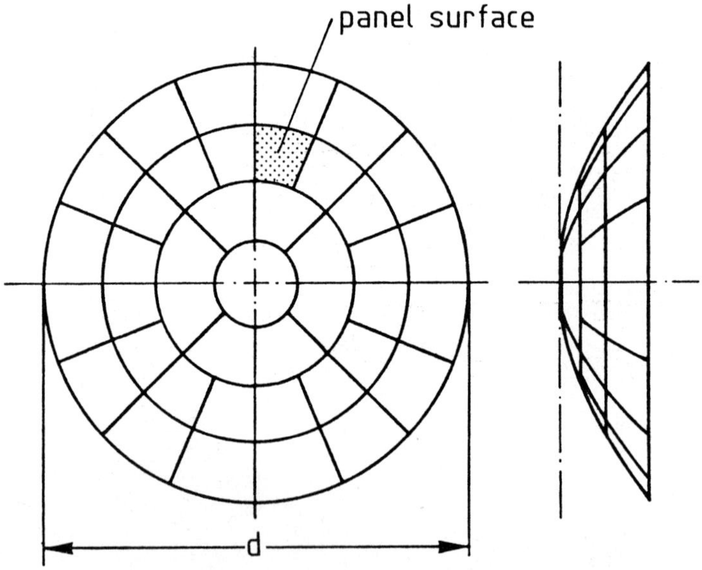

Fig. 1/13. Design of a parabolic reflector with circular aperture in panel surfaces

to Ruze, the reduced gain G can be described by a Gaussian error equation [17] :

$$\frac{G}{G_0} = e^{-(4\pi\sigma/\lambda)^2} . \tag{1-31}$$

The relation G/G_0 expresses the "efficiency" of an antenna, and

$$G_0 = \eta\left(\frac{\pi d}{\lambda}^2\right) \tag{1-32}$$

is the "gain" of an ideal parabolic antenna with d = aperture diameter, λ = wavelength, σ = standard deviation or root mean square value (rms-value), η = surface efficiency.

The rms-value σ is defined as a measure for the surface accuracy. It is determined by the method of least squares with a "bestfit"-surface being described by a set of n given points of the deformed and imperfect reflector surface [18]. As the efficiency of a parabolic antenna substantially depends on the surface accuracy, the rms-value plays the most important role besides the weight of the design of an antenna. Both objectives must be fulfilled in the best possible way. They are used as criteria in this bicriteria optimization problem.

1.3.2 Optimal Layout of a Novel Solar Energy Collector [19]

As a further example, a special type of a concentrated solar energy collector, the so-called "Rear-Focus Collector", is considered (see Fig. 1/14). It consists of several frustum-type reflector shells linked together by two intersecting ribs. The focus of the rays and accordingly the absorber are located behind the collector. The system efficiency η of the concentrating collectors depends on the geometry, the shape accuracy of the reflector, and the tracking error. These aspects have a substantial influence on the two relevant quantities of η, the concentration factor and the intercept factor. For this collector, unlike a similarly designed one developed by A. Spyridonos in the early seventies, the optimal arrangement of the single shells of the collector are determined by means of the mathematical programming. Apart from one objective function representing the system efficiency, the volume is included as a further one.

Fig. 1/14. Front view of the Rear Focus Collector (University of Calgary, Canada)

1.3.3 Shape Optimization of Satellite Tanks [20]

The supporting behaviour of shells can be considerably improved by shape optimization without increasing the weight. This is especially important for constructions in satellite technology because it is exactly this field of technology which has enormous demands on light constructions with high reliability.

Therefore, a method for the optimal layout of the middle surface and the wall thickness distribution of satellite tanks was developed. In this case the weight of the tank has to be minimized while at the same time the volume has to be maximized and the feasible stresses have to be fulfilled. In order to avoid buckling problems, negative stresses should not be permitted. Fig. 1/15 shows a view of such satellite tanks.

Fig. 1/15. View of satellite tanks (MBB-ERNO, Bremen, West Germany)

1.3.4 Optimal Shape Design of a Conveyer Belt Drum [21,22]

Efficient conveyers are necessary for extensive soil shifting operations in open mining. Here, large rubber-belt conveyers could prove to be successfully used. According to Fig. 1/16a, the conveyer belt drum, the track supporting roller, and the belt are essential components of belt conveyers. A conveyer belt drum consists of the cylindrical drum shell (1) and the bottom (2) (Fig. 1/16b). Drum bottom and shaft (4) are connected by a clamping ring (3).

Development in this field is characterized by continuously increasing demands on conveying capacity, conveying track, and operational safety which leads to enlarged distances between the axes and the conveyer belt width. As the criteria can be sufficiently realized by essentially larger tension forces, the stresses in the belt drums are inevitably enlarged, too. It was attempted to reduce these stresses by extending the wall thickness and by implementing ribs. But these measures often led to an extreme increase in weight so that damage and failure could not be avoided. The optimal shape design of a conveyer belt drum was treated with the direct method of shape optimization.

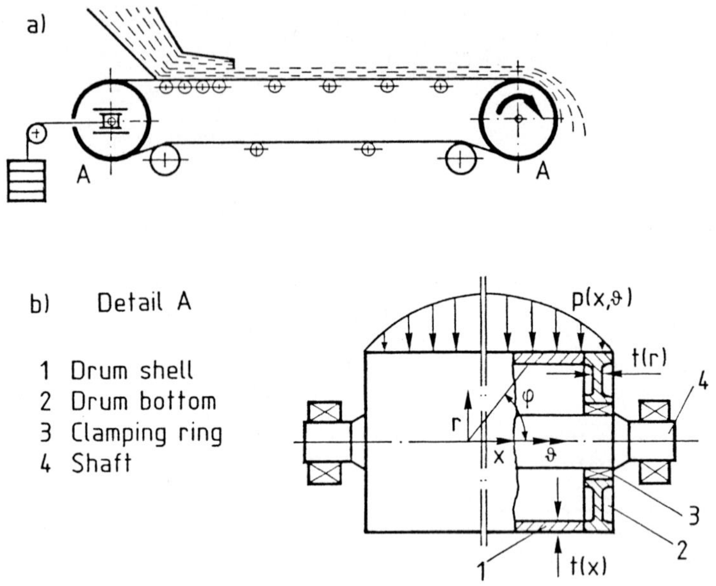

Fig. 1/16. Sketch of the belt conveyer a) complete system
b) conveyer belt cylinder

On the basis of a given midsurface contour $r(\varphi)$, the optimal wall thickness distribution $t(\varphi)$ had to be determined in a way that the criteria of optimization "minimal weight W" and "minimal reference stress $\sigma_{ref\ max}$" were fulfilled as well. Constraints were specified in terms of other limitations.

1.3.5 Optimal Layout of Tube-Flange Structures [21]

Steel engineering in various areas of "structural engineering" (crane technology, steel engineering, "offshore"-techniques, piping constructions), circular cylindrical shells are often employed as structural units connected with other elements (e.g. plates). The main parts of a slewing crane (Fig. 1/17a) are the overhang beam (1), the tubular column (2), and the slewing ball bearing (3). The bottom flange (Fig. 1/17b) which is welded to the tubular column is connected to the foundation with anchor bolts. With the force F at the overhang beam the column and also the flange connection are loaded by an axial force and a bending moment.

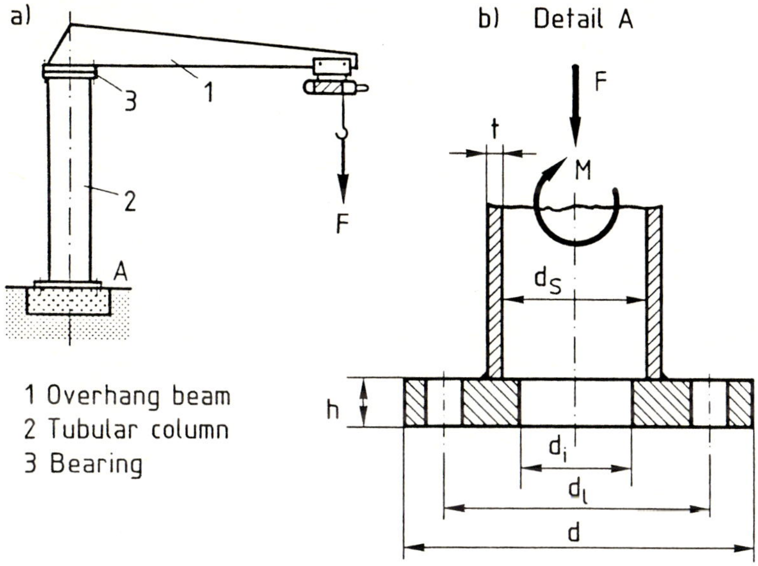

Fig. 1/17. Slewing crane a) principle sketch
b) bottom flange

In the design phase, particularly the shape of the region near the edges is important. As there are stress concentrations near the edges the problem is to find out suitable dimensions for the flange leading to a minimization of stress concentrations but additional weight reduction. These two competing objectives lead to a multicriteria optimization problem.

Some results are presented in Fig. 1/18. The conflict between the criteria $f_1(\mathbf{x}) \mathrel{\hat{=}} W(\mathbf{x})$ (weight) and $f_2(\mathbf{x}) \mathrel{\hat{=}} \sigma(\mathbf{x})$ (stress concentration) is given. Designs with low stresses give relatively high weight values. The sensitivity of the flange height $x_1 = h$ is much higher than the inner diameter $x_2 = d_i$ of the flange. During the variation only h is changed by the optimization algorithm. If $h = h_{max}$ is reached, d_j varies as well. The results discussed here are adequate and very important not only for decision making on this particular design problem of the investigated connection but also for all other examples.

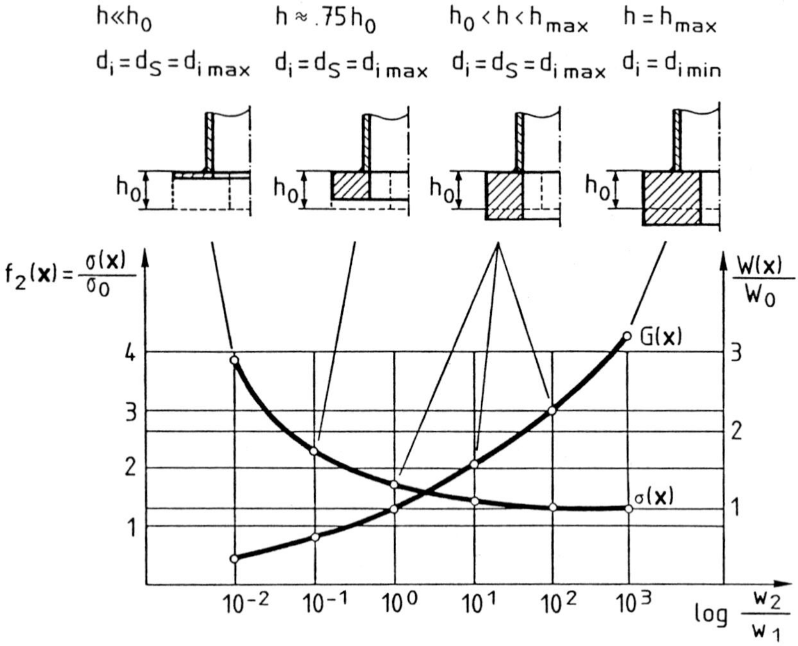

Fig. 1/18. Conflict between the two criteria weight and max. stress
σ_0, W_0 specification values
w_2/w_1 ratio of weighting factors

1.4 Conclusion

This first chapter is a presentation of the fundamentals of scalar and multicriteria optimization and illustrates the necessity of application of multicriteria optimization techniques to develop and to layout components and structures by means of some real-life examples. The applicaton of MO-techniques is primarily due to the fact that today the manufacturing of machines does not only require a minimization of costs but also observes objectives such as shape accuracy and reliability. Such problems are defined as "optimization problems with multiple objectives" (multicriteria optimization).

The objectives which are mostly competitive and nonlinear do not lead to one solution point for the optimum but rather to a "functional-efficient" (Pareto-optimal) solution set, i.e. the decision maker selects the most efficient *compromise solution* out of such a set. The use of preference functions transforms the multicriteria optimization problem into a scalar substitute problem. This so-called optimization strategy is a basic part of *optimization modelling* (see Chapter 2). For the transformation a number of preference functions such as objective weighting, distance functions, constraint-oriented transformation (trade-off method) and min-max formulation have been analysed and tested. It can be shown that the efficiency of the single preference functions depend both on the problem and on the adaptation to certain optimization algorithms.

The examples from industrial practice given in Section 1.3 show how important it is for the designer to get a tool for decision making in the design process, especially when there is more than one criterion to be fulfilled. A large number of possible multicriteria formulations which go far beyond these examples is presented in the second part of the book (Applications).

The motivation for using a multicriteria approach in engineering design can be summarized as follows:

(i) Multicriteria modelling very well reflects the design process in which usually several conflicting objectives have to be satisfied.

(ii) The designer has the possibility to explore a broader range of alternative solutions than with single criteria models for which the solution is immediately fixed after the problem-formulation.

(iii) Multicriteria formulation provides a basis for explicit trade-off between conflicting objectives or interests.

References

[1] Szabo, I.: Höhere Technische Mechanik. 5. Auflage, Berlin, Göttingen, Heidelberg: Springer 1972

[2] Szabo, I.: Geschichte der mechanischen Prinzipien. Basel, Boston, Stuttgart: Birkhäuser 1979

[3] Clausen, T.: Über die Form Architektonischer Säulen. Melanges Mathematiques et Astronomiques I (1849-1853) 279-294

[4] Bendsøe, M.P.; Olhoff, N.; Taylor, J.E.: A Variational Formulation for Multicriteria Structural Optimization. Journal of Structural Mechanics, Vol. 11, No. 4, 1983

[5] Pareto, V.: Manual of Political Economy. Translation of the French edition (1927) by A.S. Schwier. London-Basingslohe: The McMillan Press Ltd., 1971

[6] Stadler, W.: Preference Optimality and Application of Pareto – Optimality. In: Marzollo/Leitmann (eds.): Multicriterion Decision Making. CISM Courses and Lectures.Berlin, Heidelberg, New York: Springer 1975

[7] Stadler, W.: Natural Structural Shapes (The Static Case). Quarterly Journal of Mechanics and Applied Mathematics, Vol. 31, (1978) 169-217

[8] Baier, B.: Mathematische Programmierung zur Optimierung von Tragwerken insbesondere bei mehrfachen Zielen. Dissertation, TH Darmstadt, 1978

[9] Koski, J.: Truss Optimization with Vector Criterion. Tampere University of Technology Publication, Tampere 1979

[10] Eschenauer, H.: Über die Optimierung hochgenauer Tragstrukturen. Karl-Marguerre-Gedächtnisband, Schriftenreihe "THD Wissenschaft und Technik", (1980) 89-101

[11] Sattler, H.-J.: Ersatzprobleme für Vektoroptimierungaufgaben und ihre Anwendung in der Strukturmechanik. Dissertation, Universität-GH-Siegen, 1982

[12] Osyczka, A.: Multicriterion Optimization in Engineering. New York, Chichester, Brisbane, Toronto: John Wiley, 1984

[13] Radford, A.D.; Gero, J.S.; Roseman, M.A.; Balachandran, M.: Pareto Optimization as a Computer-Aided Design Tool. In: Gero, J.S. (ed.): Optimization in Computer-Aided Design. North-Holland Amsterdam, New York, Oxford: Elsevier Science Publishing Company (1984) 47-80

[14] Duckstein, L.: Multiobjective Optimization in Structural Design: The Model Choice Problem. In: Atrek, Gallagher, Ragsdell, Zienkiewicz (eds): New Directions in Optimum Structural Design. Chichester, New York, Brisbane, Toronto, Singapore: J. Wiley & Sons (1984) 459-481

[15] Koski, J.: Multicriterion Optimization in Structural Design. In: Atrek, Gallagher, Zienkiewicz (ed.): New Directions in Optimum Structural Design. Chichester, New York, Brisbane, Toronto, Singapore: J. Wiley & Sons (1984) 483-503

[16] Eschenauer, H.: Numerical and Experimental Investigations on Structural Optimization of Engineering Designs. DFG-Report of the Research Laboratory for Applied Structural Optimization, University of Siegen, May 1985

[17] Eschenauer, H.: Parabolantennen für Satellitenfunk und Radioastronomie im Millimeterwellenbereich-Forderungen und Aufgaben an den Ingenieur. In: Kreuzer, H., Bonfig, K.W.: Entwicklungen der siebziger Jahre, Gerabronn: Hohenloher Druck- und Verlagshaus, (1978) 531-549

[18] Eschenauer, H.: Multicriteria Optimization Techniques for Highly Accurate Focusing Systems. In: W. Stadler: Application of Multicriteria Optimization in Engineering and the Sciences. Plenum Publishing Corporation, (1988) 309-354

[19] Eschenauer, H.; Vermeulen, P.: Contribution to the Optimization of a Novel Solar Energy Collector. ZFW, Bd. 10, H.3, (1986) 190-198

[20] Eschenauer, H.: Shape Optimization of Satellite Tanks. Journal of Structural Optimization 1, (1989) 171-180

[21] Kneppe, G.: Direkte Lösungsstrategien zur Gestaltsoptimierung von Flächentragwerken. Dissertation, Universität-GH-Siegen, 1985

[22] Eschenauer, H.; Kneppe, G.: Min-Max-Formulierungen als Strategie in der Gestaltsoptimierung. ZAMM 6 (1985) T344-T345

[23] Pierre, D.A.; Lowe, M.J.: Mathematical Programming via Augmented Lagrangian. London: Addison-Wesley, 1975

[24] Kuhn, H.W.; Tucker, A.W.: Nonlinear Programming. Proceedings of the 2nd Berkeley Symposium on Mathematical Statistics and Probability, University of California, Berkeley, California, 1951

[25] Hettich, R.: Charakterisierung lokaler Pareto-Optima. Optimization and Operations Research. In: Oettli, W.; Ritter, K. (eds.): Lecture Notes in Economics and Mathematical Systems. No. 117. Berlin: Springer-Verlag, (1976) 127-141

[26] Isermann, H.: Strukturierung von Entscheidungsprozessen bei mehrfacher Zielsetzung. Operations Research Spectrum 1, (1979) 3-26

[27] Dück, W.: Optimierung unter mehreren Zielen. Braunschweig: Vieweg-Verlag, 1979

[28] Fandel, G.: Optimale Entscheidung bei mehrfacher Zielsetzung. In: Lecture Notes in Economics and Mathematical Systems. No. 76. Berlin: Springer-Verlag, 1972

[29] Sattler, H.J.: Eine Herleitung der Zielgewichtung in der Vektoroptimierung aus einer Abstandsfunktionsformulierung. In: Zeitschrift für Angewandte Mathematik und Mechanik. ZAMM 62 (1982) T382-T384

[30] Charnes, A.; Cooper, W.W.: Management Models and Industrial Application of Linear Programming. Vol. 1. New York: Wiley 1961

[31] Fox, R.L.: Optimization Methods for Engineering Design. London: Addison-Wesley, 1971

[32] Osyczka, A.: An Approach to Multicriterion Optimization Problems for Engineering Design, Computational Methods in Applied Mech. and Eng., Vol. 15, (1978) 309-333

[33] Osyczka, A.: An Approach to Multicriterion Optimization for Structural Design, Proceedings of International Symposium on Optimum Structural Design, University of Arizona, 1981

[34] Osyczka, A.: Multicriterion Optimization for Engineering Design, In: Gero, J.S. (ed): Design Optimization. New York: Academic Press Inc., (1985) 193-227

[35] Stadler, W.: Multicriteria Optimization in Engineering and in the Sciences. New York and London: Plenum Press, 1988

[36] Stadler, W.: Multicriteria Optimization in Mechanics (A Survey). Applied Mechanics Rewievs, Vol. 20, (1984) 1442-1471

[37] VDI-Guideline 2212: Systematisches Suchen und Optimieren Konstruktiver Lösungen. VDI-Handbuch Konstruktion, DK631: 658,512,2 (083,132)

[38] Koski, J.; Silvennoinen, R.: Norm Methods and Partial Weighting in Multicriterion Optimization of Structures. International Journal for Numerical Methods in Engineering, Vol. 24 (1987) 1101-1121

[39] Koski, J.: Defectiveness of Weighting Method in Multicriterion Optimization of Structures. Communications in Applied Numerical Methods, Vol. 1 (1985) 333-337

[40] Eschenauer, H.; Post, P.U.; Bremicker, M.: Einsatz der Optimierungsprozedur SAPOP zur Auslegung von Bauteilkomponenten. Bauingenieur 63, II (1988) 515-526

[41] Rozvany, G.I.N.: Structural Design via Optimality Criteria. Dordrecht/Boston/ London: Kluwer Academic Publishers, 1989

PART I

PROCEDURES

2 OPTIMIZATION PROCEDURE SAPOP – A GENERAL TOOL FOR MULTICRITERIA STRUCTURAL DESIGNS

M. Bremicker, H.A. Eschenauer, P.U. Post

2.1 Demands on an Optimization Procedure

As presented in Chapter 1, it is an important goal of engineering activities to improve and optimize technical designs, structural assemblies and structural components. The task of structural optimization is to support the engineer in searching for the best possible design alternatives of specific structures. The "best possible" or "optimal" structure here applies to that structure which mostly corresponds to the designer's desired concept and his objectives meeting at the same time operational, manufacturing and application demands. Compared with the "Trial and Error"-method generally used in engineering practice and based on an intuitive empirical approach, the determination of optimal solutions by applying mathematical optimization procedures is more reliable and efficient. These procedures can be expected to be more frequently applied in industrial practice. In order to apply structural optimization methods to an optimization task, both the design objectives and the relevant constraints must be expressed by means of mathematical functions. One example of a design objective is the demand for the maximum degree of stiffness of a structure which can be described by the objective "minimization of the maximum structural deformation". The design variables are the parameters of the structure, for example the cross-sectional and geometrical quantities, which should be selected in a way that the objective function can be minimized by considering additional conditions. These conditions or constraints are equality and inequality equations which include the mathematical formulation of demands such as permissible stresses, stability criteria etc. The formulation of the scalar design problem is generally given by (1-1):

$$\min_{\mathbf{x} \in \mathbb{R}^n} \{ f(\mathbf{x}) \mid \mathbf{h}(\mathbf{x}) = \mathbf{0} \; ; \; \mathbf{g}(\mathbf{x}) \le \mathbf{0} \} \; . \qquad (2\text{-}1)$$

The solution of optimization problems requires software systems which are easy to use, provide sufficient efficiency, and are available for practical application. Several optimization algorithms should be linked to structural analysis procedures in a suitable manner by means of optimization model processors [1, 2,45].

In general, a software system should meet the following requirements:

- possibility of selecting the suitable optimization algorithm for an optimization problem from a number of efficient methods,
- use of different methods for structural analysis such as finite and analytical methods,
- application of automatic design and evaluation models (pre- and post-processors) for a wide range of standard problems in optimization modelling; simple integration of special optimization models if required,
- modularized architecture with standard interfaces between the different program modules,
- extension of the program system by integrating additional modules without comprehensive implementation work,
- reduction of the numerical effort particularly for large structural optimization tasks by applying efficient algorithms (e.g. sensitivity analysis of FE-structures, solution methods for linear and nonlinear equation systems etc.),
- efficient data management for large optimization problems (large scale systems), utilization of modern programming techniques (parallel computing),
- supporting facilities for input data generation and output documentation.

2.2 Structure of the Optimization Procedure

2.2.1 Definitions

Before describing an optimization procedure and its practical realization some of the terms frequently used in this chapter shall be defined (see Figs. 2/1-2/3):

Optimization algorithm	: mathematical procedure for constrained/ unconstrained optimization (optimality criteria methods, mathematical programming methods),
Optimization procedure	: total concept involved when dealing with an optimization problem,
Optimization strategy	: approach used to reduce complex optimization problems to simplified substitute problems or smaller subproblems, respectively,
Objective function(s)	: mathematical formulation of a design objective or, in accordance with the given requirements, of several design objectives,
Preference function	: transformation of several objective functions into one scalar substitute objective function,
Constraints	: mathematically formulated design requirements which are not covered by the objective function(s),
Structural model	: mathematical description of the structural behaviour (mathematical-physical model),
Structural parameters	: parameters of the structural model,
Analysis variables	: structural parameters which can be varied during optimization computations,
State variables	: response of the structural model,
Design variables	: design quantities to be varied,
Initial design	: initial values of the design variables at the beginning of the optimization process,
Design model	: mathematical link between design variables and analysis variables,
Transformation	: transformed design variables for approximation concepts,
Evaluation model	: mathematical link between state variables and objective function and constraint values under consideration of optimization strategies,
Optimization model	: a comprehensive term for design and evaluation model(s).

2.2.2 Three-Columns Concept

When dealing with a structural optimization problem, it is recommendable to proceed following the "Three-Columns Concept" [1] (Fig. 2/2.).

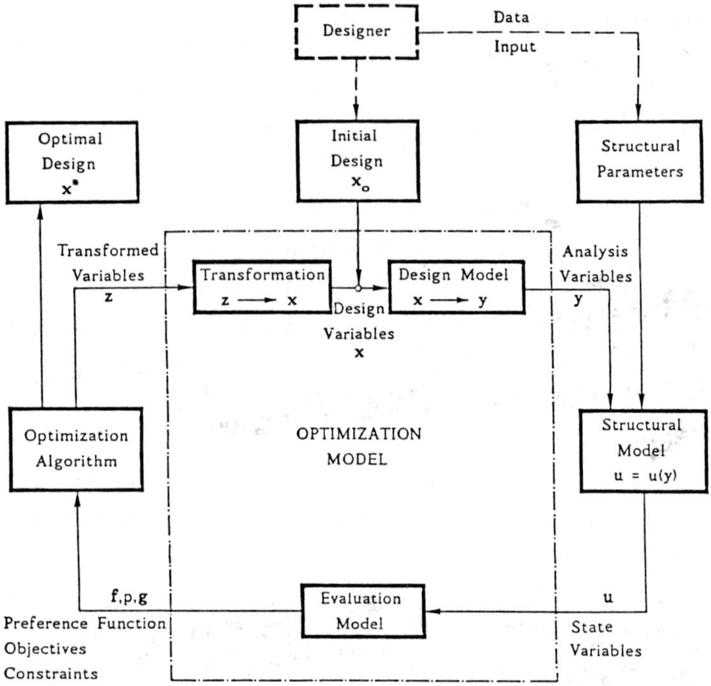

Fig 2/1. Structure of an optimization loop

The first step is the theoretical formulation of the optimization problem taking into account all relevant demands on the structure. The next step involves the solution of the subproblems "structural modelling" and "optimization modelling". From the third column an optimization algorithm is selected and linked with the structural and the optimization model to form an optimization procedure. In the following a detailed description of the columns is given.

Column 1: Structural model

Any structural optimization requires the mathematical determination of the physical behaviour of the structure. In the case of mechanical systems, this refers to the typical structural response subject to static and dynamic loading such as deformations, stresses, eigenvalues, etc. Furthermore, information on the stability behaviour (buckling loads) has to be determined. All state variables required for the objective function and constraints have to be provided. The structural calculation is carried out using efficient analysis procedures such as the finite element method or transfer matrices methods. In order to ensure a wide field of application, it should be possible to adapt several structural analysis methods.

Column 2: Optimization modelling

From an engineer's point of view, this column is the most important one of the optimization procedure. First of all, the analysis variables which are to be changed during the optimization process are selected from the structural parameters. The design model including variable linking, variable fixing, shape functions etc. provides a mathematical link between the analysis variables and the design variables. In order to increase efficiency and improve the convergence of the optimization, the optimization problem is adapted to meet the special requirements of the optimization algorithm by transforming the design variables into transformation variables. By using this approach, it is e.g. possible to almost linearize the stress constraints of a sizing optimization problem. Additionally, objective functions and constraints have to be determined by procedures that evaluate the structural response or state variables. When formulating the optimization model, the engineer has to consider the demands from the fields of design, material, manufacturing, assembly and operation.

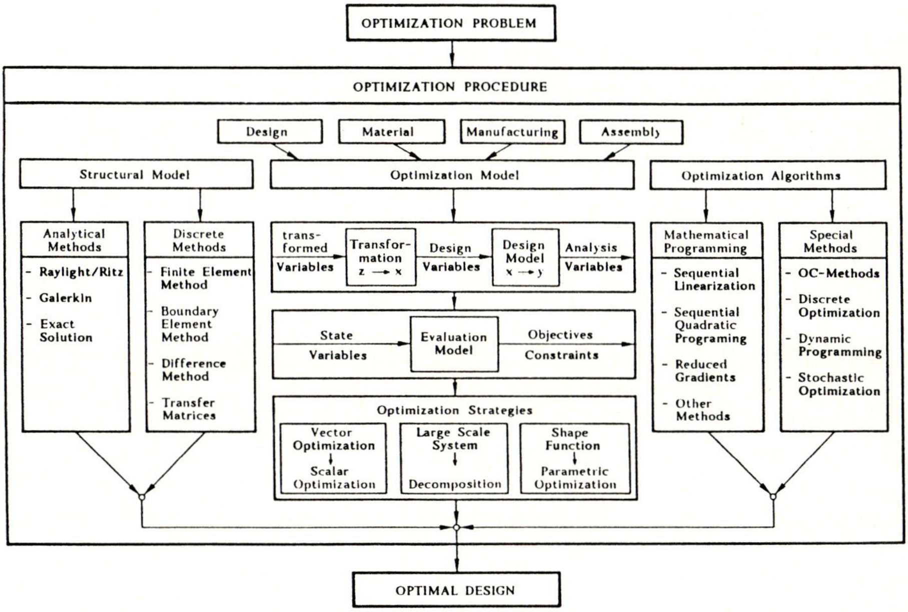

Fig. 2/2. Three-Columns Concept for solving optimization problems

Column 3: Optimization algorithms

In recent years mathematical programming algorithms have been preferred to other methods for solving nonlinear constrained optimization problems. These algorithms are iterative procedures which, proceeding from an initial design $\mathbf{x}_0$, generally provide an improved design variable vector $\mathbf{x}_k$ as a result of each iteration k. The optimization is terminated if a breaking-off criterion responds during an iteration. Numerous studies have demonstrated that the selection of the optimization algorithm is problem-dependent. This is particularly important for a reliable optimization and a high level of efficiency (computing time, rate of convergence). If, for example, all iteration results have to lie within the feasible domain, an algorithm that iterates within the feasible domain (e.g. generalized reduced gradients (GRG)) should be applied.

2.3. Basic Ideas of the Procedure SAPOP

On the basis of the "Three-Columns Concept" and on the requirements mentioned above, the software system SAPOP (**S**tructural **A**nalysis **P**rogram and **O**ptimization **P**rocedure, Fig. 2/3.) was developed. It

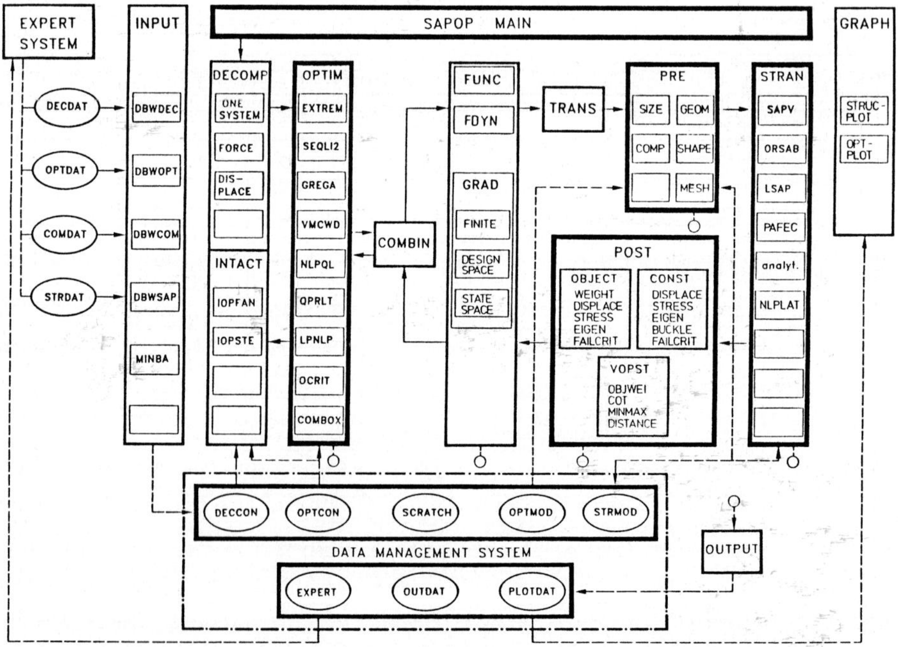

Fig. 2/3. Block diagram of the program system SAPOP

consists of three independent parts communicating with each other via a Data Management System (data base). Each of these parts is divided into individual blocks connected by standardized interfaces to ensure the largest possible modularity. Each block contains a number of interchangeable modules. When carrying out an optimization computation only those modules which are actually needed are linked together.

2.3.1 Problem Formulation and Input Data

The input system is used to prepare the input data provided by the user to be stored on a database. All quantities required to describe the structural and the optimization model as well as the parameters for controlling the optimization process are edited here. The user has to provide at least two different data items. The data file OPTDAT includes all data necessary to control the optimization process as well as the initial values of the design variables and the input quantities for the formulation of the optimization model. The data file STRDAT includes the input data for the software system which is applied to the structural analysis. The module MINBA includes a band-width-minimizer for FE-structures.

The input data for a multilevel optimization using a decomposition strategy are provided by the file DECDAT; COMDAT supplies the relevant material specifications of fibre-reinforced composites. In future, the user will be supported by an expert system EXPERT when generating the input data.

2.3.2 SAPOP Main Module

The optimization computation is actually carried out by the SAPOP main module MAIN. First of all, an initialization phase is run, and subsequently the optimization is started via the ONE-SYSTEM module of the DECOM block. Two decomposition strategies (cutting force method FORCE and deformation method DISPLACE [12]) allow to optimize large structures by optimizing substructures. A number of different optimization algorithms can be called by the DECOM-modules. Apart from the seven mathematical programming methods, an optimality criteria procedure (stress-ratio method) is available. It is also possible to couple different algorithms by means of a series connection (serial hybrid approach). Among others, the following algorithms can be applied:

COMBOX	:	Extended Version of the **COM**plex Method by Box [3],
EXTREM	:	**D**irect Search Algorithm by Jacob [4],
SEQLI2	:	**SEQ**uential **LI**near Optimization Extended Version [2-6],
VMCWD	:	**V**ariable **M**etric Method for **C**onstrained Optimization Including **W**atch **D**og Technique [7],
NLPQL	:	**N**onlinear **P**rogram with **Q**uadratic **Line** Search [8],
GREGA	:	**G**eneralized **RE**duced **G**radient **A**lgorithm [9-11],
QPRLT	:	**Q**uadratic **P**rogramming with **R**educed **L**ine-Search **T**echnique [12,13],
LPNLP	:	**L**agrange **P**enalty Method for **N**on**L**inear **P**roblems [14],
OCRIT	:	**O**ptimality **CRIT**eria Method (Stress Ratio Method) [15].

For each iteration the actual values of the objective function and constraints are required, and for most of the algorithms the gradients have to be calculated with regard to the transformation variables. The control program FUNC for structural analyses and the control programs for sensitivity analyses (gradient calculations, Section 2.4.3) are called via the interface module COMBIN. Module FDYN is an algorithm for solving time-dependent optimization problems [41].

The transformation module TRANS shifts the transformation variables into design variables. The subsequently called PRE-processor contains different design models used to determine the analysis variables from the design variables. The design model SIZE includes variable linking and variable fixing for cross-section optimization (sizing). SHAPE, GEOM and MESH modules can be used for shape and geometry optimization tasks. As far as composite designs are concerned, the module COMP transforms design variables into layer thicknesses and ply angles of a fibre composite lamina, and the corresponding material characteristics (elasticity, stiffness, thermal and hygrothermal coefficients) are calculated. If a special design model is to be used to solve an optimization problem, a corresponding program module can be included, or the entire preprocessor can be exchanged.

The structural analysis is now carried out using the updated analysis variables. These are part of the structural parameters of the mathematical-mechanical model which describes the physical behaviour of the actual structure. Systems of algebraic or differential equations are solved by using efficient numerical methods. At present, the following structural analysis methods are available in SAPOP:

SAPV-2 : **S**tructural **A**nalysis **P**rogram 5 [16], Finite-Element Method. The modules for linear displacement, stress and eigenvalue analysis are integrated in SAPOP;

ORSAB : **O**rthotrope **R**otations**s**chalen unter **a**llgemeiner **B**elastung, Transfer-Matrix Method for arbitrary loaded isotropic and orthotropic shells of revolution [1,46,47];

LSAP : **L**aminated **S**hell **A**nalysis **P**rogram [17], Finite Difference Method for anisotropic shells of fiber composite material;

NLPLAT : **N**onlinear Analysis of **L**aminated **P**lates [41], Finite Difference Method for anisotropic composite plates with time-dependent material behaviour and imperfections;

PAFEC : **P**rogram for **A**utomatic **F**inite **E**lement **C**alculation [49];

ANSYS : Finite-Element-Program of Swanson **A**naly**s**is **Sys**tems.

Apart from these programs, the user can link his own structural analysis modules to SAPOP. Thus, it is possible to deal with structures using analytical calculations or to deal with any examples from interdisciplinary fields. For the latter, however, other pre- and postprocessors are usually required in order to formulate the design and evaluation model.

The computed state variables are transferred to the postprocessor in order to determine the objective function(s) and constraints. Modules are available for computing weight as well as stress, deformation, and eigenvalue evaluations. In the case of composite structures the failure criteria of laminate composites are determined [20-23].

Multicriteria optimization problems are solved by transforming the objective function into a scalar substitute function (preference functions, see (1-10) to (1-24)).

If the range of performace of the postprocessor is not sufficient for a special application, user-defined programs for the formulation of objective functions and constraints can be linked via standardized interfaces. The actualized objective functions and constraints are transferred to the optimization algorithm via the modules FUNC and COMBIN. For a sensitivity analysis, the PRE-STRAN-POST-loop is run several times before the gradients of the objective function and constraints are available.

The data exchange between the individual blocks of the SAPOP main module is carried out by the Data Management System. Thus, all program segments have fast access to a large dataset which cannot be kept in core completely. The storage capacity is efficiently used and modularity is ensured. The Data Management System contains input and output control routines to read and write direct access files whereby the data items are identified by a character specification.

2.3.3 Evaluation of the Optimization Results

An important part of structural optimization is the critical judgement of the optimization results. In SAPOP every optimization computation is fully documented. The input module INPUT, which can be run separately, checks all input data and prints all data transferred into the Data Management System. In the main module SAPOP, the results of each optimization step are printed via the output module OUTPUT. The output for all optimization algorithms is standardized. Besides, the entire output of the structural analysis and of the pre- and postprocessor can be made available for the initial and final design.

Two graphic programs are implemented into SAPOP to get a fast survey of the optimization results. All data important for the optimization history (e.g. objective function(s) and constraints for several iterations, design variables in comparison) can be displayed with the OPTPLOT module. Furthermore, several optimization runs can be compared. For FE-structures it is possible to demonstrate the structural model using the module STRUCPLOT. These structural plots are important for the evaluation of geometrical and shape optimization.

2.4 Optimization Modelling

During the development of the SAPOP program system it was a fundamental aim to implement pre- and postprocessors for standard optimization modelling problems. First of all, this chapter gives a brief description of the design and evaluation models.

2.4.1 Design Models

The task of a design model is to calculate the analysis variables **y** from the design variables **x** by using a unique mapping rule:

$$\mathbf{y} = \mathbf{f}\,(\mathbf{x}) \;,\; \mathbf{x} \in \mathbb{R}^{n} \;,\; \mathbf{y} \in \mathbb{R}^{ny} \;. \tag{2-2}$$

The analysis variables are a subset of the structural parameters which are required to describe the physical behaviour. In structural optimization problems, the analysis variables are usually sizing quantities (e.g. thicknesses, cross-sections, moments of inertia etc.), geometrical dimensions, shape parameters or material quantities. Particularly when using discrete structural analysis methods (FE-methods, methods of finite differences etc.), it is important that all structural elements as well as the node topology are determined by the design model. One part of design modelling can be carried out by linear mapping

$$\mathbf{y}\,(\mathbf{x}) = \mathbf{A}\,\mathbf{x} + \mathbf{y}_0 \;. \tag{2-3}$$

Matrix **A** is the coordination matrix; the vector $\mathbf{y}_0 \in \mathbb{R}^{ny}$ is a constant vector. Due to a special structure of the coordination matrix, various design models can be realized. Variable linking is achieved if just one element in each row has the value "one" and if all other elements are zero. The columns of the matrix, however, contain several "one"-elements. Each analysis variable corresponds to just one design variable whereas a design variable can be allocated to several analysis variables. The addition of a constant element to a design variable allows one to consider a constant part of an analysis variable. With variable fixing a part of the analysis variable vector equals just this constant element, i.e. that all elements of the corresponding rows of the coordination matrix are zero. Another possibility is the superposition of design variables to an analysis variable where several elements of a row obtain the value "one". The more general case of this design model is the linear transformation with the analysis variables as linear functions of the design variables [45].

In SAPOP a design model with variable linking and variable fixing is employed for cross-section optimization (sizing). The moments of inertia and the section moduli of beam elements result from a linear transformation whereby each row of the coordination matrix has one non-zero element differing with a corresponding similarity factor. This ensures geometrical similarity in case of a variation of the cross-sections of a beam. The formulation of the design model for optimizing composite material components refers to the same principle. Here, the data of the material law is obtained by using the analysis variables "fibre angle" and "layer thickness".

As far as a geometrical optimization is concerned, the design variables represent the components of the vector between two given points of the structure. By adding these components and the respective constant elements, all points of a structure can be defined. This corresponds to the design model of the additive superposition. It is also possible to interpolate intermediate points. The coefficients of the coordination matrix are determined by means of simple geometrical considerations.

For shape optimization problems, SAPOP provides various shape functions to define the surface shape of a component. They allow the transformation into a parametric optimization problem [2,12,28]. Among others, the following approach functions are used: Lagrange polynomials, Bèzier curves and Bèzier surfaces, B-spline curves and surfaces, modified ellipse functions etc. In order to determine the free coefficients of most of the shape functions control points are employed. The coordinates of these control points are determined from the design variables by using the design model for geometrical optimization described above [12].

2.4.2 Evaluation Models

The expression "state variables **u**" of a structural mechanical system refers to quantities such as deformations, stresses, resulting forces, strains, eigenvalues etc. These state variables depend on the design variables via the structural model. The task of the evaluation model is to formulate the objective function vector **f** and the constraints **g** as functions of these state variables.

Generally, the constraints represent limiting demands on the state variables; i.e. the quantities of a state variable u_i should be bounded by an upper and lower limit. This demand provides two constraints which are standardized for numerical reasons:

$$u_i^l \le u_i \le u_i^u \longrightarrow g_1 = \frac{u_i}{u_i^u} - 1 \le 0 \, , \; g_2 = 1 - \frac{u_i}{u_i^l} \le 0 \, . \qquad (2\text{-}4)$$

The limiting values u_i^u and u_i^l can also depend on the design variables. According to definition (2-3), SAPOP calculates the deformation and eigenvalue constraints. The stress constraints are maximum bounds whose formulation requires stress criteria. Buckling constraints can be employed in order to optimize trusses. As far as composite optimization is concerned, various failure criteria are available (fibre break and bonding break criteria [21], max-stress criterium, max-strain criterium [22], failure criteria by Tsai-Hill [20] and Tsai-Wu [23]). For large structures it is advantageous to reduce the number of constraints. For this purpose SAPOP provides a "constraint linking strategy" which determines the most critical constraint value of a certain constraint type in a defined domain. Apart from the constraints already mentioned, upper and lower bounds of the design variables are considered.

The majority of papers published on examples of structural optimization deals with the minimization of the structural weight. However, for many applications other objectives can also be important as there are e.g. the minimization of deformations of highly accurate systems, the minimization of stresses in order to reduce stress concentrations, or the maximization of certain eigenvalues of a structure. For this reason SAPOP provides different types of objective functions, i.e. all state variables necessary for the formulation of constraints can also be used as an objective. Due to the nonlinearity of these objective functions, they are more difficult to treat than structural weight.

2.4.3 Sensitivity Analysis

Gradients of objective function(s) and constraints are determined by the sensitivity analysis with regard to the design variables (GRAD-module in Fig. 2/3.). These gradients are necessary for most optimization algorithms. They can, however, also provide the design engineer with information about the sensitivity of the structure due to changes of the design variables. Introducing the state variables $\mathbf{u} \in \mathbb{R}^{ny}$ the objective functions and constraints read as follows:

$$\mathbf{f} = \mathbf{f}\,(\mathbf{u}, \mathbf{x}) \quad \text{and} \quad \mathbf{g} = \mathbf{g}\,(\mathbf{u}, \mathbf{x}) \quad \text{with} \quad \mathbf{u} = \mathbf{u}\,(\mathbf{x}). \qquad (2\text{-}5)$$

The following explanations concentrate on the objective functions and can similarly be applied to the constraints. At first, the total differentiation yields:

$$\mathbf{df}(\mathbf{u}, \mathbf{x}) = \frac{\partial \mathbf{f}}{\partial \mathbf{u}} \mathbf{du} + \frac{\partial \mathbf{f}}{\partial \mathbf{x}} \mathbf{dx} \text{ with } \frac{\partial \mathbf{f}}{\partial \mathbf{u}} \equiv \left[\frac{\partial f_i}{\partial u_j}\right]_{m \times n_u} \tag{2-6a}$$

By eliminating **du** it is possible to determine the sensitivity matrix $\mathbf{A}_f$ whose i-th line corresponds to the derivatives $\mathbf{df}/\mathbf{dx}$:

$$\mathbf{df} = \mathbf{A}_f \, \mathbf{dx} \, . \tag{2-6b}$$

The sensitivity analysis can be carried out by the following methods:

a) Numerical sensitivity analysis (Method of forward differences)

This simple sensitivity analysis method allows to determine the coefficients of the sensitivity matrices by approximation with forward differences:

$$a_{fji} \approx \frac{f_j\,[\mathbf{u}\,(\tilde{\mathbf{x}}_i), \tilde{\mathbf{x}}_i] - f_j\,[\mathbf{u}\,(\mathbf{x}_i), \mathbf{x}_i]}{\varepsilon\, x_i} \tag{2-7}$$

with $\tilde{\mathbf{x}}_i = \mathbf{x} + \varepsilon\, x_i \mathbf{e}_i$,

ε step size factor ,

$\mathbf{e}_i$ i-th unit vector .

The numerical method can be quite easily implemented into the computer and is completely independent of the optimization and structural model. Thus, it can be used for any optimization problem even those outside the field of structural optimization. There is, however, the disadvantage that for n design variables n+1 analyses are required. This leads to very long computing times for large optimization problems.

b) Analytical methods

In order to reduce the computing time, analytical methods are used for gradient calculations [25-27]. These methods are closely linked to the structural analysis program, and their implementation requires changes of the source codes. The analytical methods were originally developed for FE-methods but can be generalized and applied to all structural analysis procedures which transform the differential equations (e.g.

transfer matrix methods, difference methods) into an ordinary equation system. In the case of linear systems the system equations are:

$$\mathbf{F}\,\mathbf{u} = \mathbf{r} \quad \text{with} \quad \mathbf{F} = \mathbf{F}(\mathbf{x}) \quad \text{and} \quad \mathbf{r} = \mathbf{r}(\mathbf{x}) \; . \tag{2-8}$$

In (2-8) $\mathbf{r}$ represents the load vector and $\mathbf{F}$ the system matrix. In most procedures the latter one is calculated from a sum or a product of submatrices of various structural elements:

$$\mathbf{F}(\mathbf{x}) = \sum_{i=1}^{ne} \mathbf{F}^{i}(\mathbf{x}) \quad \text{or} \quad \mathbf{F}(\mathbf{x}) = \prod_{i=1}^{ne} \mathbf{F}^{i}(\mathbf{x}) \; . \tag{2-9}$$

The use of the chain rule (2-6a) leads to

$$df(\mathbf{u}, \mathbf{x}) = \left[\frac{\partial f}{\partial \mathbf{u}} \frac{\partial \mathbf{u}}{\partial \mathbf{x}} + \frac{\partial f}{\partial \mathbf{x}} \right] d\mathbf{x} \; . \tag{2-10}$$

With (2-8) the derivatives of the state variables can be replaced by

$$\frac{\partial \mathbf{u}}{\partial \mathbf{x}} = \mathbf{F}^{-1}\,\mathbf{P} \quad \text{with} \quad \mathbf{P} = \frac{\partial \mathbf{r}}{\partial \mathbf{x}} - \frac{\partial \mathbf{F}}{\partial \mathbf{x}}\,\mathbf{u} \; , \tag{2-11}$$

which results in

$$\frac{\partial f}{\partial \mathbf{u}} \frac{\partial \mathbf{u}}{\partial \mathbf{x}} = \frac{\partial f}{\partial \mathbf{u}} \left(\mathbf{F}^{-1}\,\mathbf{P} \right) = \left(\frac{\partial f}{\partial \mathbf{u}}\,\mathbf{F}^{-1} \right) \mathbf{P} \, . \tag{2-12}$$

There are two ways of dealing with (2-12). When using the design space method, the derivatives of the state variables are formed from the second product according to (2-11) whereas with the state space method the first product is calculated first. For this reason an adjoint variable matrix $\mathbf{\Lambda}$ is introduced and corresponds to the product:

$$\mathbf{\Lambda}^{T} = \frac{\partial f}{\partial \mathbf{u}}\,\mathbf{F}^{-1} \longrightarrow \mathbf{F}^{T}\,\mathbf{\Lambda} = \left(\frac{\partial f}{\partial \mathbf{u}} \right)^{T} = \mathbf{P}^{*} \, . \tag{2-13}$$

Eqs. (2-11) and (2-12) then yield:

$$\frac{\partial f}{\partial \mathbf{u}} \frac{\partial \mathbf{u}}{\partial \mathbf{x}} = \left[(\mathbf{F}^{-1})^{T} \left(\frac{\partial f}{\partial \mathbf{u}} \right)^{T} \right]^{T} \mathbf{P} = \mathbf{\Lambda}^{T}\,\mathbf{P} \, . \tag{2-14}$$

The pseudo-load matrix $\mathbf{P}$ introduced in (2-11) contains $\mathbf{n}$ columns. A pseudo-load column specifies the load necessary to bring forth the same change of the state variables as a slight alteration of the corresponding design variables.

Depending on the problem, the analytical methods can reduce the computing time to a fraction of the time required by the numerical method. This mainly results from the fact that both analytical methods mentioned above carry out the sensitivity analysis by factorizing the system matrix only once.

The two analytical methods differ in the numerical effort required to solve the equation systems (2-11) and (2-13). For each load case one gets n right-hand sides when applying (2-11) (n is the number of design variables). In order to evaluate (2-12), each objective function and constraint lead to a load column. The number of load columns can be reduced in various ways. When considering an "active set"-strategy [26] only the gradients of the active constraints are required as far as the state space method is concerned. If several objective functions exist, the preference function can replace the objective function vector. With optimization algorithms based on penalty procedures or the method of multipliers, the constraints and objective functions are replaced by the penalty function so that (2-14) contains one right-hand side only. Therefore, with regard to the numerical effort required, the state space method is in most cases to be preferred to the direct method. The implementation of the state space method. however, requires more changes of the structural analysis program.

c) Semianalytical methods

The pseudo-load matrix **P** necessary for the analytical procedures is formed from the derivatives of the system matrix **F** and the load vector **r**. In several sensitivity analysis applications size optimization of FE-structures with element matrices which are simple polynomial functions of just one design variable is carried out, i.e. the derivative of these matrices can be carried out on the element level analytically.

For more general problems such as shape optimization problems, it is difficult to establish a general analytical method. Therefore, it is suggested that the problem should be approached semianalytically which means that the necessary derivatives of the submatrices in (2-9) are determined by a numerical differentiation as in (2-7):

$$\frac{\partial \mathbf{F}}{\partial x_i} \approx \sum_{j=1}^{n_e} \frac{\mathbf{F}^j(\tilde{\mathbf{x}}_i) - \mathbf{F}^j(\mathbf{x})}{\varepsilon x_i} \quad \text{or} \quad \frac{\partial \mathbf{F}}{\partial x_i} \approx \sum_{j=1}^{n_e} \frac{\mathbf{F}^j(\tilde{\mathbf{x}}_i) - \mathbf{F}^j(\mathbf{x})}{\varepsilon x_i} \prod_{\substack{k=1 \\ k \neq j}}^{n_e} \mathbf{F}^k(\mathbf{x})$$

$$\text{with} \quad \tilde{\mathbf{x}}_i = \mathbf{x} + \varepsilon x_i \mathbf{e}_i \,. \qquad (2\text{-}15)$$

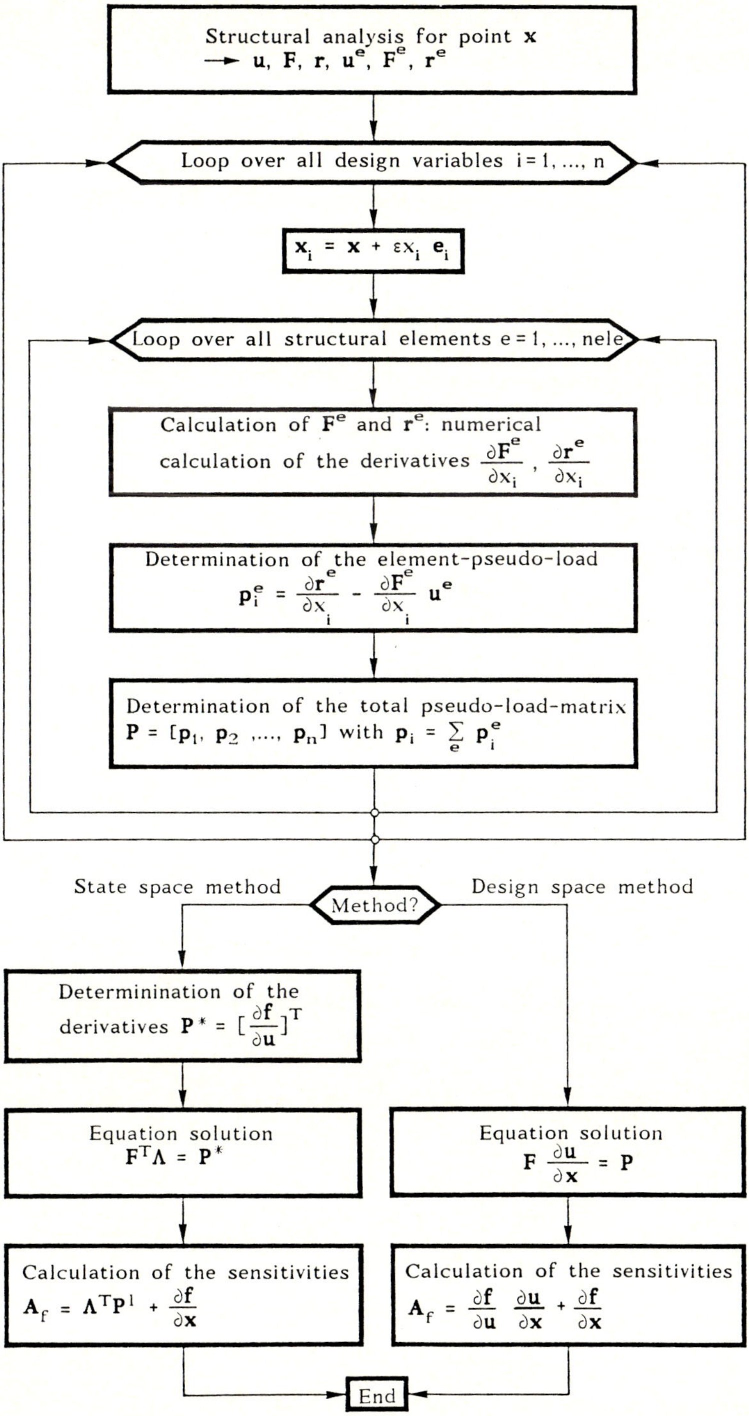

Fig. 2/4. Block diagram of the semianalytical sensitivity analysis

Here, the sum only considers those submatrices which depend on the design variable x_i. A flow chart of the semianalytical sensitivity analysis is shown in Fig.2/4. The design space method for static structural response and eigenvalues has been implemented into SAPOP.

2.5 Description of Two Optimization Algorithms

It is difficult to prefer a certain algorithm for nonlinear problems of structural optimization because their performance is problem-dependent. During the development of an optimization procedure, a large number of efficient algorithms has been established and tested in order to reduce the computing time and to increase the rate of convergence.

In the following, two optimization algorithms which successively solved various complex optimization problems will be briefly described.

2.5.1 Sequential Linear Programming Method (SEQLI) [2,6]

For the nonlinear problem, a subproblem which can be easily described in an analytical way is formulated and solved with appropriate strategies. In sequential linear programming the solution of a nonlinear optimization problem is achieved by a sequence of linearizations.

Cutting plane methods do not only take into account linearized constraints of the current iteration but also those of former iterations, i.e. with every iteration new linearized inequality constraints are added, and a part of the previous domain is cut off. The solutions of the linear substitute problems are found by the SIMPLEX-method; the objective function is linearized.

By introducing lower and upper bounds of all design variables (hypercube, move limits), Griffith and Stewart [5] extended the application to problems with solutions not lying in the intersection of constraints but generally on a hypersurface. Objective functions and constraints of the nonlinear scalar starting problem are expanded into a Taylor series at a point $\mathbf{x}^k$. Considering the linear components only, the following equations are established:

$$\begin{aligned} f\,(\mathbf{x}^k + \partial\mathbf{x}) &\approx f\,(\mathbf{x}^k) + \nabla f(\mathbf{x}^k)\partial\mathbf{x} \;, \\ h_i\,(\mathbf{x}^k + \partial\mathbf{x}) &\approx h_i\,(\mathbf{x}^k) + \nabla h_i(\mathbf{x}^k)\partial\mathbf{x} \;, \qquad i = 1,2,\ldots,q \;, \\ g_j\,(\mathbf{x}^k + \partial\mathbf{x}) &\approx g_j(\mathbf{x}^k) + \nabla g_j(\mathbf{x}^k)\partial\mathbf{x} \;, \qquad j = 1,2,\ldots,p \;. \end{aligned} \tag{2-16}$$

The design space of the linearized problem is additionally restricted by a hypercube

$$x_{il}^{k} \le x_i \le x_{iu}^{k} \qquad i = 1,2,\ldots,n \qquad (2\text{-}17)$$

because Taylor's expansion is valid for small $\partial \mathbf{x}$ only. Solutions of (2-16) are found by the SIMPLEX-method. For this, the variables need to be positive, and a linear transformation of the variables is carried out:

$$y_i = \partial x_i + (x_{iu}^{k} - x_{il}^{k}) \ , \qquad i = 1,2,\ldots,n \ . \qquad (2\text{-}18)$$

The linearized problem then reads:

$$\underset{\mathbf{y}}{\mathrm{Min}} \ \{\mathbf{c}^T\mathbf{y}\} = \mathbf{c}^T\mathbf{y}^* \ , \qquad \text{with} \quad \mathbf{c} = \nabla f(\mathbf{x}^k) \qquad (2\text{-}19)$$

as well as the linearized constraints $h_i(\mathbf{y})$, $g_j(\mathbf{y})$ and the hypercube according to (2-16) and (2-17).

The solution $\mathbf{y}^*$ of the linearized problem, found by Dantzig's SIMPLEX-algorithm, leads to an improved point $\mathbf{x}^{k+1}$ for the nonlinear model.

The hypercube is reduced by means of the correction rules:

$$x_{il}^{k+1} = x_i^{k+1} - \frac{1}{2}\alpha^{k+1}(x_{iu}^{0} - x_{il}^{0}), \qquad (2\text{-}20)$$

$$x_{iu}^{k+1} = x_i^{k+1} + \frac{1}{2}\alpha^{k+1}(x_{iu}^{0} - x_{il}^{0}), \qquad (2\text{-}21)$$

with $\alpha^0 = 1$, $\alpha^{k+1} = \dfrac{\alpha}{1+\alpha^k}$.

As $\lim \alpha^k = 0$ for $k \to \infty$, the lateral lengths of the hypercube become continually smaller within the optimization process. Fig. 2/5 shows the optimization algorithm SEQLI in the two-dimensional case. It becomes evident that the optimization process very much depends on the choice of the starting hypercube and that the convergence behaviour is faster due to the number of active constraints. Fig. 2/6 demonstrates the iterative process of the SEQLI-approximation method.

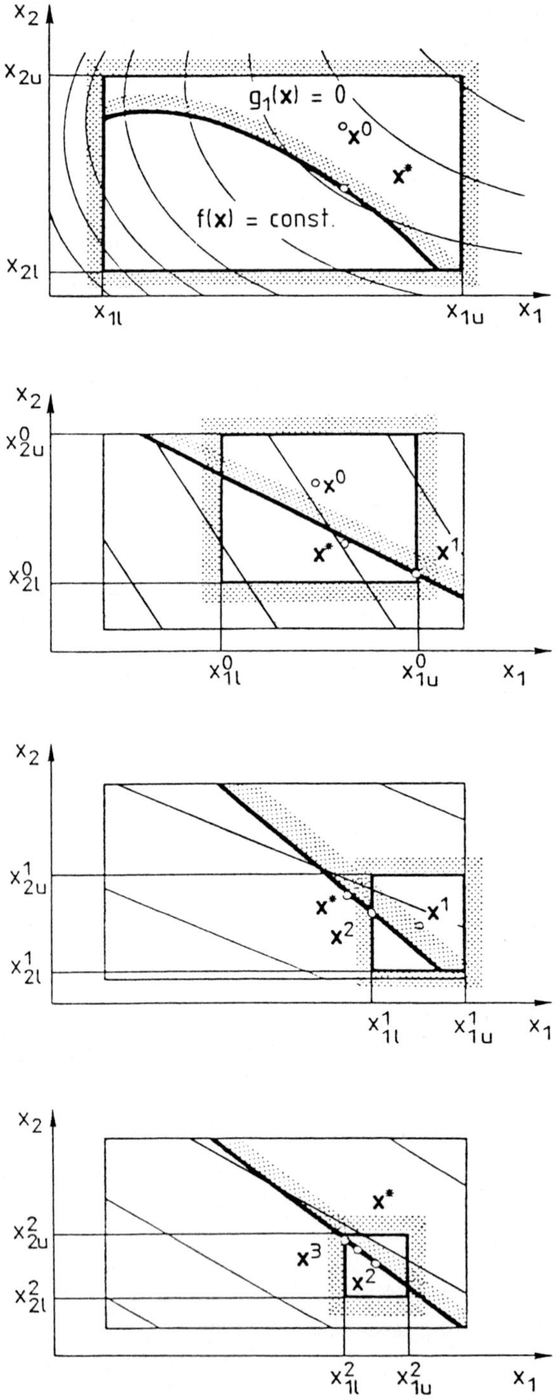

Nonlinear Starting Problem

1. Approximation

2. Approximation

3. Approximation

Fig. 2/5. Optimization procedure SEQLI in the two-dimensional case

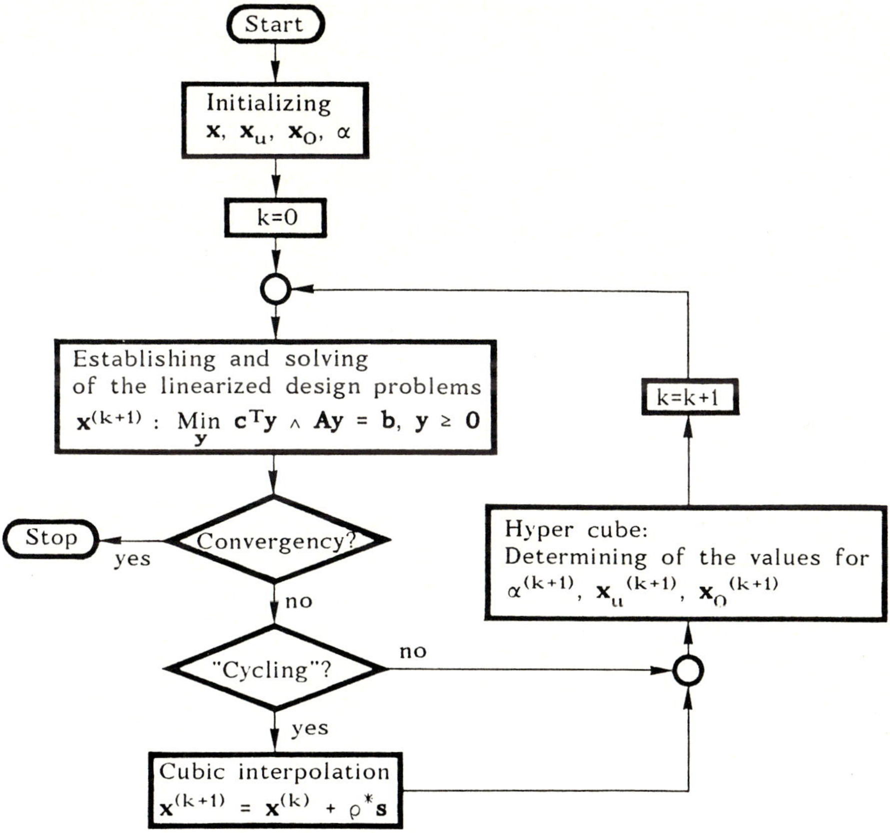

Fig. 2/6. Block diagram of SEQLI-method

2.5.2 Hybrid Method of Quadratic Programming with Reduced Line-Search Technique (QPRLT) [12]

The algorithm of the generalized reduced gradients (GRG) has proved to be an appropriate and efficient optimization procedure [9]. It belongs to the group of the Line-Search Procedures whose main operations are given in Fig. 2/7a.

Several optimization variables (basic variables) are eliminated from the objective function by means of linearized active constraints:

$$(\mathbf{x}^b - \mathbf{x}_0^b) = -(\mathbf{J}^b)^{-1}\mathbf{J}^{nb}(\mathbf{x}^{nb} - \mathbf{x}_0^{nb}) \tag{2-22}$$

with $\mathbf{x}^{b}$ basic variables (= dependent variables) ,

$\mathbf{x}^{nb}$ nonbasic variables ($\hat{=}$ independent variables) ,

$$\mathbf{J}^{b} = \left\{ \nabla_{x^b}\mathbf{h}_1, \nabla_{x^b}\mathbf{h}_2, \ldots, \nabla_{x^b}\mathbf{h}_p \right\}^T ,$$

$$\mathbf{J}^{nb} = \left\{ \nabla_{x^{nb}}\mathbf{h}_1, \nabla_{x^{nb}}\mathbf{h}_2, \ldots, \nabla_{x^{nb}}\mathbf{h}_p \right\}^T ,$$

$$h_j = g_j - x_{n+j} , \qquad x_{n+j} \geq 0 \quad \text{slack variables} .$$

By substitution one gets the "reduced objective function" that is now only a function of the nonbasic variables

$$f^R(\mathbf{x}^{nb}) = f\left\{\mathbf{x}_0^b - (\mathbf{J}^{nb})^{-1}\mathbf{J}^b(\mathbf{x}^{nb} - \mathbf{x}_0^{nb}), \mathbf{x}^{nb}\right\} . \tag{2-23}$$

The gradient of this function

$$\nabla f^R(\mathbf{x}_o) = -\mathbf{d}^{nb} = \nabla_{x^{nb}} f(\mathbf{x}_o) - \left[(\mathbf{J}^b)^{-1}\mathbf{J}^{nb}\right]^T \nabla_{x^b} f(\mathbf{x}_o) \tag{2-24}$$

is used as a search direction d^{nb} for the nonbasic variables. Eq. (2-24) gives the search direction for the basic variables

$$\mathbf{d}^b = -(\mathbf{J}^b)^{-1}\mathbf{J}^{nb}\mathbf{d}^{nb} \tag{2-25}$$

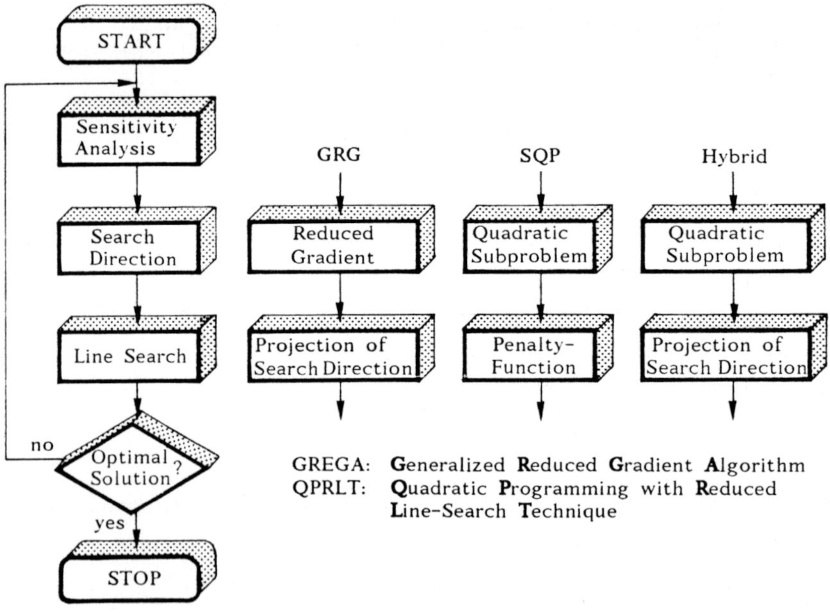

Fig. 2/7. Hybrid SQP-GRG algorithm

When carrying out a one-dimensional minimization (line search) along the complete search direction $\mathbf{d}_k^T = \{\mathbf{d}_k^{nb}, \mathbf{d}_k^b\}$ which represents a tangent line to the hypersurface of the active constraints and is thus approximately feasible, the search direction is projected to the hypersurface of the active constraints (Fig. 2/8) in order to maintain exact feasibility. In each step of the line search

$$\underset{\alpha > 0}{\text{Min}} \; f(\mathbf{x}_k + \alpha \mathbf{d}_k) \tag{2-26}$$

with the step lenght α, the nonlinear equation system

$$\mathbf{h}(\mathbf{x}_k^{nb} + \alpha \, \mathbf{d}_k^{nb}, \mathbf{x}^b) = \mathbf{0} \tag{2-27}$$

is solved for the basic variables x^b with fixed x^{nb} (e.g. using Newton's Method) which geometrically means a projection of the search direction onto the constraint hypersuface.

One of the advantages of this method is that, due to the gradient projection procedure, the excursions into the infeasible domain are relatively small; this helps prevent the structural model from degenerating. It is disadvantageous, however, that the convergence behaviour strongly depends on the choice of the basic and nonbasic variables and that only first order approximate information is used.

In order to avoid these disadvantages, the method is further developed in such a way that the search direction is not determined via the reduced gradient but by solving a quadratic subproblem as is known from the so-called sequential quadratic procedures (SQP, Fig. 2/8c) [8].

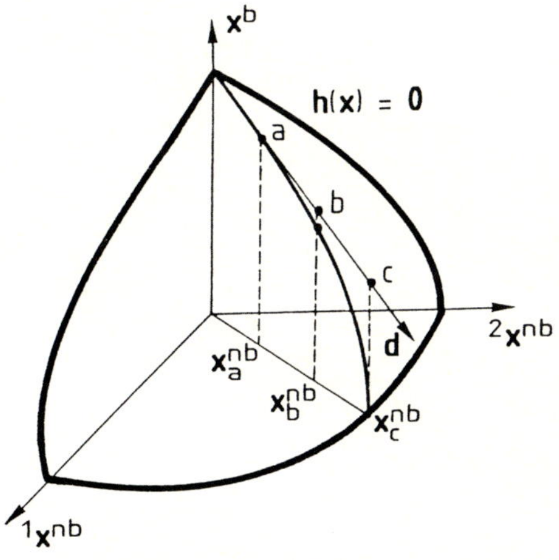

Fig. 2/8. Projection of the search direction on the hypersurface

This process starts from a feasible starting point. For determining the quadratic subproblem the following Lagrangian function is used:

$$L\,(\mathbf{x},\boldsymbol{\lambda}) = f(\mathbf{x}) - \sum_j^q \lambda_j h_j(\mathbf{x}) - \sum_j^p \lambda_{q+j}\, g_j(\mathbf{x}) \quad . \tag{2-28}$$

At the point x_k , (2-28) is expanded into a Taylor series up to the 2nd order:

$$\left.\begin{aligned} \tilde{f}\,(\mathbf{x}_k + \mathbf{d}) &= f\,(\mathbf{x}_k) + \mathbf{d}^T\,\nabla f(\mathbf{x}_k) + \frac{1}{2}\,\mathbf{d}^T\,\nabla^2 L(\mathbf{x}_k,\boldsymbol{\lambda})\,\mathbf{d}\,, \\ \tilde{g}_j(\mathbf{x}_k + \mathbf{d}) &= g_j(\mathbf{x}_k) + \mathbf{d}^T\,\nabla g_j(\mathbf{x}_k) \le 0\,, \qquad j = 1,\dots,p, \end{aligned}\right\} \tag{2-29}$$

with $\mathbf{d} = \mathbf{x}_{k+1} - \mathbf{x}_k$ search direction ,

$L\,(\mathbf{x},\boldsymbol{\lambda}) = f(\mathbf{x}) + \lambda_j \sum_{j=1}^{p} g_j(\mathbf{x})$ Lagrangian function .

The Hessian matrix $\nabla^2 L$ is calculated by means of the BFGS-Update Formula (Broyden-Fletcher-Goldfarb-Shanno). This hybrid procedure combines quantities of the SQP-algorithm (Fig. 2/8c) and the GRG-algorithm. More details are given in [12].

2.6 Comparison with other Structural Optimization Software Systems

Due to the intensive worldwide research activities in the field of structural optimization, various programming systems have been developed - many of them focusing on special purposes. Thus, they are only of limited use for general optimization problems. One or two optimization algorithms are linked to special structural analysis modules. In most cases, only a cross-section optimization with regard to minimum weight can be carried out. Some of the known software packages are described in [42-44]. and are compared in Table 2/1.

2.7 Application Example

The solution of optimization problems according to the procedure introduced here is demonstrated by a simple but interesting example of application. The cantilever truss made of steel tubes shown in Fig. 2/9 is subjected to a single load and to dead weight. For cost and functional

Name of Program	Structural Analysis	Optimization Algorithms	Objective Functions	Design Variables	Constraints
ACCESS [28,29]	FE	DM, GP, SUMT	W	S	a-d
STARS [30]	FE	OC, SR, SQP	W	S	a, b, c, f
PROSSS [31]	FE	SLP, SR, FD	W	S, G, F,	a-f
SAMCEF [28,32]	FE	DM, GP	W	S, G	a, b, c, f
PARS/EAL [33]	FE	PM, QN	W	S	a-d, f, g
FASTOP [34]	FE	OC, SR	W	S	a, f, g
LAGRANGE [35]	FE	SUMT, PM, SR, SLP, SQP, GRG	W	S, G	a-d, f, g
SAPOP [1,45]	FE, 1)	SUMT, PM, SR, SLP, SQP, GRG, SQGR	W, a-d	S, G, F	a-d, f
PANDA2 [36]	2)	FD	W	S, F	a, d, e, f
DDDU [37]	FE	SQP, DM, OC	W	S, G	a, b, f
ODYSSEY [38]	2)	FD	W	S, G	a-c
OASIS [39]	FE	LA, IA, HA, MMA	W, a	S, G	a-d, g
PASCO [40]	2)	FD	W	S, G, F	a, b, e, f
ELFINI [48]	FE, 1)	LA, GP	W	S, G	a-d, f, g

Structural analysis:
FE-Finite-Element method
1) other discrete and analytical methods
2) special analysis procedures for panel and shell components

Design Variables:
S - cross sections
G - geometrical
F - fibre angle

Objective functions/constraints:
- weight
a - stresses
b - deformations, displacements
c - eigenfrequencies
d - composite failure
e - global buckling
f - local buckling
g - aero-elastic conditions

Optimization algorithms:
OC - Optimality Criteria
SUMT - Sequential Unconstrained Minimization Technique
SR - Stress Ratio Method
SQP - Sequential Quadratic Program
SLP - Sequential Linear Program
FD - Feasible Directions Method
PM - Penalty Method
QN - Quasi Newton Method
GP - Gradient Projection Algorithm
GRG - Generalized Reduced Gradient Algorithm
SQGR - Hybrid SQP-GRG Algorithm
DM - Dual Strategy
MMA - Methods of Moving Asymptotes
LA - Linear Approximation
IA - Inverse Approximation
HA - Hybrid Approximation

Table 2/1. Some of the most important software systems in structural optimization reasons minimum weight and at the same time the largest possible structural stiffness are required. It is also important that the permissible stresses are not exceeded and that buckling stability is provided. The cross-sectional area of the tubes are optional; for the shape geometry certain bounds have to be kept.

First of all, the structural analysis procedure must be chosen, and variables as well as parameters must be defined on the basis of an initial design. To formulate the structural model, the Finite Element program SAPV-2 is used for static analyses. The structural parameters are the coordinates of the nodal points, the cross-sections and the material characteristics.

The cross-sections A_i and the nodal coordinates are the analysis variables. When defining the design variables, it must be considered that the lower nodal points are to be altered horizontally and node 2 vertically only. Furthermore, point (a+3) should not be moved. Each cross-section is defined as a design variable so that the design model for the cross-section variables can be formulated according to Section 2.4.1 as follows:

$$A_i = x_i \ , \ i = 1,\ldots,5k \qquad \text{or} \qquad \mathbf{A} = \mathbf{I}\ \mathbf{x}^S \ . \tag{2-30}$$

In this case the coordination matrix is the unity matrix **I**. The vector **x** includes the first 5k vector components of the design variable vector **x**. For geometrical optimization the distances h_i, l_{li}, and l_{ui} are defined as design variables $\mathbf{x}^g$.

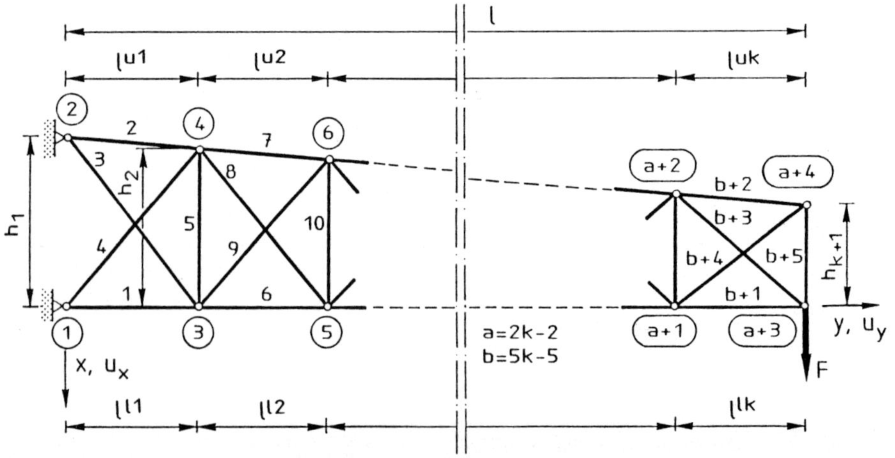

Fig. 2/9. Cantilever truss consisting of k fields

The nodal coordinates result from the following equations:

$$h_i = x^g_i \longrightarrow \lambda_{x2i} = - x^g_i , \qquad i = 1,\dots, k+1 \; ; \tag{2-31a}$$

$$l_{ui} = x^g_{k+1+i} \longrightarrow \lambda_{y2i} = \sum_{j=1}^{i} x^g_{k+1+j} , \qquad i = 1,\dots, k \; ; \tag{2-31b}$$

$$l_{li} = x^g_{2k+1+i} \longrightarrow \lambda_{y2i-1} = \sum_{j=1}^{i} x^g_{2k+1+j} , \; i = 1,\dots, k-1 \; . \tag{2-31c}$$

Together with the cross-section variables, 8k design variables are obtained. With (2-31) and (2-32) it is possible to define the initial design. In order to obtain a better solution behaviour of the optimization algorithm, the cross-section variables are transformed into reciprocal transformation variables:

$$z_i = \frac{1}{x_i} \quad \text{for } i = 1,\dots,5k \quad \text{and} \quad z_i = x_i \quad \text{for } i = 5k+1,\dots,8k \; . \tag{2-32}$$

The structural analysis program provides the state variables **u** (bar forces S_i, nodal displacements u_{xi}, u_{yi}). For the evaluation model objective functions and constraints are formulated as functions of the state variables and structural parameters. In the present case the objective function vector reads as follows:

$$\mathbf{f}(\mathbf{x}) = \begin{bmatrix} f_1(\mathbf{x}) \mathrel{\hat{=}} W(\mathbf{x}) = \sum_{i=1}^{5k} \rho \, l_i \, A_i \\ f_2(\mathbf{x}) \mathrel{\hat{=}} u_{x_{max}}(\mathbf{x}) = u_{x\,2k+1} \end{bmatrix} . \tag{2-33}$$

The evaluation model includes the transformation of the vector optimization problem into a scalar substitute problem by means of the constraint-oriented transformation using (1-14):

$$p[(\mathbf{x})] = f_i(\mathbf{x}) \text{ with } g_i(\mathbf{x}) = \frac{f_j(\mathbf{x})}{f_j^u} - 1 \le 0, \quad j = 1, 2; \quad i = 1,2, \wedge j \ne i \tag{2-34}$$

as well as the definition of stress and buckling constraints as in (2-4):

$$g_{i+1}(\mathbf{x}) = \frac{S_i(\mathbf{x})}{A_i \sigma_{max}} - 1 \le 0 \qquad \text{for } i = 1,\dots,5k \; ; \tag{2-35a}$$

$$g_{5k+1+i}(\mathbf{x}) = \frac{S_i(\mathbf{x})}{S_{i\,krit}(\mathbf{x})} - 1 \le 0 \qquad \text{for } i = 1,\dots,5k \; . \tag{2-35b}$$

The optimization algorithm uses preference functions and constraint values to calculate a new design. Apart from the formulation of the optimization model and the selection of the structural analysis program an optimization algorithm must be chosen. Different test calculations show that VMCWD [7] and QPRLT [12] have the best solution behaviour for this application. These two algorithms require gradient information. On account of its higher numerical efficiency the semianalytical procedure is used for sensitivity analysis.

The optimization model formulated above can be represented by the modules already integrated in SAPOP so that special problem routines are unnecessary. The optimization calculation is based on the following structural parameters and initial values of the design variables:

geometry	:	number of fields	$k = 3$,
		total length	$l = 5000\text{mm}$;
material	:	steel tubes	$E = 210\text{kN/mm}^2$, $\rho = 7{,}86\ \text{kg/dm}^3$,
			$\sigma_{max} = 200\text{N/mm}^2$;
load	:	single load	$F = 10\text{kN}$,
		dead weight of the truss;	

Initial values for the design variables:

cross-section variables ,	$x_{io} = 100\text{mm}^2$,	$i = 1,...,15$,
height,	$x_{io} = 1000\ \text{mm}^2$,	$i = 15,...,18$,
lengths,	$x_{io} = 1667\ \text{mm}$,	$i = 18,...,24$,
wall thickness,	$t = 2\ \text{mm}$.	

For the design problem (Chapter 1) various Pareto-optimal (functional-efficient) solution curves are presented in Fig. 2/10.

Starting from the initial design (point 1), the optimization of the cross-section variables (curve a) is carried out first. During the next step the cross-section variables remain constant and only geometrical variables are varied (curve b). Another optimization of cross-sections results in curve c. Further geometrical and cross-section optimizations do not improve the solution which is a suboptimum of the problem. The structural geometry, however, is of little practical importance (Fig. 2/10). A considerable improvement of the objective function values can be achieved by a simultaneous optimization of all design variables (curve d).

Because of the very sensitive buckling constraints it is difficult to find the optimal solutions. In this case it is particularly useful to be able to choose between different optimization algorithms. For the evaluation of the results, only a compromise solution on the bold printed section D is interesting. On the weakly efficient part of the boundary of the curve a considerable deteriation of one objective function is obtained when improving the other only slightly.

Fig. 2/11 shows a comparison of the design variables of the optimization solutions marked 2 and 3 in Fig. 2/10. With approximately the same weight the maximum deformation of the simultaneously optimized design 3 is only 23% of that of the cross-sectional optimum design 2. This results from the improved structural geometry which has a parabolic form [1]. Furthermore, in this example the independence of the semianalytical sensitivity analysis from the type of design variable (cross-sectional, geometrical variables) becomes evident. The numerical advantage can be seen by comparing the time required with the numerical procedure. The computing time required for gradient calculation using the simultaneous optimization method is reduced by 53% whereas with the cross-sectional optimization there is a time gain of 60%. These figures, however, only amount to statements of quality and cannot be transferred to other optimization problems because the time advantage of the semianalytical procedure depends on the individual problem.

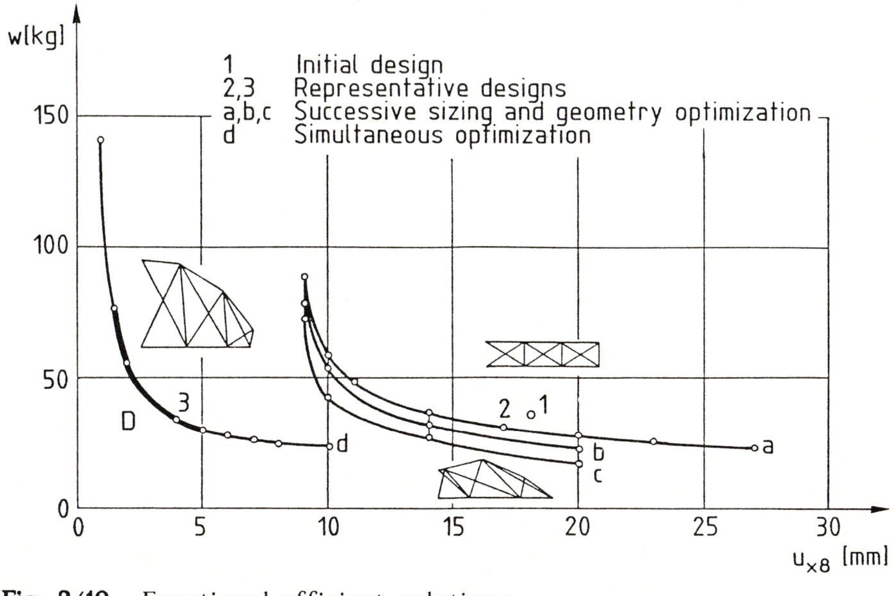

Fig. 2/10. Functional-efficient solutions

Nevertheless, it can be said that the larger the structure, the larger the number of design variables, and the smaller the influence of the design variables, the greater the reduction of computing time.

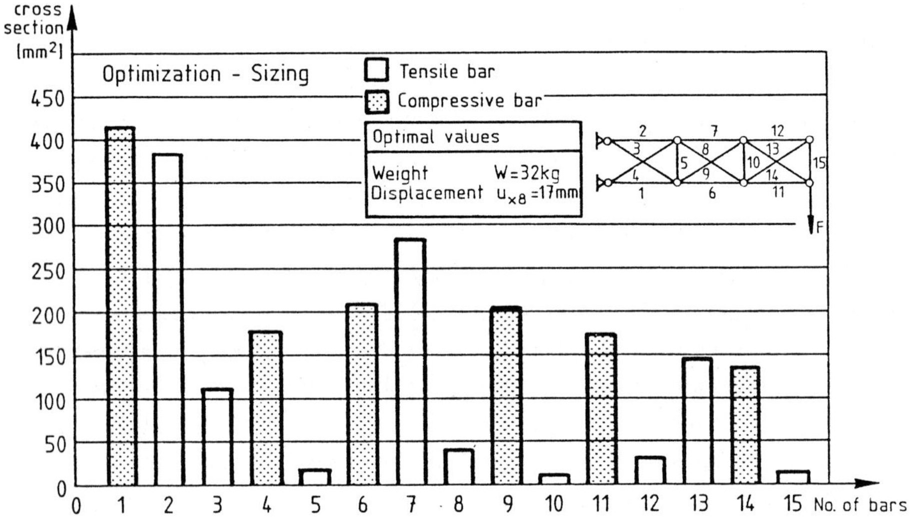

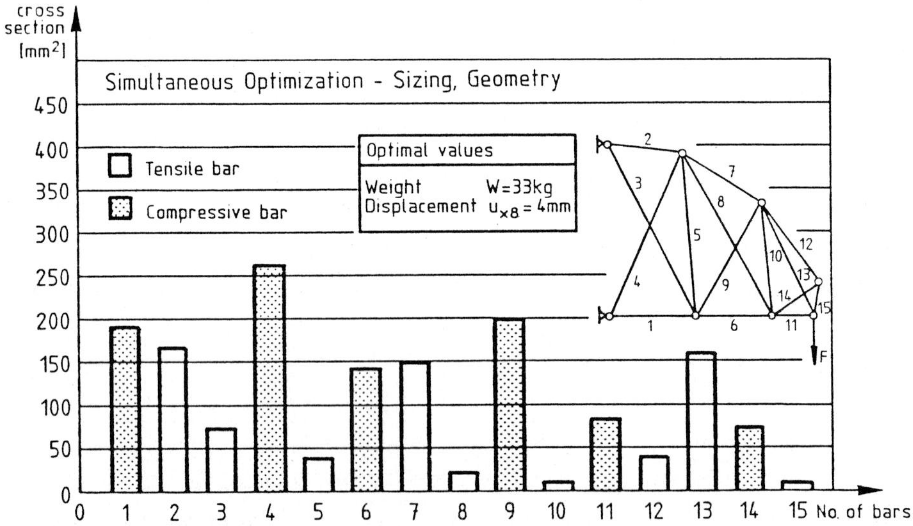

Fig. 2/11. Values of optimal design variables

2.8 Conclusion

This chapter introduces an optimization procedure for solving optimization problems with one or multiple objectives from the field of structural mechanics as well as the resultant software implementation. Starting from the Three-Columns Concept with the columns "Structural Model", "Optimization Algorithms", and "Optimization Model" and by considering other demands, the program system SAPOP (**S**tructural **A**nalysis **P**rogram and **O**ptimization **P**rocedure) was developed. Apart from an input system for data preparation and a graphic module for the evaluation of results, the main part of this program system comprises several subdomains. Each of these subdomains has program modules which are interchangeable depending on the requirements of the optimization problem. It is thus possible to link different optimization algorithms via generally formulated optimization models with several structural analysis programs. Furthermore, it is possible to implement different optimization strategies in order to solve special optimization problems (multicriteria optimization, interactive optimization, multilevel optimization).

Besides the description of the program structure the paper presents problems of optimization modelling and sensitivity analysis. The analytical procedures for sensitivity analysis used for the formulation of FE-methods have been extended to ordinary matrix methods.

In order to explain the solution concept, a combined cross-sectional and geometrical optimization of a simple bar framework is presented and discussed. This example illustrates the terminology and definitions used.

The efficiency of the SAPOP optimization procedure presented here is proved by numerous successfully solved optimization problems of different types [1,2,12,17,41]. Up to now, medium optimization problems (up to 100 design variables and 300 constraints) have been solved. However, the size of the prospective optimization problem strongly depends on the efficiency of the hardware environment. The present version of the SAPOP program is implemented on different computers, among others a MicroVAX II computer from Digital Equipment Corporation (DEC). On account of its modularity and efficient data management system, SAPOP is optimally used with workstation computers and personal computers.

References

[1] Eschenauer, H.: Numerical and Experimental Investigations on Structural Optimization of Engineering Designs. DFG-Research Report. University of Siegen, 1985. Siegen: Bonn + Fries Verlag 1986

[2] Kneppe, G.: Direkte Lösungsstrategien zur Gestaltsoptimierung von Flächentragwerken. Dissertation. Universität-GH Siegen, VDI-Fortschrittbericht, Reihe 1, Nr. 135, Düsseldorf: VDI-Verlag, 1986

[3] Sattler, H.-J.: Ersatzprobleme für Vektoroptimierungsaufgaben und ihre Anwendung in der Strukturmechanik. Dissertation. Universität-GH Siegen, VDI-Fortschrittbericht, Reihe 1, Nr. 88, Düsseldorf: VDI-Verlag, 1982

[4] Jacob, H.G.: FORTRAN-Programm zur Ermittlung eines lokalen Optimums einer beschränkten multivariablen Gütefunktion ohne Kenntnis ihrer Ableitung. PDV-Bericht 36, 1975

[5] Griffith, R.E.; Stewart, R.A.: A Nonlinear Programming Technique for the Optimization of Continuous Processing Systems. Management Science 7 (1961) 379-392

[6] Post, P.U.; Heeg, P.; Mende, B.: Erweiterte sequentielle Linearisierungsstrategie SEQLI2. Institutsbericht. Universität-GH Siegen, Institut für Mechanik und Regelungstechnik, 1985

[7] Powell, M.J.D.: VMCWD: A FORTRAN Subroutine for Constrained Optimization. Rep. DANTP 1982/NA4, University of Cambridge, 1982

[8] Schittkowski, K.: On the Convergence of a Sequential Quadratic Programming Method with an Augmented Lagrangian Line Search Function. Mathematische Operationsforschung und Statistik, Section Optimization, 4 (1983) 197-216

[9] Abadie, J.; Carpentier, J.: Generalization of the Wolfe Reduced Gradient Method to the Case of Nonlinear Constraints. In: Fletcher, R.: Optimization. New York: Academic Press, 1969

[10] Bremicker, M.: Entwicklung eines Optimierungsalgorithmus der generalisierten reduzierten Gradienten - Anwendung auf Beispiele der Strukturmechanik. Diplomarbeit. Universität-GH Siegen, Institut für Mechanik und Regelungstechnik, 1986

[11] Lasdon, L.S.; Warren, A.D.; Jain, A.; Ratner, M.: Design and Testing of a Generalized Reduced Gradient Code for Nonlinear Programming. ACM Transactions Mathematical Software 4 (1978) 34-50

[12] Bremicker, M.: Dekompositionsstrategie in Anwendung auf Probleme der Gestaltsoptimierung. Dissertation. Universität-GH Siegen. VDI-Fortschrittsbericht, Reihe 1, Nr. 173, Düsseldorf: VDI-Verlag, 1989

[13] Parkinson, A.; Wilson, M.: Development of a Hybrid SQP-GRG Algorithm for Constrained Nonlinear Programming. Proceedings of the ASME Design Engineering Technical Conference, Columbus, Ohio, Oct. 5-8, 1986 (86-DET-114)

[14] Pierre, D.A.; Lowe, M.J.: Mathematical Programming via Augmented Lagrangians. London, Amsterdam, Ontario, Sydney, Tokio: Addison Wesley Publishing Company, Inc. 1975

[15] Gallagher, R.H.: Fully Stressed Design. In: Gallagher, R.H.; Zienkiewicz, O.C.: Optimum Structural Design. London, Sydney, New York, Toronto: John Wiley and Sons, 1973

[16] N.N.: SAPV-2, A Structural Analysis Program for Static and Dynamic Response of Linear Systems. Dept. of Civil Engineering, University of Southern California, Los Angeles, 1977

[17] Fuchs, W.: Strukturanalyse und Optimierung anisotroper Schalen aus Faserverbundwerkstoff. Dissertation. Universität-GH Siegen, Institut für Mechanik und Regelungstechnik, 1986

[18] Bathe, K.J.: Finite Element Procedures in Engineering Analysis. Englewood Cliffs, New Jersey: Prentice Hall, Inc., 1982

[19] N.N.: ADINA - Automatic Dynamic Incremented Nonlinear Analysis - Theory and Modeling Guide. Watertown: ADINA Engineering Inc. 1984

[20] Adams, D.F.; Tsai, S.W.: The Influence of Random Filament Packing on the Transverse Stiffness of Unidirectional Composites. Journal of Composite Materials, 3 (1969) 368-379

[21] Puck, A.: Festigkeitsberechnung an Glasfaser/Kunststoff-Laminaten bei zusammengesetzter Beanspruchung. Bruchhypothesen und Kunststoffe, 59 (1969) 781-787

[22] Tsai, S.W.; Hahn, H.T.: Introduction to Composite Materials. Westport, Connecticut: Technomic Publishing Company 1980

[23] Tsai, S.W.; Wu, E.M.: A General Theory of Strength for Anisotropic Materials. J. Composite Materials, 5 (1971) 58-80

[24] Baier, H.: Mathematische Programmierung zur Optimierung von Tragwerken insbesondere bei mehrfachen Zielen. Dissertation. TH Darmstadt, 1978

[25] Adelman, H.M.; Haftka, T.: Sensitivity Analysis for Discrete Structural Systems - A Survey. NASA-TM-86333, 1984

[26] Haug, E.J.; Arora, J.S.: Design Sensitivity Analysis of Elastic Mechanical Systems. Computer Methods in Applied Mechanics and Engineering, 15 (1978) 35-62

[27] Haug, E.J.; Choi, K.K.; Komkov, V.: Design Sensitivity Analysis of Structural Systems. Orlando: Academic Press, Inc., 1986

[28] Fleury, C.; Schmit, L.A.: Dual Methods and Approximation Concepts in Structural Synthesis. NASA Contractor Report 3226, 1980

[29] Fleury, C.; Ramanathan, R.K.; Salama, M.; Schmit, L.A.: ACCESS Computer Program for the Synthesis of Large Structural Systems. In: ATREK, E. et al.: New Directions in Optimum Structural Design. Chichester, New York, Brisbane, Toronto, Singapore: John Wiley & Sons, 1984, 541-561

[30] Bartholomew, P.; Morris, A.J.: STARS: A Software Package for Structural Optimization. Proceedings of the International Symposium on Optimum Structural Design, University of Arizona, USA, 1981

[31] Sobieszczanski-Sobieski, J.; Rogers, J.L.: A Programming System for Research and Applications in Structural Optimization In: Atrek, E. et al.: New Directions in Optimum Structural Design. Chichester, New York, Brisbane, Toronto, Singapore: John Wiley & Sons, 1984, 563-585

[32] Fleury, C.; Braibant, V.: Application of Structural Synthesis Techniques. Proceedings of the NATO/NASA/NSF/USAF Conference of Computer Aided Optimal Design, Vol. 2, Troja, Portugal, June 29 - July 11, 1986, Center of Mechanics and Materials of the Technical University of Lisboa, 29-53

[33] Haftka, R.T.; Prasad, B.: Programs for Analysis and Resizing of Complex Structures. Computers and Structures, 10 (1979) 323-330

[34] Wilkinson, K. et al.: An Automated Procedure for Flutter and Strength Analysis and Optimization. Vol. 1 - Theory; Vol. 2 - Program User's Manual AFFDL - TR - 75 - 137, 1975

[35] Kneppe, G.; Keppeler, D.; Krammer, H.: Optimale Auslegung komplexer Luft- und Raumfahrtstrukturen mit Hilfe der Mathematischen Programmierung in Verbindung mit FE-Analysen. Vortragsmanuskript. XV. Internationaler Finite Element Kongress, 17./18. Nov. 1986, Baden-Baden, Schwarzwald

[36] Bushnell, D.: PANDA2 - A Program for Minimum Weight Design of Stiffened, Composite, Locally Buckled Panels. Computers and Structures, 25 (1987) 469-605

[37] Qian, L.X.: Structural Optimization Research in China. Proceedings of the International Conference on Finite Element Methods, Shanghai, China (1982) 16-24

[38] Botkin, M.E.; Bennet, J.A.: Shape Optimization of Three-Dimensional Folded Plate Structures. AIAA Journal, 23 (1984) 1804-1810

[39] Esping, B.; Holm, D.: Structural Shape Optimization Using OASIS. IUTAM-Symposium on Structural Optimization, Melbourne, 1988

[40] Anderson, M.S.: Application of Optimization Methods to the Practical Design of Shear and Compression Loaded Stiffened Panels. IUTAM-Symposium on Structural Optimization, Melbourne, 1988

[41] Post, P.U.: Optimierung von Verbundbauweisen unter Berücksichtigung des zeitabhängigen Materialverhaltens. Dissertation. Universität-GH Siegen. VDI-Fortschrittsbericht, Reihe 1, Nr. 172, Düsseldorf: VDI-Verlag, 1989

[42] Atrek, E.; Gallagher, R.N.; Ragsdell, K.M.; Zienkiewicz, O.C.: New Directions in Optimum Structural Design. Chichester, New York, Brisbane, Toronto, Singapore: John Wiley & Sons, 1984

[43] Haftka, T.R.; Kamat, M.P.: Elements of Structural Optimization. Maritinus Nijhoff Publishers, Netherlands, 1985

[44] Hörnlein, H.R.E.M.: Take-Off in Optimum Structural Design. Proceedings of the NATO/NASA/NSF/USAF Conference of Computer Aided Optimal Design, Vol. 3, Troja, Portugal, June 29 - July 11, 1986, Center of Mechanics and Materials of the Technical University of Lisboa, 176-199

[45] Eschenauer, H.; Post, P.U.; Bremicker M.: Einsatz der Optimierungsprozedur SAPOP zur Auslegung von Bauteilkomponenten. Bauingenieur 63, II (1988) 515-526

[46] Schiedel, E.: Die Berechnung von orthotropen Rotationsschalen beliebiger Meridianform unter allgemeiner Belastung mit Hilfe des Stufenkörperverfahrens. Dissertation TH Darmstadt, 1972

[47] Schnell, W.; Eschenauer, H.: Elastizitätstheorie 2 - Schalen. Mannheim, Wien, Zürich: BI Wissenschaftsverlag, 1984

[48] Lecina, G.; Petiau, C.: Advances in Optimal Design with Composite Materials. Proc. NATO Conf. Computer Aided Optimal Design, Vol. III. Troja, Portugal, June 29-July 11, 1986, Cent. of Mech. and Mat. Technical University of Lisboa 279-289

[49] Austin, A.: PAFEC. In: Brebbia, C.A. (ed.): Finite Element Systems - A Handbook. Berlin, Heidelberg, New York: Springer Verlag (1982) 295-305

3 INTERACTIVE MULTICRITERIA OPTIMIZATION IN DESIGN PROCESS

H.A. Eschenauer, A. Osyczka, E. Schäfer

3.1 Introduction

In this chapter MO-procedures treated in Chapter 2 will be developed into interactive procedures which integrate the decision making process into optimization algorithms. The interactive procedures provide the Decision Maker (DM) with a selection of Pareto-optimal solutions which to some extent are representative for the whole set of available solutions. This procedure consists of a sequence of decision and computation phases. In the decision phase the DM decides whether or not a solution is optimal with respect to his implicit preferences. In the latter case he must give some information about the direction in which he expects to obtain a better solution. In the computation phase the new solution is generated for the next decision phase. The procedure is stopped when the optimal solution which reflects the DM's preferences is found. Such a dialogue does not only improve the implicit preferences of the decision maker but also supports and simplifies the process of decision making.

As early as 1966, Benayon, Roy and Sussman [1] suggested an algorithm particularly designed for linear optimization problems. This algorithm limits the domain of compromise solutions until a preferred solution is achieved. The interactive procedure by Benayoun et al. [2], the so-called STEP-method, was also developed for linear problems. Further interactive optimization strategies were suggested by Geoffrion, Dyer and Feinberg [3], Fandel [4], Zionts and Wallenius [5], and Steuer [6].

The increasing number of publications on interactive procedures includes several suggestions concerning the improvement of existing procedures such as the publication by Dupre, Huckert and Jahn [7] about the STEP-method. In addition, Jahn developed an optimization strategy which is similar to the procedure by Geoffrion as far as the linearization of the nonlinear multicriteria optimization problem (MO) is concerned. Linear parametric optimization approaches were applied by Bernau [40], Brosowski [18] and Guddat et al. [8]. A great number of publications in the field of interactive MO appeared in the areas of economics and operations research in the 1970's and 1980's. A large number of reports, books and monographs on the subject *Multiple Criteria Decision Making* or *Multiobjective Decision Making* were published. Conference papers on this topic have regularly appeared since 1975 [9] , but without concentrating on mechanical problems. A survey of the current and most popular procedures can be found in the books by Hwang and Masud [10], Zeleny [11], Goicoechea, Hansen and Duckstein [12], Szidarovsky, Gershon, Duckstein [13], Sawaragi, Nakayama, Tanino [14], and Isermann [15].

For the design process interactive procedures have been mentioned or applied in only a few cases [16–19,37]. This basically results from the immense nonlinearity of the occuring problems. Furthermore, it is due to the fact that the structural analyses such as finite element methods must be carried out numerically which is a time-consuming process. In particular, such problems are regarded by Diaz [19] who presents an effective sensitivity analysis to variations in the DM's preferences based on sequential quadratic programming.

In this chapter appropriate interactive procedures for application to optimal designs are presented. A critical evaluation of the procedures is based on real-life examples. Finally, two software packages (DIALOG and CAMOS) for interactive multicriteria optimization are described.

3.2 Interactive Multicriteria Optimization Procedures

3.2.1 General Remarks on Classification

Interactive procedures can be classified according to various criteria. A distinction can be made with respect to the kind of and the stage at which preference information is required from the DM. As a further distinctive feature the particular organization scheme of an approach can be considered as it prescibes the kind of scalar substitute problems which have to be solved during the interactive optimization process. Therefore, two groups of interactive organization schemes will be distinguished here:

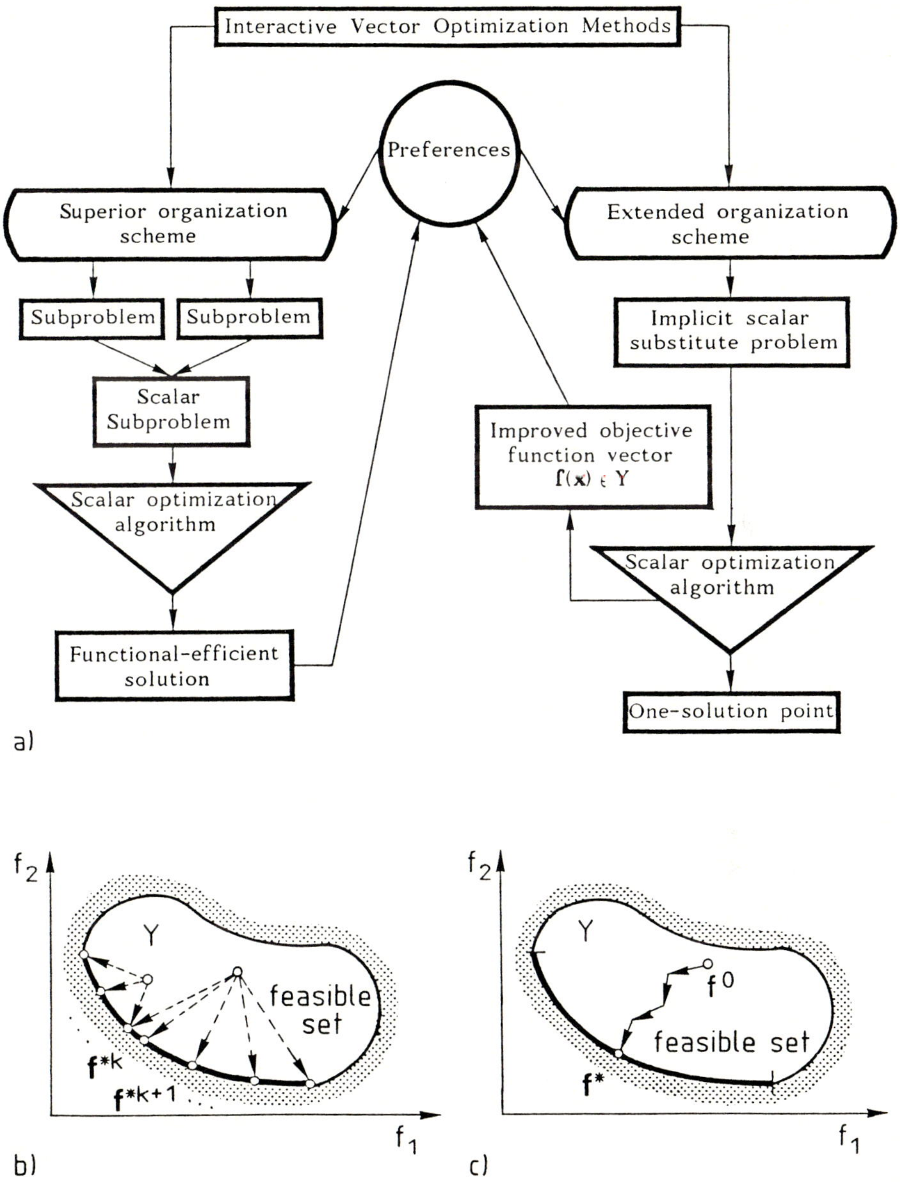

Fig. 3/1. Organization schemes of interactive vector optimization methods

a) Optimization schemes of both structures

b) Iteration of superior organization schemes in the criterion space

c) Iteration path of the extended organization schemes in the criterion space

a) Superior organization schemes (see Fig. 3/1 a,b.)

Methods with a superior organization scheme use scalar substitute problems with the preference functions described in section 1.2.2. The free parameters leading to a particular functional-efficient point can be established in a dialogue with the DM. Each substitute problem can be solved by means of any problem-adjusted mathematical programming procedure and thus leads to a functional-efficient point so that functional-efficient alternative solutions are offered to the DM. The STEP-method (see section 3.2.3)is a typical example of this group.

b) Extended organization schemes (see Fig. 3./1 a,c)

Extended organization schemes are based on the extension of particular mathematical programming algorithms so that the preference function can be changed according to the preferences of the DM before the scalar substitute problem is finally solved. This procedure provides optimization steps within the criterion space in the direction of the functional-efficient boundary whereby the DM influences the step direction. The method of the mathematical programming cannot be adjusted according to the present optimization problem. An example of this kind of procedure is the approach by Jahn (see section 3.2.4).

3.2.2 Approach by Fandel

The basic idea of Fandel's approach [4] is to support the decision maker in selecting appropriate weighting coefficients which lead to a compromise solution $\tilde{\mathbf{f}}\,(\tilde{\mathbf{x}})$ preferred by the DM after solving the substitute problem by objective weighting:

$$\min_{\mathbf{x} \in \mathbb{R}^n} \{ \mathbf{w}^T \mathbf{f}\,(\mathbf{x}) \, : \, \mathbf{g}\,(\mathbf{x}) \le \mathbf{0} \} \;, \quad \mathbf{w} \ge \mathbf{0} \;, \quad \sum_{j=1}^{m} w_j = 1 \;. \qquad (3\text{-}1)$$

By variation of the weighting coefficients w_j equation (3-1) provides all solutions of the functional-efficient boundary for convex problems. The weighting coefficients belonging to the "best" solution which will be preferred by the decision maker are still unknown. Thus, the solution set of each objective which is selected by the DM during the interactive dialogue has to be reduced successively by means of upper bounds. The subsequent steps covered iteratively characterize the approach and will be graphically described for the two-dimensional case.

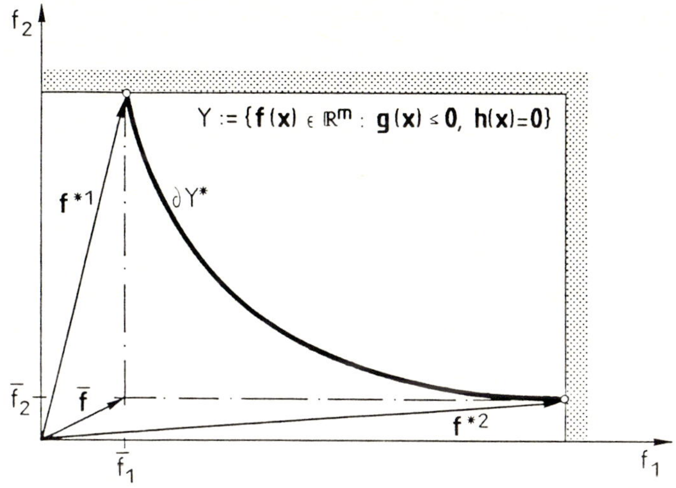

Fig. 3.2 Criterion space Y with functional-efficient boundary, individual minima $\mathbf{f}^{*i}$ and vector of ideal solutions $\bar{\mathbf{f}}$

First, the m individual minima $\bar{f}_j$ of the given criterion space Y are calculated by methods of the nonlinear mathematical programming, e.g. SEQLI2 [20] or VMCWD [21] (see Fig. 3/2):

$$\min_{\mathbf{x} \in \mathbb{R}^n} \{f_j(\mathbf{x}) : \mathbf{g}(\mathbf{x}) \leq \mathbf{0}\} \equiv \bar{f}_j \qquad \forall j = 1,\ldots,m \ . \tag{3-2}$$

They can be arranged as an ideal objective function vector $\bar{\mathbf{f}} = (\bar{f}_1, \bar{f}_2, \ldots, \bar{f}_m)^T$. The corresponding m function vectors $\mathbf{f}^{*j}$ form a (m×m)-matrix $\mathbf{B} = (\mathbf{f}^{*1}, \mathbf{f}^{*2}, \ldots, \mathbf{f}^{*m})^T$, also named pay-off table.

It is assumed that the ideal objective function vector $\bar{\mathbf{f}}$ would be the solution which is preferred by the decision maker but which is not attainable. The decision maker will then prefer that solution which comes closest to the ideal one. Therefore, a central point is calculated by means of the well-known relation

$$\mathbf{f}^M = \frac{1}{m} \sum_{j=1}^{m} \mathbf{f}^{*m} \tag{3-3}$$

and is used to reduce the remaining criterion space to a subspace $\hat{Y}$ which shall still include the preferred compromise solution. Once again, the individual minima in subspace $\hat{Y}$ are determined (Fig. 3/3)

$$\min_{\mathbf{x} \in \mathbb{R}^n} \{f_j(\mathbf{x}) \in \hat{Y}\} \equiv \hat{f}^{*j} \ \forall j=1,\ldots,m \text{ with } \hat{Y} := \{f(\mathbf{x}) \in \mathbb{R}^m : \mathbf{g}(\mathbf{x}) \leq \mathbf{0},\ \mathbf{f}(\mathbf{x}) \leq \mathbf{f}^M\} \tag{3-4}$$

and noted as pay-off table with a matrix $\hat{\mathbf{B}} = (\hat{\mathbf{f}}^{*1}, \mathbf{f}^{*2},\ldots, \mathbf{f}^{*m})^T$ containing the new individual minima $\hat{\mathbf{f}}^{*j}$.

Matrix $\hat{\mathbf{B}}$ defines a hyperplane with the plane equation

$$\hat{\mathbf{B}}\mathbf{a} = c\mathbf{e}\,, \qquad \mathbf{a} > \mathbf{0}\,, \quad \sum_{j=1}^{m} a_j = 1 \tag{3-5}$$

with $\mathbf{a} = (a_1, a_2,\ldots, a_m)^T$,
c = constant ,
$\mathbf{e} = (1,1,\ldots,1)^T$,

which is laid through the new individual minima $\hat{\mathbf{f}}^{*j}$. Vector $\mathbf{a}$ and constant c are the unknown quantities of the equation system.

The use of a hyperplane according to (3-5) considers that the dimensions of the weighting coefficients shall compensate the different physical measures and values of the objectives. In order to attain the weighting coefficients of a desirable compromise solution within the subspace $\hat{Y}$, the hyperplane is deferred in the direction of the functional-efficient boundary by minimizing the constant c until the boundary is tangented to one point only (Fig. 3/4). This linear optimization problem leads to a solution where the optimal vector $\mathbf{a}$ corresponds to the desired vector of the weighting coefficients $\mathbf{w}$.

The solution of the scalar substitute problem (3-1) can be determined now. The decision maker accepts this solution as a preferred compromise

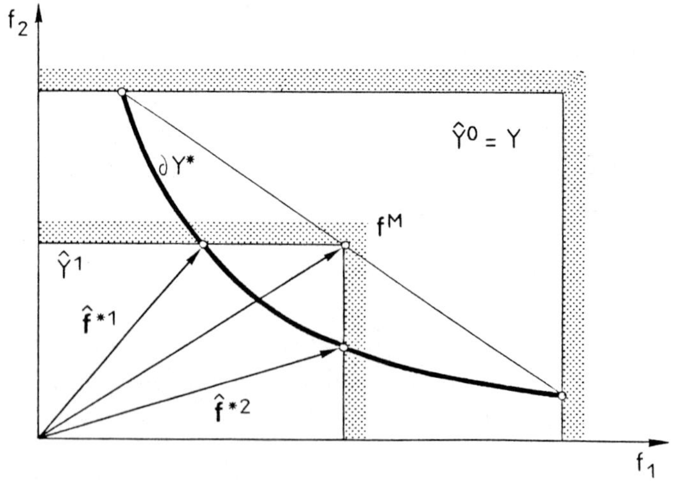

Fig. 3/3. Central point and individual minima $\hat{\mathbf{f}}^{*i}$ in the reduced subspace $\hat{Y}$

solution or starts the iteration again by restricting the criterion space according to his preferences which is done by outlining the upper bounds $\mathbf{\bar{y}}$ for the m objective functions. The individual minima of the new defined subspace $\hat{Y}$ are determined now:

$$\min_{\mathbf{x} \in \mathbb{R}^n} \{ f_j(\mathbf{x}) \in \hat{Y} \} \quad \forall j = 1,\dots,m \quad \text{with } \hat{Y} := \{ \mathbf{f}(\mathbf{x}) \in \mathbb{R}^m : \mathbf{g}(\mathbf{x}) \le \mathbf{0},\ \mathbf{f}(\mathbf{x}) \le \mathbf{\bar{y}} \}, \tag{3-6}$$

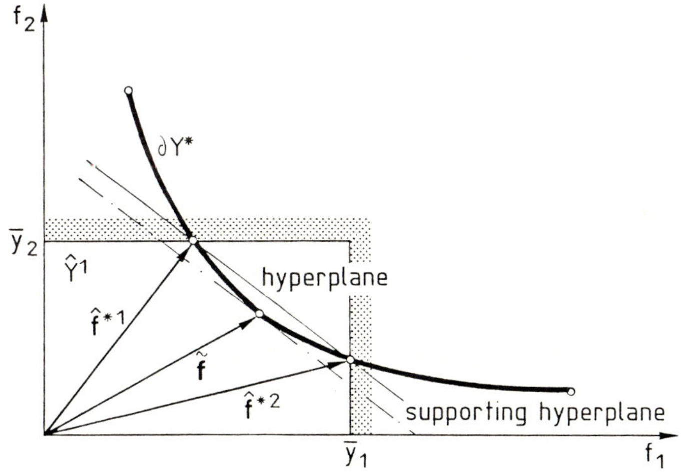

Fig. 3/4. Hyperplane and supporting hyperplane of Fandel's approach

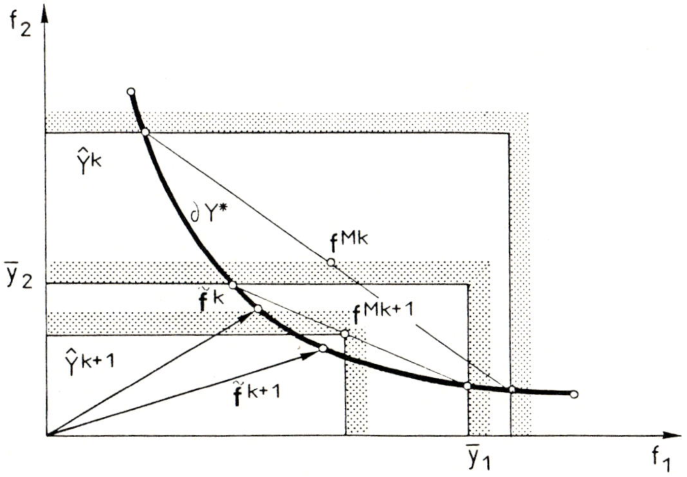

Fig. 3/5. Criterion space $\hat{Y}$ and optimal solutions $\tilde{\mathbf{f}}$ of the k-th or (k+1)-th iteration, resp.

and the iteration starts again by calculating a new central point $\mathbf{f}^M$ (see Fig. 3/5). The iteration is terminated when the breaking-off criterion

$$\| f_j(\mathbf{x}) - \bar{y}_j \| < \varepsilon \quad \forall \; j = 1,...,m \quad \text{with } \varepsilon > 0$$

is satisfied, i.e. the iteration is stopped when all subminima are placed in the ε-environment of the upper bounds $\bar{\mathbf{y}}$.

3.2.3 STEP-Method

The STEP-Method which was first proposed by Benayoun, de Montgolfier, Tergny and Laritchev [2] for solving linear MO-problems is also applicable to the nonlinear problems of structural mechanics. Similar to Fandel's approach, information from the DM about his preferences is used for reducing the criterion space Y successively. Upper bounds $\bar{\mathbf{y}}$ restrict the available criterion space. It is also assumed that the preferred compromise solution should be as close as possible to the ideal objective function vector. For this case a scalar substitute problem is applied by means of a min-max-formulation with bounded and weighted objectives (1-22)

$$\min_{\mathbf{x} \in \mathbb{R}^n} \{\beta : [\mathbf{f}(\mathbf{x}) - \bar{\mathbf{y}}]^T \mathbf{w} < \beta, \;\; \mathbf{g}(\mathbf{x}) \leq \mathbf{0}, \;\; \mathbf{f}(\mathbf{x}) \leq \bar{\mathbf{y}}\} \quad . \tag{3-7}$$

The individual steps of the procedure extended to nonlinear problems are described below.

According to Fandel's approach the m individual minima $\bar{f}_j$ of the subproblems (3-2) are determined first by using the methods of nonlinear mathematical programming. These solutions form the pay-off table (matrix **B**) containing the vector of the ideal objective functions $\bar{\mathbf{f}}$ within the diagonal elements of this matrix (see Section 3.2.2).

The weighting coefficients w_j of each objective f_j are required for solving the min-max problem (3-7) and for equalizing the different magnitudes of the weighting coefficients. Additionally, the objective functions with larger differences between the maximum and the minimum values of $\mathbf{f}^{*j}$ are given greater weight. The weighting coefficients result from

$$w_j = \frac{v_j}{\sum_{i=1}^{m} v_i} \quad , \quad j = 1,...,m \tag{3-8}$$

$$\text{with} \quad v_j = \frac{\max \{f_j^{*i}\} - f_j^{*j}}{\min \{f_j^{*i}\}} \; \frac{1}{f_j^{*j}} \; , \qquad i = 1,...,m \; .$$

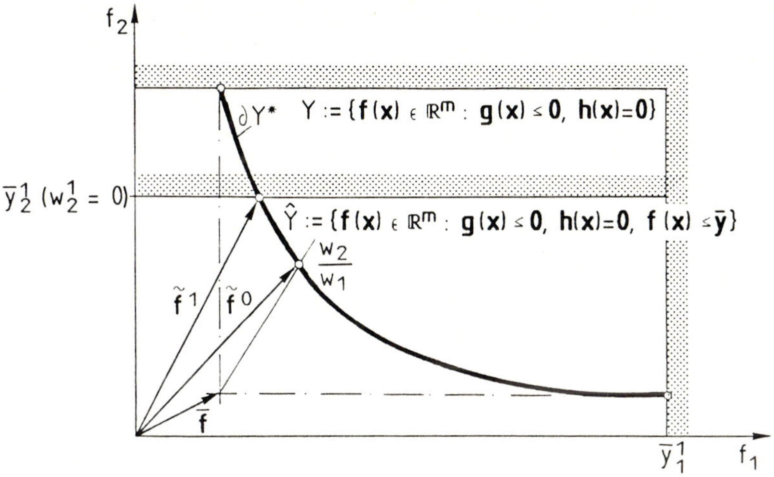

Fig. 3/6. Proposed compromise solutions of the first iteration within the initial space Y and the subspace $\hat{Y}$

The calculated weighting coefficients v_j are divided by the total sum of all v_i to fulfill the following assumptions

$$\mathbf{w} > \mathbf{0} , \quad \sum_{j=1}^{m} w_j = 1 . \tag{3-9}$$

In comparison to Fandel's method the weighting coefficients are not of an outstanding importance but only locally effective because the preferred solution is achieved by means of upper bounds for the objectives and not by a variation of the weights.

When the weighting coefficients have been determined a first compromise solution $\tilde{\mathbf{f}}^0$ is found by solving the min-max problem

$$\min_{\mathbf{x} \in \mathbb{R}^n} \{\beta : \mathbf{f}(\mathbf{x})^T \mathbf{w} < \beta , \ \mathbf{g}(\mathbf{x}) \leq \mathbf{0}\} . \tag{3-10}$$

The compromise solution is now proposed to the decision maker who compares its objective vector $\tilde{\mathbf{f}}$ with the ideal objective vector $\bar{\mathbf{f}}$. If some of the components $\tilde{\mathbf{f}}$ are satisfactory and others are not, the decision maker must indicate a certain amount of relaxation of at least one satisfactory objective to improve the unsatisfactory ones in the next iteration. This amount can be given as a differential increase Δf_j of each selected objective or as an upper bound for the j-th objective function $\bar{y}_j = \tilde{f}_j + \Delta f_j$ with $j \in \{1,...,m\}$. Otherwise the preferred solution is determined by $\tilde{\mathbf{f}}$.

According to the problem (3-10) a new compromise solution taking into account the existing upper bounds $\overline{y}_j$ is achieved. Additionally, the weighting coefficient w_j of the j-th objective is set to zero. The solution of the modified problem

$$\min_{\mathbf{x} \in \mathbb{R}^n} \{ \beta : [\mathbf{f}(\mathbf{x}) - \overline{\mathbf{y}}]^T \mathbf{w} < \beta \,,\; \mathbf{g}(\mathbf{x}) \le \mathbf{0},\; \mathbf{f}(\mathbf{x}) \le \overline{\mathbf{y}} \} \quad \text{with } w_j = 0 \,, \quad j \in \{ j = 1,\dots,m \} \tag{3-11}$$

lies in the reduced subspace $\hat{Y}$ and shows the lowest deviation of the ideal objective function vector $\overline{\mathbf{f}}$. In the two-dimensional case $\tilde{\mathbf{f}}$ is located on the intersection of a straight line with the functional-efficient boundary going through the ideal point marked by $\overline{\mathbf{f}}$ with the augmentation w_2/w_1 and the functional-efficient boundary (Fig. 3/6).

If the DM accepts this solution as a compromise solution, the iteration is stopped. Otherwise, the new iteration starts again with the relaxation of at least one objective function as described above.

3.2.4 Approach by Jahn

Jahn's approach [22] for interactive multicriteria optimization is fundamentally different from the approach by Fandel and the STEP-method just mentioned. Here, only one functional-efficient point is determined during the interactive session. Beginning with a selected starting point, the multicriteria optimization problem is solved by steps across the interior of the criterion space Y in the direction of the preferred solution on the functional-efficient boundary. Based on the scalar optimization method of feasible directions by Zoutendijk [23], also called cyclic minimization along the coordinates, the scalar substitute of the k-th direction finding problem is formulated as follows

$$\min_{\mathbf{x} \in \mathbb{R}^n} \{\beta : w_j \,[\nabla_{\mathbf{x}} f_j^k]^T \,\boldsymbol{\delta} - \beta \le 0 \,,\quad v_i \,[\nabla_{\mathbf{x}} g_i^k]^T \,\boldsymbol{\delta} - \beta \le 0 \,, -1 \le \delta_l \le 1\}$$

$$\forall\; j = 1,\dots,m \,,\quad \forall\; l = 1,\dots,n \,,\quad \forall\; i \in \{i=1,\dots,p : g_i(\mathbf{x})^k \le \mathbf{0}\} \tag{3-12}$$

with
- β - scalar preference function and auxiliary design variable,
- $\nabla_{\mathbf{x}}$ - NABLA-operator regarding the design variables $\mathbf{x}$,
- $\boldsymbol{\delta}^k$ - direction of the minimization step,
- w_j - weighting coefficient of the j-th objective function,
- v_i - weighting coefficient of the i-th active constraint.

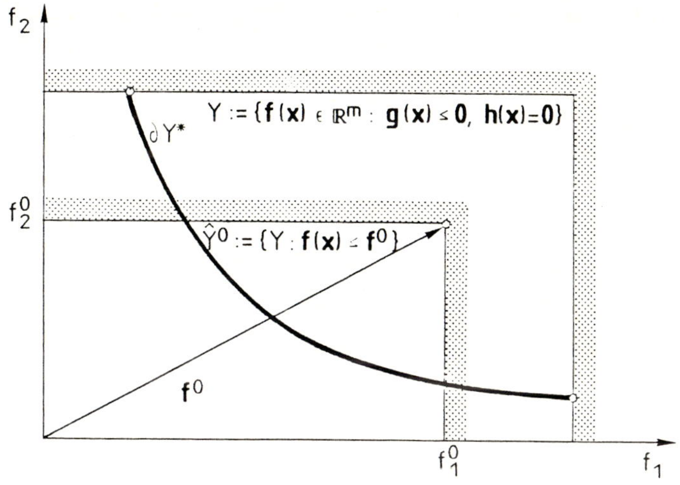

Fig. 3/7. Starting vector $\mathbf{f}^0$ and remaining criterion subspace $\hat{Y}$

The solution of the substitute problem (3-12) starts with selecting a feasible starting vector of the design variables $\mathbf{x}^0$ and with calculating the objective function vector $\mathbf{f}(\mathbf{x}^0)$. The starting point $\mathbf{x}^0$ basically influences the remaining decision alternatives (see Fig. 3/7) since the original criterion space $\hat{Y}^0$ is finally restricted to

$$\hat{Y}^0 := \{\mathbf{f}(\mathbf{x}) \in \mathbb{R}^m : \mathbf{f}(\mathbf{x}) \leq \mathbf{f}(\mathbf{x}^0)\} \quad .$$

From the presented vector $\mathbf{f}(\mathbf{x})$ the DM selects one objective function f_i which is to be minimized according to his preferences. In order to achieve this, a step direction $\boldsymbol{\delta}$ and an appropriate step length λ have to be determined which satisfy the following demands:

$$\begin{aligned} &\mathbf{x}^{k+1} = \mathbf{x}^k + \lambda^k \boldsymbol{\delta}^k , \quad \mathbf{x}^k, \mathbf{x}^{k+1} \in \mathbf{X} := \{\mathbf{x} \in \mathbb{R}^n : \mathbf{g}(\mathbf{x}) \leq \mathbf{0}\} , \\ &f_i^{k+1} < f_i^k \qquad \forall \; i \in I := \{1,\dots,m\} , \qquad (3\text{-}13) \\ &f_j^{k+1} \leq f_j^k \qquad \forall \; j \in J := \{I : j \neq i\} \end{aligned}$$

with $\mathbf{x}^k$, $\mathbf{x}^{k+1}$ - design variable vector of the k-th resp. k+1 -th iteration,

λ^k - step length of the minimization step.

The step direction is obtained by solving the linearized direction finding problem (3-12). In doing so the weighting coefficients w_j and v_i chosen by the DM influence the step direction.

The DM is now allowed to determine a preferred step length λ^k within the interval $[0, \lambda_{max}]$. The maximum step length λ_{max} is derived from a one-dimensional minimization of the following subproblem:

$$\max \{\lambda_{max} : f_i(\mathbf{x}^k + \lambda^k_{max}\, \boldsymbol{\delta}^k) < f_i(\mathbf{x}^k)\,,$$

$$f_j(\mathbf{x}^k + \lambda^k_{max}\, \boldsymbol{\delta}^k) \le f_j(\mathbf{x}^k)\,, \; (\mathbf{x}^k + \lambda^k_{max}\, \boldsymbol{\delta}^k) \in X\}$$

$$\forall \; i = 1,\ldots,m, \quad \forall \; j = 1,\ldots,m \;\text{ and } j \neq i\,. \tag{3-14}$$

The new objective function vector $\mathbf{f}^{k+1}$ for the given step direction and the step length is calculated. If the DM does not accept this vector, the iteration proceeds in determining a new step direction by solving the linearized problem (3-12). The variation of the step direction according to the preferences of the DM during several iterations provides minimization steps within the criterion space Y in the direction of the preferred functional-efficient compromise solution $\breve{\mathbf{f}}$. However, steps causing a removal of the functional-efficient boundary are not allowed (see Fig. 3/8).

If the DM accepts the last objective function vector $\mathbf{f}^k$, the iteration is stopped. The vector $\mathbf{f}^k$ in not necessarily functional-efficient but lies in the interior of the criterion space Y and $\hat{Y}^u$. Thus, a functional-efficient point within the neighbourhood of the accepted objective function

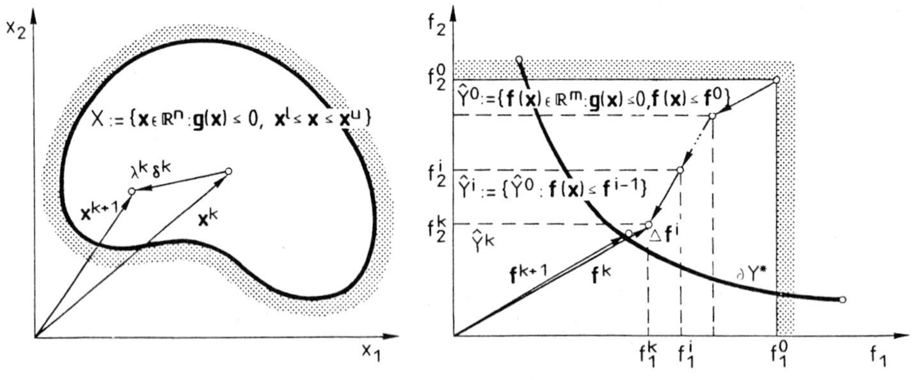

Fig. 3/8. Progress of the iteration in the design space X and in the criterion space Y resp. $\hat{Y}^k$

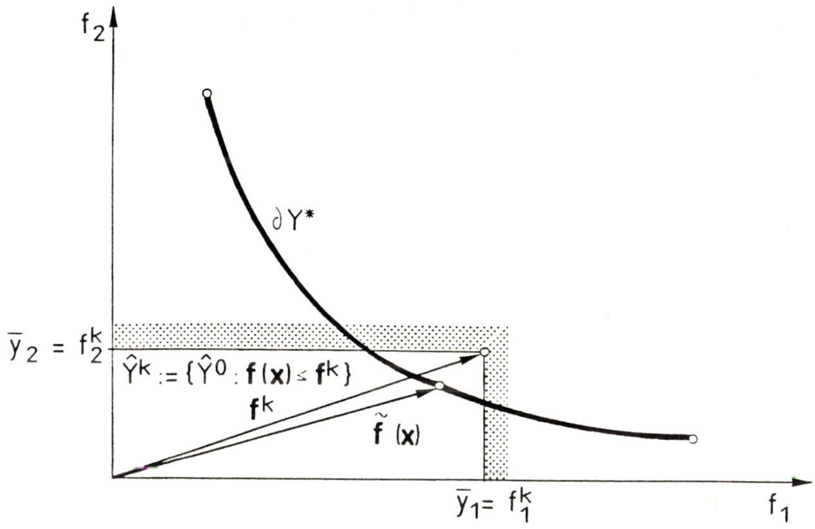

Fig. 3/9. Functional-efficient compromise solution $\tilde{\mathbf{f}}$ of the substitute problem within the neighbourhood of $\mathbf{f}^k$ and the subspace $\hat{Y}^k$

vector $\mathbf{f}^k$ is obtained by minimizing a scalar substitute problem

$$\min_{\mathbf{x} \in \mathbb{R}^n} \{ \sum_{j=1}^{m} w_j f_j(\mathbf{x}) : \mathbf{g}(\mathbf{x}) \leq \mathbf{0}, \ \mathbf{f}(\mathbf{x}) \leq \mathbf{f}(\mathbf{x}^k)\} \tag{3-15}$$

in the form of equally weighted objectives f_j (see Fig. 3/9).

3.2.5 Approach by Geoffrion

Similar to Jahn's approach, the method by Geoffrion [3] is based on a single specific mathematical programming algorithm, the Frank-Wolfe algorithm [24] with the linearized scalar substitute problem

$$\min_{\boldsymbol{\delta} \in \mathbb{R}^n} \left\{ (\nabla_{\mathbf{x}} p[\mathbf{f}(\mathbf{x})])^T \boldsymbol{\delta} : \mathbf{g}(\mathbf{x}) \leq \mathbf{0} \right\} . \tag{3-16}$$

The preference function $p[\mathbf{f}(\mathbf{x})]$ is not assumed to be explicitly known. During the interactive dialogue the decision maker is asked to assess local information about trade-off ratios between two objectives. He also has to select the required step length.

Two different ways can be used to determine the starting vector $\mathbf{x}^0$ The decision maker selects a feasible starting vector according to Jahn's approach, or the starting vector can be established by solving the substitute problem

$$\min_{\mathbf{x} \in \mathbb{R}^n} \{ \sum_{j=1}^{m} w_j \, f_j(\mathbf{x}) : \mathbf{g}(\mathbf{x}) \leq \mathbf{0} \} , \tag{3-17}$$

which leads to the functional-efficient starting point $\mathbf{x}^0 = \mathbf{x}^*$.

The direction-finding problem (3-16) has to be solved now. Information about the unknown preference function $p[\mathbf{f}(\mathbf{x})]$ is required from the DM. It is necessary to transform the original problem (3-16) by means of the chain rule

$$\nabla_{\mathbf{x}} p[\mathbf{f}(\mathbf{x})] = \sum_{j=1}^{m} \left[\frac{\partial p}{\partial f_j} \right]_{\mathbf{x}} \nabla_{\mathbf{x}} f_j(\mathbf{x}) \tag{3-18}$$

into the substitute problem of the k-th iteration

$$\min_{\boldsymbol{\delta} \in \mathbb{R}^n} \left\{ \sum_{j=1}^{m} w_j \left(\nabla_{\mathbf{x}} f_j(\mathbf{x}^k) \right)^T \boldsymbol{\delta} : \mathbf{g}(\mathbf{x}) \leq \mathbf{0} \right\}$$

with

$$w_j = \frac{\left[\frac{\partial p}{\partial f_j} \right]_{\mathbf{x}}}{\left[\frac{\partial p}{\partial f_1} \right]_{\mathbf{x}}} \quad \forall \; j = 1,\ldots,m \tag{3-19}$$

and the step direction $\boldsymbol{\delta}^k = \mathbf{x}^{k-1} - \mathbf{x}^k$. The weighting coefficients w_j reflect the trade-off of the DM between the j-th objective function f_j and a selected reference objective function as for example f_1.

There are two different methods for evaluating the required trade-off ratios to which the DM is indifferent. One way is to obtain a trade-off ratio w_j by allowing an infinitesimal change Δf_j of the j-th objective which compensates an infinitesimal change Δf_1 of the reference objective while all other objectives remain unchanged

$$w_j = - \frac{\Delta f_1}{\Delta f_j} \quad . \tag{3-20}$$

The other way to approximate the weighting factors is based on the fact that the gradient of $p[\mathbf{f}(\mathbf{x})]$ holds for the direction of the steepest ascent. Therefore, if all objectives except the first and the i-th one are held constant, the unknown preference function $p[\mathbf{f}(\mathbf{x})]$ increases most rapidly when the j-th objective changes by b_j-units in proportion to changes by b_1-units of the reference objective

$$w_j = \frac{b_j}{b_1} \quad . \tag{3-21}$$

After solving the linear direction-finding problem (3-19), the appropriate step length λ has to be determined by the DM himself who has to consider the goal

$$\min \{p[f_1(\mathbf{x}^k + \lambda^k \delta^k),\dots,f_m(\mathbf{x}^k + \lambda^k \delta^k)]: (\mathbf{x}^k + \lambda^k \delta^k) \in X\} \tag{3-22}$$

under the assumption

$$0 \le \lambda^k \le 1$$

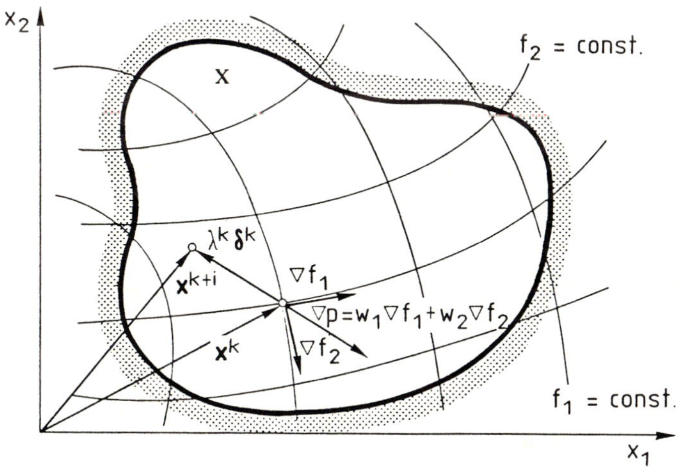

Fig. 3/10. Step direction δ and step length λ within the design space X

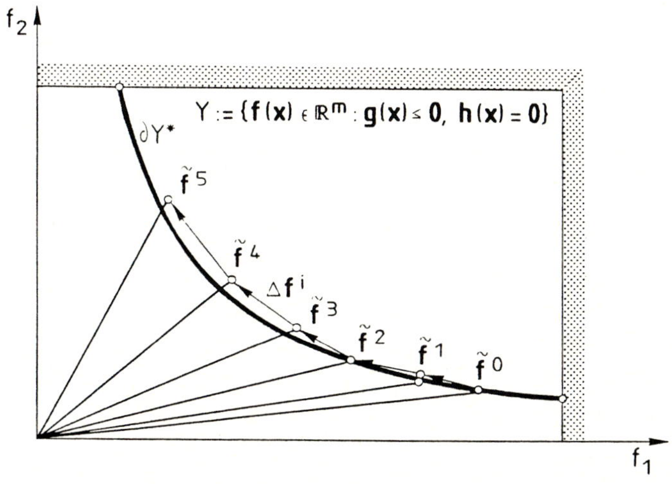

Fig. 3/11. Proposed points of the objective functions of five iterations

During the minimization (3-22) the DM is either supported by graphical representations of all objective functions in relation to the parameter λ or by a list of these dependencies given in a tabulated form.

Once the preferred step length is established, the new objective vector $\mathbf{f}^{k+1} = \mathbf{f}(\mathbf{x}^k + \lambda^k \boldsymbol{\delta}^k)$ can be calculated (see Fig. 3/10).

The decision maker evaluates the result and decides whether the procedure has to go on or whether the proposed solution (not necessarily functional-efficient) will be accepted as a compromise solution. If he decides to continue, he is able to move along the functional-efficient boundary (see Fig. 3/11).

As the approach by Geoffrion is based on the evaluation of the unknown substitute problem $p[\mathbf{f}(\mathbf{x})]$ by means of trade-off ratios within the iteration of a common scalar optimization algorithm, some other optimization algorithms may also be modified for an interactive dialogue in the way described.

3.3 Software Package DIALOG

3.3.1 Basic Structure

The program system DIALOG enables the decision maker to carry out an interactive optimization of nonlinear multicriteria optimization problems of structural mechanics. For this purpose the following requirements are met:

1. Modular structure so that the demands 2. to 5. can be fulfilled;
2. The DM can choose an interactive method without interfering in the source of the program;
3. Different structural analysis methods have to be provided;
4. The selection of an optimization method used for solving the scalar subproblems is adjustable to the problem;
5. The interfaces of the modules have to be clearly defined so that the interactive optimization program can be extended or completed by other systems at any time (e.g. SAPOP [25]);
6. In the dialogue with the DM only information about the optimization procedure and the preferences is exchanged. Constant optimization parameters and structural data are read in from a data base to relieve the DM;
7. The dialogue is documented on a data base.

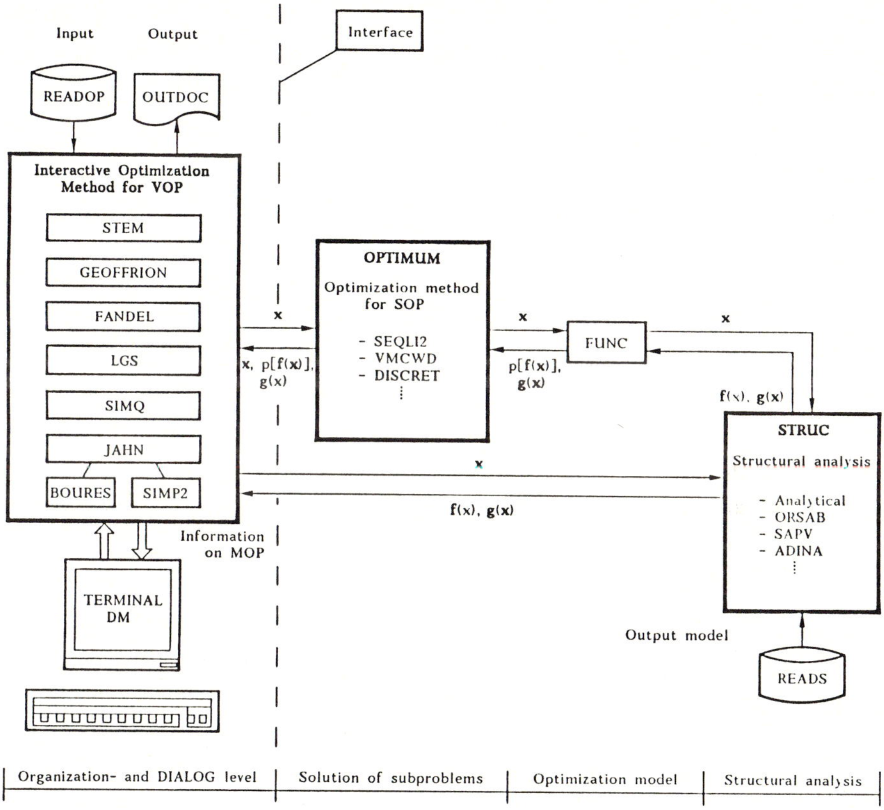

Fig. 3/12. Software Package DIALOG

Fig. 3/12 shows the program system DIALOG which fulfills these demands.

The dialogue with the DM is carried out via a terminal which allows the preparation and the exchange of information in a tabulated form or graphically. Up to now, the following interactive methods are provided:

STEM: STEP method by Benayoun et al. [2];

GEOFFRION: Method by Geoffrion [3];

FANDEL: Method by Fandel [4] with the subroutines LGS and SIMQ for solving linear equation systems;

JAHN: Method by Jahn [22] with the subroutines BOURES for the transformation of bounds into constraints and SIMP2 for the solution of a linear optimization problem.

The solution of the scalar subproblems is carried out by the module OPTIMUM which allows a selection of optimization algorithms:

SEQLI2: augmented sequential linearization strategy [20];

VMCWD: sequential quadratic programming method [21];

DISCRET: nonlinear optimization with discrete or integer design variables by limited enumeration [35].

As the formulation of the subproblems is specific for each interactive procedure, all formulations are stored in the subroutine FUNC so that they can be chosen individually.

The subroutine STRUC contains the optimization model and is varied according to the optimization problem. Here, the data of the structural analysis which is calculated by either analytical or numerical methods is used for establishing objective functions and constraints.

3.3.2 Interactive MO-Layout of a Conical Shell

This section gives a detailed description of the application of the interactive MO-procedures mentioned in 3.2. The multicriteria layout of a conical shell as the lower part of a steel tank filled with water (Fig. 3/13) illustrates the essential characteristics of the different interactive methods.

(1) *Optimization Model*

The mass M of the cone and the maximum value of the displacement w_{max} in the direction of the radius r will be minimized.

The wall thickness t(z) is given by the linear function

$$t(z) = a_1 + a_2 z = t_B + \frac{t_T - t_B}{h \cos 30^o} z \tag{3-23}$$

with the wall thicknesses t_B, t_T at the bottom and at the top of the shell and the height h of the shell as design variables. Here, only two design variables are used to describe the wall thickness because additional variables such as the wall thickness at z = h/2 will yield two different functions t(z) without having a significant effect on the optimal values of the objective functions [27]. The maximum allowable tensile and compressive stresses must not be exceeded, and the capacity of the

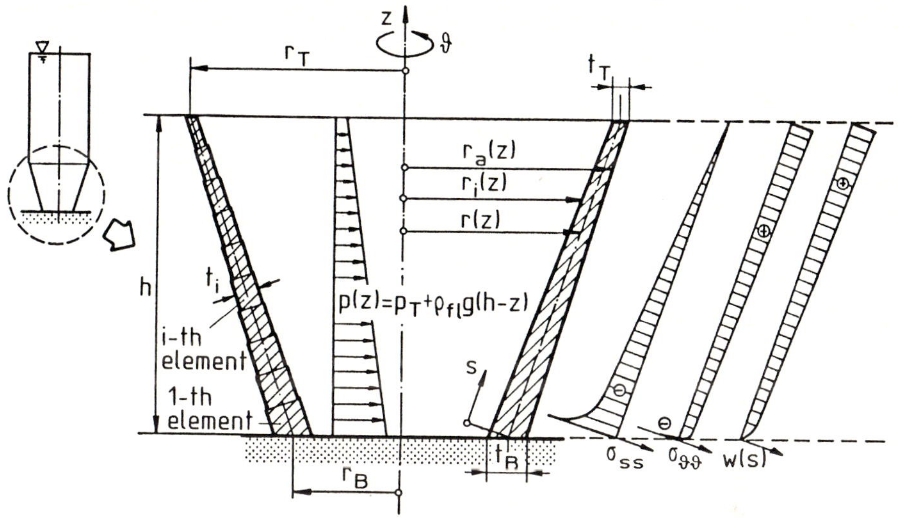

Fig. 3/13. Conical shell as the lower part of a steel tank under hydraulic pressure

fluid volume V_{fl} has to be kept constant. Hence, the two-dimensional MO-problem can be summarized as follows

$$\min_{\mathbf{x}\in X}\{\ \mathbf{f}(\mathbf{x}) \hat{=} \left(M,\ w_{max}\right)^T :\ g_1(\mathbf{x}) = \frac{\sigma_{max}}{\sigma_{tz}} - 1 \le 0\,,\ g_2(\mathbf{x}) = \frac{\sigma_{min}}{\sigma_{cz}} - 1 \le 0\,,$$

$$h_1(\mathbf{x}) = \int_0^h \pi\,(r_a^2 - r_i^2)\,dz = V_{fl}\ \} \tag{3-24}$$

with $\mathbf{x} = (t_B, t_T, h)^T$ vector of design variables,

$X := \{\ \mathbf{x}^l \le \mathbf{x} \le \mathbf{x}^u\ \}$ design space with upper and lower bounds,

σ_{tz}, σ_{cz} maximum allowable tensile and compressive stresses.

(2) *Structural Analysis via Transfer Matrices Methods*

The maximum displacement and the state quantities are determined by transfer matrices methods [26]. As shown in Fig. 3/13 the conical shell is subdivided into a finite number of smaller elements with a constant thickness $t_i(z)$ and a constant pressure $p_i(z)$ acting on the i-th element. At the bottom displacements and distortions are not allowed; at the top an initial pressure p_T is considered.

The transfer matrices method is based on equilibrium, strain-displacement relations and constitutive equations for axially symmetric loaded shells which lead to six differential equations containing six unknowns N_{ss}, Q_{ss}, M_{ss}, u, w, χ. These differential equations are solved for the i-th element by an approximation to derive a rearranged transfer equation

$$\mathbf{z}_i = \mathbf{C}_i \, \mathbf{z}_{i-1} \quad \text{with} \quad \mathbf{z}^T = (N_{ss}, \, Q_{ss} \, , \, M_{ss} , \, u, \, w, \, \beta, \, 1) \tag{3-25}$$

with a transfer matrix $\mathbf{C}_i$ and a vector of the state quantities $\mathbf{z}_i$. The boundary conditions given at the beginning of the first element $\mathbf{z}_o$ and at the end of the last ns-th element $\mathbf{z}_{ns}$ are considered by calculating the unknown quantities of the initial state vector z_o and the vector $\mathbf{z}_{ns}$ by

$$\mathbf{z}_{ns} = \mathbf{C} \, \mathbf{z}_o \quad \text{with} \quad \mathbf{C} = \mathbf{C}_o \cdot \mathbf{C}_1 \, \ldots \cdot \mathbf{C}_{ns} \quad . \tag{3-26}$$

Therefore, starting with the determined initial state vector $\mathbf{z}_o$ for each point i, all unknown quantities can be computed by (3-25). The tensile and compressive stresses are calculated by the following formulas:

$$\sigma_{ssi} = \frac{N_{ssi}}{t_i} \pm \frac{6 M_{ssi}}{t_i^2} \, , \quad \sigma_{\vartheta\vartheta i} = \frac{N_{\vartheta\vartheta i}}{t_i} \pm \frac{6 M_{\vartheta\vartheta i}}{t_i^2} \quad . \tag{3-27}$$

The stress distributions and the displacement w of the present example are shown on the right hand side of Fig. 3/13.

(3) Interactive MO-Approaches

a) Approach by FANDEL

The approach by Fandel is presented here as an example of a superior organization scheme. In order to outline the iteration using this approach, the conical shell example is considered. Fig. 3/14 shows the criterion space and the functional-efficient boundary together with all the numbered alternative solutions and points calculated during the interactive optimization.

Step 1:

Initialization of the iteration counter and the initial criterion space:

$$k = 1 \; , \quad \hat{Y}^k = Y := \{ \, \mathbf{f}(\mathbf{x}) \in \mathbb{R}^m : \mathbf{g}(\mathbf{x}) \leq \mathbf{0} \; , \; \mathbf{h}(\mathbf{x}) = \mathbf{0} \, \} \; , \qquad m = 2 \quad .$$

Step 2:
The individual minima $\overline{f}_1$, $\overline{f}_2$ (see Fig. 3/14 points 1, 2 and the corresponding vectors $\mathbf{f}^{*1}$, $\mathbf{f}^{*2}$) are determined within the current criterion space $\hat{Y}^k$ by the scalar optimization algorithm VMCWD. They form the ideal objective function vector $\overline{\mathbf{f}} = (\overline{f}_1, \overline{f}_2)^T = (168, 0.085)^T$.

Step 3:
A central point $\mathbf{f}^M$ (point 3) is calculated to reduce the remaining criterion space

$$\hat{Y}^k := \{\mathbf{f}(\mathbf{x}) \in \mathbb{R}^m : \mathbf{g}(\mathbf{x}) \le \mathbf{0}, \mathbf{h}(\mathbf{x}) = \mathbf{0}, \mathbf{f}(\mathbf{x}) \le \mathbf{f}^M\}$$

with

$$\mathbf{f}^M = \frac{1}{m} \sum_{i=1}^{m} \mathbf{f}^{*i} - (619, 0.318)^T$$

in order to improve the convergence of the procedure.

Step 4:
According to step 2, the individual minima in the *reduced criterion space* $\overline{f}_i \in \hat{Y}^k$ are determined (points 7,8).

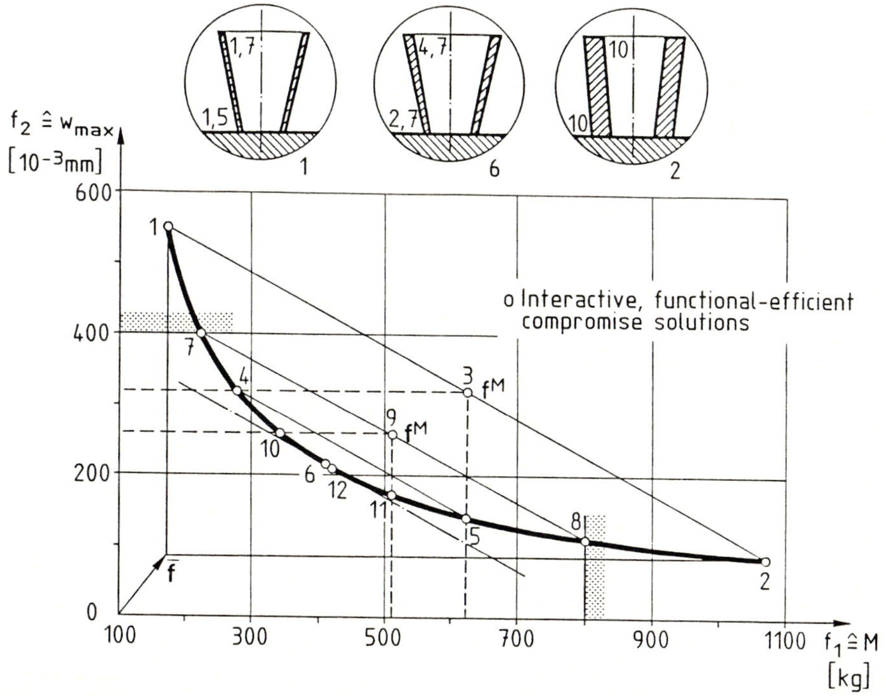

Fig. 3/14. Functional-efficient boundary determined by Fandel's approach

Step 5:
It is assumed that the preferred compromise solution $\tilde{\mathbf{f}}$ still lies within the reduced criterion space. Therefore, one functional-efficient point within $\hat{Y}^k$ has to be offered to the DM. This is done by obtaining the *weighting coefficients* w_i and by solving problem (3-1). The unknown weighting coefficients are determined by displacing a *hyperplane* with the equation (3-5) towards the functional-efficient boundary until the hyperplane is tangent to the boundary in one point only. According to Fig. 3/15 it is point 6 with the solution $\mathbf{w} = (0.34, 0.66)^T$.

Step 6:
The functional-efficient solution point (6) is offered to the DM who decides either to accept this as a preferred compromise solution and terminates the iteration or to reduce the criterion space by stating upper bounds $\bar{\mathbf{y}}$ for at least one objective function whereby the iteration counter is increased ($k = k+1$). For the present example it was considered that the
DM wants the upper bounds $\bar{y}_1 = 800$ kg and $\bar{y}_2 = 0.4$ mm to be kept

$$\hat{Y}^k := \{\mathbf{f}(\mathbf{x}) \in \mathbb{R}^m : \mathbf{g}(\mathbf{x}) \leq \mathbf{0}\ ,\ \mathbf{h}(\mathbf{x}) = \mathbf{0}\ ,\ \mathbf{f}(\mathbf{x}) \leq \bar{\mathbf{y}}\} \quad .$$

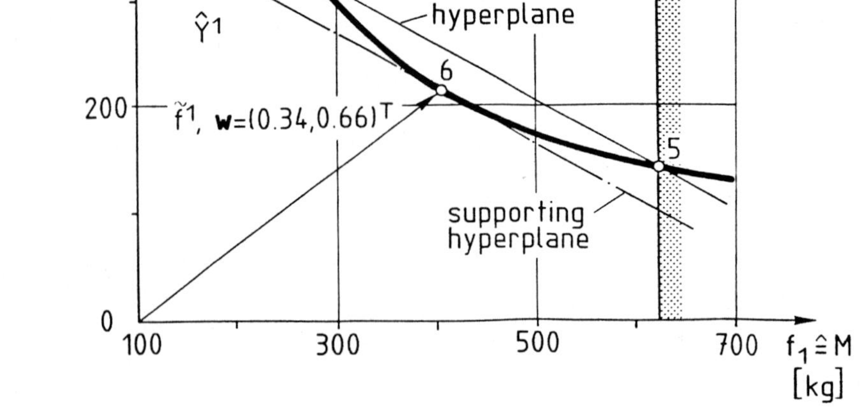

Fig. 3/15. Determination of the unknown weighting factors **w** by a hyperplane

The iteration continues with step 2. The two individual minimas 7 and 8 are calculated ($\mathbf{f}^{*1} = (219, 0.4)^T$, $\mathbf{f}^{*2} = (800, 0.109))^T$. They lie on the upper bounds $\overline{y}_1$ and $\overline{y}_2$, resp. After carrying out the described steps the iteration was terminated at point 12 (very close to point 6).

In order to get a survey of the efficient boundary, ten functional-efficient points have been obtained by means of 254 structural analyses.

b) STEP-method

Although the STEP-method can be classified as a superior organization scheme with the required preference information similar to the afore-mentioned approach, the strategy to achieve a preferred solution is different. This will be described in the following steps. Fig. 3/16 shows all functional-efficient points numbered according to their sequence.

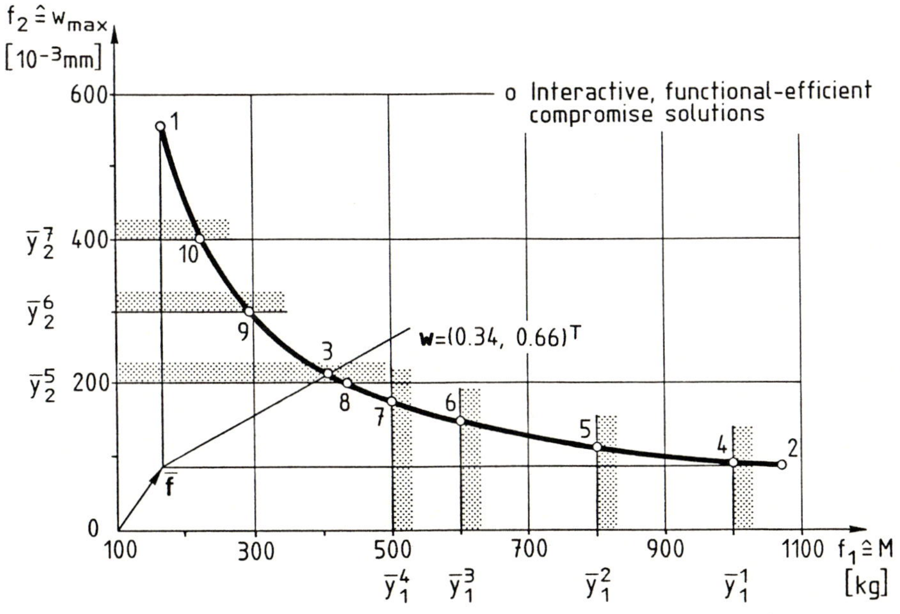

Fig. 3/16. Functional-efficient boundary ∂Y^* determined by the STEP-method

Step 1:
Initialization of the iteration counter and the initial criterion space

$$k = 1 \quad , \quad \hat{Y}^k = Y := \{ \mathbf{f}(\mathbf{x}) \in \mathbb{R}^m : \mathbf{g}(\mathbf{x}) \leq \mathbf{0} \; , \; \mathbf{h}(\mathbf{x}) = \mathbf{0} \} \; , \qquad m = 2 \; .$$

Step 2:
After determining the two vectors of the individual minima $\mathbf{f}^{*1}$, $\mathbf{f}^{*2}$, their components are used for reducing the criterion space $\hat{Y}^k = \{ Y : \mathbf{f}(\mathbf{x}) \leq \bar{\mathbf{y}} \}$ with $\bar{\mathbf{y}} = (f_1^{*2}, f_2^{*1})^T = (1070, 0.55)$.

Step 3:
To balance the different magnitudes of both objective functions, weighting coefficients which remain constant during the iterations are calculated according to (3-8).

Step 4:
A functional-efficient point (see Fig. 3/16, point 3) is obtained by solving the min-max problem (3-11) and offered to the DM.

Step 5:
The DM either terminates the iteration because his preferred solution was found, or he is asked to reduce the remaining criterion space according to his preferences by means of upper bounds $\bar{\mathbf{y}}^k$. For this application example the bounds $\bar{\mathbf{y}}^k = (1000, f_2^{*1})$ are assumed, and the criterion space is reduced

$$\hat{Y}^k := \{ Y : \mathbf{f}(\mathbf{x}) \leq \bar{\mathbf{y}}^k \}.$$

Step 6:
The DM is now asked whether he prefers that the following functional-efficient point should lie in the middle of the remaining functional-efficient boundary or whether this point should meet the l-th component of $\bar{\mathbf{y}}$. If the latter one is the case, the weighting coefficient w_l is assigned zero. For the present example $w_1 = 0$ was chosen.

The next iteration starts with step 4 solving the min-max problem in $\hat{Y}^k$. Point 4 is obtained and offered as a compromise solution. To explore the functional-efficient boundary, seven different bounds are stated with one weighting coefficient set to zero and the other one set to its initial value (see Fig. 3/16).

Altogether, ten functional-efficient alternative solutions are determined using the STEP-method for which 501 structural analyses are required.

c) Approach by Jahn

As an interactive method with an extended organization scheme the approach published by Jahn in 1982 is applied to obtain an optimal design of the conical shell. As described in Section 3.2.4 this procedure allows optimization steps within the criterion space towards the functional-efficient boundary (see Fig. 3/18b).

Step 1:
Initialization of the iteration counter k=0.

Step 2:
The DM has to select a *feasible starting point* $\mathbf{x}^0 \in X: \{\mathbf{x} \in \mathbb{R}^n: \mathbf{g}(\mathbf{x}) \le \mathbf{0}, \mathbf{h}(\mathbf{x}) = \mathbf{0}\}$ for which the vector of objective functions is evaluated and used for restricting the remaining criterion space $\hat{Y}^k: \{\mathbf{f}(\mathbf{x}) \in \mathbb{R}^m: \mathbf{g}(\mathbf{x}) \le \mathbf{0}, \mathbf{h}(\mathbf{x}) = \mathbf{0}, \mathbf{f}(\mathbf{x}) \le \mathbf{f}^0\}$ (see Fig. 3/18b, point 7').

Step 3:
The DM has to give appropriate values of the *weighting coefficients* $\mathbf{w}$ (objectives) and $\mathbf{v}$ (constraints). The optimal *step direction* $\boldsymbol{\delta}$ is determined by using a SIMPLEX algorithm. The optimal step length is still unknown.

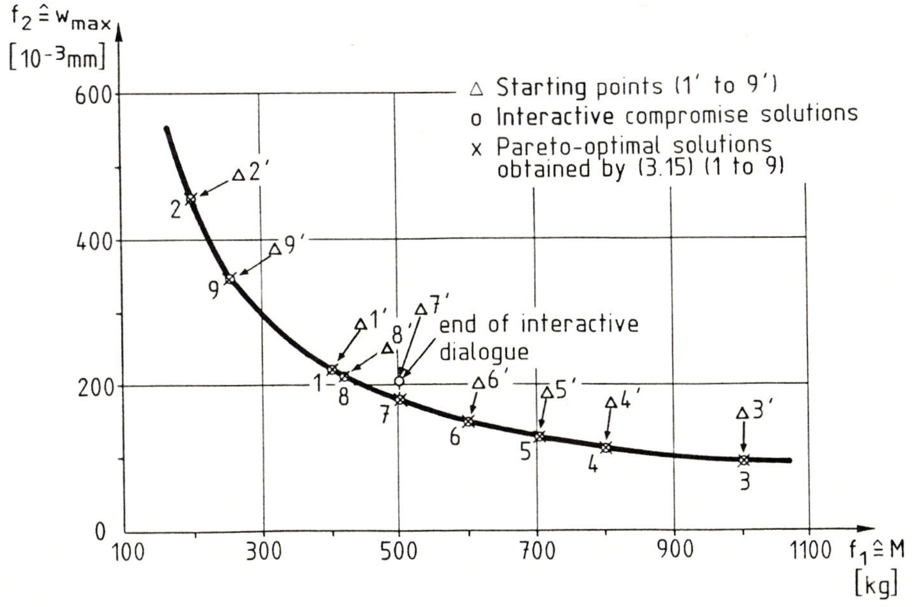

Fig. 3/17. Functional-efficient boundary ∂Y^* with points determined by Jahn's approach

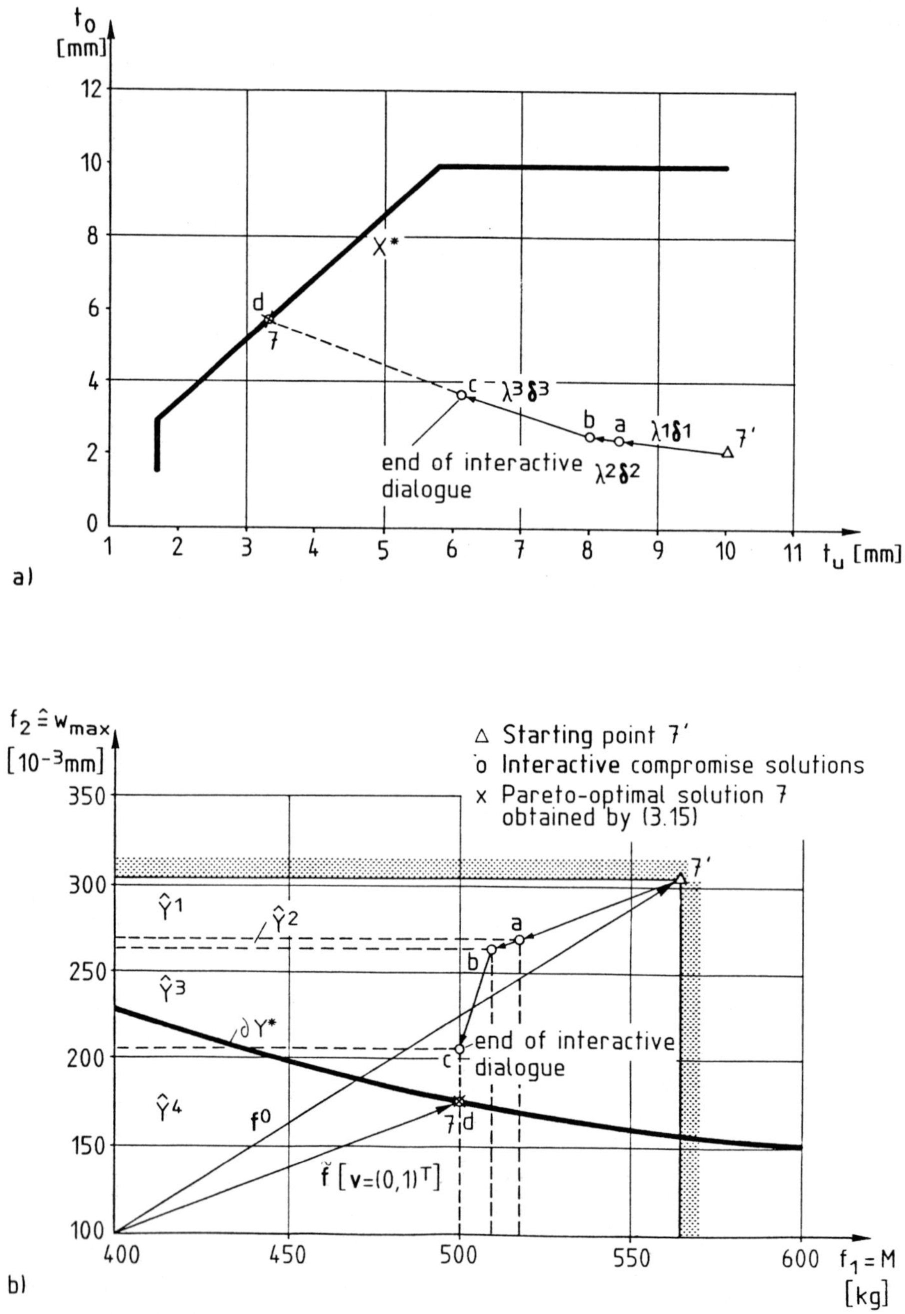

Fig. 3/18. Minimization steps starting from point (7)

a) Design space X

b) Steps towards the functional-efficient boundary ∂Y^* in the criterion space

Step 4:
To determine an optimal *step length* λ in the direction of $\boldsymbol{\delta}$, the DM selects one objective function f_i which is to be improved. The maximum allowable step length λ_{max} is either obtained by the one-dimensional minimization problem (3-14) implying that only steps within the reduced criterion space are allowed. On the other hand the DM can determine the maximum step length by trial and error.

Step 5:
Select a step length $\lambda^k < \lambda_{max}$ and evaluate $\mathbf{f}(\mathbf{x}^{k+1}) = \mathbf{f}(\mathbf{x}^k + \lambda^k \boldsymbol{\delta}^k)$ which moves a point towards the functional-efficient boundary (Fig. 3/18, point a).

Step 6:
The DM is asked either to accept the current vector of objective functions $\mathbf{f}^{k+1}$ as a compromise solution, which is generally not a functional-efficient point, or he continues the iteration with step 3 and chooses new weighting coefficients. In doing so the points b) and c) in Fig. 3/18 are obtained.

After Jahn's procedure is stopped by the DM , the iteration may be extended by determining a functional-efficient point solving (3-15). As no further functional-efficient points are evaluated, this method supplies no additional information about the functional-efficient boundary. Fig. 3/17 illustrates how Pareto-optimal points are determined using formula (3-15). In the case of our application example, nine different starting points had to be chosen to get different solutions with the amount of 490 structural analyses. This was due to the fact that the initial starting point restricting the remaining criterion space had to be feasible. Here, the feasible criterion space was only a small domain.

d) *Approach by Geoffrion*

The approach by Geoffrion with an extended organization scheme differs from Jahn's approach according to the type of preference information needed during the interactive dialogue. Additionally, only the substitute function is linearized. The solutions achieved are not necessarily functional-efficient. For the present example the solutions are shown within the criterion and the design space (Fig. 3/19, 3/20).

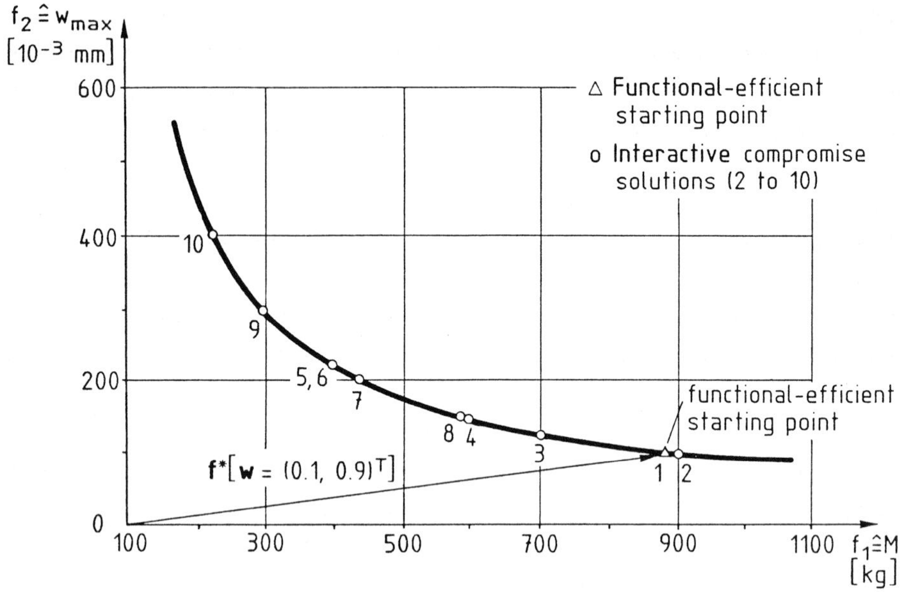

Fig. 3/19. Functional-efficient boundary ∂Y^* with solutions by Geoffrion

The following steps illustrate this approach for the conical shell.

Step 1:
Initialization of the iteration counter $k=0$.

Step 2:
A feasible starting point $\mathbf{x}^0$ has to be selected. This is done either by the DM or by calculating a functional-efficient point $\mathbf{x}^0 = \mathbf{x}^*$. In the latter case the DM states the weighting coefficients v_i, and the problem (3-17) is solved by mathematical programming. For the present example point 1 (see Fig. 3/19) was determined using $w_1 = 0.1$ and $w_2 = 0.9$.

Step 3:
The decision maker is asked to state the trade-off factors w_i $(i=1,...,m)$ between the objective f_i and the reference objective f_1 at the current point reflecting the direction of the steepest descent of the unknown substitute function with the help of (3-20) or (3-21).

Step 4:
The step direction δ is obtained by solving the direction-finding problem (3-19) which is based on a linearization of the unknown preference function p with the result $\delta^{kT} = (4.24,\ 0.)$ for the present example.

Step 5:
The decision maker is asked to state a preferred step length $\lambda^k \leq 1$ which leads to point $\mathbf{f}^k[\mathbf{x}+\lambda^k \delta^k]$ (see Fig. 3/20, point 2 with $\lambda = 0.09$). This point can either be accepted as a preferred compromise solution, or the iteration counter is increased, and the iteration continues with step 3.

Ten different solutions were obtained during one interactive optimization, shown in Fig. 3/19., for which 423 structual analyses were carried out. They show a good approximation of the functional-efficient boundary ∂Y^* although the approximation of X^* is inaccurate (see Fig. 3/20).

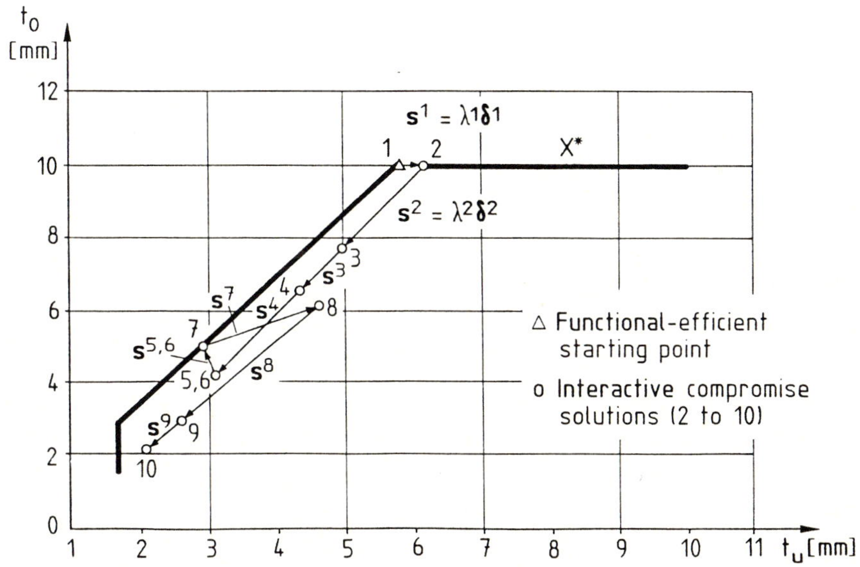

Fig. 3/20. Functional-efficient boundary X^* with solutions by Geoffrion

(4) Comparison of the Methods

By means of the application example four interactive optimization methods implemented in the program system DIALOG have been illustrated. The solution points proposed in the interactive dialogue have been included in Figures 3/14 to 3/20 to get a survey of the functional-efficient boundary. Table 3/1 shows that Fandel's approach requires the smallest number of structural analyses. With rather incomplete preferences, Fandel's method thus provides the best way to improve the information about the MO-problem. Although the method by Jahn needs only 32 structural analyses per offered solution as well, these may be too far away from the functional-efficient boundary (see Fig. 3/18). Furthermore, Jahn's approach always has to be started with a feasible starting point.

It was also investigated whether the different methods obtain a given functional-efficient point which simulates a DM who wants to reach a preferred solution as quickly as possible, taking into account his preferences. Table 3/1 shows that Jahn's method comes near the desired point with an average of 33 structural analyses. The methods with a

	FANDEL	STEM	JAHN	JAHN with (3-15)	GEOFFRION
Organization scheme	superior	superior	extended	extended	extended
Determination of the functional-efficient boundary:					
No. of offered solutions (n_l)	10	10	9	9	10
No. of determined funct.-eff. points	10	10	9	9	10
No. of structural analyses (n_{str})	254	501	290	456	423
$\frac{n_{str}}{n_l}$	25,4	50,1	32,2	50.66	42.3
Determination of a single preferred compromise solution:					
No. of structural analyses (n_{str})	185	198	33	52	57

Table 3/1. Outline of the characteristic values of the interactive optimizations superior organization scheme (Fandel, STEM) need more than 180 structural analyses. For this reason it can be concluded that methods with a superior organization scheme are suitable for a DM with rather incomplete information about the the MO-problem. With a small computation effort many solutions can be suggested here. As far as optimization problems with a time-consuming structural analysis are concerned (e.g. finite element methods), procedures with an extended organization scheme should be applied. They generally need fewer structural analyses for obtaining a preferred compromise solution than procedures with a superior organization scheme.

The advantages of both organization schemes can be combined as shown by Diaz [19]. His approach enables the DM to vary between both organization schemes during the interactive optimization dialogue. Functional-efficient points or approximate compromise solutions obtained by sensitivity analysis are presented. Starting from a functional-efficient point $\mathbf{f}^*$, the estimation of a new compromise solution is based on a sensitivity equation derived from the Lagrangian function (1-5), the sequential quadratic approximation of the MO-problem and finally the Kuhn-Tucker conditions (1-6) at $\mathbf{f}^*$. According to the variation of the DM's preferences estimate solutions are achieved by defined perturbations of the approximate quadratic problem at $\mathbf{f}^*$.

3.4 Software Package CAMOS

As mentioned in Section 3.1, interactive multicriteria methods have been developed further into decision support systems in economics and operation research [28, 29]. In engineering design, however, it is advisable that these systems should use the facilities of CAD software. Having this in mind, a **C**omputer **A**ided **M**ulticriteria **O**ptimization **S**ystem (CAMOS) was developed. The software package CAMOS is designed to

- solve single and multicriteria optimization problems for nonlinear models with continuous, integer, discrete, and mixed design variables.
- facilitate interactive processes for computer aided optimum design which includes both input and output information arrangements,
- provide an effective tool for searching interactively a preferred solution in the multicriteria decision making process.

3.4.1 Optimization Algorithms Used in CAMOS

a) Continuous Model

In order to solve nonlinear programming problems with continuous variables numerous methods and computer procedures have been developed recently (see Chapter 2). Selecting the method suitable for an optimization model is usually a difficult task, and thus most of the optimization software packages provide several methods. The following methods are used in CAMOS:

RS - **R**andom **S**earch Method,
FT - **F**lexible **T**olerance Method [31],
DS - **D**irect **S**earch Method of Hooke and Jeeves [32],
DFP - **D**avidon-**F**letcher-**P**owell Gradient Method [33],
NM - **N**elder-**M**ead Simplex Method [34].

The random search method is mainly used for

(1) generating a good starting point,
(2) solving models with discrete integer and mixed design variables,
(3) generating a set of Pareto-optimal solutions for both continuous and discrete models.

Each of the other iterative methods, i.e. FT, DS, DFP, and NM method, can be used separately or together with the RS-method in such a way that the results obtained by use of the RS-method are the starting solutions for each iterative method.

(b) Discrete Models

If all or some elements of **x** assume discrete values from a given set, then the problem is a discrete or a mixed continuous-discrete programming, respectively. The models with such variables occur quite often in engineering design. Let $\mathbf{x} = (x_1, ..., x_l, ..., x_n)^T$ be a vector of design variables for which the vector $\mathbf{x}' = (x_1, ..., x_l)^T$ defines continuous variables and the vector $\mathbf{x}'' = (x_{l+1}, ..., x_n)^T$ defines discrete variables. While establishing the optimization model it is easy to order the vector **x** in this manner. Assuming this, one can have

$l = n$ - continuous nonlinear programming,
$l = 0$ - discrete nonlinear programming,
$0<l<n$ - mixed continuous and discrete programming.

Let $X_{l+u} := \{x^1_{l+u}, ..., x^t_{l+u}\}$ be a set of discrete values which the x_{l+u} variable can assume, where u = 1, ..., n-l. If X_{l+u} for u = 1 ,..., n-l contains only integer numbers, then for l = 0 the problem is called integer nonlinear programming and for $0<l<n$ mixed continuous and integer programming.

The exact solution of the models with discrete variables encounters great difficulties even for linear models [35]. Thus, practical approaches to this problem are:

- Find the solution assuming that all variables are continuous and then round off the values of the discrete variables to the nearest values from the set X_{l+u}.
- Using the random search method one has to find out the solution for the model with discrete variables. This method can be effectively applied to examples with a small number of design variables.

(c) Combined Method

Apart from the approaches mentioned above the following combined random and iterative method is proposed and used in CAMOS.

Step 1:

Using the random search method, find the vector $\mathbf{x}^g=(x^g_1,..., x^g_l,..., x^g_n)^T$ which minimizes the objective function under the assumption that the values of each x_{l+u} variable belong to the set X_{l+u}. In other words: Find $\mathbf{x}^g$ such that

$$f(\mathbf{x}^g) = \min_{\mathbf{x} \in X} f(\mathbf{x}) \quad \text{with } x_{l+u} \in X_{l+u} \quad \text{for } u = 1,...,n-l. \tag{3-28}$$

Step 2:

Using an iterative method, find the vector $\mathbf{x}'^* = (x^*_1, ..., x^*_l)^T$ which minimizes the objective function under the assumption that $\mathbf{x}'' = (x^g_{l+1}, ..., x^g_n)^T$ is constant and contains elements with the values generated in step 1. In other words: Find $\mathbf{x}'^*$ such that

$$f(\mathbf{x}'^*) = \min_{\mathbf{x} \in X} f(\mathbf{x}) \quad \text{with } \mathbf{x}'' = \mathbf{x}''^g. \tag{3-29}$$

Step 3:

Use the vector $\mathbf{x}'^g = (x^g_1, ..., x^g_l)^T$ as the starting point for searching the vector $\mathbf{x}'^*$ and after minimizing according to step 2 substitute $\mathbf{x}''^* = \mathbf{x}''^g$.

This method is intended mainly for solving models with mixed discrete and continuous design variables, and it is also used for searching for a good feasible starting point for all iterative methods. In the latter case it is assumed that $l = n$.

3.4.2 Multicriteria Strategy Approaches

In CAMOS the following strategies are used for dealing with multicriteria optimization models:

1. Min-max method (see (1-19));
2. Global criterion method with the preference function

$$p[\mathbf{f}(\mathbf{x})] \equiv \sum_{j=1}^{m} \left[\frac{f_j(\mathbf{x}) - \bar{f}_j}{\bar{f}_j}\right]^p ; \qquad (3\text{-}30)$$

3. Weighting min-max method (see (1-22);
4. Weighting objective method (see (1-11);
5. Normed weighting objective method with the peference function

$$p[\mathbf{f}(\mathbf{x})] \equiv \sum_{j=1}^{m} w_j \frac{f_j(\mathbf{x})}{\bar{f}_j} ; \qquad (3\text{-}31)$$

6. The method for selecting the set of functional-efficient solutions from the generated set of feasible solutions [30]. This method can be used only with the random search method.

Methods which use weighting coefficients and refer to the ideal vector $\bar{\mathbf{f}}$ (the demand level) are intended to be applied interactively. Moreover, the constraint-oriented transformation method (see (1-14)) can also be employed interactively.

Multicriteria decision making is arranged in the system by means of an interactive selection of:

- multicriteria optimization methods.
- weighting coefficients for weighting min-max and objective methods.
- the ideal vector (method of displaced ideal).

3.4.3 Description of CAMOS

The software package CAMOS which is coded in FORTRAN has been prepared for a microcomputer and has been tested on an IBM PC/XT/AT.

The package is equipped with context-dependent HELP-functions which support the decision phase and the learning process.

Iterative methods for searching the minimum of a function are supported by a random search method which is used for establishing a good starting point as well as for dealing with discrete and integer variables. The optimization problem is introduced into the system by means of a problem-oriented subroutine which can reflect any desires as to the data to be introduced and the optimization results to be printed. CAMOS is the result of experience gained during several years while using the interactive multicriteria optimization system described in [30,37-39].

The general flow diagram of the system is presented in Fig. 3/21. In this system the scalar minimization algorithms are arranged in the following way:

1	- RS method which can be used for all types of optimization models;
2,3,4,5	- Iterative methods (FT, DS, DFP, NM) used for continuous programming models;
12,13,14,15	- Combined methods (RS/FT, etc.) used for continuous and mixed programming models.

The iterative methods start from the point given by the user. For the random method combined with any of the iterative methods the starting point of the iterative method is generated by the random method. If the ideal vector is calculated by the system, m separately attainable minima are to be found. In this case the random search method looks for the minimum values of all objective functions simultaneously, i.e. the values of all objective functions are calculated for each generated point. After having generated the assumed number of points, the minimum value points are stored and then used as the starting points for an iterative method. The same situation occurs while seeking the Pareto-optimal solution. However, in this case only one point is stored for an iterative method. The combined methods are also used for solving mixed continuous-integer and continuous-discrete programming models. For these models the same idea of generating a feasible starting point for the vector $\mathbf{x}'$ is applied.

The optimization problem is introduced into the system by means of the subroutine NONPRO (**Non**linear **Pro**gramming) which is a user-supplied subroutine.

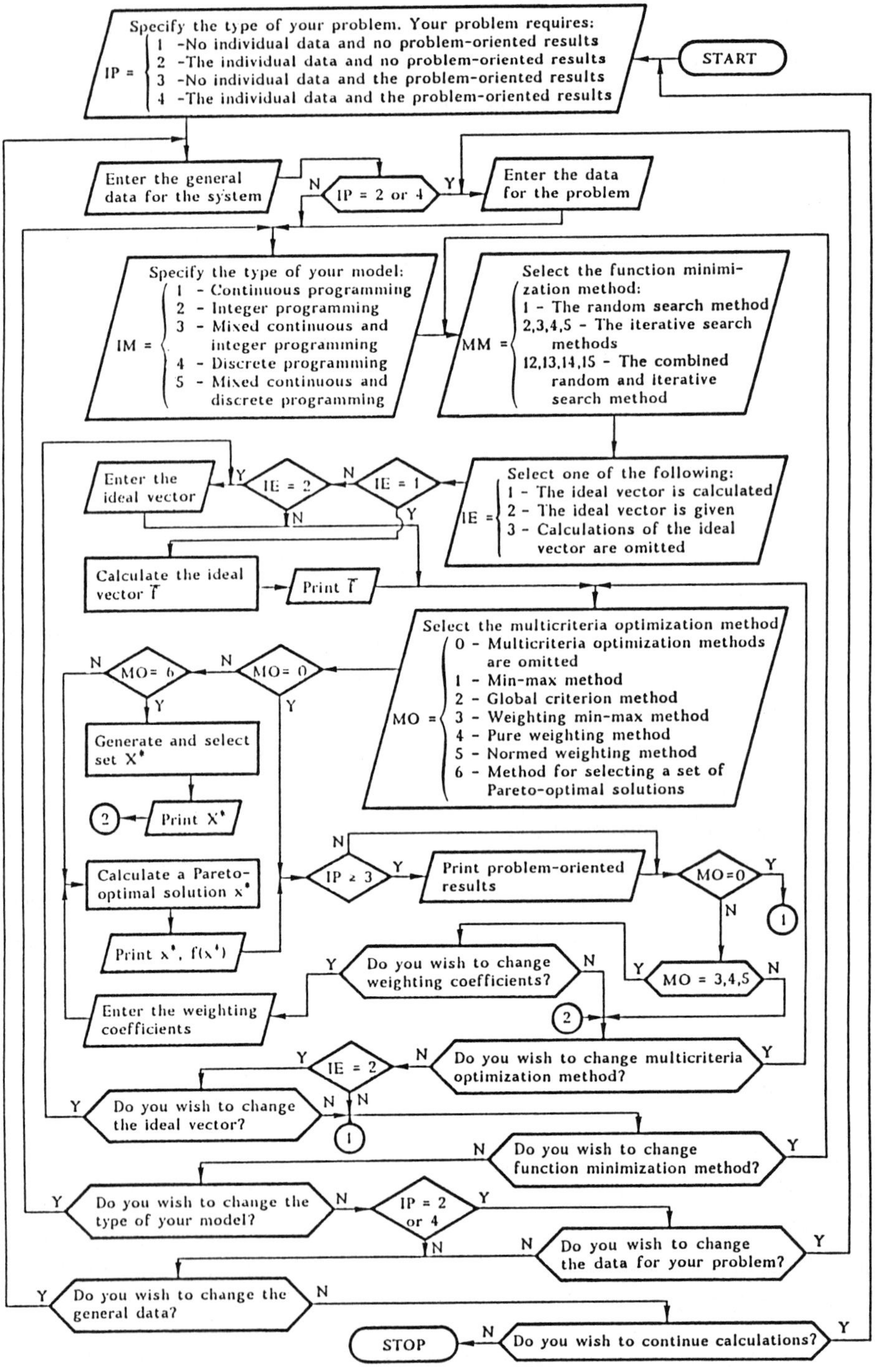

Fig. 3/21. General flow diagram of CAMOS

The structure of CAMOS and the subroutine NONPRO provide the possibility to develop computer aided optimum design. The subroutine NONPRO consists of three main parts:

- Part 1 introduces the data into the problem and/or illustrates it graphically. The constant parameters which are typical for a specific design are prepared to be entered in this part. The data for these parameters are introduced during the computer aided session (see block "Enter the data for the problem" in Fig. 3/21), and they can be modified while running the system.
- Part 2 of NONPRO is for calculating the objective and constraint functions and can be arranged in any suitable way. It is only required that the values of F(I), G(I), and H(I) have to be evaluated for given values of X(I).
- Part 3 of NONPRO is for calculating and printing the problem-oriented results and/or for the graphical illustration of the optimal solution. Note that at the decision making stage, it is usually hard to have high confidence in the solution if there is no graphical illustration.

The communication between all parts of NONPRO and CAMOS is arranged by the variables from COMMON/DAT/. Parts 1 and 3 of NONPRO are optional, and at the beginning of the computer session the user declares which of the optional parts he intends to use. The subroutine NONPRO may be prepared for each general optimum design problem as for example: optimum design of machine tool spindles, optimum design of hydrostatic bearings (see Chapter 7.2), optimum design of robot balancing mechanisms (see Chapter 5.1). All these can be treated with a computer aided design procedure.

3.4.4 Interactive MO-Layouts of a Machine Tool Spindle

A simple application example which shows the possibilities of a computer aided session with CAMOS will be presented here. More complicated examples solved by CAMOS are presented in Chapter 5.1 and 7.2. Consider the problem of a preliminary design of a machine tool spindle as presented in Fig. 3/22.

Formulation of the MO-problem:

$$\min_{\mathbf{x} \in \mathbb{R}^n} \{\mathbf{f}(\mathbf{x}) = \{ f_1(\mathbf{x}), f_2(\mathbf{x}) \} : g_j(\mathbf{x}) \le 0, \ j = 1,\dots,10\} , \qquad (3\text{-}32)$$

Vector of design variables $\mathbf{x} = (d_o, l, d_a, d_b)^T$.

Objectives:

Volume of the spindle

$$f_1(\mathbf{x}) = \frac{\pi}{4} \left[a(d_a^2 - d_o^2) + l\,(d_b^2 - d_o^2) \right]; \qquad (3\text{-}33)$$

Static displacement under the force F

$$f_2(\mathbf{x}) = \frac{Fa^3}{3EI_a} \left(1 + \frac{l}{a} \frac{I_a}{I_b}\right) + \frac{F}{c_a} \left[\left(1 + \frac{a}{l}\right)^2 + \frac{c_a\, a^2}{c_b\, l^2} \right], \qquad (3\text{-}34)$$

where I_a and I_b are the moments of inertia

$$I_a = 0.049\,(d_a^4 - d_o^4)\ , \quad I_b = 0.049(d_b^4 - d_o^4)\ ; \qquad (3\text{-}35)$$

c_a and c_b are bearing stiffnesses

$$c_a = 35400\ |\delta_{ra}|^{\frac{1}{9}}\, d_a^{\frac{10}{9}}\ , \quad c_b = 35400\ |\delta_{rb}|^{\frac{1}{9}}\, d_b^{\frac{10}{9}}\ , \qquad (3\text{-}36)$$

with δ_{ra} and δ_{rb} as preloads of the bearings.

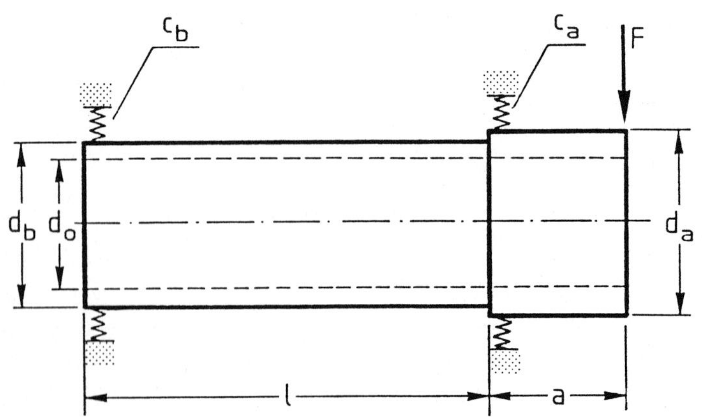

Fig. 3/22. Sketch of a machine tool spindle

Constraints:

a) side constraints (bounds)

$$g_1(\mathbf{x}) \equiv l_g - l \le 0, \qquad g_2(\mathbf{x}) \equiv l_k - l \le 0,$$
$$g_3(\mathbf{x}) \equiv d_{a1} - d_a \le 0, \qquad g_4(\mathbf{x}) \equiv d_a - d_{a2} \le 0,$$
$$g_5(\mathbf{x}) \equiv d_{b1} - d_b \le 0, \qquad g_6(\mathbf{x}) \equiv d_b - d_{b2} \le 0,$$
$$g_7(\mathbf{x}) \equiv d_{om} - d_o \le 0.$$

b) Designers' proportion requirements

$$g_8(\mathbf{x}) \equiv p_1 d_o - d_b \le 0, \qquad g_9(\mathbf{x}) \equiv p_2 d_b - d_a \le 0;$$

c) Maximal radial runout of the spindle nose Δ

$$g_{10}(\mathbf{x}) = |\Delta_a + (\Delta_a - \Delta_b)\frac{a}{l}| - \Delta \le 0,$$

where Δ_a and Δ_b are the radial runouts of the front and the end bearings.

The values of all parameters which are not design variables are introduced to the model during the computer aided session.

Since for the spindle the roller bearings can be chosen only from a given set, the optimization model is of the mixed continuous-discrete type. The variables d_a and d_b can be chosen from the set given in a bearing catalogue. For this example, it is assumed that $X_3 := \{80,85,90,95\}$ and $X_4 := \{75,80,85,90\}$.

For the above problem a printout of the computer aided optimum design session with CAMOS is shown in Fig. 3/23. The example was run on the IBM PC/XT, and the McGill Plotting Package Subroutine [36] was used for graphic output which was produced on a HP7475A plotter.

3.5 Conclusion

Multicriteria optimization strategies interacting with the DM first gained attention in the fields of management sciences, economics and operations research. These strategies, however, have been extended in such a way that they can also be applied to structural and component optimizaton problems. The nonlinearity of the optimization problem, time-consuming analytical or numerical structural analyses and the limited validity of the optimization model provide higher demands on the interactive procedure.

```
    ---PLEASE SELECT MULTICRITERION OPTIMIZATION METHOD---
3
-----------------------------------------------------------------
                        WEIGHTING MIN-MAX METHOD
-----------------------------------------------------------------
    ---PLEASE ENTER THE WEIGHTING COEFFICIENTS---
.3 .7
    *** PLEASE WAIT , OPTIMUM IS CALCULATED
    SOLUTION NO.   1
VALUE OF THE MINIMIZED QUANTITY =     .2732573E+00
VALUE OF  1 OBJECTIVE FUNCTION =     .6941010E+06
VALUE OF  2 OBJECTIVE FUNCTION =     .2307786E-01
VECTOR OF DECISION VARIABLES
    .6645399E+02     .1833649E+03     .9500000E+02     .8500000E+02
INEQUALITY CONSTRAINT VALUES
   -.1663513E+02    -.3336487E+02    -.4145399E+02    -.1500000E+02
   -.0000000E+00    -.1000000E+02    -.5000000E+01    -.1932507E+01
   -.5750000E+01    +.1119168E-03
    DO YOU WISH THE PROBLEM ORIENTED RESULTS TO BE PRINTED (Y/N) ?
Y
```

GRAPHICAL ILLUSTRATION OF THE SOLUTION

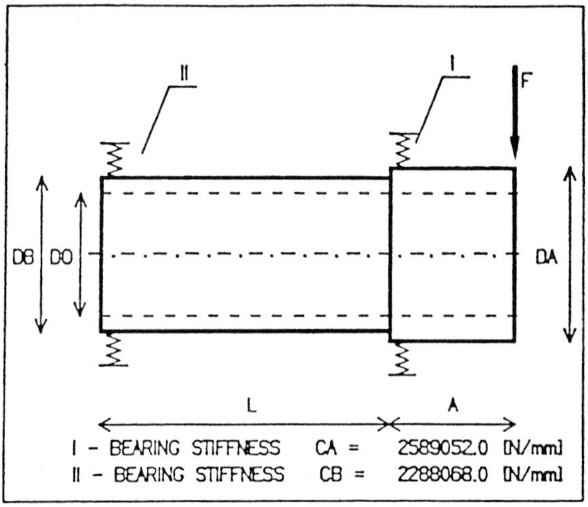

CONSTANT PARAMETERS		
DOM =	25.00	[mm]
DA1 =	80.00	[mm]
DA2 =	95.00	[mm]
DB1 =	75.00	[mm]
DB2 =	90.00	[mm]
P1 =	1.25	
P2 =	1.05	
LK =	150.00	[mm]
LG =	200.00	[mm]
A =	80.00	[mm]
E =	210000.0	[N/mm2]
F =	10000.0	[N]
DELA=	0.00540000	[mm]
DELB=	-0.00540000	[mm]
DEL =	0.01000000	[mm]
DRA =	-0.00100000	[mm]
DRB =	-0.00100000	[mm]

OBJECTIVE FUNCTION VALUES		
F(1) =	694101.00	[mm3]
F(2) =	0.02307786	[mm]

DECISION VARIABLE VALUES					
DO =	66.4	[mm]	DA =	95.0	[mm]
L =	183.3	[mm]	DB =	85.0	[mm]

```
    DO YOU WISH TO ENTER OTHER WEIGHTING COEFFICIENTS (Y/N) ?
Y
    ---PLEASE ENTER THE WEIGHTING COEFFICIENTS---
.7 .3
    *** PLEASE WAIT , OPTIMUM IS CALCULATED
    SOLUTION NO.   2
VALUE OF THE MINIMIZED QUANTITY =     .2455193E+00
VALUE OF  1 OBJECTIVE FUNCTION =     .5310598E+06
VALUE OF  2 OBJECTIVE FUNCTION =     .3018247E-01
VECTOR OF DECISION VARIABLES
    .6389444E+02     .1832864E+03     .8500000E+02     .8000000E+02
INEQUALITY CONSTRAINT VALUES
   -.1671362E+02    -.3328638E+02    -.3889444E+02    -.5000000E+01
   -.1000000E+02    -.5000000E+01    -.1000000E+02    -.1319491E+00
   -.1000004E+01    +.1139351E-03
    DO YOU WISH THE PROBLEM ORIENTED RESULTS TO BE PRINTED (Y/N) ?
```

Fig. 3/23. Example of a computer aided design cycle with CAMOS

In this chapter several appropriate interactive procedures for the application to optimal design have been presented. For all of them it is important that the articulation of preferences enables the DM to improve incomplete preferences successively by a learning process. It is advisable to distinguish the interactive approaches according to their organizaton scheme, namely the superior methods (i.e. approach by Fandel) and the extended methods (i.e. approach by Jahn). As a result of many investigations one can summarize that both groups have got advantages and disadvantages. If the DM wants to gain insight into the Pareto-optimal set ∂Y^* or X^*, he should choose an interactive method which determines many functional-efficient points combined with a small number of structural analyses. For this purpose methods with a superior organization scheme are more suitable because several functional-efficient solutions are obtained during one interactive cycle. A problem-adjusted scalar optimization algorithm can be applied as well. On the other hand, if an optimization problem with time-consuming structural analysis, i.e. FE-method, has to be carried out, the DM should consider that the total number of structural analyses has to be kept low. In this case, methods with superior organization schemes are inappropriate whereas methods with extended organization schemes need less computation time. Due to the iterations within the criterion space, the DM is able to reach his preferred solution without determining any other functional-efficient points.

Finally, two software packages (DIALOG and CAMOS) for interactive multicriteria optimization have been introduced in some detail and have been exemplified by a conical shell and a machine-tool spindle.

References

[1] Benayoun, R.; Roy, B.; Sussman, B.: ELECTRE: Une Methode pour Guider le Choix en Presence de Vue Multiple. SEMA (Metra International), Direction Scientifique, Note de Travail No. 49, Paris, June 1966

[2] Benayoun, R.; de Montgolfier, J.; Tergny, J.; Laritchev, O.: Linear Programming with Multiple Objective Functions: STEP Method (STEM). Mathematical Programming, Vol. 1 (1971) 366–375

[3] Geoffrion, A.M.; Dyer, J.S.; Feinberg, A.: An Interactive Approach for Multicriterion Optimization, with an Application to the Operation of an Academic Department. Management Science, Vol. 19, No.4 (1972) 357-368

[4] Fandel, G.: Optimale Entscheidung bei mehrfacher Zielsetzung. Berlin, Heidelberg, New York: Springer Verlag 1972

[5] Zionts, S.; Wallenius, J.: An Interactive Programming Method for Solving the Multiple Criteria Problem. Management Science, Vol. 22, No. 6 (1976) 652-663

[6] Steuer, R.E.: An Interactive Multiple Objective Programming Procedure. In: Starr, M.K.; Zeleny, M.: Studies in Management Science, North Holland Pub. Comp., Amsterdam, Vol. 6 (1977)

[7] Dupre, R.; Huckert, K.; Jahn, J.: Lösung linearer Vektormaximumprobleme durch das STEM-Verfahren. In: Späth, H. (ed.): Ausgewählte Operations Research Software in Fortran. München: Oldenburg Verlag 1979

[8] Guddat, J.; Guerra-Vasquez, F.; Tammer, K.; Wendler, K.: Multiobjective and Stochastic Optimization Based on Parametric Optimization. Berlin: Akademie-Verlag 1985

[9] Lecture Notes in Economics and Mathematical Systems: Proceedings of the International Conferences on Multiple-Criteria Decision Making. Berlin, Heidelberg, New York: Springer Verlag, annual publications

[10] Hwang, C.-L.; Masud, A.S.M.: Multiple Objective Decision Making Methods and Applications. A State of Art Survey. Lecture Notes in Economics and Mathematical Systems. Berlin, Heidelberg, New York: Springer Verlag 1979

[11] Zeleny, M.: Multiple Criteria Decision Making. New York: McGraw-Hill Book Comp. 1979

[12] Goicoechea, A.; Hansen, D.R.; Duckstein, L.: Multiobjective Decision Analysis with Engineering and Business Applications. New York, Toronto: John Wiley & Sons 1982

[13] Szidarovszky, F.; Gershon, M.E.; Duckstein, L.: Techniques for Multiobjective Decision Making in Systems Management. Amsterdam, Oxford, New York: Elsevier Pub. Comp. 1986

[14] Sawaragi, Y.; Nakayma, H.; Tanino, T.: Theory of Multiobjective Optimization. New York, London: Academic Press 1985

[15] Isermann, H.: Strukturierung von Entscheidungsprozessen bei mehrfacher Zielsetzung. OR Spektrum, Vol. 1 (1979) 3-26

[16] Koski, J.: Multicriteria Truss Optimization. In: Stadler, W.: Multicriteria Optimization in Engineering and in the Sciences. New York, London: Plenum Press 1988

[17] Carmichael, D.G.: Computation of Pareto Optima in Structural Design. Intern. Journal of Numerical Methods in Engineering, Vol. 15 (1980) 925-952

[18] Brosowski, B.: Application of Parametric Programming to the Optimal Design of Stiffened Plates. Int. Series of Num. Mathematics. Vol. 72 (1984) 47-69

[19] Diaz, A.: Interactive Solution to Multiobjective Optimization Problems. Int. Journal of Numerical Methods in Engineering, Vol. 24 (1987) 1865-1877

[20] Kneppe, G.: Direkte Lösungsstrategien zur Gestaltsoptimierung von Flächentragwerken. Dissertation Universität Siegen, Fortschritt-Berichte VDI Reihe Nr. 135. Düsseldorf: VDI- Verlag 1986

[21] Powell, M.Y.D.: A Fast Algorithm for Nonlinearly Constrained Optimization Calculations. In: Watson, G.A.: Lecture Notes in Mathematics, No. 630. Berlin: Springer Verlag 1978

[22] Jahn, J.: Ein Verfahren zur interaktiven Lösung nichtlinearer Vektoroptimierungsprobleme. OR Spektrum, Vol. 3 (1982) 239-243

[23] Zoutendijk, G.: Methods of Feasible Directions. A Study in Linear and Nonlinear Programming. Amsterdam, London, New York: Elsevier Publ. Comp. 1960

[24] Frank, M.; Wolfe, P.: An Algorithm for Quadratic Programming. Naval Research Logistics Quaterly, Vol. 3, No. 1 and 2 (1956) 95-110

[25] Eschenauer, H.; Post, P.U.; Bremicker, M.: Einsatz der Optimierungsprozedur SAPOP zur Auslegung von Bauteilkomponenten. Bauingenieur, Vol. 63 (1988) 515-526

[26] Schnell, W.; Eschenauer, H.: Elastizitätstheorie II, Schalentheorie. Mannheim, Wien, Zürich: BI-Wissenschaftsverlag 1984

[27] Thevendran, V.; Thambiratnam, D.P.: Minimum Weight Design of Conical Concrete Water Tanks. Computers & Structures, Vol. 29, No.4 (1988) 699-704

[28] Lee, J.K. and Hurst, E.G.: Decision Support Systems for Multiple Criteria Decision Making. Joint National Meeting of ORSA/TIMS, Dallas, November 1984

[29] Malakooti, B.: A Decision Support System and a Heuristic Interactive Approach for Solving Discrete Multiple Criteria Problems. IEEE Transaction Systems, Man, and Cybernetics, Vol. 18, No 2 (1988) 273-284

[30] Osyczka, A.: Multicriterion Optimization in Engineering with FORTRAN Program. New York, Chichester: John Wiley 1984

[31] Himmelblau, D.M.: Applied Nonlinear Programming. New York: McGraw-Hill Company 1972

[32] Hooke, R.; Jeeves, T.A.: Direct Search Solution of Numerical and Statistical Problems, J. Assoc. Comp. Mech., Vol. 8 (1961) 221-225

[33] Kowalik, J.; Osborne, M.R.: Methods for Unconstrained Optimization Problems. New York: American Elsevier 1968

[34] Nelder, J.A.; Mead, R.: A Simplex Method for Function Minimization. Computer J., Vol. 7 (1965) 308-313

[35] Hammer, P.; Johnson, E.L. and Korte, B.H.: Discrete Optimization. Amsterdam: North-Holland, Pub. Comp. 1979

[36] Barr, A.: McGill Plotting Package Subroutines Manual. Department of Applied Mathematics, University of Western Ontario, Canada, 1968

[37] Osyczka, A.: An Interactive Multicriterion Decision Making System. System Analysis and Simulation Symposium, Berlin, 1985

[38] Osyczka, A.: Computer Aided Multicriterion Optimization Method. Advances in Modelling and Simulation. AMSE Press, Vol. 3, No. 4 (1985) 41-52

[39] Osyczka, A.: Computer Aided Multicriterion Optimization System (CAMOS). In: Proceedings of a GAMM-Seminar "Discretization Methods and Structural Optimization". University of Siegen, FR Germany, October 5-7, 1988. Berlin, Heidelberg, New York, London, Paris, Tokyo: Springer 1989

[40] Bernau, H.: Interactive Methods for Vector Optimization. in.: Brosowski, B.; Martensen, E. (Eds): Optimization in Mathematical Physics. Frankfurt am Main, Bern, New York, Paris: Verlag Peter Lang 1987

4 KNOWLEDGE ENGINEERING AND MULTICRITERIA OPTIMIZATION

M. Balachandran, J.S. Gero

4.1 Introduction

Multicriteria optimization is a useful and challenging activity in many disciplines. It provides decision makers with tools for producing better decisions while saving time in the decision process. In the past, computers have assisted in achieving several advancements in the field of multicriteria optimization. In spite of the rapid development of multicriteria optimization techniques, their applications to real world problems have not been legion. There can be many reasons for a decision maker to avoid numerical optimization techniques and resort to conventional trial and error methods based on judgement, intuition and experience. Decision makers who are not experts in optimization techniques usually experience difficulties while using conventional optimization systems. The aim of this chapter is to show how knowledge-based systems technology can be used to overcome some of the shortcomings of earlier optimization systems. It is illustrated how knowledge-based approaches attempt to make multicriteria optimization an easier tool for decision makers to use.

First, some of the major difficulties with using conventional computer systems to multicriteria optimization are discussed:

a) Problem formulation

Most multicriteria optimization strategies require a complete formulation of the problem; that is, the objective functions, the constraint equations, and the bounds on the decision variables need to be clearly defined in mathematical terms. The most important aspect in building such a model is that the model must closely approximate the actual behaviour of the problem for the solution to be useful. This task can be time-consuming as well as a potential source of error. Moreover, it can be beyond the expertise of many decision makers as they do not usually think in terms of such mathematical models.

b) Choosing strategies

When using a conventional optimization package a decision maker has to select from different methods available for multicriteria optimization to suit the problem domain. In general, the following aspects have to be considered in selecting the appropriate procedure for the problem at hand:

- type of decision variables (discrete or continuous);
- number of objectives;
- nature of objective functions and constraints (linear or nonlinear); and
- availability of gradients of objective functions and constraint functions (differentiable or nondifferentiable).

Thus, considerable experience and mathematical background is expected from the decision maker in selecting the best possible optimization procedure.

c) Computational effort

Most of the available techniques to solve multicriteria optimization problems are usually implemented in a "black box" environment. Thus, they do not utilize many important aspects of the problem and the qualitative value of the information generated during the solution process of optimization. Hence, such programs may take excessive computational effort resulting in poor time performance.

d) Selecting initial solution (initial starting point)

Most multicriteria optimization techniques involve conversion of the multicriteria problem into a single criterion form in order to generate Pareto-optimal solutions. When dealing with nonlinear formulations the decision maker has to provide an initial solution for the problem. This requires experience and judgement from the decision maker. One major difficulty is that the initial solution given by the decision maker may be either highly infeasible or far from the optimum solution resulting in a poor performance of the algorithm.

e) Redundant and unrealistic constraints

Many of the available optimization algorithms do not recognize unrealistic or redundant constraints involved in the problem. The unrealistic constraints result in an infeasible solution to the problem whereas redundant constraints increase the computational effort.

In the following section some major advantages of systems developed using knowledge engineering concepts over the conventional computer systems are discussed.

4.2 Knowledge Engineering

4.2.1 General Concept

Conventional computer programs are deterministic and do not possess reduncancy. For any given input there is a single computational path that is always followed, and there is a single mechanism capable of producing the correct output for that input. Such algorithmic-based programs exhibit a sharp distinction between code and data.

Knowledge engineering is an emerging field which is mainly concerned with representation of knowledge and with symbolic computation. The tools and techniques available in this field allow to move away from the traditional way of "algorithmic computing" towards that of "inferential computing". A computer program which stores knowledge explicitly and manipulates that knowledge to solve problems in a specific domain is referred to as a knowledge-based system. The development of knowledge-based systems is an attempt to emulate the human process of problem-solving. Most current knowledge-based systems are divided not into code and data but into a chunk of knowledge and a controller for applying the knowledge to solve problems.

Another way to define knowledge-based systems is to compare them with conventional programs. The basic difference is that knowledge-based systems manipulate knowledge while conventional programs manipulate data. The differences can be characterized as shown below.

Conventional programs	*Knowledge-based systems*
Representation and use of data	Representation and use of knowledge
Algorithmic	Heuristic
Algebraic processes	Inferential processes
Manipulation of data bases	Manipulation of knowledge bases

Knowledge-based system methodologies now allow the incorporation of different forms of knowledge required in optimization tasks which are non-numeric. The architecture of a typical knowledge-based system is briefly discussed below. Several good discussions are available on knowledge-based systems and their development [8,13].

4.2.2 Architecture of a Knowledge-Based System

Knowledge-based systems consist of three main components: a facts base; a knowledge base representing domain knowledge; and a control strategy or an inference mechanism (see Fig. 4/1):

- The *facts base* contains facts that reflect the current state of the problem solution. The facts base builds up dynamically during the solution process of a particular problem.
- The *knowledge base* is the important part of any knowledge-based system, because the performance of the system depends on the knowledge it has. The representation of the knowlege base is dependent on the implementation of the system.
- The *inference mechanism* controls the system's activity by manipulating the facts using the knowledge base. The inference mechanism contains no domain knowledge.

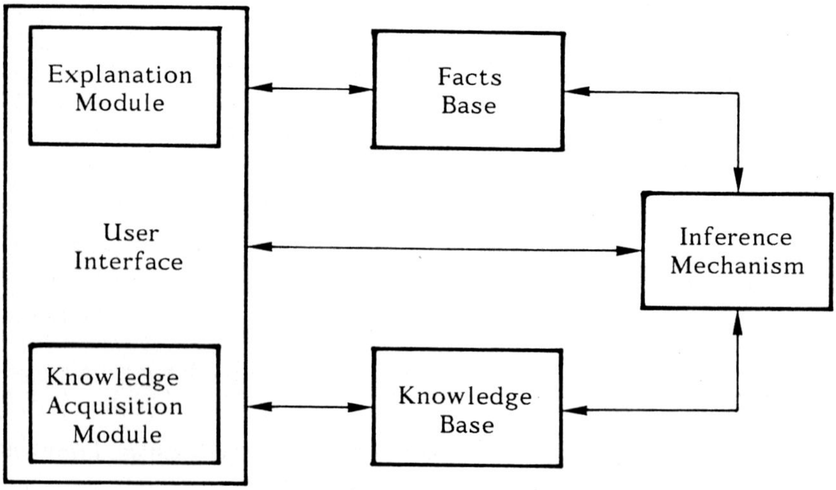

Fig. 4/1. Schematic representation of a knowledge-based system

In addition to the above primary components, there are three other secondary components to facilitate knowledge acquisition, explanation and user interface (see Fig. 4/1):

- The *knowledge acquisition module* serves as an interface between the expert(s) and the system. It provides assistance to develop the knowledge base and to structure the facts base or context for a particular application domain.
- The *explanation module* provides explanations of the inferences made by the system. Typical explanations include why certain information is requested, or how a certain conclusion was reached.
- The *user interface module* provides an interface between the user and the system. This interface usually provides a command language for directing the execution of the system. Input may be specified by a problem-oriented language or graphically.

4.3 Role of a Knowledge-Based Approach in Multicriteria Optimization

There is a number of reasons why a knowledge-based approach is suitable for multicriteria optimization:

- The human expertise necessary to perform the tasks involved in optimization problem solving can be easily incorporated using knowledge-based methodologies.
- Several human faculties can be emulated: use of heuristics and rules of thumb; performance evaluations; selection of alternative courses of action to suit the conditions involved; and symbolic reasoning.
- A knowledge-based approach attempts to extend the conventional software to bridge the gap between the software and the end user.

In recent years a few researchers have applied knowledge-based methodologies to multicriteria optimization. Radford et al. [12] predicted the use of knowledge-based methodologies in Pareto optimization. They suggest that Pareto optimization provides qualitative information for decision makers to assist in their decision and discuss the possibility of employing knowledge-based systems to extract qualitative knowledge from the Pareto optimization results. Mistree et al. [11] have investigated the use of knowledge-based systems in design decision support problems. They employ a rule-based system approach to post solution analysis of decision support problems. Knowledge-based methodologies have been proved effective to provide the solutions needed for the efficient use of optimization techniques for design problems [2,3].

In developing a knowledge-based multicriteria optimization system, the following features are desirable and useful.

- The system should provide a simple and semantically rich interface and flexible editing features.
- It should accept and represent the description of a problem in an effective and manipulable form and provide facilities for problem modification.
- It must be able to formulate the problem into a canonical form of an optimization model providing functional relationships for objectives and constraints.
- It must be capable of recognizing the types of decision variables and the functional types of objectives and constraints.
- It must be able to select an appropriate optimization algorithm and carry out the solution procedure.
- It should be able to use heuristics in the solution process itself so as to improve the performance.

It is obvious from the list that such a system should operate on information at the same level as the decision maker. Thus, it must be able to represent the problem explicitly in symbolic form and to provide functional relationships to objectives and constraints in terms of decision variables. The latter process requires the ability to understand dependency relationships of one decision variable with others. Furthermore, the system must be able to carry out symbolic manipulation such as simplifying algebraic expressions, differentiating mathematical functions, solving symbolic equations, etc. The system must also include knowledge to recognize optimization formulations and to select an appropriate algorithm(s) from an available set of algorithms.

4.4 Knowledge-Based Optimization Formulation

Optimization formulation is the process of establishing a mathematical model that reflects the reality of the physical problem at hand. The most important aspect in building such a model is that the model must closely approximate the actual behaviour of the problem for the solution to be adequate and useful. A multicriteria optimization problem is formulated in terms of the following descriptors:

- decision or design variables, respectively,
- fixed parameters,
- constraints,
- variable bounds, and
- objective functions.

Mathematical relations between the decision variables and the fixed parameters constitute an optimization model (see Chapter 2). The general multicriteria optimization problem with n decision variables, p constraints, and m criteria is formulated into the canonical form according to (1-8).

$$\min \mathbf{f}(\mathbf{x}) = (f_1(\mathbf{x}), f_2(\mathbf{x}), \dots, f_m(\mathbf{x}))^T$$

$$\begin{aligned} \text{subject to} \quad & g_i(\mathbf{x}) \le G_i && i = 1,2,\dots,p\,, \\ & h_i(\mathbf{x}) = 0, && i = p+1,\dots,q\,, \\ & x_j \ge 0 && j = 1,2,\dots,n \end{aligned} \tag{4-1}$$

with

$\mathbf{f}(\mathbf{x})$	multicriteria objective function,
$f_j(\mathbf{x})$	j-th criterion of the m individual criteria,
$\mathbf{x} = (x_1,\dots,x_n)^T$	n-component vector of decision variables,
$g_i(\mathbf{x}) \le G_i$	p constraint functions.

Formally speaking, the basic tasks involved in constructing the canonical optimization model can be described as follows:

- identification of the decision variables involved in the problem,
- provision of functional relationships in terms of the decision variables that represent the design objectives,
- provision of functional relationships in terms of the decision variables that represent the constraints and bounds,
- simplifying the mathematical expressions representing the constraints and objectives into canonical forms, and
- identifying inconsistent constraint relationships and removal of redundant constraints.

The above tasks require substantial human expertise and experience. Organizing and representing appropriate knowledge required to carry out the above tasks are important aspects in knowledge-based approaches. First, problem description and their representations are discussed.

4.4.1 Description and Representation of Optimization Problems

Several techniques developed by knowledge engineering can be utilized to represent problem descriptions. In the optimization context, the representation system should provide for representing structured objects with complex relations, constraints and goals. Although several distinct mechanisms may be suitable to handle these tasks, the one which is most appropriate is the frame-based representation [6,9]. The frame formalism exploits several important and useful ideas that are relevant to representation of problem descriptions:

- It provides a framework by which information about an entity can be grouped into a uniform structure.
- It permits the storage of expectations, defaults and procedures. The default value of an attribute of an object can normally be used when there is no value available for that attribute.
- It allows the use of hierarchies to define classes of objects and supports inheritance strategies.

a) Representing design objects

In describing objects, labels can be used to identify them. Properties of objects can be described as well. Each of these properties may be described in terms of attributes and values. An object may be related to many other objects. Therefore, the objects and their properties can be treated in terms of object-attribute value triplets where:

object - any concept involved in the design domains,
attribute - any property of an object,
value - any value of an attribute.

For example consider the following statement:

house width is 10m.

In this statement object, attribute, and value read as follows:

object	*attribute*	*value*
house	width	10

Figure 4/2 shows how the descriptions below are represented as a frame.

name:	house
slots:	width: 10m
	is-a-kind-of: building

Fig. 4/2. Objects represented as frames

The relationship between two objects can be as follows.

house is-a-kind-of building

object	*attribute*
house	is-a-kind-of
buiding	

b) Representing constraints

The representation and processing of constraints form an integral part of the optimization process. The problem constraints are normally expressed by declaring a maximum or minimum value for any object attribute or by specifying equality or inequality relationships among object attributes. Constraints can either be represented as production rules [5] or frames. Consider for example the following constraint:

house cost must not exceed \$ 100.000

Using the rule-based approach, the constraint can be represented as follows:

If	house cost is C
and	C is less than or equal to 100.000
then	the constraint is satisfied

The example illustrated above using a production rule can be represented in the form of a frame shown in Fig. 4/3.

name:	cost constraint
slots:	lhs: house cost
	rhs: 100000
	predicate: less-than-or-equal-to

Fig. 4/3. Constraints represented as frames

The "lhs" and "rhs" slots represent the "left hand side" and "right hand side" expressions of a constraint, and the "predicate" slot represents the relation involved between those expressions.

c) Representing design objectives

The multicriteria optimization process is concerned with producing solutions which satisfy the problem constraints and optimize more than one objective. The objectives are expressed as maximize or minimize some aspects of the problem domain. The general form in which an objective may be stated is as follows:

maximize f
or minimize f

where f is an object attribute or an expression of one or more object attributes.

The objectives can also be represented using the frame formalism. For example, consider the following statement:

maximize house area

This statement indicates that the objective is to maximize the "area" attribute of the object "house" and is represented as shown in Fig. 4/4.

name:	objective 1
slots:	objective function: house area
	optimality criteria: maximize

Fig. 4/4. Objectives represented as frames

4.4.2 Mathematical Symbolic Manipulation

In a mathematical modelling task a number of mathematical processes are carried out which include substitution, simplification, and differentiation. First order predicate logic has been found to be an effective tool to encapsulate this mathematical knowledge. PROLOG is one of the most popular logic programming languages based on th first order predicate logic [4]. In PROLOG, knowledge is represented as logical propositions written in well-formed formulas. The knowledge base becomes a set of PROLOG predicates encoded in a suitable way, and the inference engine becomes PROLOG's control structure. The features of pattern matching and recursive definition allow us to incorporate variety of knowledge quite easily.

a) Knowledge for mathematical simplification

Mathematical simplification is an operation during which a mathematical expression is transformed into a more simplified form in accordance with a set of rules. The rules of simplification can be encoded as PROLOG facts and rules. The following are some examples of such rules: The predicate *simp* (A,B) represents that *B* is the simplification of A.

```
simp(A-A, 0),
simp(X*0,0),
simp(0*1,X),
simp(X*1,X),
simp(1*X,X),
simp(X/1,X),
simp(X^1,X),
simp(X^0,1),
simp(X*X^2),
simp(A+B,C) :- number(A), number (B), C is A+B,!,
simp(A-B,C) :- number(A), number (B), C is A-B,!,
simp(A*B,C) :- number(A), number (B), C is A*B,!,
simp(A/B,C) :- number (A), number(B), C is A/B,!,
simp(X+0,Y) :- simp(X,Y),
simp(0+X,Y) :- simp(X,Y),
simp(X-0,Y) :- simp(X,Y).
```

```
simp(0-X,-Y):- simp(X,Y),
simp(W+X,A+B) :- simp(W,A), simp(X,B),
simp(W-X,A-B) :- simp(W,A), simp(X,B),
simp(W*X,A*B) :- simp(W,A), simp(X,B),
simp(W/X,A/B) :- simp(W,A), simp(X,B),
simp(W^X,A^B) :- simp(W,A), simp(X,B),
simp(-A,-B) :- simp(A,B),
simp(A,A).
```

b) Knowledge for symbolic differentiation

In mathematics, symbolic differentiation is an operation that converts a mathematical expression into another expression called the derivative. The following PROLOG facts represent the derivatives of some common functions in the form: derivative (function, derivative with respect to, derivative of that function):

```
derivative (x^n,x,n*x^(n-1)),
derivative (x,x,1),
derivative (sin(x), x, cos(x)),
derivative (e^x,x,e^x),
derivative (log(x), x,1/x).
```

The compound terms are differentiated using the appropriate differentiation rules as described above. The following PROLOG rules represent some basic rules of differentiation:

```
derivative (U+V,x,dU+dV)               :- derivative (U,x,dU),
                                          derivative (V,x,dV);
derivative (U-V,x,dU-dV)               :- derivative (U,x,dU),
                                          derivative (V,x,dV);
derivative (U*,V,x,U * dV+V*dU)        :- derivative (U,x,dU),
                                          derivative (V,x,dV);
derivative (U/V,x.(U*dV-V*dU)/(V*V))   :- derivative (U,x,dU),
                                          derivative (V,x,dV);
derivative (U^V,x,V*U^(V-1)*dU)        :- number (V),
                                          derivative (U,x,dU).
```

For example, the sum rule states that the derivative of a sum is the sum of derivatives. The product and quotient rules are more complicated, but the logical clauses represent their mathematical definitions easily.

4.4.3 Mathematical Formulation of Optimization Problems

In this section the process of establishing an algebraic model for multicriteria optimization is described using an example problem originally formulated by Mitchel et al. [10]. The problem concerns the optimal dimensioning of an apartment plan (see Fig. 4/5) for which a topology has been separately identified. The maximum and minimum value requirements of various rooms are specified as design constraints.

a) An example problem

Problem description

The following is a listing of the problem statements that describe the problem description in terms of relationships, constraints, and objectives. The words which appear in bold type represent the key words and symbols in a restricted language:

apartment **and** *living_room* **and** *kitchen* **and** *bathroom* **and** *bedroom* **and** *bedroom2* **and** *bedroom3* **are_kind_of** *rectangle*,
bathroom width **is** *2.45 m*,
hall width **is** *1.83 m*,
kitchen width **is** *3.05 m*,
bedroom2 width **is** *3.35 m*.

(Note: If in the following no unit is given, all dimensions are in m.)

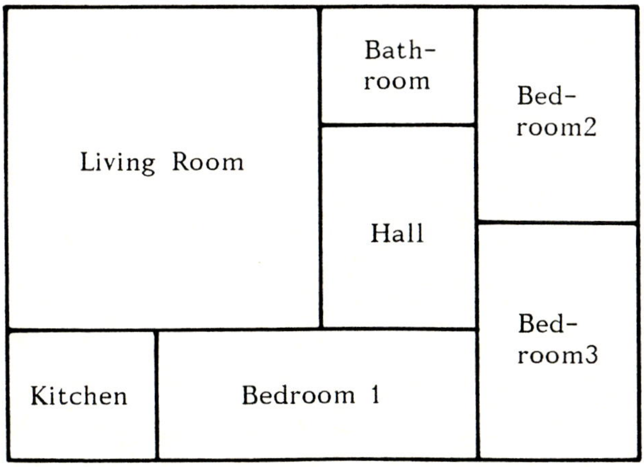

Fig. 4/5. Dimensionless representation of an apartment layout

kitchen unit_cost **is** *20,*
bedroom2 unit_cost **is** *10,*
living_room unit_cost **is** *30,*
apartment lenght = **sum_ of** *living _ room length* **and** *bathroom length* **and** *bedroom2 length,*
apartment width = **sum_ of** *living_room width* **and** *kitchen width,*
living_room width = **sum_of** *bathroom width* **and** *hall width,*
bathroom length = hall length,
bedroom1 width = kitchen width,
bedroom2 length = bedroom3 length,
sum_of *living_room length* **and** *hall length* = **sum_of** *kitchen length* **and** *bedroom1 length,*
sum_of *living_room width* **and** *kitchen width* = **sum_of** *bedroom2 width* **and** *bedroom 3 width,*
apartment cost = **sum_of** *living_room cost* **and** *kitchen cost* **and** *bathroom cost* **and** *hall cost* **and** *bedroom1 cost* **and** *bedroom2 cost* **and** *bedroom3 cost,*
maximum *living_room area* **is** *30* m^2,
minimum *living_room are* **is** *15* m^2,
maximum *kitchen area* **is** *12* m^2,
minimum *kitchen area* **is** *5* m^2,
maximum *bathroom area* **is** *6.5* m^2,
minimum *bathroom area* **is** *4.5* m^2,
maximum *hall area* **is** *7.2* m^2,
minimum *bedroom1 area* **is** *10* m^2,
maximum *bedroom area* **is** *18* m^2,
minimum *bedroom3 area* **is** *10* m^2,
maximize *apartment area.*
minimize *apartment cost.*

Interpretation of the problem

Information gathered from the problem description is interpreted and transformed into a format suitable for subsequent processes involved in the formulation. Firstly, the objects involved in the design are identified and their descriptions including relationships to other objects are set up in appropriate frames. Secondly, the constraints and objectives specified are interpreted and transformed into a set of frames as discussed previously.

Problem representation

Figures 4/6 a, b, c) present some example frames showing how the objects, objectives, and constraints of this problem are represented in the system.

```
name:   living room
slots:  a-kind-of
                   value      :  rectangle
        width
                   expression :  (bathroom width)
                                        +
                                 (kitchen width)
        unit-cost
                   value      :  30
```

```
name:   constraint-1
slots:  lhs
                   expression :  (apartment area)
        rhs
                   value      :  30.0
                   unit       :  m^2
        predicate
                   value      :  less-than-or-equal-to
```

```
name:   objective-1
slots:  objective-function
                   expression :  (apartment area)
        rhs
        optimality-criteria
                   value      :  maximize
```

Fig. 4/6. a) Design object represented in a frame
b) Design constraint represented in a frame
c) Design objective represented in a frame

b) Identifying decision variables

In identifying the decision variables the system starts with a set of variables for which fixed values have been provided by the decision maker and tries to find values for unknown variables using appropriate mathematical relationships. For example consider the mathematical relationship given below:

living_room width = **sum_of** *bathroom width* **and** *hall width*

The above relationship is transformed by the interpreter into a mathematical equation as follows:

living_room width = (bathroom width) + (hall width) .

It is then identified by the system that there are three variables involved in this equality relationship, namely living_room width, bathroom width, and hall width. If the values for any two of these variables are known the value for the third variable is found by simply solving this equation. The system therefore is continually scanning all available mathematical equations and attempting to find values for unknown parameters with known parameters. For example, from the above relationship the following result is deduced:

living_room width = 2.45 m + 1.83 m ,
i.e. living_room width = 4.28 m.

This process is continued until there are no more parameters for which values can be found. At this stage the system will be able to know which variables are to be treated as decision variables as distinct from fixed parameters. The mathematical equalities which have more than one unknown variable will be treated as equality constraints.

c) Formulating problem constraints

The problem constraints simply describe dependencies among decision variables and are represented as mathematical inequalities or equalities. Each constraint is encoded as a frame with three slots, namely left-hand side part, right-hand side part, and relationship type. In formulating a constraint into a mathematical form, the system starts with the description of that constraint obtained from the appropriate frame representation. An example will be used to illustrate the basic process involved in transforming the constraint description into a canonical form. Consider the following constraint:

maximum living_room area **is** 30 m^2 .

This is interpreted as the upper limit value for living_room area is 30.0 m^2. Initially, a constraint is transformed into mathematical inequality or equality form based on its nature of relationship; thus, it can be rearranged into one of the following standard forms:

$$Lhs - Rhs \leq 0,$$
$$Lhs - Rhs \geq 0,$$
$$Lhs - Rhs = 0.$$

In case of the example given above the result would be:

(living_room area) - 30.0 $\leq$ 0 .

At this stage the variables involved in the constraint are determined by scanning the left-hand side expression of the constraint. A variable may have a value or some functional relationship with other variables. If such relationship or value exists for a variable, it is obtained directly from the frame representing that variable by inheriting from another related frame. The information obtained from the frame is then substituted for the variable in the constraint. This procedure is repeated recursively for every variable until no further substitution is possible. The variables which remain in the current state of the constraint are denoted as x_i $i=1,2,...,n$. Thus, x_i is a variable representing the ith quantity. After replacing the variables with the appropriate algebraic symbols, the constraint is algebraically simplified to reduce it to a canonical form.

In the present example the only variable which is involved in the constraint "living_room area". An expression for living_room area is inherited from the frame "rectangle"

living_room area = (living_room lenght)*(living_room width).

The following relationships are obtained from the problem statements:

living_room width = (bathroom width) + (hall width),
bathroom width = 2.45 m,
hall width = 1.83 m.

After the substitution process the constraint is expressed in the following form:

(living_room length)*(2.45+1.83) $\leq$ 30.0 .

At this stage algebraic symbols are introduced to replace the variables, and the constraint becomes:

$x_1*(2.45+1.83) \leq 30.0$.

where x_1 = living_room length .

The above form of the constraints is then algebraically simplified to reduce it to a canonical form as follows:

$x_1 \leq 7.025$.

d) Formulating objectives

In the optimization model the objectives must be expressed as computable functions of the decision variables. The objectives are stated as maximize or minimize one or more aspects of the problem. For example consider the following statement:

maximize *apartment area* .

This statement indicates that the "area" attribute of the object "apartment" is to be maximized. Initially, a functional relationship describing the area of the apartment is obtained. The function describing the area of the apartment is obtained from the frame "rectangle". Initially, the objective function will be as follows:

maximize (apartment length)*(apartment width) .

As illustrated in the previous section the substitution process is applied to each variable in turn. The objective function is then transformed as follows:

maximize ((living room_length) + (bathroom length) + (bedroom2 length))*
((living room_width) + (kitchen width)) ,
maximize ((living_room length) + (bathroom length) + (bedroom2 length))*
((2.45 + 1.83) + (3.05)) .

Now, algebraic symbols for the variables are introduced, and the objective function is algebraically simplified to reduce it to a canonical form:

$$\text{maximize } (x_1+x_2+x_3)*(2.45+1.83+3.05) \ ,$$
$$\text{maximize } 7.33*x_1+7.33*x_2+7.33*x_3 \ ,$$

where x_1 = living_room length,
x_2 = bathroom length,
x_3 = hall length.

At this stage the decision variables have been identified, the constraints expressed in terms of these have been constructed and the objective functions written in terms of those variables. They will now be in an algebraic form.

4.5 Knowledge-Based Optimization Control

This section deals with two other important aspects involved in multicriteria optimization processes. One aspect is about knowledge representation for recognition of optimization formulations and selection of appropriate algorithms. The other aspect is concerned with applying knowledge-based control mechanisms in solution generation processes. It is argued that by utilizing some knowledge-based control mechanisms the performance of the solution processes can be improved significantly.

4.5.1 Recognition of Optimization Formulation

This section discusses the knowledge required to recognize optimization formulations expressed in canonical forms. It is shown that the first order predicate logic formalism can easily encapsulate this kind of knowledge. Mathematical programming models are vital for solving optimization problems. There is no ideal mathematical programming technique which can solve any type of optimization problem. The problem each technique can solve depends on the type of variables, the nature of constraints, the objective functions, and availability of the gradients of constraints and objective functions. Thus, in general, the selection of the right algorithm depends on a number of aspects:

- type of decision variables (discrete or binary or continuous),
- number of objectives (single objective or multiple objectives),
- nature of objective function(s) and constraints (linear or nonlinear), and
- availability of gradients of objective function(s) and constraint functions (differentiable or nondifferentiable).

The canonical form of the optimization model represents the objectives and constraints of the problem as mathematical functions expressed in terms of the decision variables. In order to identify the structure of an optimization model, a computer system needs to have the knowledge to recognize the variable and function types involved in the model.

a) Knowledge for algebraic functions recognition

An algebraic function, in its reduced form, is the sum of positive and/or negative terms, each of which is expressed in terms of one or more variables. Common types of algebraic terms are linear, quadratic, polynomial, and nonlinear. An algebraic expression is linear if all of its terms are of the linear form. Thus, the type of an algebraic function depends on the type of every term involved in that function. Such knowledge can be encoded using PROLOG's features such as pattern matching and recursive definition. A compound expression, for example, is linear if all of its terms are of the linear form. This condition can be encoded as follows:

expression (A+B, linear) :- term (B, linear), expression (A, linear),
expression (A−B, linear) :- term (B, linear), expression (A, linear),

where "expression (E,T)" means that T is the type of expression E. The first rule simply states that an expression of the form A+B, in which B represents its last term, is linear if both B and the rest of the expression A are linear. The second rule deals with the case where the last term is a negative term. These two rules are applicable only when the expression has at least two terms. Hence, it is necessary to define further rules to handle expressions which have only one term. For this purpose, the following rules are added.

expression (A, linear) :- term (A, linear).
expression (−A, linear) :- term (A, linear).

These rules simply define that an expression which has a single term is linear if that term is linear.

Now, one needs to define the necessary rules to determine whether a given term is linear. An algebraic term is linear if it is represented by a variable or by a product of a variable and a number. The following rules encode these conditions:

term (A, constant) :- *integer (A); float (A),*
term (A , linear) :- *symbol (A),*
*term (A*B, linear)* :- *term (A, constant), term (B, linear),*

where "term(t,T)" means that term t is of type T. Likewise, similar logic programs can be written to recognize other types of functions such as quadratic, polynominal, nonlinear, etc. These logic programs are a representation of the knowledge used in recognizing algebraic functions.

4.5.2 Optimization Algorithm Selection

Depending upon the problem formulation, optimization techniques provide a number of distinct algorithms from which the designer can select. Recognizing canonical forms of optimization models and selecting appropriate methods for their solutions require knowledge about various optimization models and the appropriate methods for solving them. This kind of knowledge can be conveniently represented by using IF-THEN rules.

a) Knowledge for optimization algorithm selection

The system selects the algorithm which is most appropriate to the problem based on the following criteria:

- nature and type of design variables,
- nature and type of objective functions,
- nature and type of design constraints,
- number of objective functions.

Some of the rules are presented below for illustration:

If all the variables are of continuous type
and all the constraints are linear
and the objective function is linear
then conclude that the model is linear programming
and execute linear programming algorithm.

If all the variables are of binary type
and all the constraints are linear
and the objective function is linear
then conclude that the model is zero one linear programming
and execute zero one linear programming algorithm.

If all the variables are of integer type
and all the constraints are linear
and the objective function is linear
then conclude that the model is integer linear programming
and execute integer linear programming algorithm.

If all the variables are of continuous type
and the constraints are nonlinear or
the objective function is nonlinear
then conclude that the model is nonlinear programming
and execute nonlinear programming algorithm.

If all the variables are of continuous type
and all the constraints are linear
and the number of objectives is greater than 1
and all the objective functions are linear
then conclude that the model is multicriteria linear programming
and execute multicriteria linear programming algorithm.

When the rule interpreter correctly indentifies the appropriate rule whose antecedents are satisfied, the type of optimization model is inferred, and an appropriate algorithm to solve the optimization model is selected. This kind of production system can act as a "front end" consultant to an optimization package which contains a set of algorithms to solve various optimization models.

4.5.3 Knowledge-Based Control in Pareto-Optimal Set Generation

a) Pareto-optimal set generation

Here, three most commonly used methods for Pareto optimal set generation, namely the *weighting method*, the *noninferior set estimation (NISE)* method, and the *constraint* method are considered. All of these generating techniques have one element in common; they all involve conversion of the multicriteria problem into a single objective form in order to generate Pareto solutions. The *weighting method* transforms the multicriteria function to a single criterion function through a parameterization of the relative weighting of the criteria (1-11). The *NISE method* is an extension of the weighting method with the mechanism for quickly converging onto the Pareto set. However, since this method works on the concept of the weighting method, Pareto-optimal solutions in nonconvex parts of the set will not be generated. The *constraint method* is another technique which transforms a multicriteria function of m criteria into a single criterion one by retaining one selected criterion as the primary criterion to be optimized and treating the remaining m-1 criteria as constraints. Balachandran and Gero [3] have discussed the relative merits and demerits of these three methods.

b) Knowledge as control mechanisms

In generating the Pareto-optimal set, or an approximation to it, the generation of an individual solution is carried out independently of the control of the process used to generate it. The three methods mentioned above simply transform a multicriteria problem into a single criterion optimization process. This process does not make use of the knowledge of the overall Pareto-optimal set in order to perform more efficiently. For example, in the NISE method the decision about which segment to use for the generation of the next solution is taken on the basis of an error criterion. Using this strategy, considerable effort can be expended generating closely spaced solutions which do not enhance the overall Pareto set. Appropriate rules to carry out this control more usefully in order to provide an outline of the whole Pareto set are given (see Fig. 4/7), where S_i is solution i:

If $\overline{S_iS_{i+1}}$ is the maximum unexplored segment length
and $\overline{S_iS_{i+1}}$ is greater than the minimum acceptable distance
then generate next solution between S_i and S_{i+1}.

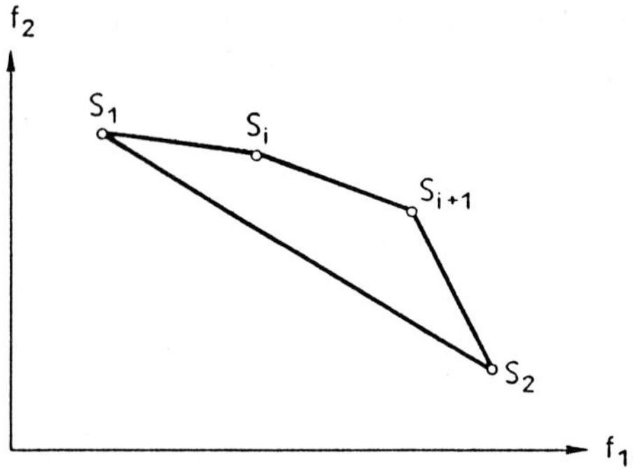

Fig. 4/7. Points S_1 and S_2 represent the optimal solutions of f_1 and f_2 individually [1)]

If $\overline{S_i S_{i+1}}$ is the maximum unexplored segment length
and $\overline{S_i S_{i+1}}$ is less than the minimum acceptable distance
then terminate the solution generation process.

If all segments have been explored
then there is no other solution outside the current boundary.

If same solution optimizes both objectives
then there is no conflict between those criteria.

The value of the minimum acceptable distance is designated as a percentage of the length of the primary line segment $\overline{S_1S_2}$ (see Fig. 4/7). The first rule is a declarative statement of the procedural control knowledge which will produce a better approximation to the shape of the Pareto-optimal set for the same computational load. The second rule is simply the procedural terminating condition. The third rule adds the knowledge that another way to terminate is to exhaust all possible Pareto-optimal solutions. The final rule adds the domain specific knowledge concerned with one of the boundary conditions in multicriteria optimization.

c) Knowledge to select an alternate generation process

The NISE method with the knowledge stated above would require fewer solutions to provide the overall outline of the Pareto-optimal set. However, the NISE-method is incapable of generating the Pareto-optimal solutions in either nonconvex or flat portions of the set. On

1) In contrast to chapter 1, the optimization problem is here defined as a maximization of criteria. Thus, the curvature of the p-efficient boundary differs from that of the other chapters.

the other hand, the constraint method will generate convex portions as well as nonconvex portions of the Pareto-optimal set and can thus be used to generate solutions in the "gaps" created in the Pareto-optimal set by the NISE method. The knowledge required to perform this task can be encoded as a rule as follows:

If the NISE method has failed to generate solutions between S_i and S_{i+1}
and the length of $\overline{S_iS_{i+1}}$ is greater than the minimum acceptable distance
then generate solutions between S_i and S_{i+1} using the constraint method.

This rule simply encodes the knowledge used to decide when an alternate method of generation can be used.

d) Knowledge from the Pareto-optimal set

The purpose of this section is to discuss the qualitative knowledge contained in the generated Pareto-optimal set. The knowledge is derived by recognizing the characteristic features of Pareto-optimal sets.

In dealing with conflicting objectives, the Pareto-optimal solutions are generated and analysed in order to make compromise decisions. The Pareto-optimal solutions are usually plotted as 2-D graphs since the graphical representations are more convenient to understand than the numerical tables. Furthermore, the graphical representation of the Pareto-optimal set provides qualitative knowledge of importance to the decision maker. Gero and Radford [7] have discussed the role of such knowledge in design decision making. This knowledge is mainly dependent on the shape of the Pareto-optimal boundary.

The knowledge required to identify the shape of Pareto-optimal boundaries and the qualitative knowledge that can be obtained from those curves can be encoded as a set of rules. The computer system can use this form of knowledge to assist the designer in making decisions efficiently. Some example rules employed to identify the best compromise solution from the 2-D Pareto-optimal set are given below:

If the shape of Pareto-optimal curve is heavily convex
then there is less conflict likely for the Pareto set
and best compromise solution lies in the north-east corner of the curve.

If the shape of Pareto-optimal curve is heavily concave
then there is considerable conflict for the Pareto set
and compromise solution depends on user's preference.

If the shape of Pareto optimal curve is planar
then the performance trade-offs between all criteria are linear
and further information is needed before choosing a solution.

In the preceding sections it was discussed how knowledge about the results can be used in the efficient generation of those results and how knowledge about processes can be used to select appropriate processes. In the following section these aspects are illustrated by an example problem.

4.6 Illustration Example

4.6.1 System Implementation

A system has been developed and implemented on SUN Microsystems workstations. It implements the notions described in the previous sections. The production rules and the inference engine are encoded in LISP. The pattern matching knowledge used in the simplification and the recognition of algebraic functions are encoded in PROLOG. The optimization algorithms are all written in C. The system is composed of four major modules as describes below:

- A *user interface* that provides various facilities for the designer to communicate with the system. It includes a semantic interpreter to translate the user input into a form that can be used by the knowledge-based system.
- An *optimization problem formulator* which builds a canonical optimization model for the given problem.
- An *optimization model recognizer* which identifies the nature of the optimization formulation and selects an appropriate algorithm.
- An optimization problem solver which carries out the necessary numeric computation to find the solution of the problem.

A high level controller for the entire system to allow all four components to communicate with each other is written in LISP. Example sessions with the systems are given below to illustrate the ideas and concepts discussed in this chapter.

4.6.2 Knowledge-Based Optimization Formulation and Recognition

Figures 4/8a) and 4/8b) show a session with the system during which the multicriteria optimization problem described in Section 4.4.3a) was formulated and identified by the system. Figure 4/8a) shows the mathematical model constructed for the problem and Fig. 4/8b) shows the information generated when attempting to recognize the algebraic model of the problem.

```
MATHEMATICAL MODEL FOR THE APARTMENT PROBLEM

maximize 7.33 * X1 + 7.33 * X2 + 7.33 * X3
minimize 128.4 * X1 + 79.55 * X2 + 73.3 * X3 + 61 * X4 + 30.5 * X5

subject to

4.28 * X1 <= 30
4.28 * X1 => 15
3.05 * X4 <= 12
3.05 * X4 => 5
1.83 * X2 <= 7.2
2.45 * X2 <= 6.5
2.45 * X2 => 4.5
3.05 * X5 <= 18
3.05 * X5 => 10
3.35 * X3 <= 18
3.35 * X3 => 10
3.98 * X3 <= 18
3.98 * X3 => 10
X1 + X2 - X4 - X5 = 0.0

continuous variables X1 to X5

X1 = living_room_length
X2 = hall_length
X3 = bedroom3_length
X4 = kitchen_length
X5 = bedroom_length

enter command
==>
```

Fig. 4/8a). Mathematical model constructed by the system for the apartment problem of section 4.4.3a)

OPTIMIZATION MODEL RECOGNITION

The following information has been identified

Variables

number of continuous variables = 5,
all the variables are continuous.

Objectives

number of objective functions = 2,
all the objective functions are linear.

Constraints

number of less-than-or-equal-to constraints = 7,
number of greater-than-or-equal-to constraints = 6,
number of equal-to constraints = 1,
all the constraints are linear.

The model has the following properties:

all the variables are continuous,
the number of objectives is 2,
all the objective functions are linear,
all the constraints are linear.

This is a mulitcriteria linear programming formulation.

enter command
==>

Fig. 4/8 b). Information generated by the system when attempting to recognize the algebraic model of the problem

4.6.3 Knowledge-Based Control in Pareto-Optimal Set Generation

In order to illustrate the knowledge-based control in Pareto-optimal set generation the example problem formulated in the above section will be used. The system first generated a set of Pareto-optimal solutions using the NISE method incorporated with the knowledge described in Section 4.5.3b).

a) Computation of the Pareto-optimal solution set

Fig. 4/9 and Table 4/1 show the Pareto-optimal solutions generated by the NISE-method. The minimum acceptable distance between any two solutions was specified as 15% of the primary line segment.

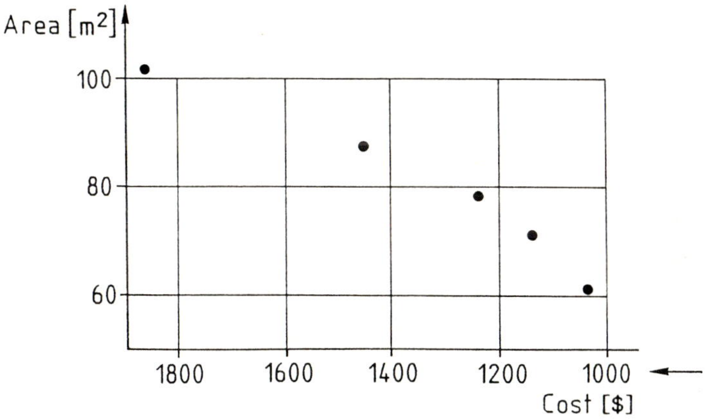

Fig. 4/9. Pareto-optimal solutions generated by the NISE-method using the rules given in section 4.5.3b)

Pareto-optimal Solutions					
Solution No.	***Area [m]²***		***Costs [units]***		
1	103.98		1857.26		
2	88.43		1450.17		
3	78.29		1230.37		
4	72.30		1140.53		
5	61.03		1027.83		
Design Variables [m]					
Solution No.	x_1	x_2	x_3	x_4	x_5
1	7.01	2.65	4.52	3.93	5.73
2	4.89	2.65	4.52	1.64	5.9
3	3.50	2.65	4.52	1.64	4.52
4	3.50	1.84	4.52	1.64	3.70
5	3.50	1.84	2.99	1.64	3.70

Table 4/1. Pareto-optimal solution set generated by the NISE method (1) Maximize floor area; (2) Minimize construction cost

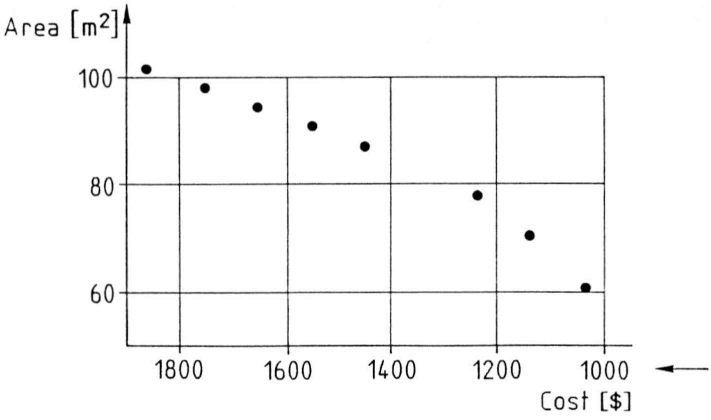

Fig. 4/10. Constraint method in the largest "gap" shown in Fig. 4/9

Using the rules given in Section 4.5.3c) the system generated additional Pareto solutions by applying the constraint method in the region where the NISE method has failed to generate. Some solutions are shown in Fig. 4/10 and Table 4/2.

The first set of results shown in Table 4/1 and Fig. 4/9 was obtained by the NISE-method with the knowledge described in Section 4.5.3b). In this case only five Pareto-optimal solutions were generated. The NISE method, however, could not generate the Pareto-optimal solutions in either nonconvex or flat portion of the set. In the knowledge-based approach the system automatically switched the algorithm to the constraint method and produced a better approximation of the Pareto-optimal set by generating additional solutions in the largest "gap" left by the NISE method (see Fig. 4/10).

Although no single generating technique is likely to be efficient for all types of multiple criteria problems, knowledge-based control mechanisms allow the system to use the knowledge about the results produced by an algorithm in order to select alternate processes if required. Hence, the overall performance of the system and the resultant decisions are improved using a range of techniques with a knowledge-based controller.

Pareto-optimal Solution					
Solution No.	Area $[m]^2$		Costs [unit]		
1	103.98		1857.26		
2	100.10		1755.49		
3	96.21		1653.72		
4	92.32		1551.95		
5	88.43		1450.17		
6	78.29		1230.37		
7	72.30		1140.53		
8	61.03		1027.83		
Decision Variable [m]					
Solution No.	x_1	x_2	x_3	x_4	x_5
1	7.01	2.65	4.52	3.93	5.73
2	6.48	2.65	4.52	3.37	5.76
3	5.95	2.65	4.52	2.80	5.80
4	5.42	2.65	4.52	2.20	5.87
5	4.89	2.65	4.52	1.64	5.90
6	3.50	2.65	4.52	1.64	4.52
7	3.50	1.84	4.52	1.64	3.70
8	3.50	1.84	2.99	1.64	3.70

Table 4/2. Complete set of Pareto-optimal solutions
Objectives: (1) Maximize floor area
(2) Minimize construction cost

4.7 Conclusion

The structure and operation of a knowledge-based system used in conjunction with numerical optimization techniques for multicriteria decision making has been described. The introduction of knowledge-based systems in this area is significant. Multicriteria optimization processes involve several types of knowledge. The system which was developed incorporating those distinct types of knowledge is more versatile and useful than conventional optimization systems. Examples illustrated in this chapter obviously show the potential of this approach to multicriteria optimization.

The nature and difficulties associated with formulating multicriteria problems into mathematical models have been discussed, and the use of particular symbolic computation and knowledge-based methodologies to handle these tasks has been demonstrated. A symbolic manipulation system was developed to carry out the mathematical operations involved in the formulation process. The frame formalism was shown to be particularly useful for representing design information in terms of design objects, objectives and constraints in a concise and comprehensive manner. The reasoning capabilities of a frame system have been utilized to assist in the optimization formulation process.

The capability of recognizing optimization formulations and selecting appropriate algorithms was another important aspect addressed in detail. The predicate logic and production rule formalisms have been utilized to encapsulate the necessary knowledge to carry out these tasks. Furthermore, the use of explicit knowledge to guide and control Pareto-optimal set generation processes has been illustrated. It was shown that, by incorporating some better control mechanisms based on explicit knowledge, the performance of Pareto-optimal set generating processes can be upgraded significantly.

References

[1] Balachandran, M.; Gero, J.S.: A Comparison of Three Methods for Generating the Pareto Optimal Set. In: Engineering Optimization 7 (1984) 319-336

[2] Balachandran, M.; Gero, J.S.: A Knowledge-Based Approach to Mathematical Design Modelling and Optimization. In: Engineering Optimization 12 (1987a) 99-115

[3] Balachandran, M.; Gero, J.S.: Use of Knowledge in Selection and Control of Optimization Algorithms. In: Engineering Optimization 12(2) (1987b) 163-173

[4] Clocksin, W.F.; Mellish, C.S.: Programing in PROLOG. Berlin: Springer-Verlag 1981

[5] Davis, R.; King, J.: An Overview of Production Systems. In: Elcock, E.W.; Michie, D.; Horwood E. (ed): Machine Intelligence 8, Chichester (1977) 300-332

[6] Fikes, R.; Kehler, T.: The Role of Frame-Based Representation in Reasoning. In: Communication of the ACM 28(9) (1985) 904-920

[7] Gero, J.S.; Radford, A.D.: The Place of Multicriteria Optimization in Design. In: Design Policy Conference Proceedings, Design Council, London: (1984) 3:81-85

[8] Hayes-Roth, F.; Waterman, D.A.; Lenat, D.B. (eds): Building Expert Systems. Reading: Addison-Wesley 1983

[9] Minsky, M.: A Framework for Representing Knowledge. In: P.H. Winston (ed): The Psychology of Computer Vision. New York: McGraw-Hill (1975) 211-277

[10] Mitchell, W.J.; Steadman, J.P.; Liggett, R.S.: Synthesis and Optimization of Small Rectangular Floor Plans. In: Environment and Planning B, 3 (1976) 37-70

[11] Mistree, F.; Karandikar, H.; Kamal, S.: Rule-Based Post Solution Analysis of Decision Support Problems: Some Preliminary Results. In: D. Sriram; R. Adey (ed): Applications of Artificial Intelligence in Engineering Problems (1986) 301-316

[12] Radford, A.D.; Gero, J.S.; Rosenman, M.A.; Balachandran, M.: Pareto Optimization as a Computer Aided Design Tool. In: Gero, J.S. (ed): Optimization in Computer Aided Design. Amsterdam: North-Holland (1985) 47-69

[13] Watermann, D.: A Guide to Expert Systems. Reading: Addison-Wesley 1986

PART II

APPLICATIONS

5 MECHANISMS AND DYNAMIC SYSTEMS

5.1 Optimal Counterweight Balancing of Robot Arms Using Multicriteria Approach

J. Koski, A. Osyczka

5.1.1 Introduction

The increasing application of industrial robots in different fields of technology augments the demand for further improvement of their performance. Energy consumption and working accuracy especially are becoming more and more important in assessing the efficiency of a robot. One way to improve these factors is the proper balancing of a robot manipulator. There are two main methods of balancing a robot manipulator: 1) by spring mechanisms or 2) by counterweights [1,2]. The problem of the optimum design of a robot spring balancing mechanism is discussed in Chapter 5.2. In this chapter, counterweight balancing of robot arms is the subject of investigation.

Counterweight balancing is often used for establishing better mass distributions of mechanisms, and some approaches to the optimum design of different kinds of mechanisms have been discussed in the literature [3,4,5]. Since the balancing of a robot manipulator involves the minimization of driving forces or torques as well as the support reactions at joints, it seems reasonable to apply the multicriteria approach. Using this approach, an optimization model based on the rigid-body dynamics of robot arms is presented. The solution of the multicriteria optimization model provides the designer with a set of Pareto-optimal solutions which enables him to make the right decision in a conflicting situation.

5.1.2 Kinetic Model of a Robot Arm

Two alternative approaches are generally used in deriving the equations of motion for robot manipulators. These are the Lagrangian and the Newton-Euler approaches. By using angular coordinates for the PUMA–560 robot, schematically shown in Fig. 5.1/1, it is possible to calculate the generalized torques at each joint applying the following equation:

$$M_{ti} = \frac{d}{dt}\left(\frac{\partial L}{\partial \dot{\theta}_i}\right) - \frac{\partial L}{\partial \theta_i}, \tag{5.1-1}$$

where θ_i is the rotation at joint i and $\dot{\theta}_i$ is the corresponding angular velocity. The term

$$L = T - V \tag{5.1-2}$$

represents the Lagrangian function of the mechanical system. Here, T is the total kinetic energy of the system and V is the total potential energy. The application of (5.1-1) to a fully articulated robot arm results in the following nonlinear second-order system of differential equations

$$\mathbf{A}\ddot{\boldsymbol{\theta}} + \mathbf{B}\dot{\boldsymbol{\theta}}^2 + \mathbf{c} - \mathbf{m} = \mathbf{0}\,. \tag{5.1-3}$$

Here, the vector of angular accelerations is given by

$$\ddot{\boldsymbol{\theta}} = (\ddot{\theta}_1, \ddot{\theta}_2, \ldots, \ddot{\theta}_N)^T, \tag{5.1-4}$$

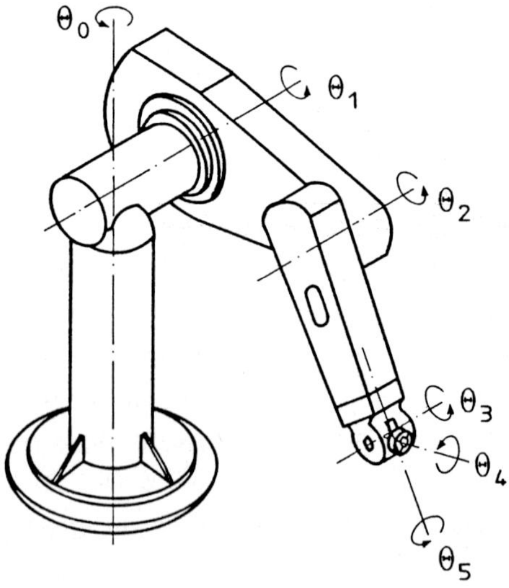

Fig. 5.1/1. PUMA-560 robot and schematic representation of coordinate angles θ_i

and the vector of squared angular velocities by

$$\dot{\boldsymbol{\theta}}^2 = (\dot\theta_1\dot\theta_1, \dot\theta_1\dot\theta_2, \dots, \dot\theta_1\dot\theta_N \mid \dot\theta_2\dot\theta_1, \dot\theta_2\dot\theta_2, \dots, \dot\theta_2\dot\theta_N \mid \dots \dot\theta_k\dot\theta_1, \dot\theta_k\dot\theta_2, \dots, \dot\theta_k\dot\theta_N \mid \dots \dot\theta_N\dot\theta_1, \dot\theta_N\dot\theta_2, \dots, \dot\theta_N\dot\theta_N)^T , \tag{5.1-5}$$

where N is the number of joints. The corresponding matrices are

$$\mathbf{A} = \begin{bmatrix} D_{11} & D_{12} & \cdots & D_{1N} \\ D_{21} & D_{22} & \cdots & D_{2N} \\ \cdot & & & \\ \cdot & & & \\ \cdot & & & \\ D_{N1} & D_{N2} & \cdots & D_{NN} \end{bmatrix}$$

and

$$\mathbf{B} = \left[\begin{array}{cccc|cccc|cccc} D^1_{11} & D^1_{12} & \cdots & D^1_{1N} & D^2_{11} & D^2_{12} & \cdots & D^2_{1N} & D^N_{11} & D^N_{12} & \cdots & D^N_{1N} \\ D^1_{21} & D^1_{22} & \cdots & D^1_{2N} & D^2_{21} & D^2_{22} & \cdots & D^2_{2N} & D^N_{21} & D^N_{22} & \cdots & D^N_{2N} \\ & \cdot & & & & \cdot & & & & \cdot & & \\ & \cdot & & & & \cdot & & & & \cdot & & \\ & \cdot & & & & \cdot & & & & \cdot & & \\ D^1_{N1} & D^1_{N2} & \cdots & D^1_{NN} & D^2_{N1} & D^2_{N2} & \cdots & D^2_{NN} & D^N_{N1} & D^N_{N2} & \cdots & D^N_{NN} \end{array} \right] , \tag{5.1-6}$$

where it is assumed that each joint has one degree of freedom. The elements of matrix **A** are the inertia terms, and the elements of matrix **B** represent the centripetal and Coriolis terms. All these terms depend on the position of the arm, i.e. $D_{ij} = D_{ij}(\theta_i)$. Vector $\mathbf{c} = (D_1, D_2, \dots, D_N)^T$ includes the gravitational terms D_i , and $\mathbf{m}$ is the vector of torques. Kinetic equations (5.1-3) represent the rigid-body motion of the arm, and they are geometrically nonlinear because of large rotations θ_i.

The manipulator is an isostatic structure, and thus it is possible to get explicit expressions for all forces and moments in the system. The friction in the joints as well as the flexibility of the arm are not included in the following design model. For the application of optimization methods, a two-member robot arm, which corresponds to the two arms of PUMA-560 robot in a plane motion, is considered. This arm is assumed to move in the xy-plane only (corresponding angular coordinates θ_i are shown in Fig. 5.1/2). The masses of the members are m_1 and m_2. They are located as point masses at distances e_1 and e_2 from the joints. The external load is represented by the point mass m_3. In the sequel, only the counterweight masses m_4 and m_5 as well as their distances from the joints x_1 and x_2 are treated as design variables, whereas all the other quantities are fixed.

The torques of this two-member robot are obtained from the Eq. (5.1-1) and are expressed as follows:

$$M_{t1} = D_{11}\ddot{\theta}_1 + D_{12}\ddot{\theta}_2 + D^1_{11}\dot{\theta}^2_1 + D^2_{12}\dot{\theta}^2_2 + (D^1_{12} + D^2_{11})\,\dot{\theta}_1\,\dot{\theta}_2 + D_1\,,$$

$$M_{t2} = D_{21}\ddot{\theta}_1 + D_{22}\ddot{\theta}_2 + D^1_{21}\dot{\theta}^2_1 + D^2_{22}\dot{\theta}^2_2 + (D^1_{22} + D^1_{21})\,\dot{\theta}_1\,\dot{\theta}_2 + D_2\,. \tag{5.1-7}$$

The terms appearing in these equations are defined in the Appendix.

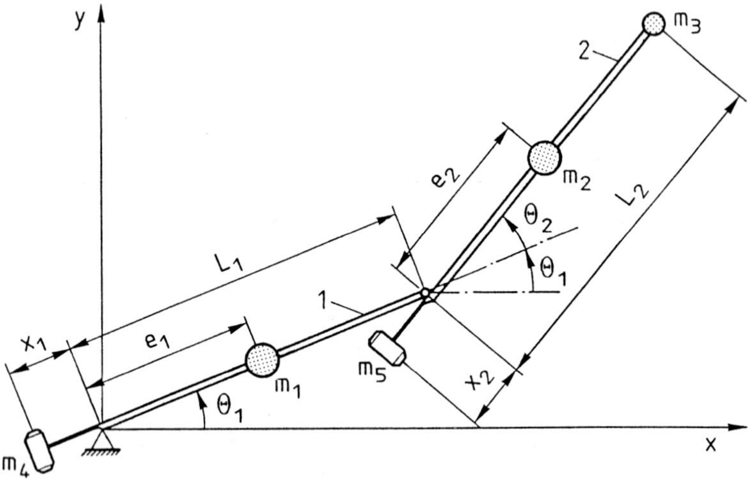

Fig. 5.1/2. Mechanical model of a robot arm for optimization

In addition to the torques, the joint forces are considered in the optimization process. In the present application, the most convenient way of solving them is to use the force equilibrium conditions in both coordinate directions x and y. For this purpose, the free-body diagrams of both members have been depicted in Fig. 5.1/3. The positive directions in the figure are associated with the global xy-axes, and the positive rotation direction is counterclockwise. By computing the accelerations from the well-known kinematic equation

$$\mathbf{a}_P = \mathbf{a}_Q + \boldsymbol{\alpha} \times \mathbf{r}_{P/Q} + \boldsymbol{\omega} \times (\boldsymbol{\omega} \times \mathbf{r}_{P/Q}) \ , \tag{5.1-8}$$

analytic expressions for $\mathbf{a}_i$ (1,...,5) can be obtained. Here, $\mathbf{a}_Q$ is the acceleration vector of the comparison point, $\boldsymbol{\alpha}$ the angular acceleration vector of the member, $\mathbf{r}_{P/Q}$ the position vector from point P to point Q along the member, and $\boldsymbol{\omega}$ the angular velocity of the member. For member 1 point Q is the support point and $\boldsymbol{\alpha} = (0, 0, \ddot{\theta}_1)^T$, $\boldsymbol{\omega} = (0, 0, \dot{\theta}_1)^T$. For member 2 point Q is at the joint and $\boldsymbol{\alpha} = (0, 0, \ddot{\theta}_0)^T$, $\boldsymbol{\omega} = (0, 0, \dot{\theta}_0)^T$ where $\theta_0 = \theta_1 + \theta_2$. Vector $\mathbf{r}_{P/Q}$ depends on the 5 selected calculation points. Here, the detailed expressions for the accelerations are presented separately, and they also appear as a part of the terms in (5.1-7). The corresponding inertia forces $m_i a_i$ ($i = 1,...,5$) and the moments $J_j \alpha_j$, ($j = 1,2$) with the complete free-body diagrams are shown in Fig. 5.1/3.

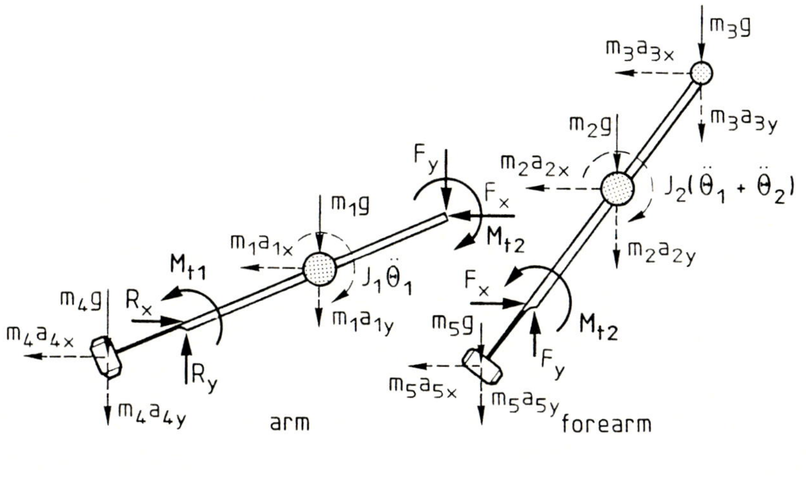

Fig. 5.1/3. Free-body diagrams of robot arms

The accelerations of the points at which the point masses are located have the following explicit expressions:

$$\mathbf{a}_1 = \ddot{\theta}_1 e_1 \begin{bmatrix} -\sin\theta_1 \\ \cos\theta_1 \end{bmatrix} - \dot{\theta}_1^2 e_1 \begin{bmatrix} \cos\theta_1 \\ \sin\theta_1 \end{bmatrix} ,$$

$$\mathbf{a}_2 = \ddot{\theta}_1 L_1 \begin{bmatrix} -\sin\theta_1 \\ \cos\theta_1 \end{bmatrix} - \dot{\theta}_1^2 L_1 \begin{bmatrix} \cos\theta_1 \\ \sin\theta_1 \end{bmatrix} + e_2\ddot{\theta}_o \begin{bmatrix} -\sin\theta_o \\ \cos\theta_o \end{bmatrix} +$$

$$e_2\dot{\theta}_o^2 \begin{bmatrix} -\cos\theta_o \\ -\sin\theta_o \end{bmatrix} ,$$

$$\mathbf{a}_3 = \ddot{\theta}_1 L_1 \begin{bmatrix} -\sin\theta_1 \\ \cos\theta_1 \end{bmatrix} - \dot{\theta}_1^2 L \begin{bmatrix} \cos\theta_1 \\ \sin\theta_1 \end{bmatrix} + L_2\ddot{\theta}_o \begin{bmatrix} -\sin\theta_o \\ \cos\theta_o \end{bmatrix} + \tag{5.1-9}$$

$$+ L_2\dot{\theta}_o^2 \begin{bmatrix} -\cos\theta \\ -\sin\theta_o \end{bmatrix} ,$$

$$\mathbf{a}_4 = \ddot{\theta}_1 x_1 \begin{bmatrix} \sin\theta_1 \\ -\cos\theta \end{bmatrix} + \dot{\theta}_1^2 x_1 \begin{bmatrix} \cos\theta_1 \\ \sin\theta_1 \end{bmatrix} ,$$

$$\mathbf{a}_5 = \ddot{\theta}_1 L_1 \begin{bmatrix} -\sin\theta_1 \\ \cos\theta_1 \end{bmatrix} - \dot{\theta}_1^2 L_1 \begin{bmatrix} \cos\theta \\ \sin\theta_1 \end{bmatrix} + x_2\ddot{\theta}_o \begin{bmatrix} \sin\theta_o \\ -\cos\theta_o \end{bmatrix} +$$

$$+ x_2\dot{\theta}_o^2 \begin{bmatrix} \cos\theta_o \\ -\sin\theta_o \end{bmatrix} .$$

Here again the notations $\theta_o = \theta_1 + \theta_2$, $\dot{\theta}_o = \dot{\theta}_1 + \dot{\theta}_2$ and $\ddot{\theta}_o = \ddot{\theta}_1 + \ddot{\theta}_2$ have been used. By applying the force equilibrium conditions in the coordinate directions, the following joint reactions (see Fig. 5.1/3)

$$\begin{aligned} R_{2x} &= m_2 a_{2x} + m_3 a_{3x} + m_5 a_{5x} , \\ R_{2y} &= m_2 a_{2y} + m_3 a_{3y} + m_5 a_{5y} + (m_2 + m_3 + m_5) g \end{aligned} \tag{5.1-10}$$

to member 2 are obtained. The torque M_{t2} at the joint can be calculated from the moment equilibrium condition. By applying the same routine to member 1, it is possible to derive expressions for the support reactions:

$$R_{1x} = m_1 a_{1x} + m_2 a_{2x} + m_3 a_{3x} + m_4 a_{4x} + m_5 a_{5x} ,$$

$$R_{1y} = m_1 a_{1y} + m_2 a_{2y} + m_3 a_{3y} + m_4 a_{4y} + m_5 a_{5y} + \qquad (5.1\text{-}11)$$

$$+ (m_1 + m_2 + m_3 + m_4 + m_5) g$$

The torque M_{t1} is obtained again from the moment equilibrium condition. Corresponding to the standard Lagrangian approach, the torques M_{t1} and M_{t2} must be the same as those computed from (5.1-7). The resulting support reactions are:

$$R_1 = \sqrt{R_{1x}^2 + R_{1y}^2} , \qquad R_2 = \sqrt{R_{2x}^2 + R_{2y}^2} . \qquad (5.1\text{-}12)$$

The torques M_{ti} and the forces R_i are chosen as criteria in the optimization model. It is important to present the detailed expressions for M_{ti} and R_i because the choice of the design variables as well as the general complexity of the optimization problem are associated with these formulae.

5.1.3 Formulation of the Optimization Problem

Next, a multicriteria optimization problem based on the rigid-body kinetic model of the robot arm (see Fig. 5.1/2) will be formulated, and the numerical design data for this case will be given. The objective is to find such masses m_4 and m_5 for the counterweights and such joint distances x_1 and x_2 which will minimize the chosen four design criteria. Consequently, the design variable vector is

$$\mathbf{x} = (x_1, x_2, x_3, x_4)^T , \qquad (5.1\text{-}13)$$

where the first two are the distances shown in Fig. 5.1/2, $x_3 = m_4$ and $x_4 = m_5$. The upper and the lower limits for all these four design variables can be given in the form

$$x_i^l \leq x_i \leq x_i^u , \qquad i = 1,\dots,4 . \qquad (5.1\text{-}14)$$

The torques M_{t1} and M_{t2} at the arm joints are chosen as the first two criteria of the vector objective function. Their minimization is important because it is then possible to use smaller motors, and the energy consumption is lower if the variation ranges of the torques are small. In the explicit expressions (5.1-7), terms m_4x_1, $m_4x_1^2$, m_5x_2, and $m_5x_2^2$ appear, and thus it is reasonable to choose the design variables in the way presented.

The torques do not depend on the design variables alone but also on the position of the robot arm (θ_1, θ_2), on the angular velocities $(\dot{\theta}_1, \dot{\theta}_2)$, and on the angular accelerations $(\ddot{\theta}_1, \ddot{\theta}_2)$. Usually, the working space of the robot arm is restricted, and thus constraints of the form

$$\theta_i^l \le \theta_i \le \theta_i^u \ , \qquad i = 1, 2, \tag{5.1-15}$$

are needed. Here, θ_i^l and θ_i^u are the lower and the upper limits of the angles θ_i. In each position of the arm, the angular velocities and accelerations may be different. In order to optimize the performance of the robot, the torques should be as small as possible at all working positions and at all existing angular velocity acceleration combinations. Thus, the first two criteria are chosen as follows:

$$\begin{aligned} f_1(\mathbf{x}) &= \max_{\theta_1} \ \max_{\theta_2} \ \max_{\dot{\theta}_i, \ddot{\theta}_i} M_{t1} \ , \\ f_2(\mathbf{x}) &= \max_{\theta_1} \ \max_{\theta_2} \ \max_{\dot{\theta}_i, \ddot{\theta}_i} M_{t2} \ , \end{aligned} \tag{5.1-16}$$

where notation $\dot{\theta}_i$, $\ddot{\theta}_i$ is associated with the chosen angular velocity profile. This is shown in Fig. 5.1/4 where a trapezoidal profile, typical of robot applications, has been depicted for both members. The corresponding angular accelerations are also presented in this figure.

The construction of joints, especially the choice of bearings, depends largely on the reaction forces at the joints. Thus, it seems reasonable to choose the maximum values of the joint forces as two additional criteria. By using the fixed trapezoidal velocity profiles (see Fig. 5.1/4) and every feasible position of the arm, these criteria can be expressed in the form

$$f_3(\mathbf{x}) = \max_{\theta_1} \max_{\theta_2} \max_{\dot\theta_i, \ddot\theta_i} R_1 ,$$

$$f_4(\mathbf{x}) = \max_{\theta_1} \max_{\theta_2} \max_{\dot\theta_i, \ddot\theta_i} R_2 . \qquad (5.1\text{-}17)$$

The geometrical interpretation of all the four criteria is the following: a small movement at every positon (θ_1, θ_2) of the arms is performed by using the fixed trapezoidal profiles $(\dot\theta_1, \dot\theta_2, \ddot\theta_1, \ddot\theta_2$ shown in Fig. 5.1/4), and the maximum values of the torques and the joint forces during the movement are selected.

By using the design variables x_i given in (5.1-13), the criteria presented in (5.1-16) and (5.1-17), the side constraints of (5.1-14), and the state constraints of (5.1-15), it is now possible to formulate the multicriteria optimization problem

$$\min (f_1(\mathbf{x}), f_2(\mathbf{x}), f_3(\mathbf{x}), f_4(\mathbf{x}))^T ,$$

subject to

$$\theta_i^l \le \theta_i \le \theta_i^u , \qquad i = 1, 2, \qquad (5.1\text{-}18)$$

$$x_i^l \le x_i \le x_i^u , \qquad i = 1, 2, 3, 4.$$

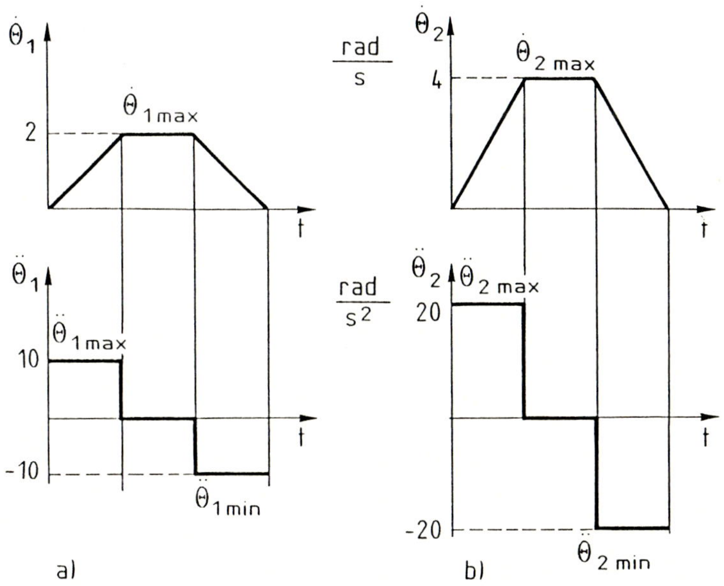

Fig. 5.1/4. Angular velocities and corresponding angular accelerations of robot arms

The minimum values of conflicting criteria cannot be obtained at the same design point. In multicriteria problems it is reasonable to apply the Pareto-optimality concept which is described in Chapter 1. Usually, not just one, but a set of Pareto-optima is characteristic of a multicriteria problem. The corresponding vectors $\mathbf{f}(\mathbf{x}^*) = (f_1(\mathbf{x}^*), f_2(\mathbf{x}^*), f_3(\mathbf{x}^*), f_4(\mathbf{x}^*))^T$ in the criterion space R^4 are called "minimal solutions".

Next, the numerical design data for the design problem is given. These values are close to the arms of the PUMA-560 robot shown in Fig. 5.1/1 [6]. By using the notations defined in Fig. 5.1/2 and in (5.1-13), the following data are used in the optimization calculations:

$$m_1 = 17 \text{ kg}, \quad m_2 = 6 \text{ kg}, \quad m_3 = 2 \text{ kg},$$

$$L_1 = L_2 = 0.43 \text{ m}, \quad e_1 = 0.07 \text{ m}, \quad e_2 = 0.14 \text{ m},$$

$$\theta_1^l = -40^0, \quad \theta_1^u = 220^0, \quad \theta_2^l = -140^0, \quad \theta_2^u = 140^0,$$

$$\dot{\theta}_{1\,max} = 2\,\frac{\text{rad}}{\text{s}}, \quad \dot{\theta}_{2\,max} = 4\,\frac{\text{rad}}{\text{s}},$$

$$\ddot{\theta}_{1\,max} = 10\frac{\text{rad}}{\text{s}^2}, \quad \ddot{\theta}_{2\,max} = 20\frac{\text{rad}}{\text{s}^2},$$

$$x_1^l = x_2^l = 0, \quad x_1^u = x_2^u = 0.2 \text{ m}, \quad x_3^l = x_4^l = 0,$$

$$x_3^u = 35 \text{ kg}, \quad x_4^u = 15 \text{ kg},$$

$$J_1 = 0.2619 \text{ kgm}^2, \quad J_2 = 0.0924 \text{ kgm}^2 .$$

5.1.4 Solution Method

The numerical solution of the robot arm balancing problem (5.1-18) is obtained on the basis of the Computer Aided Multicriteria Optimization System (CAMOS) which uses several scalar and multicriteria optimization methods (see Chapter 3 and [7]). For the problem treated here the method of combined random and sequential search has been used for the generation of Parero-optima. The choice of the method was mainly based on the expectation of several local minima which proved to be true during the calculations. First, some points were generated by the random search method, and the best of them were stored and used as the starting points for the sequential search procedure. By minimizing each criterion separately in the feasible set, the ideal vector $\bar{f} = (\bar{f}_1, \bar{f}_2, \bar{f}_3, \bar{f}_4)^T$ was obtained. After the numerical solution of these four scalar optimization problems, it was possible to use the weighting min-max method (see (1.22)) for generating several Pareto-optimal solutions. Here, the weights w_i $(i = 1,...,4)$ were used as parameters which could be normalized by using $w_1 + w_2 + w_3 + w_4 = 1$. Each parameter combination corresponds to one Pareto-optimum of the problem (5.1-18). While seeking both, the ideal vector and the other Pareto-optima, the random search method was used in combination with the Nedler-Mead simplex method [8] with a penalty function.

For the numerical computation of the criteria values, the term $\max_{\theta_1}(\cdot)$ can be obtained in the following way. The value of each torque and of each joint force is calculated at the positions θ_1^l, $\theta_1^l + \Delta\theta_1$, $\theta_1^l + 2\Delta\theta_1,...,\theta_1^u$ where the increment $\Delta\theta_1$ is properly chosen. Then the maximum value of these positions is selected separately for each criterion. The same calculations can be carried out for the term $\max_{\theta_2}(\cdot)$ after choosing increment $\Delta\theta_2$. The smaller the increments $\Delta\theta_i$, the better the accuracy. However, the computation time is increased. The terms $\max_{\dot{\theta}_i\ddot{\theta}_i}(\cdot)$ require the computation of torques and joint forces for all expected combinations of $\dot{\theta}_i$ and $\ddot{\theta}_i$ which is impossible to realize. Thus, only some chosen combinations of the angular velocities and accelerations are considered for given θ_1 and θ_2. These can be found in Table 5.1/1 which presents the seven calculation points. After calculating M_{ti} and R_i for all the rows in Table 5.1/1 the maximum values can be chosen.

It is obvious that the numerical evaluation of the criteria values is computationally expensive. Using the IBM PC/AT computer, the calculation of the ideal vector $\bar{f}$ needs about three hours, and the subsequent calculation of any other Pareto-optimum about one hour each. The application of the method presented in Chapter 5.2 can make the computation process more efficient. For some robot applications, the evaluation of the objective functions can be simplified by assuming for example that calculations of M_{ti} and R_i are carried out only for certain positions of the robot arms as well as for selected angular velocities and accelerations (e.g. points 3 and 5 in Table 5.1/1)). Such alternatives are available when using the computer program for the presented optimization model.

Point	$\dot{\theta}_1$	$\dot{\theta}_2$	$\ddot{\theta}_1$	$\ddot{\theta}_2$
1	0	0	$\ddot{\theta}_{1\,max}$	$\ddot{\theta}_{2\,max}$
2	$\frac{1}{2}\dot{\theta}_{1\,max}$	$\frac{1}{2}\dot{\theta}_{2max}$	$\ddot{\theta}_{1\,max}$	$\ddot{\theta}_{2\,max}$
3	$\dot{\theta}_{1\,max}$	$\dot{\theta}_{2max}$	$\ddot{\theta}_{1\,max}$	$\ddot{\theta}_{2\,max}$
4	$\dot{\theta}_{1\,max}$	$\dot{\theta}_{2\,max}$	0	0
5	$\dot{\theta}_{1\,max}$	$\dot{\theta}_{2\,max}$	$-\ddot{\theta}_{1\,max}$	$-\ddot{\theta}_{2\,max}$
6	$\frac{1}{2}\dot{\theta}_{1\,max}$	$\frac{1}{2}\dot{\theta}_{2\,max}$	$-\ddot{\theta}_{1\,max}$	$-\ddot{\theta}_{2\,max}$
7	0	0	$-\ddot{\theta}_{1\,max}$	$-\ddot{\theta}_{2\,max}$

Table 5.1/1. Angular velocities and accelerations at the calculation points shown in Figure 5.1/4

5.1.5 Pareto-Optimal Designs

Fifteen Pareto-optimal robots have been generated by using the module of CAMOS which calculates one Pareto-optimum for each weighting coefficient combination. The weights w_i ($\sum w_i = 1$), the Pareto-optimal design variable vectors $\mathbf{x}^*$, and the corresponding minimal solutions $\mathbf{f}^*$ are shown in Table 5.1/2. The results are illustrated graphically in Fig. 5.1/5 which presents four Pareto-optimal robot arms.

In Fig. 5.1/6 the minimal solutions have been presented graphically. Pareto-optima 1...15 are in such an order that criterion f_1 is increasing continuously. Thus, $\mathbf{x}_i^*$ and $\mathbf{x}_{i+1}^*$ are not necessarily neighbouring points in the design space. If the minimal set is represented sufficiently, it is possible to obtain a satisfactory solution, i.e. the best Pareto-optimum from this kind of a figure. Otherwise, one can apply the procedure given in CAMOS where $\mathbf{w}$ is changed until a satisfactory solution is found.

$w = (w_1, w_2, w_3, w_4)^T$	$x = (x_1, x_2, x_3, x_4)^T$	$f(x) = (f_1(x), f_2(x), f_3(x), f_4(x))^T$	No.
min $f_1(x)$	(0.199, 0.199, 34.98, 5.77)	(85.7, 32.2, 750.6, 302.8)	1
min $f_2(x)$	(0.175*, 0.114, 10,24*, 14.86)	(162.7, 23.2, 711.1, 450.7)	15
min $f_3(x)$	(0.198*, 0.14 ,* 0.001, 0.002)	(132.2, 41.9, 373.8, 194.5)	12
min $f_4(x)$	(0.191,* 0.198*, 14.3,* 0.001)	(111.3, 41.9, 419.4, 194.5)	4
(0,25, 0.25, 0.25, 0.25)	(0.186, 0.198, 7.95, 4.06)	(125.5, 35.0, 509.8, 268.5)	10
(0.3, 0.3, 0.2, 0.2)	(0.171, 0.184, 16.9, 5.66)	(117.8, 32.5, 599.1, 299.1)	7
(0.35, 0.35, 0.15, 0.15)	(0.194, 0.182, 19.6, 7.59)	(114.2, 29.5, 666.9, 336.19)	6
(0.4, 0.4, 0.1, 0.1)	(0.130, 0.193, 32.9, 7.84)	(107.9, 28.6, 769.6, 345.6)	3
(0.2, 0.2, 0.3, 0.3)	(0.190, 0.197, 8.05, 3.82)	(125.0, 35.5, 505.2, 263.3)	9
(0.15, 0.15, 0.35, 0.35)	(0.126, 0.175, 3.89, 2.93)	(132.9, 37.2, 457.3, 244.8)	13
(0.1, 0.1, 0.4, 0.4)	(0.103, 0.114, 0.138, 2.08)	(138.5, 39.3, 408.9, 228.2)	14
(0.5, 0.1, 0.2, 0.2)	(0.172, 0.093, 26.5, 0.45)	(99.0, 41.4, 592.3, 201.6)	2
(0.1, 0.5, 0.2, 0.2)	(0.198, 0.157, 17.0, 7.84)	(121.9, 29.9, 645.9, 335.2)	8
(0.4, 0.2, 0.2, 0.2)	(0.148, 0.182, 20.6, 3.6)	(112.3, 36.0, 606.6, 257.9)	5
(0.2, 0.4, 0.2, 0.2)	(0.152, 0.198, 10.3, 6.6)	(127.7, 30.7, 584.8, 320.7)	11

Table 5.1/2. Pareto-optima and minimal solutions corresponding to the robot problem

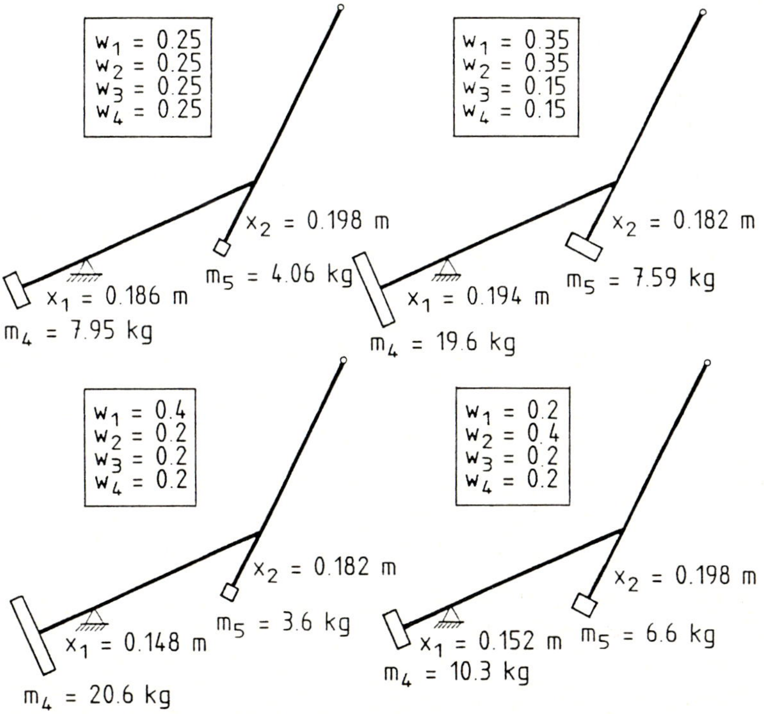

Fig. 5.1/5. Minimal solutions of Pareto-optimal robot arms

From these results it is obvious that the criteria f_1 and f_2 are in conflict with each other as well as with f_3 and f_4. The last two criteria, however, are more or less nonconflicting. This observation suggests that a combined criterion $\lambda_1 f_3 + \lambda_2 f_4$ could be used to decrease the dimension of the multicriteria problem. Here, it is also possible to choose $\lambda_1 = 1$, $\lambda_2 = 0$, i.e. one reaction represents both. This seems reasonable also from a physical point of view. It is interesting to notice the large variation in ranges of the criteria values, even if all solutions are already Pareto-optimal. The solutions seem to be very sensitive to the design variables which are not active but located at the separately attainable minima. At these points some variables can be chosen freely, and they are depicted by an asterisk (*) in Table 5.1/2. Note that when the mass of the counterweight is close to zero, variables x_1 or x_2 do not influence the solution.

Variables x_1 and x_3 have no influence on the minimum of $f_2(\mathbf{x})$ and $f_4(\mathbf{x})$ which is the case when weak Pareto-optima occur [9]. Thus, the values of x_1 and x_3 in the rows two and four of Table 5.1/2 were obtained while additionally minimizing $f_1(\mathbf{x})$ and $f_2(\mathbf{x})$ with equal importance for the given values of x_2 and x_4, established at the separately attainable minima of $f_3(\mathbf{x})$ and $f_4(\mathbf{x})$.

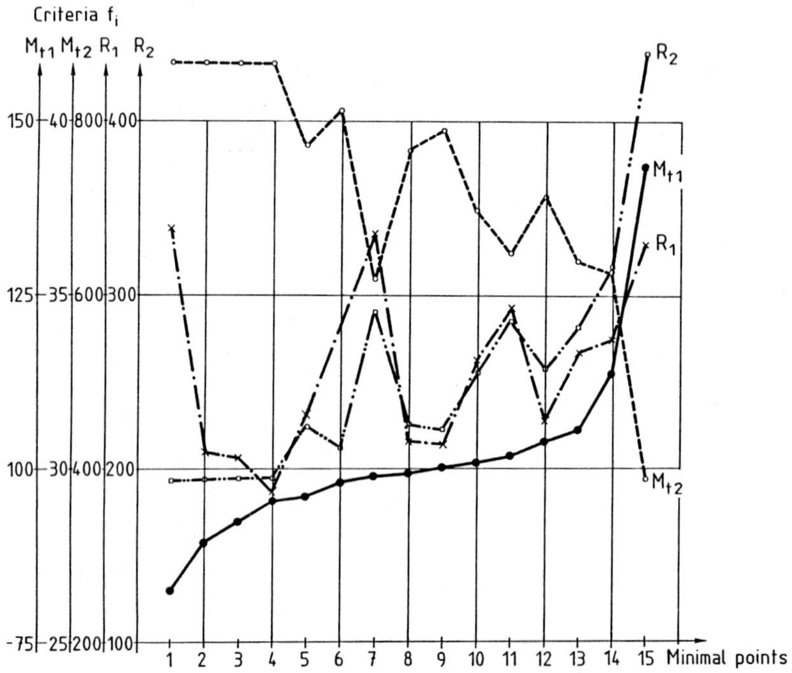

Fig. 5.1/6. Minimal solutions of the multicriteria robot problem

5.1.6 Conclusion

A multicriteria optimization problem for robot arm balancing has been formulated by using dynamic support reactions and torques as criteria. Lengths and masses of balancing mechanisms have been used as design variables, and side constraints have been applied to these variables. The mathematical solution to this problem, called the Pareto-optimal set , is represented by fifteen Pareto-optima obtained by the optimization system CAMOS. These correspond to fifteen different robot designs and points in the criterion space.

These results show that the optimal structures differ considerably depending on the importance given to each criterion. Also, the criterion values vary from one optimal solution to another. This suggests that plenty of Pareto-optima should be available for the designer who can bring in his personal experience in choosing the best available structure. The numerical computation involved is costly and time-consuming. Another problem is the probable existence of local minima in this very nonconvex problem. The results also show that it is difficult to predict the optimal mass distribution for these dynamic criteria without applying the optimization approach.

The design of robot arms is a complex problem including the aspects of strength, flexibility, and vibration. The overall optimization problem is very extensive, and thus it is difficult to solve it numerically even for one Pareto-optimum. Furthermore, the aspects associated with robot control should be considered in the design process. This will unavoidably result in the application of several criteria such as mass, energy consumption, stiffness, controlindex. It seems practical to proceed into a hierarchical optimization process where several subproblems such as dynamic behaviour, structural design, and control design are treated as separate problems. One subproblem is that of balancing; it can be based completely on the rigid-body motion of the arms, thus resulting in a problem of practical size and complexity. Now that a set of optimal alternatives is available to the designer, it is possible to make the structural design of the arms more flexible than if all data values were to be fixed.

The module which generates the Pareto-optima for this dynamic problem is an important part of an interactive design system. This system is under development, and sensitivity information between the subsystems will be needed to create a working computer aided design system for robot arms. In addition to this, problems will also arise in the numerical solution of the min-max-max-max-type problem when more arms are considered and more Pareto-optima are needed.

References

[1] Herve, I.M.: Design of Spring Mechanism for Balancing the Weight of Robots. Proceedings of the 6th Symposium of Theory and Practice of Robots and Manipulators. Cracow, Poland: MIT Press 1986

[2] Subbiah, M.; Sharan, A.M.: The Optimal Balancing of the Robotic Manipulation. IEEE. CH2282-2 (1986) 828-835

[3] Hockey, B.A.: The Minimization of the Fluctuation of Input Shaft Torque in Plane Mechanisms. Mechanisms and Machine Theory Vol. 7 (1972) 335-346

[4] Lee, T.W.; Cheng, C.: Optimum Balancing of Combined Shaking Force, Shaking Moment and Torque Fluctuations in High-Speed Linkages. ASME Paper No. 83-DET-9, Design and Production Engineering. Technical Conference, September 1983

[5] Tricano, S.J.; Lowen, G.G.: Simultaneous Optimization and Dynamic Reactions of Four-Bar Linkage with Prescribed Maximum Shaking Force. ASME Journal of Mechanisms. Transmission and Automation in Design, Vol. 105 (1983) 520-525

[6] Armstrong, B.; Oussama, K; Burdick, J.: The Explicit Dynamic Model and Inertial Parameters of the PUMA 560 Arm. IEEE (1986) 510-521

[7] Osyczka, A.: Computer Aided Multicriterion Optimization System (CAMOS). In: Proceedings of a GAMM-Seminar, October 5-7, 1988, Siegen, FRG. Berlin, Heidelberg, New York, London, Paris, Tokyo 1989

[8] Nelder, J.A.; Mead, R.A.: Simplex Method for Function Minimization. Computer J., Vol. 7 (1965) 308-313

[9] Koski, J.: Bicriterion Optimum Design Method for Elastic Trusses. Dissertation, Acta Polytecnica Scandinavica, Mechanical Engineering Series No. 86, Helsinki 1984

Appendix

The terms appearing in the condensed equation (5.1-7) are:

Coefficients for torque M_{t1}:

$$D_{11} = m_1e_1^2 + m_4x_1^2 + m_2e_2^2 + m_3L_2^2 + m_5x_2^2 + (m_2 + m_3 + m_5)L_1^2 + \\ + 2m_2e_2L_1\cos\Theta_2 + 2m_3L_1L_2\cos\Theta_2 - 2m_5x_2L_1\cos\Theta_2 + J_1 + J_2 ,$$

$$D_{12} = m_2e_2^2 + m_3L_2^2 + m_5x_2^2 + m_2e_2L_1\cos\Theta_2 + m_3L_1L_2\cos\Theta_2 - \\ - m_5x_2L_1\cos\Theta_2 + J_2 ,$$

$$D_{11}^1 = - m_5x_2L_1\left[\sin(\Theta_0 + \Theta_1) + \sin\Theta_2)\right] + m_5x_2^2\sin 2\Theta_0 ,$$

$$D_{12}^2 = - m_2e_2L_1\sin\Theta_2 - m_3L_1L_2\sin\Theta_2 - m_5x_2L_1\sin(\Theta_0 + \Theta_1) + \\ + m_5x_2^2\sin 2\Theta_0 ,$$

$$D_{12}^1 + D_{11}^2 = - 2m_2e_2L_1\sin\Theta_2 - 2m_3L_1L_2\sin\Theta_2 - 2m_5x_2L_1\sin(\Theta_0 + \Theta_1) + \\ + 2m_5x_2^2\sin 2\Theta_0 ,$$

$$D_1 = m_1ge_1\cos\Theta_1 - m_4gx_1\cos\Theta_1 + m_2ge_2\cos\Theta_0 + m_3gL_2\cos\Theta_0 - \\ - m_5gx_2\cos\Theta_0 + (m_2 + m_3 + m_5)gL_1\cos\Theta_1 .$$

Coefficients for torque M_{t2}:

$$D_{21} = m_2(L_1e_2\cos\Theta_2 + e_2^2) + m_3(L_1L_2\cos\Theta_2 + L_2^2) - \\ - m_5(L_1x_2\cos\Theta_2 - x_2^2) + J_2 ,$$

$$D_{22} = m_2e_2^2 + m_3L_2^2 + m_5x_2^2 + J_2 ,$$

$$D_{21}^1 = m_2L_1e_2\sin\Theta_2 + m_3L_1L_2\sin\Theta_2 + m_5(2x_2^2\sin\Theta_0\cos\Theta_0 - L_1x_2\sin\Theta_2) ,$$

$$D_{22}^2 = 2m_5x_2^2\sin\Theta_0\cos\Theta_0 ,$$

$$D_{22}^1 + D_{21}^2 = 4m_5x_2^2\sin\Theta_0\cos\Theta_0 ,$$

$$D_2 = m_2ge_2\cos\Theta_0 + m_3gL_2\cos\Theta_0 - m_5gx_2\cos\Theta_0 ,$$

where J_1 and J_2 are the rotary inertias of members 1 and 2, respectively. Notation $\Theta_0 = \Theta_1 + \Theta_2$ is used for convenience.

5.2 Multicriteria Optimization of Computationally Expensive Functions and its Application to Robot Spring Balancing Mechanism Design

A. Osyczka, J. Zajac

5.2.1 Introduction

Modelling of most engineering problems involves the use of time-consuming complex numerical methods such as simulation techniques, numerical integration, finite element analyses and others. Optimization of problems modelled in this way is often prohibitive considering the amount of computer time required when commonly used methods are applied. Thus recently, great efforts have been made to develop some more effective methods and a first step has been made for solving single criteria optimization models [1-3]. These methods use the approximation of analysis functions by low order polynomials over some small region for reducing the number of function evaluations. The optimization methods involve calculations of a computationally expensive function at a sufficient number of points to construct a low order polynomial approximation. Then, an iterative optimization algorithm is used for finding the minimum of the approximate function. At the point obtained the optimization model is replaced by a new approximate model, and the process continues until an improvement in the objective function cannot be distinguished.

A much more complicated problem occurs while solving multicriteria optimization models because several objective functions are to be minimized and a set of Pareto-optimal solutions is to be found. For multicriteria models the methods mentioned above become ineffective since the number of computationally expensive function evaluations will still be large and it will increase with both the number of objective functions considered in the model and the number of Pareto-optimal solutins to be generated in order to make the right decision. This paper proposes a new method in which the number of function evaluations is correlated to the number of design variables.

The method presented here is applied to the optimum design of a spring balancing mechanism of industrial robots. The optimization problem consists in seeking such design variables as minimize the following objective functions: static torque to drive the arm, work performed by the motor, and static force at the joint. The design variables are: the characteristics of the spring and the location of the spring mechanism in relation to the robot arm. In this model the objective functions not only depend on the decision variables but also on parametric variables such as the position of the robot arms and the mass of the payload. These are not design variables but variables which influence the objective functions.

5.2.2 General Description of the Method

The method is based on a statistical experimental approach [4-6], and its main idea is to substitute computationally expensive functions by approximate functions which reflect the behaviour of real functions. A constrained optimization is then performed on the approximate functions using any suitable optimization technique.

Statistical experimental methods involve the following steps:

- mathematical modelling,
- experimental design,
- statistical analysis of the experiments.

Mathematical modelling

A mathematical model of the process with computationally expensive functions can be written in a general form as presented in Fig. 5.2/1 where:

$\mathbf{x} = (x_1, ..., x_n)^T$ - vector of input variables which represent design variables,

$\mathbf{y} = (y_1, ..., y_m)^T$ - vector of output variables which represent objective functions and/or functions considered as constraints,

$\mathbf{r} = (r_1, ..., r_t)^T$ - vector of random or uncontrollable variables,

$\mathbf{c} = (c_1, ..., c_s)^T$ - vector of constant parameters.

In the problem considered here the output variables $\mathbf{y}$ can be evaluated in terms of input variables $\mathbf{x}$, but the process of evaluation is com-

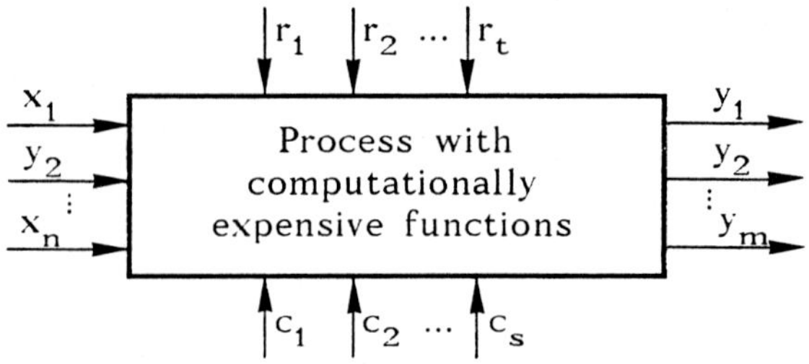

Fig. 5.2/1. General model of the process with computationally expensive functions

putationally expensive. A substitute model of the process can be created and written in the form:

$$\begin{aligned} y_1 &= F_1\,(x_1\,,\dots,\,x_n)\;, \\ y_2 &= F_2\,(x_1\,,\dots,\,x_n)\;, \\ &\vdots \\ &\vdots \\ y_m &= F_m\,(x_1\,,\dots,\,x_n)\;. \end{aligned} \tag{5.2-1}$$

The functions F_j describe the dependence of input and output variables for the given vectors **c** and for the neglected vectors **r**. The choice of the functions F_j approximating the behaviour of the real function is one of the most important parts of modelling and should be based on a theoretical analysis of the process. If this analysis provides no answer, a second order polynomial model is generally used. The frequent use of second order polynomials is justified by the fact that they reflect the nonlinear character of the process and can be effectively applied at the next stage, i.e. at the stage of the experimental design. The approximation model for three input variables would for example have the form:

$$\begin{aligned} y = b_o + b_1\,x_1 + b_2\,x_2 + b_3\,x_3 + b_{11}\,x_1^2 + b_{22}\,x_2^2 + b_{33}\,x_3^2 \,+ \\ + b_{12}\,x_1\,x_2 + b_{13}\,x_1\,x_3 + b_{23}\,x_2\,x_3\;, \end{aligned} \tag{5.2-2}$$

where

b_o estimate of mean value,
b_i estimate of 1st order coefficients,
b_{ii} estimate of 2nd order coefficients,
b_{ij} estimate of 2nd order cross products coefficients.

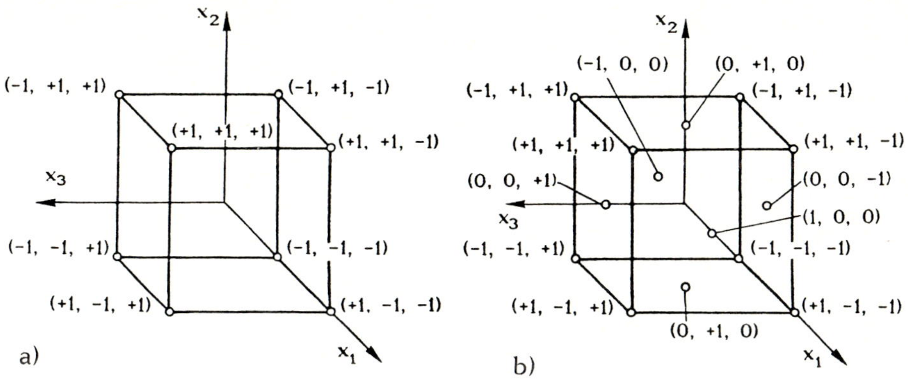

Fig. 5.2/2. Examples of experimental designs

Experimental designs

Experimental designs, also called test plans, specify the set of analysis runs required for input variables. For this set of variables a corresponding set of output variables is evaluated after running a series of experiments. For models with computationally expensive functions this series is simulated on a computer. The selection of a specific experimental design depends on the goals to be achieved in experimentation and on the features of the process. The most common experimental design is a full factorial design as presented for three input variables in Fig. 5.2/2a). This design consists of running all combinations of the variables at their upper and lower bounds denoted by plus and minus one, respectively. Another design, the so-called B_n design [6] is presented in Fig. 5.2/2b). In this case the additional points for running experiments are placed in the middle of the face of a polyhedron, and the zero value refers to the arithmetic mean of upper and lower bounds of the variable.

Statistical analysis of experiments

The statistical analysis of experiments is used for approximating the results of experiments, evaluating the error of this approximation and also evaluating the level of significance and adequacy of the function obtained. The least square method is generally used for function approximation. A less frequently used method is the Chebyshev method [8] which seeks an approximate function which minimizes the maximum error.

Objective functions approximated by using either of these methods are then verified as to their adequacy. This is usually done by means of an F Snedecor test [9] using the statistics called the variance quotient of the form:

$$F = \frac{S^2(y)_2}{S^2(y)_1} , \tag{5.2-3}$$

where $S^2(y)_2$ - variance of approximation errors,
$S^2(y)_1$ - variance of measurement errors.

Then, F is compared with the critical value of the variance quotient F_{α, f_1, f_2} where:

α - level of significance,

f_1, f_2 - number of degrees of freedom referred to the variances $S^2(y)_1$ and $S^2(y)_2$, respectively.

Note that most optimum design models are assumed to be deterministic, even if they are not. For deterministic models the value of the variance $S^2(y)_1$ is equal zero. Thus, the influence of random variables on the process of evaluation of the objective functions has to be considered separately for each problem while estimating the value of $S^2(y)_1$.

It is assumed that the approximate function is adequate if

$$F < F_{\alpha, f_1, f_2} . \tag{5.2-4}$$

In the computer simulation the following formula is generally used to test the adequacy of the function

$$\alpha(F) \geq \alpha^* , \tag{5.2-5}$$

where

$\alpha(F) = \alpha(F,f_1,f_2)$ - level of significance, evaluated on the basis of formula (5.2-3),

α^* - assumed critical value of the level of significance.

The model is adequate if (5.2-5) is satisfied and then well-known multicriteria optimization methods are used for solving this simplified model. In the case that the approximate functions are inadequate, the whole process is reconsidered from the beginning. The main reason why the model may be inadequate is the fact that the approximate functions do not reflect the physical behaviour of the system.

The steps of the method described above are shown in Fig. 5.2/3.

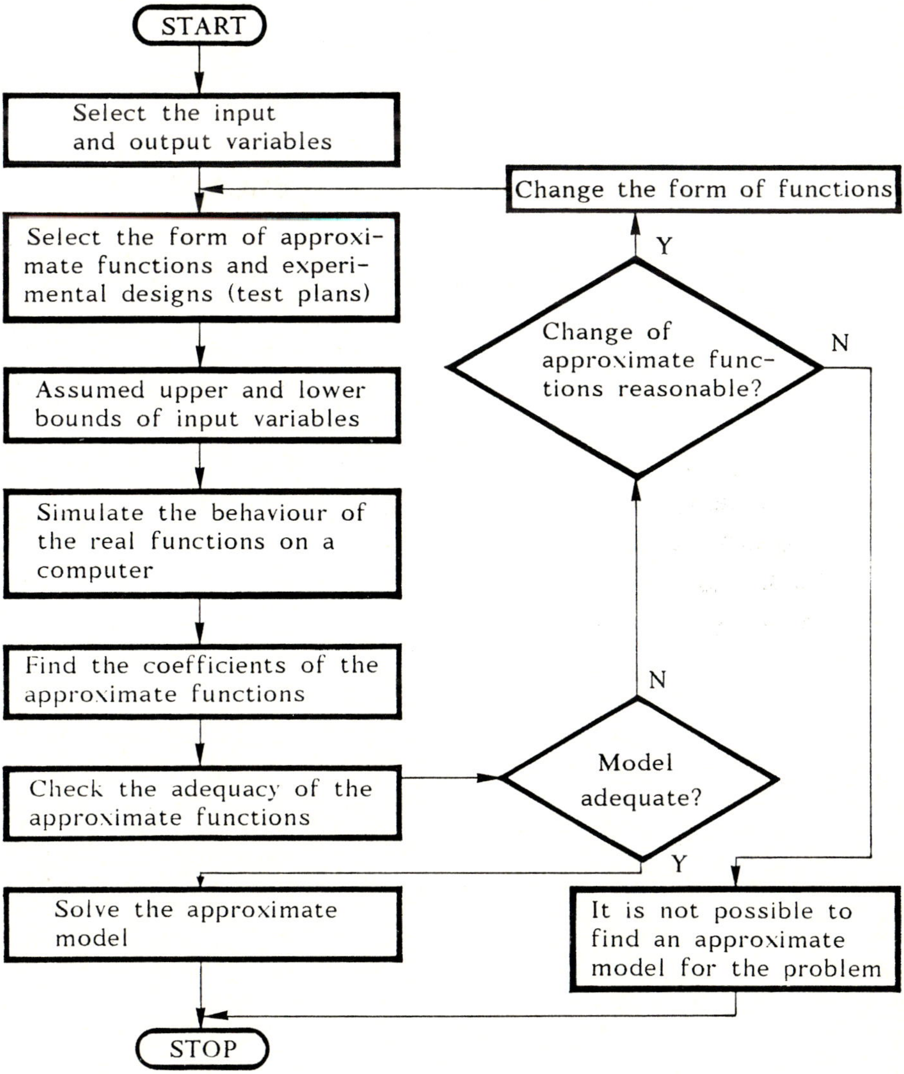

Fig. 5.2/3. Scheme of the method

5.2.3 Optimum Design of Spring Balancing Mechanisms of Industrial Robots

Robot balancing mechanisms are used for decreasing both the load of the arm driving system and the net power required for the robot. There are various ways of balancing robot arms [11]. Here, a spring balancing mechanism has been chosen as the subject of investigation. Models discussed in literature [10-12] for the design and optimization of mechanisms of this kind are simplified and do not consider various configurations the robot can have during the rotation of the arms. The optimization model in which all possible positions of the robot arms are taken into account is the model with computationally expensive functions. The solution of such a model by means of the method described in section 5.2.2 will be discussed here.

a) Formulation of the optimization problem

Consider a rigid body model of the robot as presented in Fig. 5.2/4 in which a hydraulic spring mechanism is used for balancing the arm. For industrial robots of general applications the forearm spring balancing mechanism is rarely used since it does not provide a great balancing effect. The forearm balancing mechanism is sometimes used in robots for special applications [12].

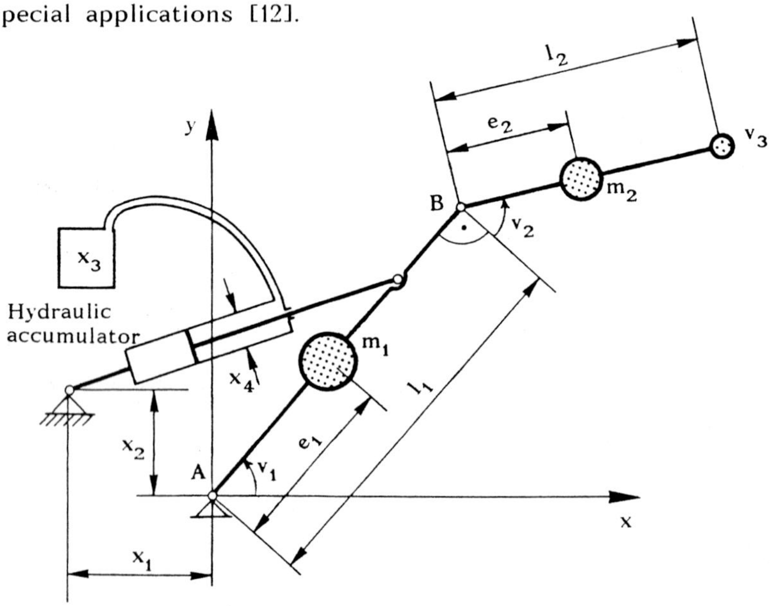

Fig. 5.2/4. Rigid body model of the robot with a hydraulic spring mechanism

The quantities which influence the balancing effect are

- location of the spring mechanism in relation to the robot base,
- spring characteristics.

These quantities form the vector of design variables $\mathbf{x} = (x_1, ..., x_4)^T$ with

x_1, x_2 - coordinates of the mounted end of the spring referred to the joint A,

x_3 - pressure in the hydraulic accumulator,

x_4 - diameter of the hydraulic cylinder.

The constraints imposed on design variables are side constraints of the form:

$$x_i^l \leq x_i \leq x_i^u \quad , \qquad i = 1, ..., 4 \tag{5.2-6}$$

and behaviour constraints which refer to the capacity of the hydraulic accumulator and its strength condition.

A good performance of the robot depends on the following criteria:

- static torque on the shaft of the driving motor,
- work performed by the driving motor,
- static force at the joint A.

These three criteria form a vector function:

$$\mathbf{f}(\mathbf{x},\mathbf{v}) = [f_1(\mathbf{x},\mathbf{v}), f_2(\mathbf{x},\mathbf{v}), f_3(\mathbf{x},\mathbf{v})]^T , \tag{5.2-7}$$

where $\mathbf{v} = (v_1, v_2, v_3)^T$ is a vector of parametric variables which influence the vector function (5.2-7) although they are not design variables. In this model the positions of the arm and the forearm, and the mass of the payload are these parametric variables. The elements of the vector $\mathbf{v}$ vary within certain limits. These variations have to be considered while evaluating the values of the objective functions.

The static torque M_A at the point A is the function of the design variable vector $\mathbf{x}$ and the parametric variable vector $\mathbf{v}$ and can be expressed as:

$$M_A = \vartheta_1 (\mathbf{x},\mathbf{v}) \quad . \tag{5.2-8}$$

Thus, the first objective function has the form:

$$f_1(\mathbf{x}, \mathbf{v}) = \max_{v_i^l \le v_i \le v_i^u} \vartheta_1(\mathbf{x},\mathbf{v}) , \quad i = 1,2,3 , \qquad (5.2\text{-}9)$$

where v_i^l and v_i^u are lower and upper bounds of the parametric variables, respectively.

The second objective function has the form:

$$f_2(\mathbf{x}, \mathbf{v}) = \int_{v_1^l}^{v_1^u} \max_{v_i^l \le v_i \le v_i^u} \vartheta_1(\mathbf{x},\mathbf{v}) , \quad i = 2,3 . \qquad (5.2\text{-}10)$$

The reaction force R at the joint A is also the function of the vectors **x** and **v** which can be expressed as:

$$R = \vartheta_2(\mathbf{x},\mathbf{v}) , \qquad (5.2\text{-}11)$$

and the third objective function has the form:

$$f_3(\mathbf{x},\mathbf{v}) = \max_{v_i^l \le v_i \le v_i^u} \vartheta_2(\mathbf{x},\mathbf{v}) , \quad i = 1,2,3 . \qquad (5.2\text{-}12)$$

All objective functions cannot be expressed in explicit forms. An algorithm for calculating the values of the objective functions for the given vector **x** has to be worked out. This algorithm, however simple, makes the process of evaluation of the objective functions computationally expensive, since one has to consider all possible combinations of elements of the vector **v**, i.e. all possible positions of robot arms and loads.

Note that the situation is similar when a single criterion optimization model is considered. If the first objective function is to be minimized and the second and third functions are considered as constraint functions, the process of evaluation of these functions is still computationally expensive.

The static behaviour of the robot manipulator is considered here. Its dynamic behaviour may only have influence on the constraints imposed on the spring, i.e. the additional constraints referring to the spring stiffness are to be introduced into the optimization model.

b) Solution Method of the Optimization Problem

In order to solve the optimization problem formulated, in a reasonable computing time, a simplified model has been built using the approach described in section 5.2.2.

Quantities which describe the spring balancing mechanism are:

(i) Input variables (design variables) x_1, x_2, x_3, x_4;

(ii) Output variables (objective functions)

$$y_1 = f_1(\mathbf{x}), \quad y_2 = f_2(\mathbf{x}), \quad y_3 = f_3(\mathbf{x}); \tag{5.2-13}$$

(iii) Random variables
- friction in the hydraulic spring,
- changes in the wrist position,
- calculation errors in estimating robot arm masses and their coordinates;

(iv) Constant parameters
- extreme positions of the robot arms,
- maximum payload,
- type and volume of hydraulic accumulator.

Since there is no theoretical evidence as to the form of approximate functions, second order polynomial functions with first order interactions are considered here. Thus,

$$\begin{aligned} y_j = {} & b_o + b_1 x_1 + b_2 x_2 + b_3 x_3 + b_4 x_4 + b_{11} x_1^2 + b_{22} x_2^2 + b_{33} x_3^2 + \\ & + b_{44} x_4^2 + b_{12} x_1 x_2 + b_{13} x_1 x_3 + b_{14} x_1 x_4 + b_{23} x_2 x_3 + \\ & + b_{24} x_2 x_4 + b_{34} x_3 x_4 \end{aligned}$$

with $j = 1,2,3$. (5.2-14)

Among various types of experimental designs which may be applied to this problem, the B_4-design has been used here. For four design variables this design requires 24 experiments (see the example) which are performed by means of computer simulation.

The results of simulation experiments are the basis for approximating objective functions by means of polynomial functions which are then verified as to their adequacy.

At this point, the characteristics of the simulation model of robot manipulation should be explained. For reasons of simplicity, it is usually assumed that the rigid-body model of the robot is of a deterministic type, and in this case simulation experiments for the given values of parameters always provide the same results. This model is used at the initial stage of robot design at which it is not possible to determine robot parameters such as masses of the robot arms and coordinates of their gravity centres with great accuracy. These parameters are estimated only roughly, and mean values are used in the simulation algorithm. Assuming a normal distribution of the parameters with a known mean value, only the variance of the distribution function should be established. This can be done using the following equation:

$$S^2(y) = \left(\frac{\Delta}{3}\right)^2, \qquad (5.2\text{-}15)$$

where Δ is the estimated maximum error of the parameter.

Formula (5.2-15) reflects the ratio of the maximum error to three standard deviations. Knowing the distribution functions of the robot parameters, their values can be generated at random. After making a series of simulation experiments in the centre of the experimental design, the variance of measurement errors can be calculated from the formula:

$$S^2(y)_1 = \frac{\sum_{i=1}^{r} (y^{(i)} - \bar{y}^{(i)})^2}{r-1}, \qquad (5.2\text{-}16)$$

where r is the number of simulation experiments made in the centre of the experimental design.

The optimization model of the spring balancing mechanism obtained in this way consists of objective functions of the form of the second order polynomials with first order interactions (see formula 5.2-14), and any conventional method will solve this model in a short computation time. For the problem discussed, the interactive multicriteria optimization system has been adapted to solve the model [13].

c) Numerical Example

Consider the problem of designing a hydraulic spring balancing mechanism of a robot as presented in Fig. 5.2/4 for the following data:

mass of the arm	m_1 = 254.4 kg,
mass of the forearm	m_2 = 105.3 kg,
maximum mass of payload	v_3^u = 80 kg,
length of the arm	l_1 = 1.0 m,
length of the forearm	l_2 = 1.35 m,
arm gravity centre distance	e_1 = 0.52 m,
forearm gravity centre distance	e_2 = 0.53 m,
number of springs:	2 .

The lower and upper bounds of the parametric variables are:

$$6 \le v_1 \le 120, \qquad -30 \le v_2 \le 150, \qquad 0 \le v_3 \le 80 .$$

The values of v_1 and v_2 are given in deg, and v_3 in kg. The assumed lower and upper bounds of the decision variables are:

$$-0.01 \le x_1 \le 0.01 , \qquad 0.26 \le x_2 \le 0.34 ,$$
$$3.0 \le x_3 \le 3.6, \qquad 0.034 \le x_4 \le 0.04 .$$

The values for x_1, x_2, and x_4 are given in m, and x_3 in MPa.

For B_4 experimental design, the list of required experiments with coded and natural input variables is presented in Table 5.2/1. The last three columns of Table 5.2/1 contain the responses, i.e. the results of computer evaluations of the output variables at the given points. The coefficients of approximate functions have been obtained by using the Chebyshev method. These coefficients for the coded input variables are presented in Table 5.2/2. In this way the optimization model with computationally expensive objective functions is replaced by a simple one which is solved in short computation time. For this example the computation time for seeking one Pareto-optimal solution for the simplified model was almost equal to the evaluation time of an objective function at a given point.

The results of seeking some of the Pareto-optimal solutions are presented in Table 5.2/3. The weights given in this table refer to the weighting min-max method (see 1-22) used for solving the multicriteria model.

No of exper	Coded input variables				Natural input variables				Response output variables		
	x_1	x_2	x_3	x_4	x_1	x_2	x_3	x_4	y_1	y_2	y_3
1	1	1	1	1	.01	.34	3.6	.040	3546.7	16652.3	4405.4
2	−1	1	1	1	−.01	.34	3.6	.040	3684.1	17443.4	4552.4
3	1	−1	1	1	.01	.26	3.6	.040	1960.5	13648.9	3042.8
4	−1	−1	1	1	−.01	.26	3.6	.040	1727.8	14200.7	2736.7
5	1	1	−1	1	.01	.34	3.0	.040	2606.3	13788.4	3510.8
6	−1	1	−1	1	−.01	.34	3.0	.040	2720.9	14447.8	3611.5
7	1	−1	−1	1	.01	.26	3.0	.040	1828.5	11856.7	3433.9
8	−1	−1	−1	1	−.01	.26	3.0	.040	1746.0	11884.0	3168.5
9	1	1	1	−1	.01	.34	3.6	.034	2053.8	10851.3	3034.0
10	−1	1	1	−1	−.01	.34	3.6	.034	1886.5	10851.3	2824.2
11	1	−1	1	−1	.01	.26	3.6	.034	2452.1	10851.8	3895.2
12	−1	−1	1	−1	−.01	.26	3.6	.034	2438.7	10851.8	3752.3
13	1	1	−1	−1	.01	.34	3.0	.034	1977.5	9761.6	3456.0
14	−1	1	−1	−1	−.01	.34	3.0	.034	1954.6	9761.6	3205.6
15	1	−1	−1	−1	.01	.26	3.0	.034	2827.7	9762.0	4270.2
16	−1	−1	−1	−1	−.01	.26	3.0	.034	2816.5	9762.0	4169.2
17	1	0	0	0	.01	.30	3.3	.037	1880.4	11411.5	3178.2
18	−1	0	0	0	−.01	.30	3.3	.037	1694.0	11411.5	2857.2
19	0	1	0	0	.00	.34	3.3	.037	2154.7	11685.6	3053.5
20	0	−1	0	0	.00	.26	3.3	.037	2084.2	11409.8	3528.7
21	0	0	1	0	.00	.30	3.6	.037	1935.6	12054.9	2813.7
22	0	0	−1	0	.00	.30	3.0	.037	1785.7	10764.7	3248.1
23	0	0	0	1	.00	.30	3.3	.040	2087.1	13966.0	3066.5
24	0	0	0	−1	.00	.30	3.3	.034	2189.5	10305.7	3526.9

Table 5.2/1. Computer simulation results for B_4 experimental design

	b_0	b_1	b_2	b_3	b_4	b_{11}	b_{22}	b_{33}
y_1	180.0	10.712	160.64	78.673	74.202	29.192	318.52	-62.009
y_2	11399	-137.94	474.24	948.49	1943.7	210.82	148.69	-22.185
y_3	3019.7	19.738	10.149	-97.091	0.0522	105.07	271.43	-116.41
	b_{44}	b_{12}	b_{13}	b_{14}	b_{23}	b_{24}	b_{34}	—
y_1	337.37	-50.253	57.707	-2.3589	140. 14	501.70	195.12	—
y_2	514.03	-59.821	-59.820	-134.94	-0.0116	750.96	343.78	—
y_3	259.58	-91.490	-11.609	9.8353	102.96	470.53	205.27	—

Table 5.2/2. Polynomial coefficients for coded input variables

No	$\mathbf{w} = (w_1,w_2,w_3)^T$	$\mathbf{x} = (x_1, x_2, x_3, x_4)^T$	$f(\mathbf{x}) = (f_1(\mathbf{x}), f_2(\mathbf{x}), f_3(\mathbf{x}))^T$
1	min $f_1(\mathbf{x})$	$(0.0054, 0.2872, 3.00, 0.0382)^T$	$(1618.9, 10949.9, 3021.3)^T$
2	min $f_2(\mathbf{x})$	$(-0.001, 0.3371, 3.00, 0.0349)^T$	$(1947.3, 9213.8, 3176.7)^T$
3	min $f_3(\mathbf{x})$	$(0.0032, 0.3284, 3.59, 0.0340)^T$	$(1918.2, 10419.0, 2733.0)^T$
4	$(0.33, 0.33, 0.33)^T$	$(0.0000, 0.3079, 3.00, 0.0364)^T$	$(1694.5, 10199.5, 3025.3)^T$
5	$(0.25, 0.25, 0.50)^T$	$(0.0000, 0.2984, 3.00, 0.0376)^T$	$(1654.3, 10784.1, 2965,9)^T$
6	$(0.33, 0.0, 0.67)^T$	$(-0.0017, 0.2984, 3.00, 0.0376)^T$	$(1707.7, 12072.0, 2807.9)^T$
7	$(0.67, 0.0, 0.33)^T$	$(-0.0099, 0.2774, 3.59, 0.0370)^T$	$(1681.6, 12514.2, 2872.9)^T$

Table 5.2/3. Optimization results for spring balancing mechanism

5.2.4 Computer Aided Optimum Design of Robot Spring Balancing Mechanism

On the basis of the method presented in sections 5.2.2 and 5.2.3 a computer program for the design automation and optimization of the robot spring balancing mechanism has been developed. The program is coded in FORTRAN 77 and implemented on an IBM PC/XT/AT computer. The program is fully interactive, and it is equipped with a graphic menu. There are two types of graphic menu. The first type refers to the selection of the path from the list displayed on the screen using cursors only. The second type is used for introducing the data for calculations. Both, verbal notation and graphical illustration are used for describing each quantity to be introduced. The main menu contains the following parts:

- selection of the type of mechanism,
- introduction/modification of the data,
- selection of the approximation method,
- processing the calculations,
- printing the results.

The decision phase, i.e. the selection of the Pareto-optimal solution is also supported by a graphical output on the screen or plotter. For the optimal solution the designer may get a view graph of the balancing mechanism for all positions of the robot arms.

The program can be used for solving both single and multicriteria optimization models for designing hydraulic and mechanical spring balancing mechanisms of the arm of an industrial robot. While the program is running, a great amount of data is processed. Thus, the most important data is shown on the screen whereas all data referring to the calculations processed by the computer are recorded on a disk file. The program contains an option "view of the output data" which enables the designer to look through the "history" of the calculations.

5.2.5 Conclusion

The problem of multicriteria optimization with computationally expensive functions is quite common for design procedures. In this chapter an approximate analysis model which provides explicit expressions of the behaviour of the computationally expensive functions in terms of the design variables has been presented. Such an approach may improve the efficiency of the calculations. Moreover, once the explicit model has been established, it can be used for efficient multiple reanalyses. This becomes important for the application of optimization methods to CAD procedures. These procedures are interactive, and thus none of the methods requiring long computations can be implemented. The method proposed provides the basis for an efficient computer procedure for designing the spring balancing mechanism of industrial robots which has a computer aided optimum design form.

On the other hand, the method presented here will usually be less accurate than other conventional methods and can handle the optimization problems with a only few design variables. The nature of the statistical experimental method causes these disadvantages. Although the method is less accurate, it will locate the solution near the global minimum. Thus, it can also be used for quick evaluations of a good starting point for conventional methods. In the case of multicriteria optimization, each objective function will have a different starting point. The method will find these points in one series of experiments. Due to the reason mentioned above, further investigations and testing are planned.

References

[1] Kardis, J.P.; Turns, S.R.: "Efficient Optimization of Computationally Expensive Objective Functions", Proceedings of ASME, 18th Mechanisms Conference, 10th Design Automation Conference, Oct. 1984

[2] Bryce, G.R.; Collette, D.R.: "Statistical Design Experiments for Process Optimization", Microelectronic Manufacturing and Testing, Oct. 1984

[3] Fee, I.W; Parkinson A.R.; Bryce, G.R.; Balling, R.I.: "Approximation of Computationally Expensive and Noisy Functions for Constrained Nonlinear Optimization", Transactions of ASME, Journal of Mechanisms, Transmissions and Automation in Design, Vol. 109, (1987) 528-532

[4] Myers, R.H.: "Response Surface Methodology". Boston: Allyn and Bacon Inc. 1971

[5] Akhnazarova, S., Kafarov, V.: "Experiment Design and Optimization in Chemistry and Chemical Engineering". Moscow: Mir Publishers, 1982

[6] Polański, Z.: Planowanie doświadczeń w technice (Experiment Design in Engineering). Warsaw: PWN 1984

[7] Diamond, W.J.: Practical Experiment Design. Belmont, California: Lifetime Learning Publications 1981

[8] Damels, R.W.: An Introduction to Numerical Methods and Optimization Techniques. New York: North-Holland 1978

[9] Snedecor, G.W., Cochran, T.E.: Statistical Methods. 6th Edition. Iowa State University Press, 1967

[10] Herve, J.M.: Design of Spring Mechanism for Balancing the Weight of Robots. Proceedings of the 6th Symposium on Theory and Practice of Robots and Manipulators. Cracow, Poland: MIT Press 1986

[11] Volmer, I. (ed.): Industrieroboter. Berlin: VEb Verlag Technik 1983

[12] Subbiah, M; Sharan, A.M.: The Optimal Balancing of the Robotic Manipulators. IEEE, CH 2282-2/86, 828-835

[13] Osyczka, A.: Multicriterion Optimization in Engineering with FORTRAN Programs. Chichester: John Wiley & Sons 1984

5.3 On the Optimal Synthesis of a Automotive Drive Train

F. Pfeiffer

5.3.1 The Automotive Drive Train

An automotive drive train usually comprises components like motor, disk flywheel, transmission coupling, transmission gears, elastic coupling, damper, Cardan shaft, differential gear, and the driving shafts with the wheels. The vibration behaviour of such a drive system is essentially influenced by the following external and internal excitation sources:

- The gas forces as produced within the motor act as the most penetrating excitation for the drive system. They represent a type of external excitation and depend on the engine design (number of cylinders).
- The time-variant moment of inertia of the crankshaft with all cylinders and connecting rods leads to a parametric excitation.
- The time-variant behaviour of stiffness and damping produces parametric excitation in all meshes of gears. Moreover, small errors in profile grinding of the gear teeth may lead to additional internal excitation. [3].
- Errors of Cardan joints produce parametric excitation.

The character of drive train vibrations is generally nonlinear due to the nonlinear characteristics of all joints and couplings. However, we will see later that a linear approximation gives useful results for a large class of practical applications.

One important goal of any drive train design is to keep vibrations as small as possible. As vibrations in machine dynamics always represent a certain form of unwanted and vagabond energy the design requirements are twofold. Firstly, excitations have to be suppressed wherever possible. Secondly, vibration energy must be dissipated as efficiently as possible. Both steps depend on all design parameters of the drive system, which leads directly to an optimization problem of considerable complexity.

In order to establish an approximate solution, the drive train system is reduced to the minimum number of degrees of freedom without loosing its important physical properties. Furthermore, the system is linearized with reference to a chosen working state. In order to cut down the number of design parameters, a sensitivity analysis is performed which allows a selection of the most important parameters with regard to some multicriteria functions. The reduced drive train model, together with the reduced set of design parameters to be optimized, forms the basis for an optimization process yielding Pareto-optimal solutions in the sense of Section 1.2.2.

5.3.2 The Mechanical Model

A typical drive train system is shown in Figure 5.3/1. A motion of the car at a constant engine speed Ω is considered, where the motion within the drive train system might possess either a pushing or a pulling character. A mechanical multibody-system model with all possible Degrees Of Freedom (DOF) would result in a description with more than 100 DOF which is too large to serve as a basis for a multicriteria optimization. For model reduction extensive investigation with a large model (~100 DOF) and a reduced model (24 DOF) has been performed with the following results [1]:

- Low frequency excitations penetrate the complete drive train system. The most important source stems from the engine's gas force.
- High frequency excitations act more or less locally. Their influence disappeares very quickly with the increasing distance from the source. (parametric excitation in all gears).
- All other excitations like time-variant crankshaft properties as well as all nonlinearities in the joints have only a very small influence on vibration amplitudes and thus on vibration energies.

With these results the mechanical model can be reduced to $f = 24$ rational degrees of freedom (Fig. 5.3/1). Additionally, the mathematical model can be established in a linearized form. Only the motor excitation was taken into consideration and represented by four harmonics.

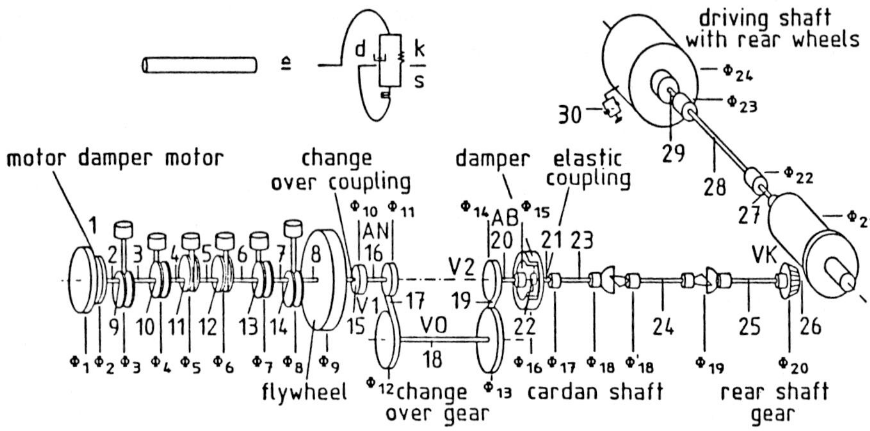

Fig. 5.3/1. The drive train of a car

The reduction process of the mechanical model results in saving about 98% of the computing time as compared with that of a complete nonlinear calculation. The equations of motion of the linear system are

$$\mathbf{M}\ddot{\mathbf{q}} + \mathbf{D}\dot{\mathbf{q}} + \mathbf{K}\mathbf{q} = \mathbf{r}\ . \tag{5.3-1}$$

where

$$\mathbf{r} = \frac{1}{2}\mathbf{r}_0 + \sum_{k \in K} (\mathbf{r}_k e^{\frac{1}{2}ik\Omega} + \bar{\mathbf{r}}_k e^{-\frac{1}{2}ik\Omega t})\ ,$$

$$\mathbf{r}_k = \frac{1}{2}(\mathbf{r}_{k1} + i\mathbf{r}_{k2}) \qquad (i = \sqrt{-1})\ .$$

The vector $\mathbf{q} \in \mathbb{R}^f$ contains all n rational degrees of freedom. **M**, **D**, and **K** (all in $\mathbb{R}^{f,f}$) are the mass, damping and stiffness matrices, respectively , and $\mathbf{r} = \mathbf{r}(t) \in \mathbb{R}^f$ is the excitation vector. The vector $\bar{\mathbf{r}}_k$ is the conjugate complex of $\mathbf{r}_k$, Ω the rotating speed of the engine, t the time, and $K := \{3,6,9,12\}$ for a six-cylinder engine. The solution of the system then reads as follows:

$$\mathbf{q}(t) = \frac{1}{2}\mathbf{q}_0 + \sum_{k \in K} (\mathbf{q}_k e^{\frac{1}{2}ik\Omega t} + \bar{\mathbf{q}}_k e^{-\frac{1}{2}ik\Omega t})\ ,$$

$$\mathbf{q}_k = \left[-\left(\frac{k^2\Omega^2}{4}\right)\mathbf{M} + i\left(\frac{k\Omega}{2}\right)\mathbf{D} + \mathbf{K}\right]^{-1}\mathbf{r}_k = \tag{5.3-2}$$

$$= \frac{1}{2}(\mathbf{q}_{k1} - i\mathbf{q}_{k2})\ .$$

System (5.3-1) with the solution (5.3-2) represents the mechanical analysis model for all further optimization steps.

5.3.3 The Optimization Model

The aim of the optimization consists in reducing noise and vibration within the drive train for a given motor excitation. The main emphasis will be laid on three aspects: a low noise level, small vibration amplitudes of the flywheel and the rear wheel, and a low rattling noise level in the transmission gears. These properties must be attained for a driving speed range from Ω_1 = 750 rpm to Ω_2 = 5750 rpm, where the necessary averaging follows the formula:

$$f_j = \frac{1}{f_{jR}} \left(\frac{1}{\Omega_2 - \Omega_1} \int_{\Omega_1}^{\Omega_2} F_j \cdot d\Omega \right) \qquad (5.3\text{-}3)$$

with f_{jR} = reference value for the standard set of design variables, j = 1,2,3.

Considering a low noise level, the noise in a drive train system may be represented approximately by the total kinetic energy as the sum of all f degrees of freedom and of all excitation modes K, where K := {3, 6, 9, 12} for a six-cylinder engine which is defined as the first part of the objective

$$F_1 = \Omega^2 \sum_{m=1}^{24} \Phi_m \sum_{k \in K} k^2 (a_{km}^2 + b_{km}^2) , \qquad (5.3\text{–}4)$$

and the a_{km} and b_{km} are the Fourier-coefficients of the series expansion for the angular degree of freedom Φ_m. They can be derived from equations (5.3-2) [1]. The sum m includes all n = 24 rational degrees of freedom.

The vibrations of the flywheel and the rear wheels can be measured by the maximum mean amplitudes of the corresponding angular degrees of freedom (Fig. 5.3/1). The following sum leads to the second part of the objective:

$$F_2 = \hat{\Phi}_9 + \hat{\Phi}_{24} , \qquad (5.3\text{-}5)$$

where

$$\hat{\Phi}_m := \sqrt{\sum_{k \in K} (a_{km}^2 + b_{km}^2)} \quad . \qquad (5.3\text{-}6)$$

Rattling in transmission gears is produced in those gears of the transmission which run loosely in the mesh without a load. For a given excitation from the motor, the rattling noise essentially depends on the angular vibrations of the countershaft [4]. Therefore, its mean amplitudes should be as small as possible. According to Fig. 5.3/1 this yields the third part of the objective:

$$F_3 = \hat{\Phi}_{12} + \hat{\Phi}_{13} \; . \tag{5.3-7}$$

In order to obtain the dimensionless objective function, f_j is divided by a reference value f_{jR} (5.3-3) which is evaluated for a standard set $\mathbf{x}_0 \in \mathbb{R}^{n_0}$ of design variables. In our special case, this set $\mathbf{x}_0$ is determined by the design of an existing car. For the purpose of evaluating a better combination of design variables, it is usually impossible to modify a given set $\mathbf{x}_0$ in an arbitrary way. One is constrained by the condition $\mathbf{x}_i^l \le \mathbf{x}_i \le \mathbf{x}_i^u$ which may also be expressed as inequality constraints

$$\mathbf{g}(\mathbf{x}) \le 0 \quad \text{with} \quad \mathbf{g} \in \mathbb{R}^{2n_0}, \quad \mathbf{x} \in \mathbb{R}^{n_0} \; ,$$

$$\mathbf{g}_i = \mathbf{x}_i^l - \mathbf{x}_i \; , \qquad \mathbf{g}_{i+n_0} = \mathbf{x}_i - \mathbf{x}_i^u \; . \tag{5.3-8}$$

As in most cases of machine design, the deviations from a standard reference level must be small due to principal design considerations. In these cases, one deals with a linear synthesis problem [2]. With equations (5.3-3) to (5.3-8) and with respect to (1-8), the following multicriteria problem is obtained:

$$\min_{\mathbf{x} \in \mathbb{R}^{n_0}} \{\mathbf{f}(\mathbf{x}) : \mathbf{g}(\mathbf{x}) \le \mathbf{0}\} \qquad \mathbf{f} \in \mathbb{R}^3, \; \mathbf{g} \in \mathbb{R}^{2n} \; . \tag{5.3-9}$$

5.3.4 The Solution Procedure

A complete amplitude-frequency-analysis dealing with the vibration problems yields only one point of the design variable space. In our case it is even worse. In order to evaluate one point of the design variable space the complete dynamics for the given speed range must be evaluated according to (5.3-3). In the first step the basic mechanical model has already been reduced. In the second step the large number of design variables must be reduced to a reasonable amount. This can be achieved by performing a sensitivity analysis. If a slightly modified sensitivity matrix is defined as

$$\mathbf{S} := \left(\frac{\partial \mathbf{f}}{\partial \mathbf{x}}\right) \operatorname{diag}\,(\mathbf{x}^u - \mathbf{x}^l), \quad \mathbf{S} \in \mathbb{R}^{3,n} \quad , \tag{5.3-10}$$

then a combined measure of sensitivity $\frac{\partial \mathbf{f}}{\partial \mathbf{x}}$ and of the possible range of design variable modifications $(\mathbf{x}^u - \mathbf{x}^l)$, the latter having a higher practical relevancy, can be obtained. Equation (5.3-10) allows us to select those design variables which have the most significant influence on the objective functions. These design variables might be modified to a large extent, or they can possess both of these properties. All design variables with any $S_{ji} < \varepsilon_i$ will be deleted. They do not enter into the optimization process which again results in considerable savings of computing time. The reduction process may be expressed formally as follows:

$$\text{If} \quad (\max S_{ji} \le \varepsilon_i) \quad \text{then} \quad (x_i = x_{0i}), \quad (i = 1,2,\ldots,n_0) \ ,$$

$$(\mathbf{x}_0 \in \mathbb{R}^{n_0}) \longrightarrow (\mathbf{x} \in \mathbb{R}^{n}) \ , \quad \text{where } n < n_0 .$$

The multicriteria optimization problem (5.3-9) is further simplified by choosing the method of objective weighting in a modified form, according to equations (1-11), (1-13b):

$$p[\mathbf{f}(\mathbf{x})] = \sqrt{\sum_{j-1}^{3} w_j \,[f_j(\mathbf{x})]^2} \ , \tag{5.3-12}$$

$$\sum_{j=1}^{3} w_j = 1, \quad 0 \le w_j \le 1 \ .$$

By using these equations, the multicriteria optimization problem (5.3-9) can be written in the simpler form (1-10):

$$\min_{\mathbf{x} \in \mathbb{R}^{n'}} \{p[\mathbf{f}(\mathbf{x})] : \mathbf{g}(\mathbf{x}) \le \mathbf{0}\} \ . \tag{5.3-13}$$

According to section 1.2.2, a min-max formulation represents the best compromise of the optimal design variables $\mathbf{x}^*$ with respect to the objective functions $\mathbf{f}(\mathbf{x}^*)$.

In our case, one can define acccording to (1-13a)

$$\max f_i(\mathbf{x}^*) \longrightarrow \min \ . \tag{5.3-14}$$

As an optimization algorithm for this example a gradient method was used. The preference function was applied as a strategy for obtaining the geometrical properties of the functional efficient surface [1].

5.3.5 Results

The optimization method presented above has been introduced for the drive train of a German car, the design variables of which were taken as the reference or starting vector $x_0 \in \mathbb{R}^{n_0}$ with $n_0 = 66$. The mechanical model of the system (Fig. 5.3/1) was reduced to 24 rational degrees of freedom; only the excitation by motor imbalance was taken into consideration by using an approach with four harmonics. In section 5.3.3, the three objective functions were defined ((5.3-3) to (5.3-7)) as measures of noise, of vibrations of the flywheel and the rear wheels, and of rattling. The allowed design tolerances were given in accordance with the principal design criteria for the car under consideration.

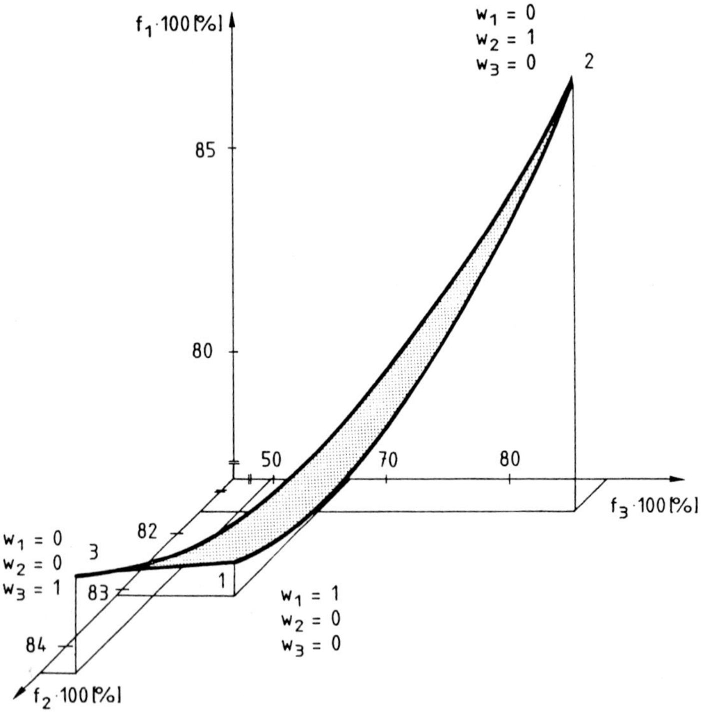

Fig. 5.3/2. Functional-efficient surface

Applying the sensitivity analysis as specified in Section 5.3.4, one could reduce the number of design variables from n = 66 to n = 18 (5.3-11). Values of ε_i, equal to 0.05, were chosen for all i = 1, ... ,n. The most important of the remaining design variables are (in the order of maximum sensitivity (5.3-10)) the flywheel dimension, transmission coupling dimensions, motor damper size, and the elastic coupling dimensions. All other components of the drive train have minor influence on the chosen criteria.

On the basis of the reduced mechanical model and of the reduced number of design variables, the optimization procedure defined by (5.3-13) has been evaluated, yielding one optimal point in the criterion space for each combination of weighting coefficients w_j (5.3-12). Each of these optimal points comprise a set of optimal design variables x^*. All these optimal points form the functional-efficient surface in the criterion space which is shown in Fig. 5.3/2. In order to determine the best compromise, (5.3-14) is applied [1]. For the given data set the best compromise was achieved for the weighting coefficients w_1 = 0.108 (noise), w_2 = 0 (vibrations flywheel, rear wheels), w_3 = 0.892 (rattling). The properties of the original system could be improved by approximately 18% with regard to the objective functions.

5.3.6 Conclusion

In this chapter a method for the optimization of the vibrational behaviour of an automotive drive train was presented. When dealing with a dynamical problem of this type, it is necessary to reduce the mechanical model and the number of design variables as far as possible. The mechanical model is cut down to 24 rational degrees of freedom, and the system is linearized. A comparison with a nonlinear model having nearly 100 degrees of freedom does not show any significant loss of information. The number of design variables is reduced from 66 to 18 by applying a sensitivity analysis and by retaining only those variables of which the sensitivity is more than 5% with regard to the substitute scalar objective function (5.3-14). These objective functions serve as measures for noise, flywheel and rear wheel vibrations, and rattling. A Pareto-optimal solution is determined on the basis of the reduced mechanical model and the reduced number of design variables. Considering the three objective functions, an improvement of about 18%, compared with the original system, was achieved as the best compromise solution.

References

[1] Gosdin, M.: Analyse und Optimierung des dynamischen Verhaltens eines Pkw-Antriebsstranges. Fortschritt-Berichte VDI. Reihe 11, Nr. 69. Düsseldorf: VDI-Verlag GmbH 1985

[2] Gosdin, M.; Pfeiffer, F.: Über eine optimale Synthese von Schwingungssystemen. ZAMM 67, Nr.5 (1987) T 401 - T 402

[3] Kücükay, F.: Dynamik der Zahnradgetriebe, Modelle, Verfahren, Verhalten. Berlin, Heidelberg, New York: Springer Verlag 1987

[4] Pfeiffer, F.; Kücükay, F.: Eine erweiterte mechanische Stoßtheorie und ihre Anwendung in der Getriebedynamik. VDI-Z Bd. 127 (1985) Nr. 9, 341-349

5.4 Modelling of Multibody Systems by Means of Optimization Procedures

H.H. Müller-Slany

5.4.1 Adapted Models for Elasto-Mechanical Systems

Dynamic analysis and simulation of elasto-mechanical systems require, for many applications, simplified physical models. In general, these can be ascertained by a reduction of the number of degrees of freedom. For complex simulation problems such as the design of control systems or the modelling of large-scale systems, these substitute systems, e.g. multibody systems, represent the dynamic behaviour of real elasto-mechanical subsystems. The advantage of reducing the degrees of freedom is that the system dynamics can be clearly presented and carried out by mini-computer systems. Such models are thus the basis for the design or variation of kinematic constraints in complex systems.

The modelling of complex elasto-mechanical systems is only reasonable when the essential properties of the condensed model are identical with those of the real system. In order to establish lumped mass systems the distributed mass and stiffness properties of the real system are substituted by concentrated parameters regarding the linear discretization rule. This leads to a multibody system which is of limited use for simulation problems because the original dynamics of the real system is not sufficiently modelled. Thus, heuristic modelling is replaced by an adapted model which is a multibody system that allows one to adapt the dynamic properties to those of the real system by means of optimization algorithms. This very accurate model is of low order and thus appropriate for any kind of simulation.

The properties to be adapted for representing the dynamics of a real system are provided by the mathematical model for free or forced vibrations. The equation of motion for the undamped, free or forced mechanical system reads:

$$\mathbf{M}\ddot{\mathbf{q}} + \mathbf{K}\mathbf{q} = \begin{cases} \mathbf{0} \\ \mathbf{f}(x) \end{cases} , \tag{5.4-1}$$

with the position vector $\mathbf{q} \in \mathbb{R}^f$, the mass matrix $\mathbf{M}$, the stiffness matrix $\mathbf{K}$ (all in $\mathbb{R}^{f \times f}$, f = number of the degrees of freedom). For complex systems one obtains a differential equation of high order n=2f; for continuous systems f becomes infinite.

The free and forced vibrations of the system can be described in terms of free vibration modes. The eigenvalue problem

$$[-\omega_i^2 \mathbf{M} + \mathbf{K}]\, \boldsymbol{\varphi}_i = \mathbf{0} \ , \quad i = 1,\dots,f \tag{5.4-2}$$

with the eigenfrequencies ω_i and the eigenmodes $\boldsymbol{\varphi}_i$ yields the modal matrix $\boldsymbol{\Phi} = [\ \boldsymbol{\varphi}_i,\dots,\boldsymbol{\varphi}_f]$. If the modal coordinate ξ_i, the generalized mass m_i as well as the stiffness k_i are introduced into (5.4-1):

$$\begin{aligned} \xi_i &= (\boldsymbol{\Phi}^{-1}\,\mathbf{q})_i \ , \\ m_i &= \boldsymbol{\varphi}_i^{\,T} \mathbf{M}\, \boldsymbol{\varphi}_i \ , \\ k_i &= \omega_i^2\, m_i = \boldsymbol{\varphi}_i^T\, \mathbf{K}\, \boldsymbol{\varphi}_i \ , \quad i = 1,\dots,f \ , \end{aligned} \tag{5.4-3}$$

f decoupled differential equations are required to describe the free and forced vibrations:

$$\ddot{\xi}_i + \omega_i^2\, \xi_i = \begin{cases} 0, \\ \dfrac{1}{m_i}\,(\boldsymbol{\Phi}^T \mathbf{f}(\mathbf{x}))_i \ , \quad i = 1,\dots,f \ . \end{cases} \tag{5.4-4}$$

For practical modelling it is sufficient to consider only the lower modal equations of (5.4-4). This is due to the fact that higher eigenvibrations with correspondingly high eigenfrequencies need a great amount of energy for excitation. Furthermore, the higher eigenmodes of structures made of conventional material damp out faster than the lower ones. There are no definite rules for the selection of the necessary number of eigenmodes. Here, the engineer has to decide with regard to the particular problem.

The physical modelling of a complex, elasto-mechanical system by means of a simplified but a highly accurate substitute model of low order requires the correlation of the first eigenfrequencies and eigenmodes. If the physical model also describes internal and external forces which influence the real system, all mass-specific system quantities such as mass of the system, elements of the inertia tensor, and the position of the centre of mass have to be described accurately as well.

The adapted model for a complex elasto-mechanical system is a simple but a highly accurate physically substitute system whose mass-specific and modal parameters correspond to those of the real system. It is only a residual error (Table 5.4/1) which remains.

Property	Real system	Adapted model
Total mass	M^R	$M^R \approx M^M = \sum_i m_i$
Centre of mass	$\mathbf{r}_S^R$	$\mathbf{r}_S^R \approx \mathbf{r}_S^M = [r_x, r_y, r_z]^T$
Inertia tensor	$\boldsymbol{\Theta}_S^R$	$\boldsymbol{\Theta}_S^R \approx \boldsymbol{\Theta}_S^M = \begin{matrix} \Theta_{xx} & 0 & 0 \\ 0 & \Theta_{yy} & 0 \\ 0 & 0 & \Theta_{zz} \end{matrix}$
Eigenfrequency	ω_i^R, i=1,...,p	$\omega_j^R \approx \omega_j^M$, j=1,...,f, f<p
Eigenmode	$\boldsymbol{\varphi}_i^R$, i=1,...,p	$\boldsymbol{\varphi}_j^R \approx \boldsymbol{\varphi}_j^M = [x_1, y_1, z_1, \ldots, x_n, y_n, z_n]_j^T$ j=1,...,f, f<p

Table 5.4/1. Dynamic properties for the adapted model (R = real system, M = model)

In this context it is very important that even experimental system data can be implemented into the adapted model. Thus, a model is available for simulation problems which are difficult to describe analytically. For system maintenance, the implementation of measurement results in an adaptation process makes it possible to identify changes of parameters due to machine damage very early [1].

It is very important for control systems that the eigenmodes of the model correspond to the free vibrations of the real system. By this method it is possible to identify the correct state of motion for given system points by means of sensors. With the small number of degrees of freedom the adapted models make efficient simulation of complex elasto-mechanical systems possible, because the given dynamic properties are realized precisely.

5.4.2 Adaptation of the Model as an Optimization Procedure

The adaptation of a model aims at generating a simplified, physical model as a lumped-mass system or multibody system with given dynamic properties. Two problems have to be solved here:

1) Designing an appropriate structure for the model which makes it possible to solve the given simulation task,
2) Searching the system parameters to minimize the difference between the assumed dynamic properties of the model and the real system.

The solution of the first problem requires the designer's experience as a decision maker. Formal solution strategies can be applied to a small extent only. The second problem, however, can be formulated as a nonlinear, constrained multicriteria optimization problem. It can be solved by applying the direct methods of mathematical programming. In a program-controlled iteration process the parameters of the adapted model are changed until the given system properties correspond to those of the real system.

For the initial design of the reduced model which is a lumped-mass system, the following conditions have to be observed:

- the kinematics of the model with a limited number of degrees of freedom should correspond to the eigenmodes of the real system,
- the important points of the system, e.g. the sensor places, should be described in the model,
- the interfaces of substructure elements should be modelled in a realistic way,
- the free vibrations of the model should not show eigenmodes which cannot be carried out by the real system.

The parameters of this model are design variables $\mathbf{x} \in \mathbb{R}^n$ of the optimization problem:

- geometrical data (point coordinates, cross-section areas, moments of inertia),
- discrete mass points.

The feasible data range of these design variables have to fulfill particular constraints so that the desired model can realize the real system efficiently (e.g. no overlap in the field of FE-modelling):

$$x_i^l \le x_i \le x_i^u , \qquad i = 1,\dots,n. \qquad (5.4\text{-}5)$$

According to (2-4), the appropriate inequality constraints read as follows:

$$\mathbf{g}(\mathbf{x}) \le \mathbf{0}, \quad \text{with } \mathbf{g} \in \mathbb{R}^{2n} , \qquad \mathbf{x} \in \mathbb{R}^n ,$$

$$g_i = \frac{x_i}{x_i^u} - 1 \le 0 , \qquad (5.4\text{-}6)$$

$$g_{i+n} = 1 - \frac{x_i}{x_i^l} \le 0 .$$

The adapted model should have some of the given dynamic properties of the real system. The realization of each system quality $S_i(\mathbf{x})$ is considered as an objective $f_i(\mathbf{x})$ and is described by an error function depending on the design variable vector. Altogether, the objectives $f_i(\mathbf{x})$, $i = 1,\dots,k$ formulate the objective function vector $\mathbf{f}(\mathbf{x}) \in \mathbb{R}^k$:

$$f_i(\mathbf{x}) = 1 - \frac{S_i^M(x)}{S_i^R(x)} \quad , \qquad i = 1,\dots,k \tag{5.4-7}$$

with M = substitute model,
R = real system.

For the adapted model the following dynamic properties $S_i(\mathbf{x})$ are the most important objectives:

- mass of the system $\quad S_1 = \sum_i m_i$,
- centre of mass $\quad S_{2,3,4} = r_{x,y,z}$,
- elements of the inertia tensor $\quad S_{5,6,7} = \Theta_{xx,yy,zz}$, (5.4-8)
- eigenfrequencies $\quad S_{8,\dots,7+f} = \omega_{1,\dots,f}$,
- eigenmodes $\quad S_{8+f,\dots,7+2f} = \boldsymbol{\varphi}_{1,\dots,f}$.

The adaptation of eigenmodes requires a modification of the error function (5.4-7). The vector of a particular mode is normalized for both the real system and the adapted model.

$$\left|\boldsymbol{\varphi}_i^R(\mathbf{x})\right| = \left|\boldsymbol{\varphi}_i^M(\mathbf{x})\right| = 1 \ . \tag{5.4-9}$$

The formulation of the error function depends on the difference vector

$$f_\varphi(\mathbf{x}) = \left|\boldsymbol{\varphi}_i^R(\mathbf{x}) - \boldsymbol{\varphi}_i^M(\mathbf{x})\right| = \sqrt{\sum_j (\varphi_{ij}^R - \varphi_{ij}^M)^2} \ . \tag{5.4-10}$$

As the angle of the phase difference γ_i between $\boldsymbol{\varphi}_i^R$ and $\boldsymbol{\varphi}_i^M$ in the design space ranges from $-\frac{\pi}{2} \le \gamma_i \le \frac{\pi}{2}$, it yields:

$$0 \le f_\varphi(\mathbf{x}) \le \sqrt{2} \ . \tag{5.4-11}$$

Equations (5.4-5) to (5.4-10) form the multicriteria optimization problem (1-8)

$$\min_{\mathbf{x} \in \mathbb{R}^n} \{\mathbf{f}(\mathbf{x}) : \mathbf{g}(\mathbf{x}) \le \mathbf{0}\} \ . \tag{5.4-12}$$

A modified method of objective weighting (see Section 1.2.2) proved to be efficient for the transformation of the MO-problem (5.4-12) into a scalar optimization problem:

$$p[\mathbf{f}(\mathbf{x})] = \sqrt{\sum_i w_i [f_i(\mathbf{x})]^2} \;, \qquad w_i = \{0 \text{ or } 1\} \;. \tag{5.4-13}$$

The preference function $p[\mathbf{f}(\mathbf{x})]$ to be evaluated has the form of a cost or a utility function with an overlap of all quadratic errors. The weighting factor w_i determines whether or not the error of a particular dynamic system feature is considered. During the optimization process the total optimization calculation between the initial and the final design is controlled by the modification of the weighting factors. Until the adapted model is established, the modelling is carried out by an hierarchical optimization procedure which modifies the numbers w_i in three steps. Each of the steps is characterized by taking into account additional inequality constraints whereby the dynamic properties adapted in the step before are observed:

1st step

The objective function contains the errors of the mass-specific values: mass, centre of mass, elements of the inertia tensor.

2nd step

The adapted mass-specific values are introduced as constraints; the objective function contains the errors of the eigenfrequencies ω_i.

3rd step

The adapted eigenfrequencies are additionally introduced as constraints; the objective function contains the errors of the eigenmode $\boldsymbol{\varphi}_i$.

The result of the third optimization step is the adapted model as a final design. The additional constraints to be formulated for the system quantity S_i already adapted yields:

$$g_k(\mathbf{x}) = \left[1 - \frac{S_i^M(\mathbf{x})}{S_i^R(\mathbf{x})}\right]^2 - \varepsilon_i \le 0 \;. \tag{5.4-14}$$

According to the optimization procedure, it is guaranteed that the relative error belonging to a special dynamic property S_i between the real system and the adapted model will be less than $\sqrt{\varepsilon_i}$. As the optimization algorithm a hybrid method of the quadratic programming with reduced line search technique [2,3] was applied to determine models adapted in this way.

5.4.3 Example: Adapted Model for the Reflector of a Parabolic Antenna

A physical model realizing essential dynamic properties accurately will be established for simulation studies of the reflector of a parabolic antenna with an aperture diameter of 10 m. This model should also allow a short computing time as well as a clear presentation of vibrations. The reflector is a truss structure made of bars, beams, and membrane elements (Fig. 5.4/1) covered with panels. The FE-model of the structure shows more than 900 degrees of freedom.

The dynamic properties (eigenfrequencies and eigenmodes) are compiled in Table 5.4/2 as a result of computations of the reflector with a complete FE-model.

The aim of the modelling process is a simple lumped-mass model for the reflector representing the mass-specific data (mass, position of the centre of mass, moments of inertia) and the frequencies of the first eight eigenvibrations with high accuracy. The eigenmodes should be qualitatively precise.

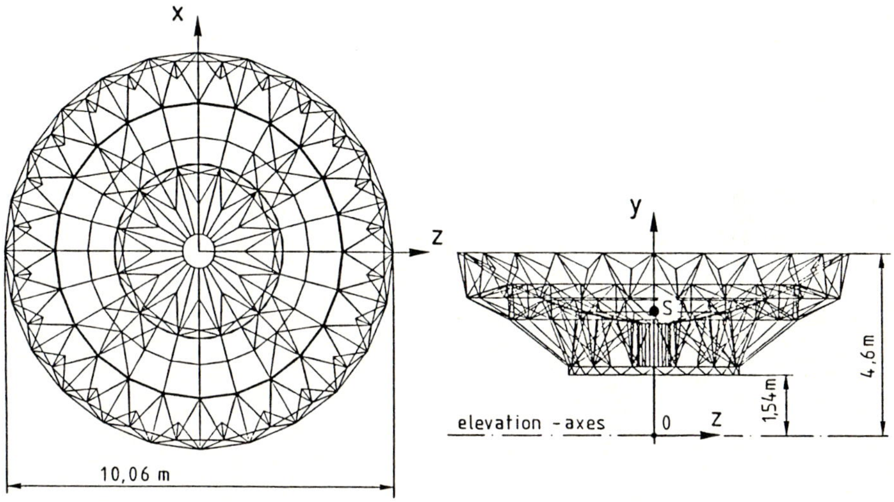

Fig. 5.4/1. Reflector of a parabolic antenna

Reflector values:

mass	$m = 4627$ kg ,
centre of mass	$y_S = 3050$ mm ,
moments of inertia	$\Theta_{xx0} = \Theta_{zz0} = 68180$ kg m^2,
	$\Theta_{yy0} = 42760$ kg m^2.

Mode i	Frequency ω(Hz)	Vibration mode
1	22.88	torsional vibration around axis of symmetry
2 3	37.03 37.03	bending vibration around lateral axis (double eigenvalue, two different modes)
4 5	42.18 42.33	shell vibration with 4 nodal lines (different eigenvalues and modes)
6 7	50.30 50.30	shell vibration with 6 nodal lines (double eigenvalue, 2 diff. modes)
8	55.56	bending vibration in direction of axis of symmetry

Table 5.4/2. Vibration modes 1 to 8 of the reflector

According to Fig. 5.4/2 the beam structure with discrete mass points is designed as an initial design. The FE-model consists of 40 beam elements with 144 degrees of freedom. All elements can be correlated with seven different beam groups taking into account the fact that according to the real structure the diagonal ring-lacer are stiffer than the other four ones. This constructive design leads to slightly different frequencies of the 4th and 5th eigenvibrations.

In order to determine design variables, necessary constructive connection points have to be established first, like the diameter of the inner ring and the additional points on the ring-lacer. The diameter of the inner ring describes the connection points of the reflector at the elevation frame, and the points on the ring-lacer refer to the couple points of the subreflector construction. All other system values have to be adapted until the final design corresponds to the objectives mentioned before. As the model, with the exception of the ring-lacer, is axially symmetric, the system variables can be combined in groups.

Thus, an optimization problem with n=35 design variables is obtained:

$\mathbf{x} = (x_1,\ldots,x_{35})^T$,

$x_{1\ldots3}$ three position coordinates,

$x_{4\ldots10}$ seven beam cross-sections,

$x_{11\ldots31}$ 21 moments of inertia of the beams,

$x_{32\ldots35}$ four mass points.

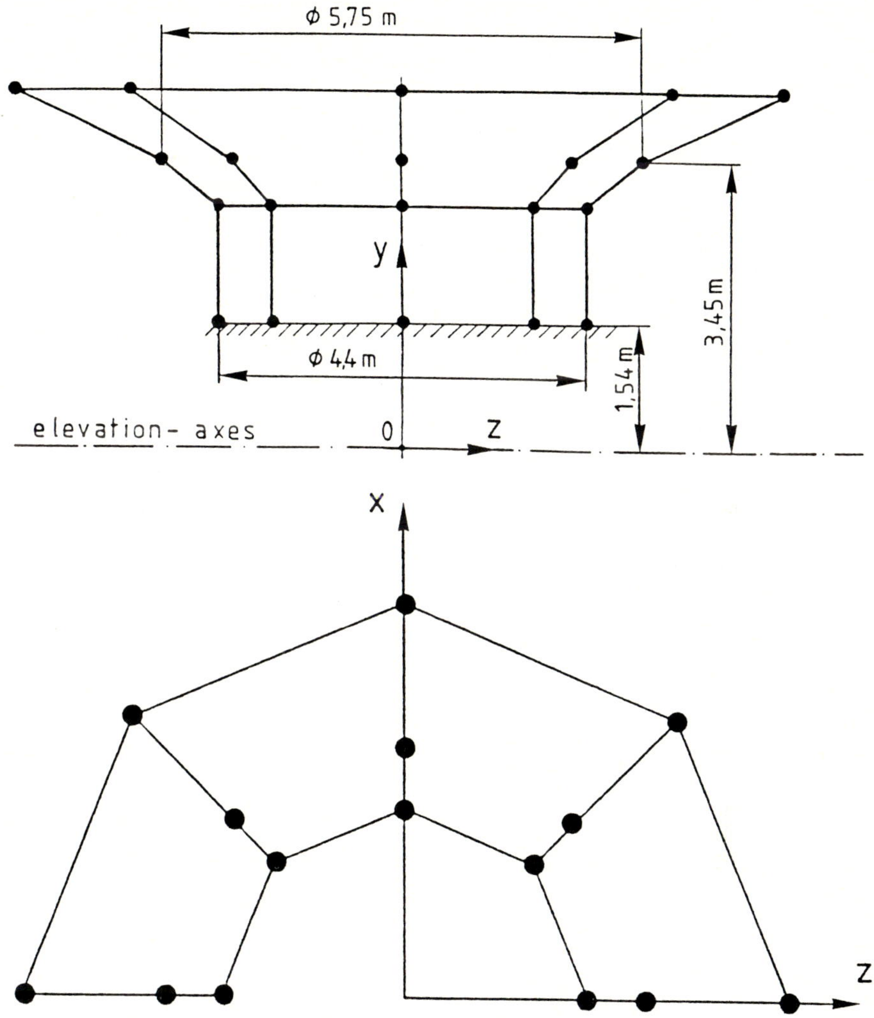

Fig. 5.4/2. Lumped-mass system for the reflector

The optimization problem for adapting the model to the modal data (Table 5.4/2) and to the mass-specific system features yields:

$$\min_{\mathbf{x} \in \mathbb{R}^n} \{ p[\mathbf{f}(\mathbf{x})] \; : \; \mathbf{g}(\mathbf{x}) \le \mathbf{0} \} \; . \tag{5.4-15}$$

The adaptation of the model is carried out in two optimization steps:

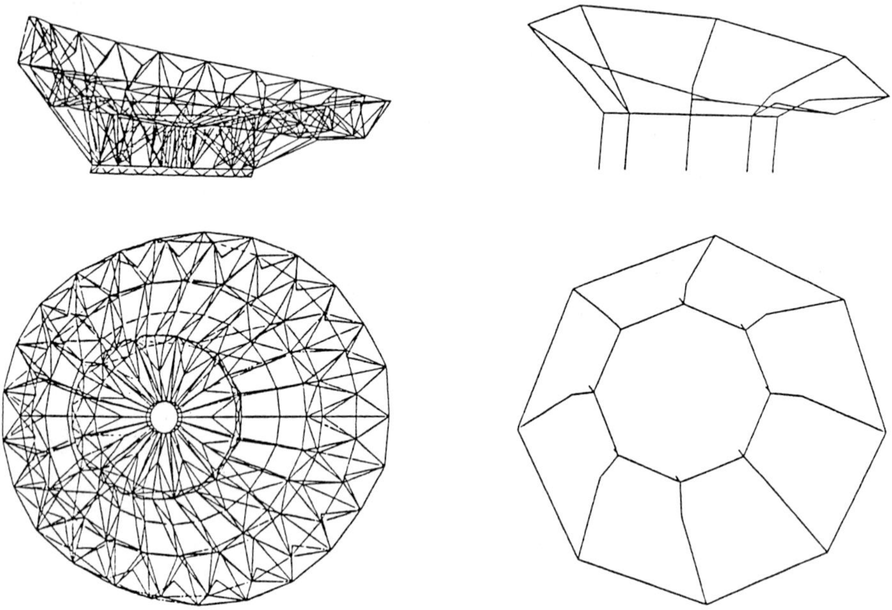

Fig. 5.4/3. Second eigenmode: ω_2 = 37.03 Hz

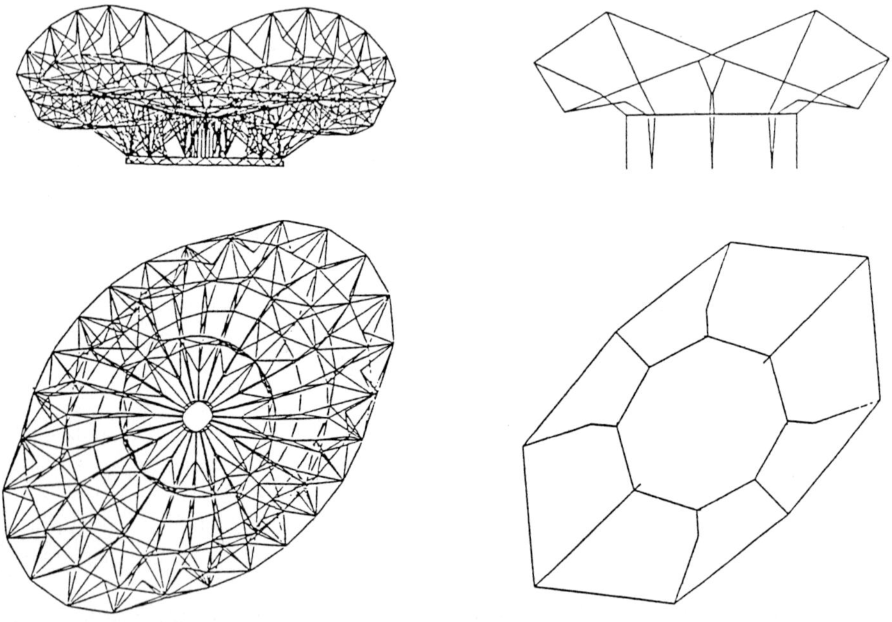

Fig. 5.4/4. Fifth eigenmode: ω_5 = 42.33 Hz

1st step

$$p[\mathbf{f}(\mathbf{x})] = \sqrt{\left(1 - \frac{m^M}{m^R}\right)^2 + \left(1 - \frac{\Theta^M_{xx}}{\Theta^R_{xx}}\right)^2 + \left(1 - \frac{\Theta^M_{yy}}{\Theta^R_{yy}}\right)^2 + \left(1 - \frac{y^M_S}{y^R_S}\right)^2},$$

$$\mathbf{g}(\mathbf{x}) = [g_1,\dots,g_{70}]^T \text{ according to (5.4-6)} . \qquad (5.4\text{-}16)$$

2nd step

The final design of the 1st step will be taken as the initial design of the 2nd step of the optimization procedure.

$$p[\mathbf{f}(\mathbf{x})] = \sqrt{\sum_{i=1}^{8} \left(1 - \frac{\omega^M_i}{\omega^R_i}\right)^2},$$

$$\mathbf{g}(\mathbf{x}) = [g_1,\dots,g_{74}]^T, \qquad (5.4\text{-}17)$$

$g_1(\mathbf{x}) \dots g_{70}(\mathbf{x})$ according to (5.4-6).

$$g_{71}(\mathbf{x}) = \left|1 - \frac{m^M}{m^R}\right| - \varepsilon \le 0,$$

$$g_{72}(\mathbf{x}) = \left|1 - \frac{\Theta^M_{xx}}{\Theta^R_{xx}}\right| - \varepsilon \le 0,$$

$$g_{73}(\mathbf{x}) = \left|1 - \frac{\Theta^M_{yy}}{\Theta^R_{yy}}\right| - \varepsilon \le 0,$$

$$g_{74}(\mathbf{x}) = \left|1 - \frac{y^M_S}{y^R_S}\right| - \varepsilon \le 0, \quad \varepsilon = 10^{-8} .$$

The result of the optimization calculation is an adapted model which realizes the given dynamic properties $\omega_1,\dots,\omega_8$ and the mass geometry with an error of only $\varepsilon < 10^{-8}$.

The correct mode sequence could be secured by an interactive search of the appropriate initial design. Figs. 5.4/3 and 5.4/4 show the modes $\boldsymbol{\varphi}_2(\mathbf{x})$ and $\boldsymbol{\varphi}_5(\mathbf{x})$ of the real system and the adapted model.

5.4.4 Conclusion

This chapter deals with the problem of computing simplified but useful models for the analysis and simulation of complex elastomechanical structures. The question is how to construct a physical model with given mass-specific and spectral data. In literature this problem is known as "inverse" or "reconstruction problem" in system dynamics [4]. The heuristic discretization methods, which are normally applied, provide models with a reduced order of freedom but with large errors in the dynamic system behaviour. Another modelling task occurs if the simplified model is required to have special measured dynamic properties of a real complex system.

The problem of generating simplified adapted models with given dynamic properties is formulated as a constraint vector optimization problem. A vector objective function with error expressions as single objectives depending on the theoretical, measured or desired dynamic properties of the substitute model has to be minimized. The multicriteria optimization problem is solved by transforming it into a scalar optimization problem by means of a cost functional with weighting factors that change stepwise. Starting with an initial design, the adapted model is determined as a final design with minimized errors under given dynamic properties by means of a program controlled hierarchical optimization procedure.

The modelling concept is illustrated by the design of an adapted substitute model for a complex reflector structure of a parabolic antenna.

References

[1] Müller-Slany, H.H.: Angepaßte Schwingungsmodelle zum Bestimmen von Parameteränderungen. ZAMM 69 (1989) T312-T314

[2] Eschenauer, H.; Bremicker, M.: Über die Leistungsfähigkeit einiger MP-Algorithmen im Gestaltoptimierungsprozeß. ZAMM 69 (1989) T85-T87

[3] Bremicker, M.: Dekompositionsstrategie in Anwendung auf Probleme der Gestaltsoptimierung. Dissertation, Universität Siegen. VDI-Fortschrittsbericht, Reihe 1, Nr. 173, Düsseldorf: VDI-Verlag, 1989

[4] Gladwell, G.M.L.: Inverse Problems in Vibration. Dordrecht. Boston, Lancaster: Martinus Nijhoff Publishers, 1986

5.5 Approximation Methods in Structural Optimization Using Experimental Designs for Multiple Responses

D.H. van Campen, R. Nagtegaal, A.J.G. Schoofs

5.5.1 Introduction

For the analytical task in structural optimization problems, the finite element method (FEM) is predominantly used due to its accuracy and its serviceable and flexible modelling facilities. However, FEM-analyses are often very time-consuming which is a serious drawback in their application to iterative optimization. One of the biggest problems encountered here is that of getting trapped in a local optimum. In order to find the global optimum several optimization runs are necessary, and this results in high computing costs.

The *Experimental Design Theory* (EDT), often referred to as "Design of Experiments", provides an alternative method which, under certain circumstances, is more efficient [1]. Instead of repeatedly analysing the structure and subsequently updating the design variables towards an optimum, several structures chosen a proiri are analysed. The FEM-analysis of a structure will be termed as "experiment" and the set of structures to be analysed as "Experimental Design" (ED). In order to run the experiments, values, so-called levels, have to be chosen for each design variable. By assigning a level to each of the design variables, we "create" a structure. The space spanned by the design variables is called the design space. Thus, a point in the design space represents the set of levels assigned to the design variables, and the experimental design is represented by a set of points in the design space. To the results of the experiments, simple mathematical models are fitted which are subsequently used as problem functions in the optimization.

We will denote n as the number of design variables in the optimization problem. These variables will be stored in a column matrix **x**:

$$\mathbf{x} = (x_1, x_2, \ldots, x_n)^T . \tag{5.5-1}$$

A particular response or state variable of the experiments will be named y_i. In general, several response variables, e.g. a number m, are relevant to the optimization problem functions. Hence, we are interested in modelling the following relationships g_i between response variables y_i and design variables (see Section 2.4.1):

$$y_i = g_i(\mathbf{x}) \quad , \qquad i = 1,\dots,m . \tag{5.5-2}$$

In order to determine the experimental design the following questions have to be answered:

1. Which design and response variables are of importance?
2. What is their range of interest?
3. Which form of the functions g_i (see(5.5-2)) must be chosen?
4. How many and which levels have to be chosen for each design variable?
5. How many and which points in the design space, i.e. which ED should be chosen?
6. How should the parameters be estimated and tested in the functions g_i ?

A most valuable aspect of EDT is its guiding the experimentation in a structured way. It is impossible to answer all of the above questions at once, and it is unwise to do so. The questions constitute a circular problem: for instance, in order to define the ED, the type of relationship (5.5-2) has to be known. Yet, the number of parameters in a relationship can only be estimated after an ED has been defined and the experimentation carried out. The solution to this problem is to start with a preliminary ED of moderate extent. Analysis of the data, emerging from this ED, will provide better answers to the questions mentioned above, and a more adequate ED can be defined. The whole procedure is repeated until finally all questions are answered in a satisfactory way.

The discretization of the design space, by assigning a certain number of levels to each design variable, results in a set of candidate points which is denoted by X, and the number of elements of X, i.e. the number of design points, is denoted by r. The number of candidate points in X will rapidly become too large as the number of design variables and the number of levels increase. Analysing all candidate points in X will then become expensive and inefficient. In this case, the restriction of the experimental design to a subset of X is advisable.

The experimental design theory provides optimality criteria and algorithms by means of which optimal experimental designs can be generated. These criteria are based on inferential properties of the assumed relationship (model) between response variables and design variables. The properties may be the variance of parameter estimators or response estimators (predictions).

In the next section optimality criteria will be derived for one response variable. In Section 5.5.3 these criteria will be derived for multiple response variables. A computer program has been written for designing experiments and fitting regression models in the case of multiple response variables (Section 5.5.5). In Section 5.5.6 we illustrate the proposed method by optimizing a link of a continuously variable transmission system.

5.5.2 Experimental Designs in the Case of a Single Response Variable

a) Regression equations

In the first step, only one response variable y is considered. It is presumed that there is a true physical relationship between y and the input quantities $\boldsymbol{\xi}$ which depends on the physical constants $\boldsymbol{\theta}$

$$y = \varkappa(\boldsymbol{\xi}, \boldsymbol{\theta}) \quad . \tag{5.5-3}$$

If the true physical relationship is known, then only the vector $\boldsymbol{\theta}$ needs to be estimated. Often, the relation (5.5-3) is unknown or is too complex and (5.5-3) is approximated locally by an interpolating function $g(\mathbf{x},\boldsymbol{\beta})$

$$y = g(\mathbf{x},\boldsymbol{\beta}) + b(\mathbf{x}) \quad , \tag{5.5-4}$$

where $\mathbf{x}$ contains the set of n design variables obtained by coding or standardizing the input quantities $\boldsymbol{\xi}$; $\boldsymbol{\beta}$ are the unknown coefficients in the interpolating function; $b(\mathbf{x})$ is the systematic departure (bias) from the true physical relationship.

If y contains random errors (i.e. measurement errors), then y is said to be stochastic. In this case, a tilde ($\sim$) is denoted. The model then reads

$$\underset{\sim}{y} = g(\mathbf{x},\boldsymbol{\beta}) + b(\mathbf{x}) + \underset{\sim}{e} \ , \tag{5.5-5a}$$

where $\underset{\sim}{e}$ is the random error.

For several reasons, it will be assumed here that there are random errors in the computer experiments:

1. If the computer experiments are repeated, the output will remain the same: in practice there is no random error, and the model should include only the systematic error. Research on designing and analysing computer experiments without random errors has recently been conducted by Welch [2] and Currin et al. [3]. Welch reports that for larger problems, design optimization requires the solution of a large system of equations which results in very high computing costs. Often, these equations are ill-conditioned. Currin et. al. use a Bayesian approach in which the design and analysis are driven by a so-called prior process. In the latter case, problems arise regarding the choice of the prior function. The traditional ED-methods used here do not show these disadvantages and are, for the time being, the only suitable tools for usually large problems arising in structural optimization. These methods, however, require the existence of random errors.
2. Normally, computer experiments are not repeated completely. A complete repetition of an experiment would require
   - choosing new coordinates and material properties from statistical distributions,
   - generating a new element mesh which introduces a random error into the output. We believe that this approach to computer experimentation, in which experiments are repeated by using complete statistical distributions, is far more realistic. Furthermore, this approach will enable us to give statements of uncertainty about the predictions of the model.

In this chapter it will be assumed that there is no systematic error. A way of dealing with this type of error can be found in Welch [10]. In fact, errors of this kind may be handled in the program to be discussed later in this chapter. Our experience has shown that assuming random errors only leads to a good design if reasonable computing time has been invested. Therefore, all departures from the true relationship will be regarded as random errors. Additionally, this serves the aim of simplicity.

Our model then reads

$$\underset{\sim}{y} = g(\mathbf{x}, \boldsymbol{\beta}) + \underset{\sim}{e} \quad . \tag{5.5-5b}$$

Furthermore, it will be assumed that the interpolating function is linear:

$$\underset{\sim}{y} = \mathbf{f}^T(\mathbf{x})\boldsymbol{\beta} + \underset{\sim}{e} = \beta_1 f_1(\mathbf{x}) + \ldots + \beta_k f_k(\mathbf{x}) + \underset{\sim}{e}, \tag{5.5-6}$$

where the components $f_1, f_2,\ldots,f_k$ of the column $\mathbf{f}$ are the known postulated mutually independent functions of $\mathbf{x}$ while the components $\beta_1, \beta_2,\ldots,\beta_k$ of the column $\boldsymbol{\beta}$ are unknown parameters. Often the value of $f_1(\mathbf{x})$ is chosen to be 1 for all $\mathbf{x}$. Relation (5.5-6) is called a regression equation. Only linear regression equations will be considered in this chapter (i.e. linear in the components of $\boldsymbol{\beta}$).

b) Parameter and response estimation

The unknown parameters $\boldsymbol{\beta}$ have to be determined from the results of a sample of N experiments characterized by the design points $\mathbf{x}_1$, $\mathbf{x}_2,\ldots,\mathbf{x}_N$. The regression equations (5.5-6) for these design points can be written in matrix notation:

$$\underset{\sim}{\mathbf{y}} = \mathbf{X}\boldsymbol{\beta} + \underset{\sim}{\mathbf{e}}, \tag{5.5-7}$$

where $\underset{\sim}{\mathbf{y}} = [\underset{\sim}{y}_1, \underset{\sim}{y}_2,\ldots,\underset{\sim}{y}_N]^T$ is the column of response values, $\underset{\sim}{\mathbf{e}}$ is the column of errors and $\mathbf{X}$ is an (N*k)-matrix, called the "design matrix", which is expressed as:

$$\mathbf{X} = [\mathbf{f}(\mathbf{x}_1), \mathbf{f}(\mathbf{x}_2),\ldots, \mathbf{f}(\mathbf{x}_N)]^T \tag{5.5-8}$$

Unbiased estimates of $\boldsymbol{\beta}$ can be computed if the expected values $E(\underset{\sim}{\mathbf{e}})$ are zero, and the variances of the errors are the same for all design points:

$$E(\underset{\sim}{\mathbf{e}}\,\underset{\sim}{\mathbf{e}}^T) = V(\underset{\sim}{\mathbf{e}}) = \underset{\sim}{\sigma}^2\mathbf{I}, \tag{5.5-9}$$

where $\underset{\sim}{\sigma}$ is the standard deviation.

In general, the number of design points in the experiment should exceed the number of the parameters $\boldsymbol{\beta}$. Then $\boldsymbol{\beta}$ can be estimated by using a least-square technique. $\hat{\boldsymbol{\beta}}$ (^ indicates an estimate) is calculated from

$$\hat{\boldsymbol{\beta}} = (\mathbf{X}^T\mathbf{X})^{-1}\mathbf{X}^T\mathbf{y} \,. \tag{5.5-10}$$

Subsequently, for each design point $\mathbf{x}$ an estimation $\hat{y}$ of the response variable can be calculated from

$$\hat{y}(\mathbf{x}) = \mathbf{f}^T(\mathbf{x})\,\hat{\beta} \;. \tag{5.5-11}$$

c) The accuracy of parameter and response estimators

A measure of the accuracy of the column of estimators $\hat{\underset{\sim}{\beta}}$ is the variance-covariance matrix $V(\hat{\underset{\sim}{\beta}})$ which is defined as

$$V(\hat{\underset{\sim}{\beta}}) = E[(\hat{\underset{\sim}{\beta}} - \beta)(\hat{\underset{\sim}{\beta}} - \beta)^T] = \underset{\sim}{\sigma}^2 (\mathbf{X}^T\mathbf{X})^{-1} \quad . \tag{5.5-12}$$

For the response estimators $\hat{\underset{\sim}{\mathbf{y}}}(\mathbf{x})$, the variances $V[\hat{\underset{\sim}{\mathbf{y}}}(\mathbf{x})]$ are used as a measure of their accuracy. From (5.5-11) and (5.5-12) follows

$$V[\hat{\underset{\sim}{\mathbf{y}}}(\mathbf{x})] = \mathbf{f}^T(\mathbf{x})(\mathbf{X}^T\mathbf{X})^{-1}\,\mathbf{f}(\mathbf{x})\,\underset{\sim}{\sigma}^2 \,. \tag{5.5-13}$$

Often, the variance $\underset{\sim}{\sigma}^2$ of the measured responses is unknown. An estimate for $\underset{\sim}{\sigma}^2$ can be calculated from

$$\hat{\sigma}^2 = \frac{1}{n-k}\,(\mathbf{y} - \mathbf{X}\hat{\beta})^T(\mathbf{y} - \mathbf{X}\hat{\beta}) \quad . \tag{5.5-14}$$

Using this result, the estimates $\hat{V}(\hat{\underset{\sim}{\beta}})$ and $\hat{V}[\hat{\underset{\sim}{\mathbf{y}}}(\mathbf{x})]$ can be calculated.

d) Optimization criteria for experimental designs

The major aspect of EDT is the formulation of an ED in a way that an accurate regression model can be determined with a minimum amount of experimental effort. For this purpose, the optimization criteria are based on the variance of the parameter estimators or the variance of the response estimators. The objective is to select the best N design points $\mathbf{x}_1,\ldots,\mathbf{x}_N$ (which do not necessarily have to be distinct) from a fixed set χ of r candidate points chosen a priori. It should be emphasized that the optimality of the ED depends on the choice of the set of candidate points.

As mentioned above, EDs are evaluated using the variances of the parameter estimator $V(\hat{\underset{\sim}{\beta}})$ or the variances of the response estimator $V[\hat{y}(\mathbf{x})]$ as a measure (see (5.5-11) and (5.5-12), respectively). In both cases the quality of the ED is a function of the matrix $(\mathbf{X}^T\mathbf{X})^{-1}$, and the objective is to determine, among all possible N-point EDs, that ED which makes $(\mathbf{X}^T\mathbf{X})^{-1}$ minimal. However, the minimum of a matrix is not a well defined concept, and a number of operational criteria have been developed. The most important criteria are:

1. D-optimality: minimize det $(\mathbf{X}^T\mathbf{X})^{-1}$. (5.5-15a)
2. G-optimality: minimize the max. response variance. (5.5-15b)
3. V-optimality: minimize the average response variance. (5.5-15c)

The choice of the criterion depends on
1. the purpose of the model usage, i.e.
   - parameter estimation,
   - response estimation;
2. the scope of the optimization problem.

The D-criterion gives accurate parameter estimations and requires the least computing time. This criterion may be used if
- the parameters are of main interest or if
- the scope of the optimization problem, i.e. the number of design variables and/or the number of levels is very large, or if multiple responses are involved.

The G and V-criteria may be used if the main purpose of the model is response estimation:
- the G-criterion may be used only if a few response estimations are to be made and if the deviation between the estimation and the observation is not too large,
- the V-criterion may be used if many response estimations are to be made.

e) Optimization algorithms

Four classes of optimization algorithms can be distinguished. The first class contains algorithms which generate and evaluate all possible N-point designs. This is very time-consuming and only feasible for small values of N. Welch [9] has developed a "branch and bound"-algorithm as a variant in which a binary tree of minimization problems is generated. Not all designs are generated and evaluated. By exploiting bounds on the minimization, only the branches which might contain D-optimal designs are created. Although the algorithm guarantees global D-optimal designs, the computing costs are extremely high.

The second class of algorithms uses mathematical optimization techniques for a direct maximization of det $(\mathbf{X}^T\mathbf{X})$, a procedure which is equivalent to the minimization of det $(\mathbf{X}^T\mathbf{X})^{-1}$. Due to the complexity of det $(\mathbf{X}^T\mathbf{X})$, these methods are not successful.

The third class of algorithms neglects the integer character of the number of replications of the design points and consequently results in a so-called discrete ED. An exact design is created from the discrete design by rounding off the real numbers of replications to integers. This may result in an appropriate approximation if the number of points N is sufficiently higher than the number of parameters k. For this reason, discrete designs are sometimes called "approximate designs". Such a design can be used as an initial design in exchange algorithms. The algorithms for the construction of optimal discrete designs depend on the optimality criterion used [4].

Finally, the fourth class concerns the exchange algorithms. Mitchell [5] developed an efficient algorithm called DETMAX as the most popular of all exchange algorithms. The algorithm starts with an initial ED; the final goal is an optimal N-point ED. During each iteration step, the candidate point which results in the largest increase of det $(\mathbf{X}^T\mathbf{X})$ is added to the design, and, subsequently, the point which results in the smallest decrease of det $(\mathbf{X}^T\mathbf{X})$ is removed from the design. The number of points in the initial design may be larger or smaller than N. If necessary, the algorithm first adds or rejects points until the number of points in the ED is equal to N. In order to avoid local optima, the algorithm is able to perform "excursions" in which several points are added in one single step, and, subsequently, a number of points is reduced to N. If the resulting N-point ED has not been improved, another excursion will be made from the same initial design. If the excursion is successful, the resulting ED will be used as the starting ED in a further attempt to maximize det $(\mathbf{X}^T\mathbf{X})$. The algorithm terminates if no better ED can be found. The algorithm generates high quality EDs at relatively low computing costs.

f) Use of partial derivatives of response variables

The differentiation of the regression equation (5.5-6) with respect to the design variable x_i yields:

$$\frac{\partial \underset{\sim}{y}}{\partial x_i} = \beta_1 \frac{\partial f_1}{\partial x_i} + \ldots \beta_k \frac{\partial f_k}{\partial x_i} + \frac{\partial \underset{\sim}{e}}{\partial x_i}, \qquad i = 1,\ldots,n \quad . \tag{5.5-16}$$

In FEM-formulations such partial derivatives of y can be computed efficiently, and thus (5.5-16) can be used successefully, together with (5.5-6), to estimate the parameters $\boldsymbol{\beta}$. Furthermore, the accuracy of the partial derivatives of the resulting regression model will be increased. Subsequently, the partial derivatives may be used in optimization algorithms.

5.5.3 Experimental Designs in the Case of Multiple Responses

- *Multiple regression equations*

In general, more than one response variable is of interest. Each of these variables will have its own functional relationship to the design variables. Often, they correlate, because they originate from the same system.

In the case of multiple responses, a possible approach would be to construct experimental designs for each response variable separately. From the economical point of view, it is much more favourable to construct one optimal ED which takes into account all response variables simultaneously.

Furthermore,in the case of multiple responses, generalization makes it possible to use the partial derivatives of a response variable in the experimental design. The regression equations for the partial derivatives are obtained by differentiating the regression equation of the response variable (see (5.5-16).

In the following equation, m is the index of the response variables $y_1,\ldots,y_m$, and $\mathbf{u}$ is the column containing these variables and their partial derivatives in the following order:

$$\mathbf{u} = [y_1, \frac{\partial y_1}{\partial x_1},\ldots, \frac{\partial y_1}{\partial x_n},\ldots, y_m, \frac{\partial y_m}{\partial x_1},\ldots, \frac{\partial y_m}{\partial x_n}]^T \quad . \qquad (5.5\text{-}17)$$

Since only linear models are used, y_j, $j = 1,2,\ldots,m$ is assumed to depend linearly on k_j parameters as the elements of $\boldsymbol{\beta}_j$. Then, the regression equations are given by:

$$y_j(\mathbf{x}) = \mathbf{f}_j^T(\mathbf{x})\,\boldsymbol{\beta}_j \qquad \text{for} \quad j = 1,2,\ldots,m \qquad (5.5\text{-}18a)$$

and

$$\frac{\partial y_j(\mathbf{x})}{\partial x_i} = \frac{\partial \mathbf{f}_j^T(\mathbf{x})}{\partial x_i}\boldsymbol{\beta}_j \qquad \text{for} \quad j = 1,2,\ldots,m \quad \text{and } i = 1,2,\ldots,n, \qquad (5.5\text{-}18b)$$

where $\mathbf{f}_j(\mathbf{x})$ is a column vector whose elements are postulated functions.

The postulated models (5.5-18) can be represented in a matrix notation as follows. The columns containing the parameters β_1, β_2,..,β_m are filed in a column β expressed by

$$\beta = \left[\beta_1^T, \ \beta_2^T, \ldots, \beta_m^T\right]^T . \tag{5.5-19}$$

The column vector β contains k elements, where

$$k = \sum_{j=1}^{m} k_j \quad . \tag{5.5-20}$$

Next, the columns $\mathbf{f}_j(\mathbf{x})$ and $\dfrac{\partial \mathbf{f}(\mathbf{x})}{\partial x_l}$ are extended with m-1 columns $\mathbf{z}_l$, $l = 1, 2, \ldots, m$, $l \neq j$, each containing k_l zeros. The extended columns will be denoted by a dash ('), and they are given by

$$\mathbf{f}_j'(\mathbf{x}) = \left[\mathbf{z}_1^T, \mathbf{z}_2^T, \ldots, \mathbf{z}_{j-1}^T, \ \mathbf{f}_j^T(\mathbf{x}), \ \mathbf{z}_{j+1}^T, \ldots, \mathbf{z}_m^T\right]^T \quad , \tag{5.5-21a}$$

$$\frac{\partial \mathbf{f}_j'(\mathbf{x})}{\partial x_i} = \left[\mathbf{z}_1^T, \mathbf{z}_2^T, \ldots, \mathbf{z}_{j-1}^T, \ \frac{\partial \mathbf{f}_j^T(\mathbf{x})}{\partial x_i}, \ \mathbf{z}_{j+1}^T, \ldots, \mathbf{z}_m^T\right]^T . \tag{5.5-21b}$$

Using these definitions, the regression equations (5.5-18) can be written as:

$$\mathbf{u}(\mathbf{x}) = \mathbf{F}^T(\mathbf{x})\beta \quad , \tag{5.5-22}$$

where $\mathbf{F}(\mathbf{x})$ is a $k*m(n+1)$ matrix given by:

$$\mathbf{F}(\mathbf{x}) = \left[\mathbf{f}_1', \ \frac{\partial \mathbf{f}_1'}{\partial x_1}, \ldots, \frac{\partial \mathbf{f}_1'}{\partial x_n}, \ldots, \mathbf{f}_m', \ \frac{\partial \mathbf{f}_m'}{\partial x_1}, \ldots, \frac{\partial \mathbf{f}_m'}{\partial x_n}\right]. \tag{5.5-23}$$

- *Weighted least squares estimation*

As in the previous section, N design points $\mathbf{x}_1$, $\mathbf{x}_2$,...., $\mathbf{x}_N$ are to be selected from a set χ of r candidate points. In these design points the FEM-analysis results in the values $\mathbf{u}(\mathbf{x}_1),\ldots,\mathbf{u}(\mathbf{x}_N)$ for the responses $\mathbf{u}(\mathbf{x}_1),\ldots,\mathbf{u}(\mathbf{x}_n)$. Let the difference between $\underset{\sim}{\mathbf{u}}(\mathbf{x}_i)$ and $\mathbf{u}(\mathbf{x})$ be denoted by $\underset{\sim}{\mathbf{e}}_i$, then

$$\underset{\sim}{\mathbf{u}}(\mathbf{x}_i) = \mathbf{u}(\mathbf{x}_i) + \underset{\sim}{\mathbf{e}}_i = \mathbf{F}^T(\mathbf{x}_i)\beta + \underset{\sim}{\mathbf{e}}_i \quad \text{for} \quad i = 1,2,\ldots,N \ , \tag{5.5-24}$$

where $\mathbf{F}(\mathbf{x})$ is given by (5.5-23).

We will again assume that the elements of $\underset{\sim}{\mathbf{e}}_i$ are noncorrelating random errors with an expected value zero and a known constant variance. Furthermore, it is assumed that $\underset{\sim}{\mathbf{e}}_i$ and $\underset{\sim}{\mathbf{e}}_j$ do not correlate if $i \neq j$:

$$E(\underset{\sim}{\mathbf{e}}_i) = \mathbf{0} \; ; \quad E(\underset{\sim}{\mathbf{e}}_i \, \underset{\sim}{\mathbf{e}}_j^T) = \delta_{ij}\Sigma^2 \quad \text{for } i,j = 1,2,\ldots,N \quad . \tag{5.5-25}$$

Here, δ_{ij} is the Kronecker symbol and Σ is a known $m(n+1) * m(n+1)$ diagonal matrix whose diagonal elements $\sigma_1,\ldots,\sigma_{m(n+1)}$ are the standard deviations of the elements of $\mathbf{u}$ (see (5.5-17)).

As the error variances may be unequal, $\boldsymbol{\beta}$ should be estimated by using the method of weighted least squares. By premultiplying (5.5-24) with Σ^{-1}, the new regression equation takes the form of

$$\underset{\sim}{\mathbf{w}}(\mathbf{x}_i) = \mathbf{W}^T(\mathbf{x}_i)\boldsymbol{\beta} + \underset{\sim}{\eta}_i \qquad \text{for } i = 1,2,\ldots, N \quad , \tag{5.5-26}$$

where

$$\underset{\sim}{\mathbf{w}}(\mathbf{x}_i) = \Sigma^{-1}\underset{\sim}{\mathbf{u}}(\mathbf{x}), \quad \mathbf{W}^T(\mathbf{x}_i) = \Sigma^{-1}\mathbf{F}^T(\mathbf{x}_i) \text{ and } \underset{\sim}{\eta}_i = \Sigma^{-1}\underset{\sim}{\mathbf{e}}_i \quad .$$

The errors $\underset{\sim}{\eta}_i$ still have zero expectation: $E(\underset{\sim}{\eta}_i) = \mathbf{0}$. However, the error variance matrix of $\underset{\sim}{\eta}_i$, $V(\underset{\sim}{\eta}_i)$, now becomes a unity

$$E(\underset{\sim}{\eta}_i \underset{\sim}{\eta}_i^T) = E(\Sigma^{-1}\underset{\sim}{\mathbf{e}}_i\underset{\sim}{\mathbf{e}}_i^T\Sigma^{-1}) = \Sigma^{-1}E(\underset{\sim}{\mathbf{e}}_i\underset{\sim}{\mathbf{e}}_i^T)\,\Sigma^{-1} = \Sigma^{-1}\Sigma\Sigma\Sigma^{-1} = I \; .$$

Therefore, the model (5.5-26) satisfies the standard least squares assumptions, and $\boldsymbol{\beta}$ can be estimated by using an ordinary least squares fit of the new model.

Estimates $\hat{\boldsymbol{\beta}}$ for the parameters $\boldsymbol{\beta}$ can be determined from the requirement that the weighted error sum of squares SS_e is minimal, i.e.

$$SS_e = \sum_{i=1}^{N}\Big[\mathbf{w}(\mathbf{x}_i) - \mathbf{W}^T(\mathbf{x}_i)\,\hat{\boldsymbol{\beta}}\,)^T(\mathbf{w}(\mathbf{x}_i) - \mathbf{W}^T(\mathbf{x}_i)\,\hat{\boldsymbol{\beta}}\Big]. \tag{5.5-27}$$

This results in a set of k equations. $\hat{\boldsymbol{\beta}}$ is obtained from

$$\hat{\boldsymbol{\beta}} = \Big[\sum_{i=1}^{N}\mathbf{W}(\mathbf{x}_i)\,\mathbf{W}^T(\mathbf{x}_i)\Big]^{-1}\;\sum_{i=1}^{N}\Big[\mathbf{W}(\mathbf{x}_i)\mathbf{w}(\mathbf{x}_i)\Big] \quad . \tag{5.5-28}$$

It can be shown that the variance-covariance matrix of the parameter estimators is

$$V(\underset{\sim}{\hat{\boldsymbol{\beta}}}) = N\mathbf{M}^{-1} \quad , \tag{5.5-29}$$

where $\mathbf{M} = \frac{1}{N}\sum_{i=1}^{N}\Big[\mathbf{W}(\mathbf{x}_i)\,\mathbf{W}^T(\mathbf{x}_i)\Big]$ is the weighted design matrix.

In order to estimate $\underset{\sim}{\beta}$, the error variance matrix Σ^2 should be known beforehand. When Σ^2 is unknown, Σ^2 and $\underset{\sim}{\beta}$ can be estimated by using an iterative reweighted least squares procedure [1].

Using these results, the unbiased estimators $\hat{\mathbf{u}}(\mathbf{x})$ for the response quantities can be derived from:

$$\hat{\mathbf{u}}(\mathbf{x}) = \mathbf{F}^T(\mathbf{x})\,\hat{\beta} \quad . \tag{5.5-30}$$

The variance-covariance matrix $\mathbf{V}(\hat{\mathbf{u}}(\mathbf{x}))$ is expressed as:

$$\mathbf{V}(\hat{\underset{\sim}{\mathbf{u}}}(\mathbf{x})) = \mathbf{F}^T(\mathbf{x})\ \mathbf{V}(\hat{\underset{\sim}{\beta}})\mathbf{F}(\mathbf{x}) \quad . \tag{5.5-31}$$

- *Generalized optimization criteria*

Equation (5.5-24) shows that $\mathbf{V}(\hat{\underset{\sim}{\beta}})$ is determined completely by the statistical properties Σ of the errors, the model functions $\mathbf{F}(\mathbf{x})$, and the design points $\mathbf{x}_1,\ldots,\mathbf{x}_N$. Assuming that the error variance matrix Σ is known, then, for a given number N of design points, the matrix $\mathbf{V}(\hat{\underset{\sim}{\beta}})$ can only be influenced by the choice of the design points $\mathbf{x}_1,\ldots,\mathbf{x}_N$ from the set X of candidate points $\chi_1,\ldots,\chi_r$. As stated in the previous section, the objective of the optimal design of experiments for multiple responses is to determine the number $N_i (N_i \geq 0)$ of observations at candidate point χ_i $(i = 1,2,\ldots,r)$ in such a way that $N_1 + N_2 + \ldots N_r$ equals the number N of experiments. The use of Equation (5.5-29) shows distinctly that the variance-covariance matrix $\mathbf{V}(\hat{\underset{\sim}{\beta}})$ can be considered a function of the column $\mathbf{N} = \left[N_1, N_2, \ldots, N_r\right]^T$

$$\mathbf{V}(\hat{\underset{\sim}{\beta}},\mathbf{N}) = N\ \mathbf{M}^{-1}(\mathbf{N}) \ . \tag{5.5-32}$$

It has been shown that the accuracies of the parameter estimators $\hat{\underset{\sim}{\beta}}$ and of the response estimators $\hat{\underset{\sim}{\mathbf{u}}}(\mathbf{x})$ are determined by the variance-covariance matrix $\mathbf{V}(\hat{\underset{\sim}{\beta}}, \mathbf{N})$ and therefore by the elements of the column $\mathbf{N}$. These elements may be subject to constraints which lead to the following generalized definition of an optimal experimental design for multiple responses:

> The experimental design characterized by $\mathbf{N}^*$ is optimal if the elements of $\mathbf{N}^*$ satisfy all specified constraints, and a specified criterion $C = C(\mathbf{N})$ is minimal for $\mathbf{N} = \mathbf{N}^*$.

The D-criterion remains the same as in (5.5-15)

$$C(\mathbf{N}) = \det\ (\hat{\mathbf{V}}(\hat{\underset{\sim}{\beta}},\mathbf{N}) \quad \text{for D-optimality.} \tag{5.5-33a}$$

Nagtegaal [8] proposes the following G and V-criteria

$$C(\mathbf{N}) = \max_{i=1,\dots,r} \left[\text{trace}\,(\hat{V}(\hat{\underset{\sim}{u}}(\chi_i))\right] \quad \text{for G-optimality,} \tag{5.5-33b}$$

$$C(\mathbf{N}) = \frac{1}{r} \sum_{i=1,\dots,r} \left[\text{trace}\,(\hat{V}(\hat{\underset{\sim}{u}}(\chi_i)\right] \quad \text{for V-optimality.} \tag{5.5-33c}$$

Generalizations of the optimization algorithms can be found in the works by Fedorov [4] and Nagtegaal [8].

5.5.4 Model Fitting and Testing

The building of an accurate regression model for a given system or structure is an iterative process. At the start of each iteration step the model functions $f_i(\mathbf{x})$, $i = 1,2,\dots,k$, must be available. The iteration step then involves the design of an experiment to collect data followed by estimation of the parameters from the collected data and evaluation of the model. Evaluation implies answering questions like:

- Is the model valid?
- Are the estimated parameters accurate enough?
- Are the model predictions accurate enough for all relevant values of $\mathbf{x}$?
- Which parameters are significantly different from zero, and which parameters can be dropped without affecting the predictive ability of the model too much?

For the selection of the significant parameters, several algorithms have been developed. The backward elemination algorithm starts with the full model and subsequently removes the nonsignificant parameters in successive steps. The forward selection algorithm starts with the most significant parameter and adds significant parameters successively. A popular and effective algorithm is the stepwise regression algorithm developed by Efroymson [6]. Stepwise regression is a combination of forward selection and backward elimination.

Several tests and criteria for judging the model adequacy are available. We emphasize that the model should be tested by using a new set of data. Tests based on data which was used to derive the model can be deceptive and dangerous. By carefully inspecting the residuals, i.e. the differences between observed and predicted values, we can learn much about the validity of the model. Assumptions about distributions of errors and the significance of the parameters can be tested. Draper et al. [7] present a thorough investigation into this subject. Plots of residuals against predicted values or against design variables are very illustrative.

If the results of the model testing require further model improvement, it is necessary to perform another model building cycle consisting of design data collection, model fitting, and retesting.

5.5.5 Computer Program for Experimental Design

Nagtegaal [8] developed an interactive computer program called CADE (**C**omputer **A**ided **D**esign of **E**xperiments). Additionally, facilities for the analysis of experiments have been implemented.

CADE consists of three main modules, namely model input, design of experiments, and model fitting. For the experimental design part, the core of the program ACED (Welch [9]) has been used. In CADE, the optimality criteria and algorithms of ACED have been generalized to multiple responses. Facilities for the selection of parameters have been implemented. The selection procedures are capable of handling partial derivatives by means of an iterative reweighted least squares procedure. CADE has been written in Fortran 77 and runs on Apollo D3000 work stations and on VAX systems.

5.5.6 Application

The iterative model building process will be illustrated by a stress concentration problem in a link of a chain of a continuous variable transmission system (see Fig. 5.5/1). Each section of the chain contains a number of links of about 0.5 mm in thickness. The pins which transmit the driving force to the conical discs are locked up by the links in subsequent sections of the chain. Figure 5.5/2a shows the geometry of the link. The loading force F is 268 N. Only a symmetric loading case was considered, allowing us to use only one quarter of the link in the finite element model (Fig. 5.5/2b).

The links are manufactured using a punching machine leading to an increased danger of crack initiation. In order to reduce the chance of crack-growth, tensile stresses along the contour of the link had to be minimized. In Fig. 5.5./3 three areas A,B, and C are indicated with potentially high tensile stresses. The level of these stresses can be influenced by variation of the geometry parameters x_1, x_2, and x_3. Hence, these parameters were used as design variables.

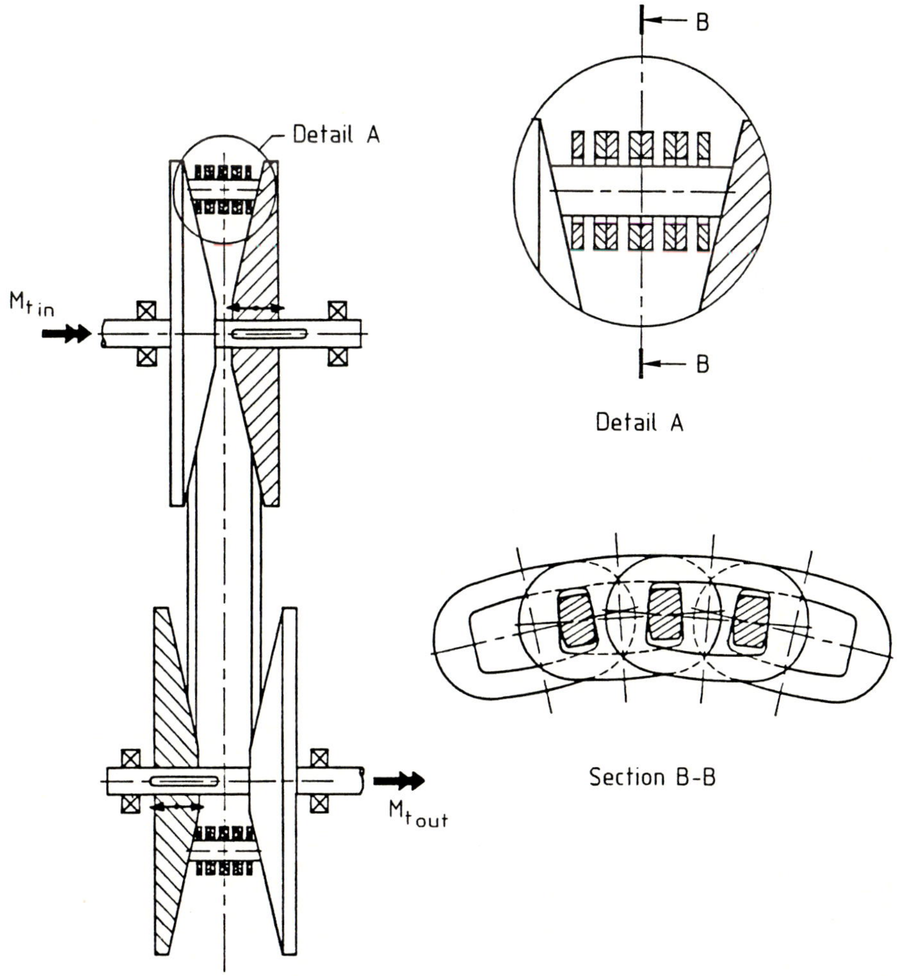

Fig. 5.5/1. Continuous variable transmission system

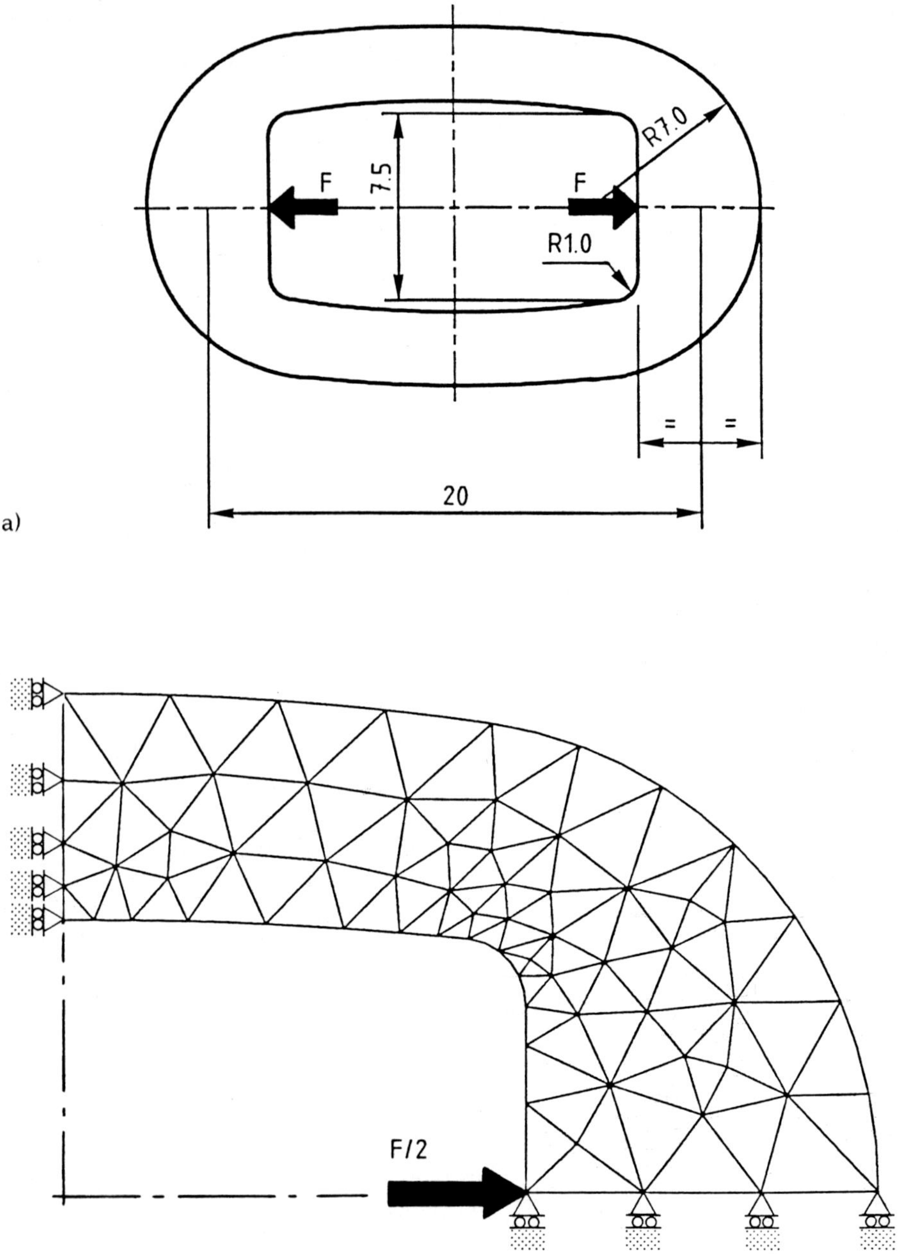

a)

b)

Fig. 5.5/2. Models of the link
a) Real geometry model
b) FE-model

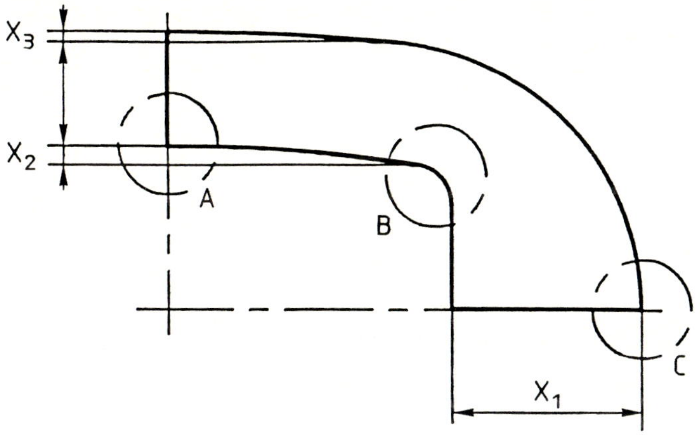

Fig. 5.5/3. Areas with potentially high tensile stresses

a) Model building

We build regression models for the maximum tensile stresses in the areas A, B, and C as functions of the variables x_1, x_2, and x_3.

The upper and lower limits of the design variables are given by constructive constraints:

$$4.5 \le x_1 \le 6.0 \text{ [mm]} \;,$$
$$0.0 \le x_2 \le 0.6 \text{ [mm]} \;, \qquad (5.5\text{-}34)$$
$$0.0 \le x_3 \le 0.6 \text{ [mm]} \;.$$

Each design variable was varied on four equidistant levels covering the design area given in (5.5-34) (see Table 5.5/1).

The maximum tensile stresses in the areas A, B, and C were denoted by y_A, y_B, and y_C. In the first model building cycle the same regression model was used for the maximum tensile stresses in each area:

$$\underset{\sim}{y}_i = \beta_1 + \beta_2 x_1 + \beta_3 x_1^2 + \beta_4 x_2 + \beta_5 x_2^2 + \beta_6 x_3 + \beta_7 x_3^2 + \beta_8 x_1 x_2 +$$
$$+ \beta_9 x_1 x_3 + \beta_{10} x_2 x_3 + \beta_{11} x_1 x_2 x_3 + \underset{\sim}{e}_i \;, \qquad (i = A,B,C) \,. \quad (5.5\text{-}35)$$

For the set of candidate points from which the experimental design had to be selected, we used all possible combinations of the levels given in Table 5.5./1. This resulted in 4x4x4 = 64 candidate points.

Model (5.5-35) contains 11 unknown parameters requiring at least 11 observations to estimate the parameters. One FEM-analysis provided us

Variable	Level 1	Level 2	Level 3	Level 4
x_1	4.5	5.0	5.5	6.0
x_2	0.0	0.2	0.4	0.6
x_3	0.0	0.2	0.4	0.6

Table 5.5/1. Levels of the design variables

with 4 observations, namely one value of the stress and three values of its partial derivatives with respect to the design variables:

$$y_i\ ,\ \frac{\partial y_i}{\partial x_1}\ ,\ \frac{\partial y_i}{\partial x_3}\ ,\ \frac{\partial y_i}{\partial x_3}\ , \qquad (i = A,B,C) \quad .$$

Hence, a minimum of 3 FEM-runs was required; we decided to carry out 5 FEM-runs. These 5 points were selected from the 64 candidate points using the DETMAX-algorithm of the program CADE. Table 5.5/2 shows the optimized 5-point experimental design which was obtained. For this design the trace of the variance-covariance matrix, averaged over all design points, is minimal.

In order to obtain a good fit we had to extend our regression model (5.5-35) with higher order interactions. The model fitting resulted in the following regression models for the maximum tensile stress in the areas A,B, and C:

$$y_A = 540.3 - 110.1x_1 + 7.7x_1^2 + 201.6x_2 + 10.5x_2^2 - 76.3x_3^2 - $$
$$- 17.1x_1x_2 - 3.0x_1x_2x_3 + 12.6x_1x_3 \ , \qquad (5.5\text{-}36a)$$

Design points	x_1	x_2	x_3
2	4.5	0.0	0.2
16	4.5	0.6	0.6
45	5.5	0.6	0.0
50	6.0	0.0	0.2
64	6.0	0.6	0.6

Table 5.5/2: Initial optimal experimental design

$$y_B = 870.9 - 199.3x_1 + 15.2x_1^2 - 263.6x_2 - 46.1x_2^2 + 41.2x_1x_3 + \\ + 3.1x_1^2x_3 - 7.0x_1^2x_3 \quad , \qquad (5.5\text{-}36b)$$

$$y_C = 1311.6 - 335.9x_1 + 23.7x_1^2 + 44.2x_2 + 7.7x_2^2 - 39.2x_3^2 - \\ - 0.9x_1^2x_2 + 6.0x_1x_3^2 \quad . \qquad (5.5\text{-}36c)$$

Table 5.5/3 gives two measures of the accuracy of these models:

- The fraction of the variation about the mean that is explained by the fitted model. This is an overall measure of the fit attained and is called "adjusted R-squared". For a perfect fit the fraction explained variation is 1. For a stochastic response variable, however, the fraction of explained variation will always be less then 1. This is due to the existence of a random error.
- The maximum absolute deviation between observations and predictions.

From table 5.5/3 we see that the models predict the tensile stresses very well. However, the measures in Table 5.5/3 were calculated using the same data as was used for the estimation of the parameters. A more valid test requires new data. We enlarged the models with third order terms. With the program CADE we augmented the 5-point design with three extra points, using the new models simultaneously. These three points should indicate whether or not models with third order terms will fit better and therefore show possible shortcomings of the original models (5.5-36). Minimizing the maximum trace of the variance-covariance matrix with the DETMAX-algorithm resulted in an augmented design for which all other optimization criteria were good as well. Table 5.5/4 gives the extra design points.

Area	Explained variation	Maximum absolute deviation
A	90.2%	0.33%
B	86.7%	1.9 %
C	98.2%	0.5 %

Table 5.5/3. Explained variation and maximum residuals of the estimated models

Design points	x_1	x_2	x_3
12	4.5	0.4	0.6
52	6.0	0.0	0.6
57	6.0	0.4	0.0

Table 5.5/4. Design points for model testing

Area	Residuals in %		
	Point 12	Point 52	Point 57
A	-0.73	-0.99	-0.25
B	-0.83	-0.14	3.35
C	-0.15	0.11	0.13

Table 5.5/5. Residuals in the test points

At the test points FEM-analyses were made and predictions with the original models (5.5-36) were calculated. Table 5.5/5 shows the deviations between prediction and observation of the maximum tensile stresses.

If deviations of 5% and less are considered to be satisfactory, we can stop the model building process by fitting the models again. For this we use the observations of all 8 design points and change to the optimization of the link. If for example deviations of 1% or less are required, we have to do one more extra model building cycle.

In order to evaluate the test above, we carried out a much more rigorous test of the regression models. Here, we chose one hundred points in the design space at random and compared FEM-observations in these points with the predictions of the models (5.5-36). Fig. 5.5/4 shows the histograms of the residuals.

Note that the DETMAX-algorithm had chosen critical design points for the models A and B for which the deviations appear to be the largest. Our testing method, using only a few but well chosen design points, appears to work very well.

We may conclude that, based on as little as five FEM-analyses using partial derivatives, regression models of good overall fit could be derived. Moreover, without the use of partial derivatives it would have been impossible to estimate the parameters of the models (5.5-36).

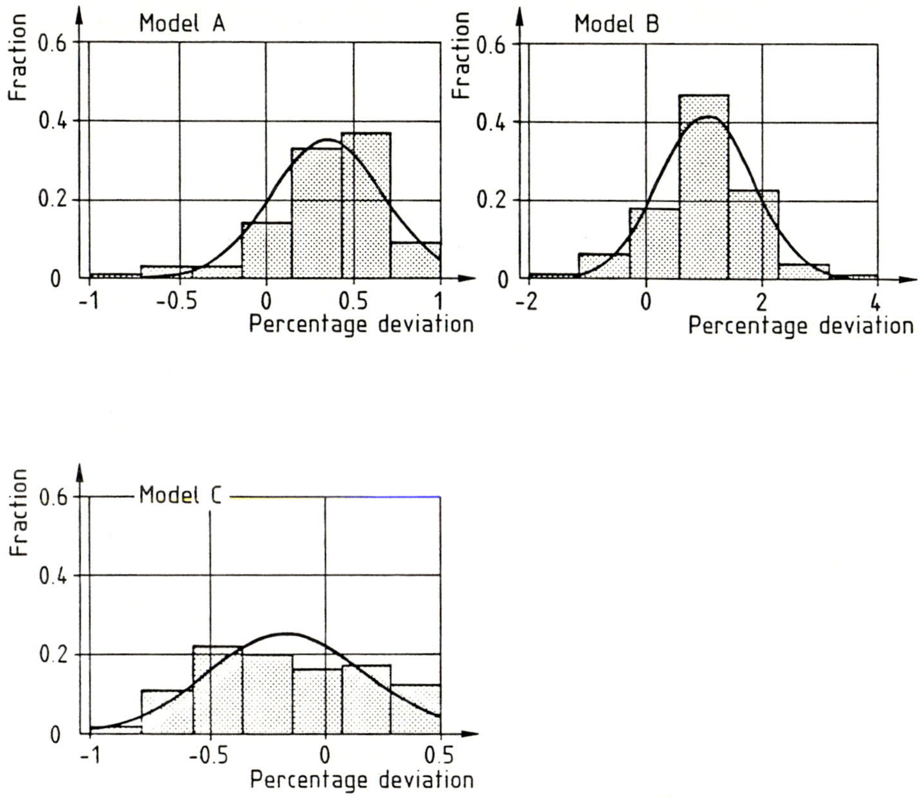

Fig. 5.5/4. Distribution of residuals for 100 randomly chosen test points

b) Optimization

Optimization using regression models can be carried out very simply and cheaply and can be guaranteed to find a global optimum. We optimized the link by stepping through the design space following a narrow spaced 20 x 20 x 20 three-dimensional grid computing the maximum tensile stresses with the models (5.5-36) and recording the best combination of design variables encountered. This procedure resulted in the following global optimum for which the maximum tensile stresses are minimal:

$$x_1 = 6.0\,, \qquad x_2 = 0.6\,, \qquad x_3 = 0.6\ .$$

Note that the program CADE had selected this optimum as a point in our 5-point experimental design.

The maximum tensile stresses according to the models were:

$$y_A = 214.50 \left[N/mm^2\right], \quad y_B = 112.10 \left[N/mm^2\right], \quad x_C = 157.80 \left[N/mm^2\right].$$

FEM-analyses in this optimum resulted in almost the same values for the stresses because the models were fitted using this optimum:

$$y_A = 213.12 \left[N/mm^2\right], \quad y_B = 112.14 \left[N/mm^2\right], \quad y_C = 157.89 \left[N/mm^2\right].$$

The optimization required only 28 sec of CPU-time on an Apollo DN3000 workstation.

c) A remark on the use of derivatives

We frequently experience that using derivatives results in models that predict the observations, used to fit the parameters, slightly worse than models that have been built without the use of derivatives. However, using derivatives will result in models with a better overall fit in the whole design area.

5.5.7 Conclusion

We have introduced a method for finding a relationship (regression model) between input and output of a FEM-analysis. Once this relationship has been established, the regression model can be used instead of the FEM-analysis. The advantages are:

- FEM-analyses and optimization can be done separately; FEM-programs do not have to be extended with an optimization algorithm.
- Computations with a regression model require very little CPU-time. Thus, optimization can be done in a straightforward way by searching through a grid in the design space; this guarantees finding the global optimum if the grid is sufficiently narrow spaced.
- When constraints are changed, the regression model may still be used for optimization. This may be done when the design space is narrowed down. When the design space is extended in some direction, the regression model has to be adapted by carrying out extra FEM-

analysis, perhaps dropping observations in areas that are no longer of interest and fitting the model again. This is sometimes called "response surface design" [11].

However, in comparison to the iterative optimization, this method has the following disadvantages:

- There is a lot of data transport and thinking involved which takes more time of the scientist than the iterative optimization: he or she has to make an experimental design, perform the FEM-analysis, fit the regression model(s) and perhaps test them several times until the models are accurate enough.
- There is a loss in accuracy, especially when the design space is large. This may be overcome by narrowing down the design space to the area where the optimum is expected to be found and by building a regression model in this area.

We have shown that partial derivatives can be used with great benefit in building regression models with a good overall fit.

The proposed method can be applied to single as well as to multi-criteria optimization in the field of experimental design and structural optimization.

References

[1] Schoofs, A.J.G.: Experimental Design and Structural Optimization Ph.D. Thesis. Eindhoven University of Technology. Enschede, The Netherlands: Tiso Press 1987

[2] Sacks, J.; Schiller, S.B.; Welch, W.J.: Designs for Computer Experiments. Technometrics, 31 (1989) 41-47

[3] Currin, C.; Mitchell. M.: Morris, M.; Ylvisaker, D.: A Bayesian Approach to the Design and Analysis of Computer Experiments. Report ORNL-6498. Oak Ridge National Laboratory. Oak Ridge, Tennessee 37831

[4] Fedorov, V.V.: Theory of Optimal Experiments. New York: Academic Press, 1972

[5] Mitchell, T.J.: An Algorithm for the Construction of D-optimal Experimental Designs. Technometrics, 16 (1974) 203-210

[6] Efroymson, M.A.: Multiple Regression Analysis. In: Ralston, A.; Wilf, H.S.: Mathematical Methods for Digital Computers. New York: John Wiley 1960

[7] Draper, N.R.; Smith, H.: Applied Regression Analysis. New York: John Wiley 1981

[8] Nagtegaal, R.: Computer Aided Design of Experiments. A Program for Experimental Design and Model Building (in Dutch). Report WFW 87.005, Eindhoven: Univ. Press Eindhoven 1987

[9] Welch, W.J.: Algorithm for the Construction of Experimental Designs. User's Manual ACED, Version 1.6.1. Univ. of British Columbia, Vancouver, Canada, 1985

[10] Welch, W.J.: A Mean Squared Error Criterion for the Design of Experiments. Biometrica, 70 (1983) 205-223

[11] Box, G.E.P.; Draper, N.R.: Empirical Model Building and Response Surfaces. New York: John Wiley 1987

6 AIRCRAFT AND SPACE TECHNOLOGY

6.1 Multicriteria Optimal Layouts of Aircraft and Spacecraft Structures

G. Kneppe

6.1.1 Introduction

In the aerospace industry the methods of structural optimization have been integrated into the process of engineering design in many cases. Especially for complex design problems, their application leads to optimal layouts which fulfill all requirements in the best possible manner. Fundamental suggestions for this use in industrial practice were made by L.A. Schmit [1]. Nowadays, mathematical optimization algorithms and finite element methods set the basis for optimization computations with a high rate of generality and efficiency [2]. The additional inclusion of optimization models leads not only to a very modular architecture but also to the direct consideration of all relevant practical demands [3]. A variety of examples of aircraft and spacecraft structures shows the advantages of this procedure [4].

So far, the structural weight has often been the most essential and the only objective function for optimization. But today, more and more problems occur where other objectives have to be optimized with the same priority as well. For the design of aircraft, the aerodynamic and aeroelastic features are required in addition to the structural weight. Different objectives have to be fulfilled for the spacecraft structures during the launching phase and during the operation in orbit. Here, the vector optimization methods are suitable for determining unique and optimal compromise solutions.

Two examples of applications - a composite fin of an advanced supersonic aircraft and a satellite structure - are chosen to show the application of the multicriteria approach. The optimization models are formulated and the context to the real structures is explained. Some remarks on the theoretical foundations of aeroelastics, heat transfer, and dynamics surveying the corresponding structural and sensitivity analysis are given. The vector optimization results are summarized.

Before going into more detail, however, the following section will briefly outline the optimization procedure used.

6.1.2 Optimization Procedure

For spacecraft and aircraft, the design of high performance structures involves large scale optimization problems with objectives and constraints from different disciplines. These objectives and constraints have to be fulfilled simultaneously, e.g. statics, buckling, aeroelastics, and dynamics. In order to treat such problems, the optimization procedure MBB-LAGRANGE has been developed since 1984. It has already led to considerable improvements in solving many design problems [4,5]. The program architecture is organized according to the concept of H.A. Eschenauer [3] with the main parts, being: optimization algorithm, optimization model, and structural analysis including sensitivity analysis.

The corresponding *optimization models* are based on the general nonlinear programming problem according to (1-1). The design variables $\mathbf{x}$ are cross sectional areas of rods and bars, wall thicknesses of membrane and shell elements, laminate thicknesses for every single layer in composite elements or nodal coordinates for geometry optimization problems. The inequality constraints in (1-1) include limitations of displacements, stresses, strains, buckling, aeroelastic efficiencies, flutter speed, divergence speed, natural frequencies, dynamic response, and design variables.

In the case of scalar optimization, the objective function $f(\mathbf{x})$ often includes the structural weight or another linear combination of the design variables. However, it is also possible to define one of the constraint functions as an objective and to introduce the weight as a constraint at the same time. If vector optimization problems are under consideration, then optimization strategies $p[\mathbf{f}(\mathbf{x})]$ ensure the transformation to scalar substitute problems according to Section 1.3.2 .

It is necessary to provide several different *optimization algorithms* because there is no single algorithm known which can be adapted to any type of problem. The following algorithms are implemented:

- IBF : Inverse Barrier Function,
- MOM : Method of Multipliers,
- SLP : Sequential Linear Programming,
- SRM : Stress Ratio Method,
- RQP1, RQP2 : Recursive Quadratic Programming,
- GRG : Generalized Reduced Gradients.

The *structural and sensitivity analysis* are based on finite element methods (FEM). Static, buckling, dynamic, aeroelastic, and flutter modules have been incorporated. It is possible to treat homogeneous as well as fibre reinforced composite materials with isotropic, orthotropic or anisotropic behaviour. The element library contains the types: rod, bar, membrane (3, 4 and 8 nodes), shell (3 and 4 nodes), and volume elements. In addition, shell structures can be analyzed with a special transfer procedure. This transfer matrix procedure transforms the transfer matrices into a stiffness matrix that can be assembled together with the remaining finite element stiffness matrices. This mixed procedure allows very efficient analysis of large shell structures with complex boundary conditions.

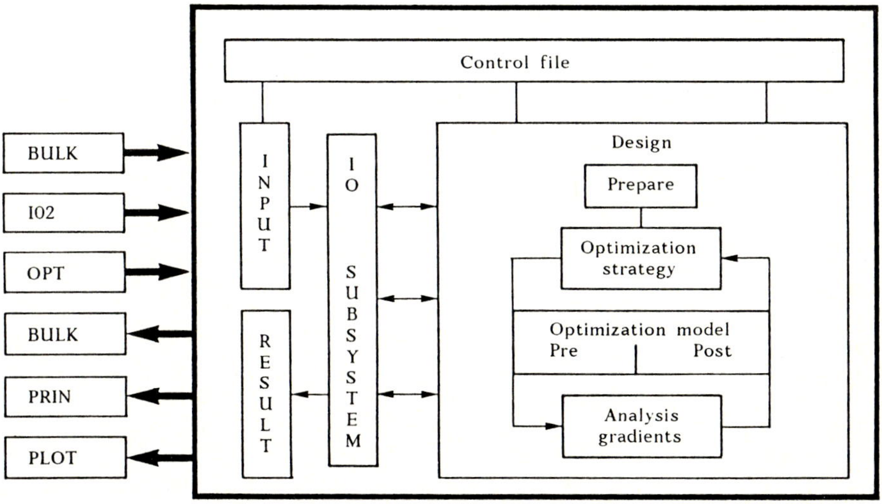

Fig. 6.1/1. Program architecture of MBB-LAGRANGE

The program architecture of MBB-LAGRANGE, shown in Fig. 6.1/1, has a modular set-up with defined interfaces. While the modules INPUT and RESULT are used to enter data or to process the results, the real optimization calculations take place in the DESIGN module. There is a strong separation between optimization and analysis. The "mathematics" is mostly located in the part "optimization algorithm". The "physics" which stands for the structural response and its derivatives is realized in the analysis and gradient module. Both are linked by the "optimization model" (similar to SAPOP-Procedure, Chapter 2).

An interactive user exit with an integrated knowledge based system supports all phases of optimization runs. Further possibilities such as the automatic preparation of batch procedures for different hardware systems or the automatic linking dependent on the problem size ensure a high user comfort. Standard interfaces enable the integration into the CAE-environment (NASTRAN, I-DEAS, PATRAN). The graphical input of some optimization data, especially, saves a lot of time and is really helpful for variable linking, buckling fields and displacement constraints.

6.1.3 Multicriteria Optimization of a Composite Fin

Aircraft structures are often characterized by minimum weight designs. Additional demands include feasible stresses or strains, allowable deformations, critical buckling loads, aeroelastic efficiencies, flutter speed, and manufacturing constraints. Here, the optimization of a composite fin is presented as one example from a wide range of possible applications e.g. frames, bulk heads, fuselages, wings and fins. The optimization of a composite fin is presented.

This composite fin is part of an advanced supersonic aircraft. The two objectives "minimum weight" and "maximum aeroelastic efficiency" are most important for the optimal layout. Hence, the following concept study has the task to fulfill these two objectives in an optimal way and at the same time to consider all further demands as constraint functions. Since the two objectives cannot be fulfilled simultaneously, an objective conflict arises which is typical of vector optimization problems.

The corresponding FE-model is illustrated in Fig. 6.1/2. The cover skins are made of carbon fibre composite material. These laminates

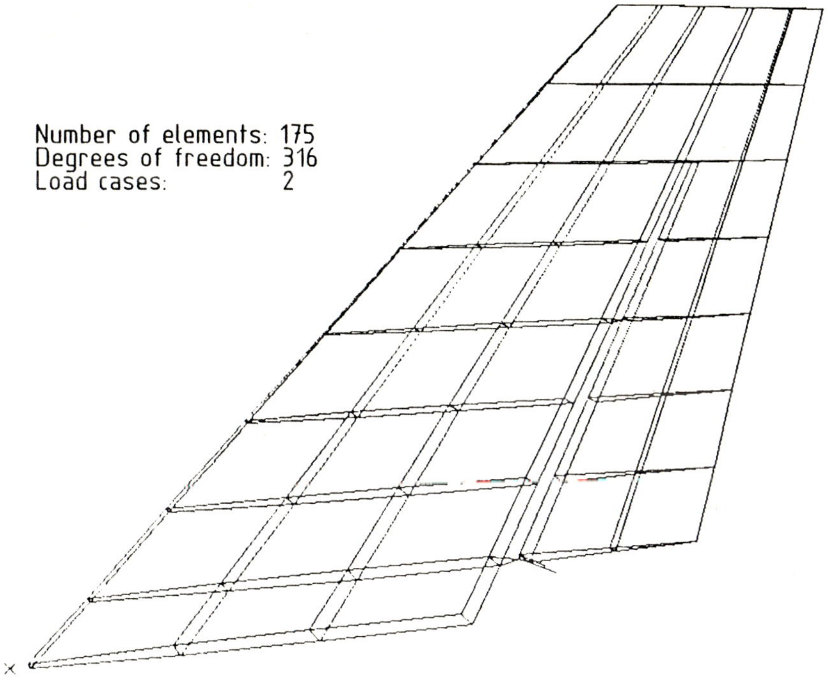

Fig. 6.1/2. Finite element model of a composite fin

consist of arrangements of layers with four different fibre angles for the stabilizer (α_1=70^0, α_2=-65^0, α_3=-20^0, α_4=25^0) and three angles for the rudder (α_1=-70^0, α_2=-25^0, α_3=65^0). The substructure is realized by an aluminium honeycomb core. Two significant load cases for a high supersonic flight condition have been selected (Ma = 1.8, q = 12 psi). The first load case represents a yawing condition of 9° and the second a rudder deflection of 20°. The structure is supported at the connection points to the fuselage.

For every element of the skin the corresponding thickness of every single layer is defined as an independent variable. After variable linking with regard to the symmetry, n = 102 design variables remain for the optimization. The tip of the fin and the substructure are fixed during the optimization process:

$$x_i = t_i , \qquad i = 1,\dots,n . \tag{6.1-1}$$

The objective function vector $\mathbf{f}(\mathbf{x})$ includes the structural weight $W(\mathbf{x})$ and the aeroelastic efficiency $\eta(\mathbf{x})$:

$$\mathbf{f}(\mathbf{x}) = \begin{bmatrix} f_1(\mathbf{x}) \\ f_2(\mathbf{x}) \end{bmatrix} = \begin{bmatrix} W(\mathbf{x}) \\ -\eta(\mathbf{x}) \end{bmatrix} , \tag{6.1-2}$$

where

$$\max \eta(\mathbf{x}) = - \min [-\eta(\mathbf{x})]$$

is chosen.

With the element density ρ_i and the element area A_i the structural weight and its derivatives can be calculated with respect to the design variables:

$$W(\mathbf{x}) = W_O + \sum_{i=1}^{n} \rho_i A_i x_i , \qquad (6.1\text{-}4)$$

$$\frac{\partial W}{\partial x_i} = \rho_i A_i . \qquad (6.1\text{-}5)$$

In order to determine the aeroelastic efficiency, first of all, the elastic displacement vector $\mathbf{u}$ due to the aerodynamic loading has to be calculated. By assuming small deformations, the finite element formulation yields [6]:

$$\mathbf{K}(\mathbf{x})\ \mathbf{u}(\mathbf{x}) = \mathbf{Q}\ \mathbf{u}(\mathbf{x}) + \mathbf{p}_A(\mathbf{x}) \qquad (6.1\text{-}6)$$

with $\mathbf{K}$ - stiffness matrix,
$\mathbf{u}$ - displacement vector,
$\mathbf{Q}$ - aerodynamic influence coefficient matrix (transformed to FE-system),
$\mathbf{p}_A$ - load vector for aeroelastics.

In these equations the actual aerodynamic pressure distribution is estimated by means of the aerodynamic influence coefficient matrix which mainly depends on the geometry of the aerodynamic system (see Fig. 6.1/3). First, this matrix is calculated by means of the aerodynamic model, then it is transformed into the FE-system. $\mathbf{Q}$ is a non-symmetrical matrix as opposed to the symmetrical matrix $\mathbf{K}$. The linear equations (6.1-6) are solved with an iterative solution method [7].

The aeroelastic efficiency is defined by the following affine linear relationship [7]:

$$\eta(\mathbf{x}) = 1 + \mathbf{b}^T \mathbf{u}(\mathbf{x}) . \qquad (6.1\text{-}7)$$

The vector $\mathbf{b}$ contains the behaviour of the geometry and the aerodynamics for the rigid structure. The aeroelastic efficiency is a measure for the ratio of aerodynamic effects between the ideal rigid and the real elastic structure. It can be interpreted as an error estimation of the aerodynamic results due to elastic deformations because the aerodynamic calculations are performed for the rigid structure.

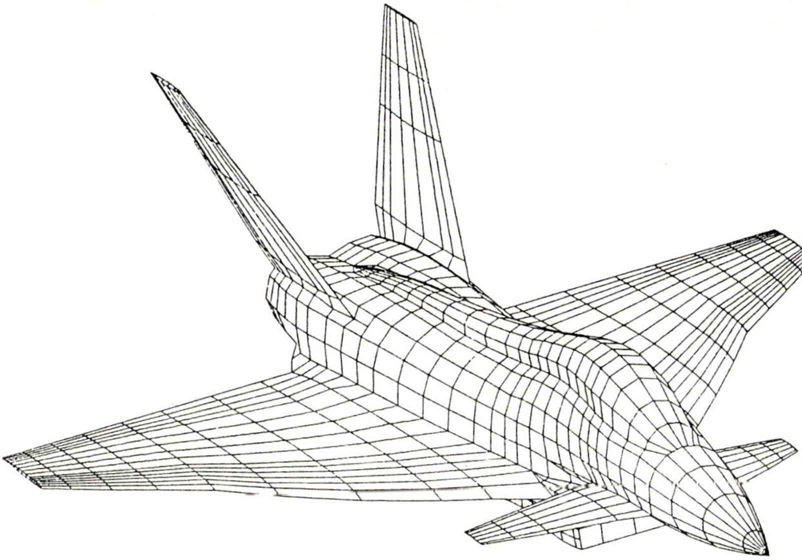

Fig. 6.1/3. Aerodynamic system for advanced supersonic aircraft

The derivatives of the aeroelastic efficiency with respect to the design variables are derived from (6.1-6) and (6.1-7):

$$\frac{\partial \eta}{\partial x_i} = \mathbf{b}^T \, [\mathbf{K} - \mathbf{Q}]^{-1} \left[- \frac{\partial \mathbf{K}}{\partial x_i} \mathbf{u} + \frac{\partial \mathbf{p}_A}{\partial x_i} \right] . \qquad (6.1\text{-}8)$$

In addition, a maximum strain criterion is imposed to satisfy strength requirements. Accordingly, one inequality constraint is formulated for every layer, every element and every load case which leads to the total number of $m_g = 408$ constraints:

$$g_j(\mathbf{x}) = \frac{|\varepsilon_{max,j}(\mathbf{x})|}{\varepsilon_{0,j}} - 1 , \qquad j = 1,\ldots,m_g , \qquad (6.1\text{-}9)$$

where

ε_{max} - maximum strain,

ε_0 - allowable strain.

Here, the strains

$$\boldsymbol{\varepsilon}(\mathbf{x}) = \mathbf{B}(\mathbf{x}) \; \mathbf{u}(\mathbf{x}) \qquad (6.1\text{-}10)$$

result from a static analysis of the FE-model:

$$\mathbf{K}(\mathbf{x}) \; \mathbf{u}(\mathbf{x}) = \mathbf{p}_S(\mathbf{x}) \qquad (6.1\text{-}11)$$

with

$\mathbf{B}$ - displacement strain matrix,

$\mathbf{p}_s$ - static load vector.

With (6.1.-10) and (6.1-11) the derivatives of the constraints (6.1-9) are given by:

$$\frac{\partial g_j}{\partial x_i} = \frac{\partial g_j}{\partial \varepsilon}\frac{\partial \varepsilon}{\partial x_i} + \frac{\partial g_j}{\partial \varepsilon}\frac{\partial \varepsilon}{\partial \mathbf{u}}\,\mathbf{K}^{-1}\left[-\frac{\partial \mathbf{K}}{\partial x_i}\mathbf{u} + \frac{\partial \mathbf{p_S}}{\partial x_i}\right]. \qquad (6.1\text{-}12)$$

The side constraints of the design variables complete the optimization model:

$$x_i^l \le x_i \le x_i^u \quad , \qquad i = 1,\ldots,n\ . \qquad (6.1\text{-}13)$$

Using the constraint oriented transformation (1-14), the multicriteria optimization problem (6.1-2) is reduced to a scalar substitute problem. The structural weight is minimized as the main objective and the aeroelastic efficiency has to achieve different constraint levels. Sequential linear programming SLP as well as recursive quadratic programming RQP are employed as optimization algorithms.

The functional-efficient solutions are illustrated in Fig. 6.1/4 where the structural weight is normalized by means of the reference weight W_0. The initial design has the aeroelastic efficiency $\eta = 0.73$ and the structural weight $W/W_0 = 2.08$. The influence of different constraint levels upon the structural weight is apparent. There, the lower and upper bounds of the efficiency $0.85 < \eta < 0.95$ represent the range of practical importance. Because η is maximized and not minimized, the curvature of the current functional-efficient boundary differs from those boundaries which are obtained in Section 1.3. In this case, the design with $\eta = 0.89$ and $W/W_0 = 2.83$ is chosen as the optimal compromise solution (Fig. 6.1/4). For designs with $\eta > 0.89$, the increase of the structural weight is too large in comparison to the relatively small improvements of η. On the other hand, the designs with $\eta < 0.89$ provide no essential reductions in weight. The functional-efficient boundary of this composite fin represents a good device for making the decision which of the designs is more appropriate for the respective demands.

For the chosen compromise solution, the optimal thickness distribution of the skin is illustrated in Fig. 6.1/5. The contour plots show the thicknesses of the first and the third layer. There are higher magnitudes close to the connecting points, to the fuselage, and to the rudder. The leading edge and wide areas of the first, second, and fourth layer get smaller thicknesses which tend to meet the lower constraints (6.1-13).

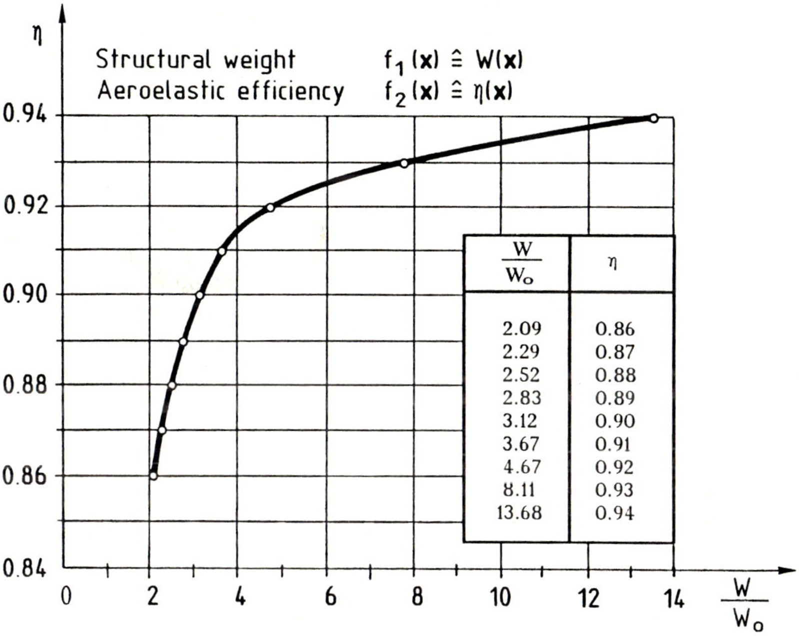

$\frac{W}{W_o}$	η
2.09	0.86
2.29	0.87
2.52	0.88
2.83	0.89
3.12	0.90
3.67	0.91
4.67	0.92
8.11	0.93
13.68	0.94

Fig. 6.1./4. Functional-efficient boundary for a composite fin

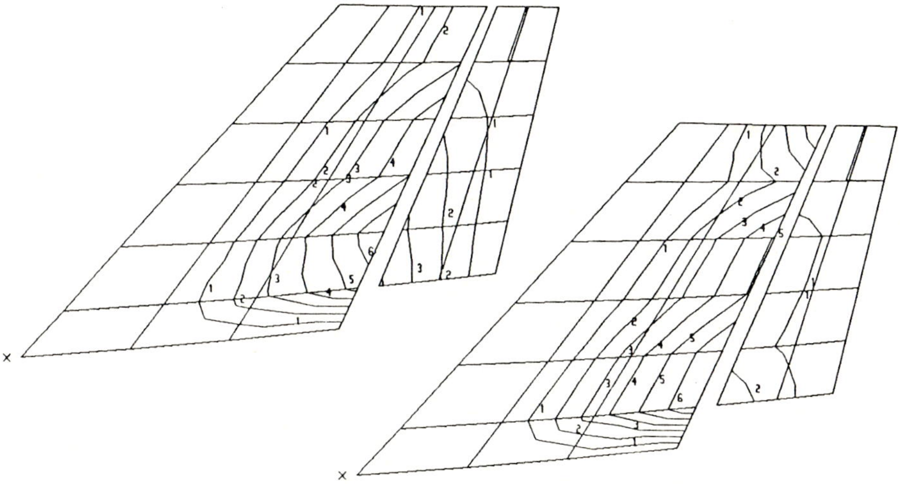

Fig. 6.1/5. Optimal thickness distribution of the first and the third layer

6.1.4 Heat Flux and Frequency Optimization of a Satellite Structure

Two different operating phases are usually most decisive for spacecraft structures. The dynamic loads dominate during the launching phase. On the other hand, the following mission in orbit is characterized by thermal effects. Thus, an objective conflict occurs because none of the possible designs permits a simultaneous optimal fulfillment of dynamic and thermal objectives. That means, optimization methods have to be applied for determining an optimal multicriteria compromise solution.

The ISO-satellite, the European infrared space observatory, will be used for exploring cosmic infrared radiation (Fig. 6.1./6). The inner part of the satellite contains a telescope with a sensor which has to be cooled down to an operating temperature of about 0^0 Kelvin. For this purpose, a cooling medium is stored in special tanks. Thus, it is evident that the availability of this medium limits the lifetime of the satellite.

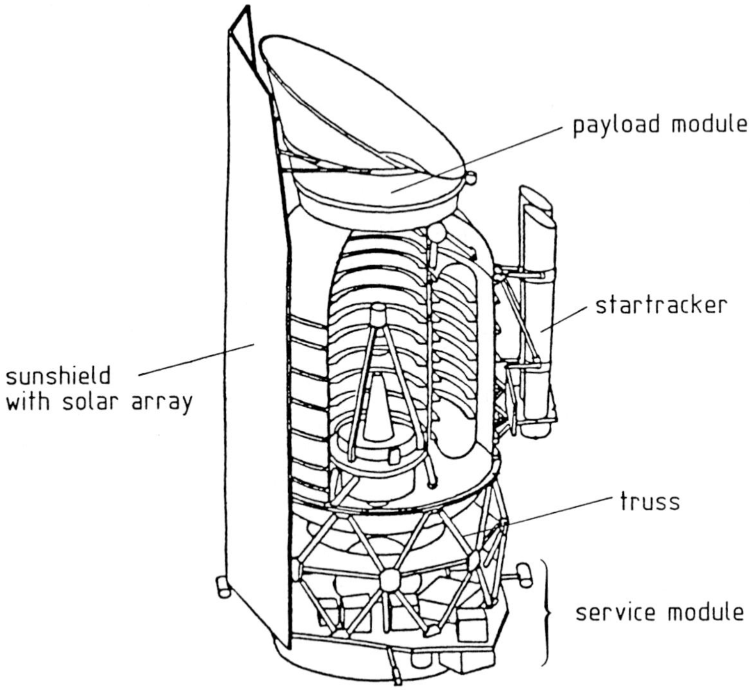

Fig. 6.1./6. ISO-satellite (infrared space observatory)

In order to maximize the lifetime, the heat flux from the outer to the inner structure will be minimized as the first objective function. The second objective function is directed to the dynamic behaviour of the structure during the launching phase. The natural frequency of the dominant axial vibration mode is to be maximized.

The principal sketch of the structural model presented in Fig. 6.1/7 shows that the inner and outer structure are linked by means of a spatial framework and by suspensions. Only a small number of suspensions is included in order to reduce the heat flux from the inner structure. Furthermore, each of these suspensions consists of four loops made of glass or carbon fibre composite material with small heat conductivities as well as with sufficient elastic characteristics. The inner and outer structure are modelled by shell elements which approximate the mass distribution and the elastic behaviour of the real components. The FE-model involves 32 shell elements, 148 bars, 32 rods, and 672 degrees of freedom.

The design variable vector **x** includes the cross sectional areas of the spatial framework and the suspensions (Fig. 6.1/7). For the suspensions of both the lower and the upper part of the structure, respectively, one design variable corresponds to one of the loops. Two other design variables are cross-sectional areas of the framework, but here, the lower and upper variables are linked together. The wall thicknesses of the shells will not be changed. Finally, it is necessary to ensure an

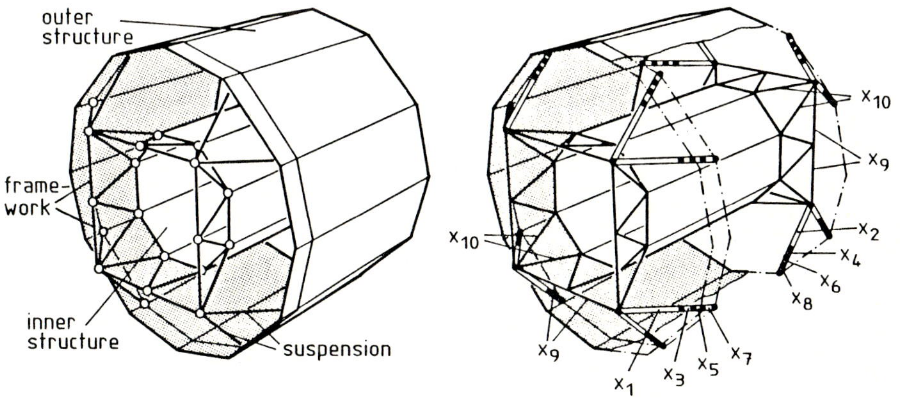

Fig. 6.1/7. Finite element model and design variables of the ISO-satellite structure

axisymmetric structure which is achieved by variable linking. Consequently, we obtain $n = 10$ independent design variables:

$$x_i = A_i , \qquad i = 1,\ldots,n . \qquad (6.1\text{-}14)$$

These variables are restricted due to design and manufacturing demands:

$$x_i^l \le x_i \le x_i^u , \qquad i = 1,\ldots,n . \qquad (6.1\text{-}15)$$

In this case, the multicriteria optimization problem consists of finding the minimum heat flux $\Phi(\mathbf{x})$ and the maximum natural frequency $\omega_A(\mathbf{x})$ (compare (6.1-2) and (6.1-3)):

$$\mathbf{f}(\mathbf{x}) = \begin{bmatrix} f_1(\mathbf{x}) \\ f_2(\mathbf{x}) \end{bmatrix} = \begin{bmatrix} \Phi(\mathbf{x}) \\ -\omega_A(\mathbf{x}) \end{bmatrix} . \qquad (6.1\text{-}16)$$

The heat flux Φ contains conduction, convection and radiation. Here, the relationship between the heat flux Φ and the design variables $\mathbf{x}$ is implicit and nonlinear.

Its calculation requires the costly determination of view factors and the numerical solution of the corresponding differential equations. Previous studies have shown, however, that the heat flux Φ can be linearized with respect to the design variables (6.1-14) and their move limits (6.1-14). Corresponding checks for the initial and optimal designs confirm this approach:

$$\Phi(\mathbf{x}) \approx \Phi_l(\mathbf{x}) = c_o + \sum_{i=1}^{n} c_i x_i , \qquad (6.1\text{-}17)$$

$$\frac{\partial \Phi}{\partial x_i} \approx \frac{\partial \Phi_l}{\partial x_i} = c_i , \qquad (6.1\text{-}18)$$

where

$$c_o = \Phi(\mathbf{x}_o) - \frac{\partial \Phi(\mathbf{x}_o)}{\partial \mathbf{x}} \mathbf{x}_o , \qquad c_i = \frac{\partial \Phi(\mathbf{x}_o)}{\partial \mathbf{x}_i} ,$$

$\mathbf{x}_o$ - reference design variable vector .

The coefficients c_o and c_i are calculated in advance by a heat transfer program based on a node method.

The natural frequency ω_A appears in the eigenproblem which characterizes the structural modal analysis:

$$\left[\mathbf{K}(\mathbf{x}) - \omega_k^2(\mathbf{x})\, \mathbf{M}(\mathbf{x})\right] \mathbf{q}_k(\mathbf{x}) = 0 . \qquad (6.1\text{-}19)$$

In this equation **K** and **M** represent the stiffness and mass matrices, and $\mathbf{q}_k$ are the modal displacements, e.g. the eigenmodes associated with the eigenvalues ω_k^2.

The natural frequency $\omega_k(\mathbf{x})$ is chosen as an objective function $\omega_A(\mathbf{x})$ out of $k=1,\dots,n_f$ which belongs to the dominant axial vibration mode $\mathbf{q}_A(\mathbf{x})$. As this connection between the axial mode and the sequence in the frequency spectrum can change during the optimization process, it has to be checked within every iteration and to be adapted if necessary. This demand is satisfied by applying a special method [8] which is based on a difference metric defined for modes of the ν-th and the $(\nu+1)$-th iteration.

If the eigenvectors are scaled by $\mathbf{q}_k^T \mathbf{M} \mathbf{q}_k = 1$, then the derivatives of the natural frequency ω_A are as follows:

$$\frac{\partial \omega_A}{\partial x_i} = \frac{1}{2\omega_A} \mathbf{q}_A^T \left(\frac{\partial \mathbf{K}}{\partial x_i} - \omega_A^2 \frac{\partial \mathbf{M}}{\partial x_i} \right) \mathbf{q}_A . \tag{6.1-20}$$

According to the constraint oriented transformation (1-14), the heat flux is minimized as the main objective. Due to the selection of appropriate constraint levels for the frequency, one gets different functional-efficient solutions. For this scalar substitute problem with one dominant constraint the optimization algorithm RQP shows a good efficiency.

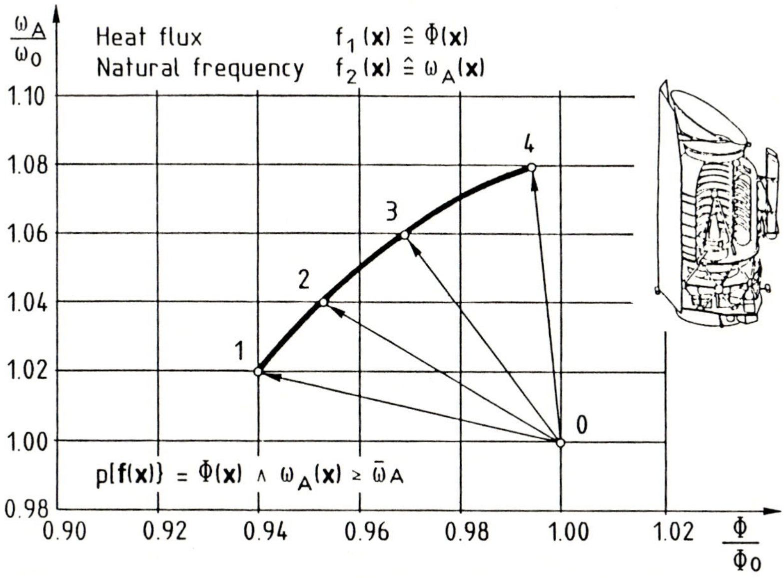

Fig. 6.1/8. Functional-efficient boundary for ISO-satellite

	Initial Design	Optimal Compromise Solutions			
	0	1	2	3	4
$\Phi(\mathbf{x})$	1.0	0.94	0.95	0.97	0.99
ω_A	1.0	1.02	1.04	1.06	1.08
x_1	1.0	1.82	1.82	1.82	1.82
x_2	1.0	0.92	0.97	1.08	1.20
x_3	1.0	0.79	0.82	0.88	0.94
x_4	1.0	0.42	0.43	0.48	0.53
x_5	1.0	0.88	0.91	0.98	1.05
x_6	1.0	0.47	0.49	0.55	0.61
x_7	1.0	1.00	1.00	1.00	1.00
x_8	1.0	0.74	0.78	0.87	0.97
x_9	1.0	1.65	1.65	1.65	1.65
x_{10}	1.0	2.99	3.08	3.24	3.42

Table 6.1/1. Initial design and functional-efficient solutions of ISO-satellite

The optimization results are summarized in Fig. 6.1/8 and Table 6.1/1 where the heat flux Φ, the frequency ω_A, and the design variables $\mathbf{x}$ are normalized by means of their initial values Φ_0, ω_0 and $\mathbf{x}_0$. The functional-efficient boundary shown in Fig. 6.1/8 provides the corresponding optimal heat flux within the eigenfrequency range $1.02 \le \omega_A \le 1.08$ which is important in practice. In this case, the functional efficient boundary has a nearly linear contour. Finally, the design with $\Phi = 0.99$ and $\omega_A = 1.08$ (point (4)) is chosen as the optimal compromise solution.

6.1.5 Conclusion

The previous section contributes to the application of the multicriteria approach to aircraft and spacecraft structures. The optimal design of a composite fin and a satellite structure are presented as real life examples. For the fin, an objective conflict occurs between the structural weight and the aeroelastic efficiency while the heat flux and a natural frequency are the competing objectives for the satellite. Since none of the possible designs permit a simultaneous optimal fulfillment of all objectives in either of the two cases, optimization strategies presented in Chapter 1 are used for transforming these multicriteria optimization problems to scalar substitute problems. Finally, the determination of the corresponding functional-efficient boundaries is a good device for the selection of the optimal compromise solution which is more appropriate for the respective demands.

The optimization procedure MBB-LAGRANGE is used for solving these problems. It is especially suited for large-scale problems with a high number of design variables and constraints. An important feature is the simultaneous treatment of several demands resulting from different disciplines. By means of defined interfaces and a user shell of pre- and postprocessors, broad intergration of CAE supports and improves the application of this procedure.

References

[1] Schmit, L.A.: Structural Design by Synthesis, Proc. 2nd Conf. Elet. Comp. ASCE, New York, (1960) 105-122

[2] Atrek, E.; Gallagher, R.H.; Ragsdell, K.M.; Zienkiewicz, O.C.: New Directions in Optimum Structural Design. Chichester, New York, Brisbane, Toronto, Singapore: John Wiley 1984

[3] Eschenauer, H.A.: Rechnerische und experimentelle Untersuchungen zur Strukturoptimierung von Bauteilen, DFG-Forschungsbericht, Forschungslaboratorium für Angewandte Strukturoptimierung, Universität-GH Siegen, 1985

[4] Kneppe, G.; Krammer, J.; Winkler, F.: Structural Optimization of Large Scale Problems Using MBB-LAGRANGE, 5th World Congress and Exhibition on FEM, Salzburg (Austria), 1987

[5] Kneppe, G.; Hartzheim, W.; Zimmermann, G.: Development and Application of an Optimization Procedure for Space and Aircraft Structures, Proc. GAMM-Seminar, Discretization Methods and Structural Optimization, Siegen (FR Germany), 1988

[6] Bisplinghoff, R.L.; Ashley, H.: Principles of Aeroelasticity, New York: Dover Publication 1962

[7] Gödel, H.; Hörnlein, H.: Aeroelasticity and Optimization in Fighter Aircraft Design, AGARD-Report No. 740, Special Course on Fundamentals of Fighter Aircraft Design, 1986

[8] Ross, C.: Strukturoptimierung mit Nebenbedingungen aus der Dynamik. Dissertation. Technische Universität München, to appear 1990

6.2 Multicriteria Design of Spacecraft Structures with Special Emphasis on Mass and Stiffness

H. Baier

6.2.1 Introduction and Technical Background

Spacecraft structures should generally have small mass, high stiffness, and sufficient safety margins. Their overall design heavily depends on the type of mission planned. whether it be a communication spacecraft, a scientific spaceprobe, or a science lab like Spacelab or the future Space Station. In any case, the spacecraft has to survive the launch loads with sufficient safety margin. They occur because of static and especially dynamic interaction with the launcher system.

In-orbit performance requirements are mainly related to (partly very stringent) thermal distortion limitations for communication platforms and scientific telescopes or to the dynamic interaction of flexible appendages such as solar arrays or large reflectors with an in-orbit flight control system. For larger structures, deployment from the stowed launch condition has to be considered. Therefore, the typical design problem for spacecraft structures is to minimize mass, dynamic response, (thermal) distortions and to satisfy constraints on frequencies, stresses, displacements and fabrication. In most cases, the overall design problem is governed by stiffness requirements while stress constraints usually become relevant locally in areas where loads are introduced. These very different demands are to be met by effective design approaches, by choosing specific materials such as fibre composites and, especially, by intensive applicaton of analysis and optimization tools. Undoubtedly, these tools have to be rather general and flexible [1]. On the other hand, special and important types of problems require the development and application of special tools. Some of these general and special approaches as well as their applications are outlined in the following.

6.2.2 Problem Statement and Discussion

Optimal design problems are seldom unique and may be formulated in different ways. Moreover, various mathematical statements for a given design problem may exist. This applies even more to multicriteria design where one can choose between different ways of transforming multicriteria problems into scalar ones. Thus, the problem statements and remarks given below should be considered as typical but not necessarily complete.

The optimal design problem of spacecraft structures is characterized by

objective functions:	minimize mass, strain energy, function of static and dynamic distortion or maximize eigenfrequencies;
constraints:	failure criteria, displacements, frequencies, mass, fabrication;
design variables:	cross-sectional areas, plate thicknesses, laminate and sandwich structure, dimensions and positions of stiffeners, shape, etc.;
system equations:	finite element method; special approaches in special classes (e.g. for axisymmetric shells);
loads:	static, thermal, dynamic (harmonic; lower frequency transients; higher frequency random loads).

This nonlinear, usually nonconvex and implicit multicriteria optimization problem puts severe demands on computational efficiency, especially when dynamics have to be taken into account. This problem is discussed in more detail in [2].

For the sake of briefness, some more precise and simple problems are given below. Large scale applications, related to those given above, are outlined in Section 6.2.5. For a large and relatively heavy spacecraft, such as the multimirror X-ray satellite XMM (see Fig. 6.2/1), ensuring that a lower bound for the fundamental (bending) eigenfrequency is exceeded, is a design task to be solved already in earlier design stages. This is neccessary to avoid severe dynamic interaction with the launcher. The corresponding mathematical statement reads as follows:

- minimize: mass, dynamic response at specific degrees of freedom;
- maximize: fundamental eigenfrequency;
- constraints: lower/upper bounds of design variables.

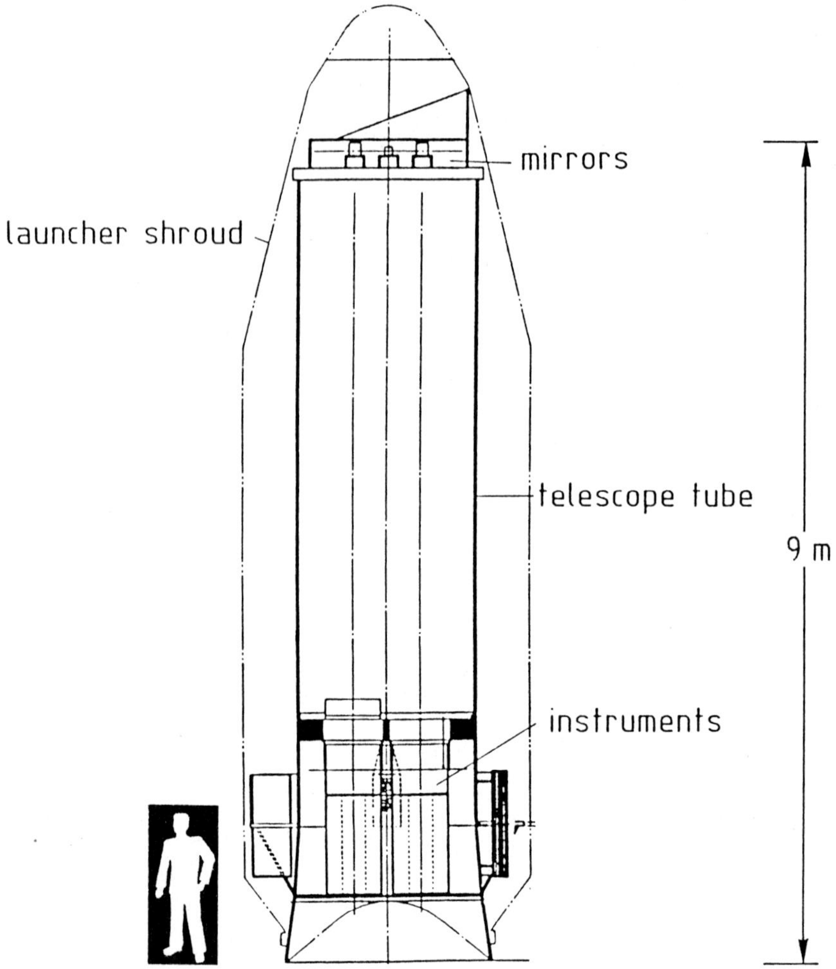

Fig. 6.2/1. Configuration of an X-ray satellite XMM

The constraints are introduced to keep the design variables (laminate build-up, thicknesses) within reasonable bounds.

For large scientific space telescopes, high shape accuracy with low mass is to be achieved. Thereby, shape accuracy is defined by a bestfit rms-deviation of the real (e.g. thermally) deformed shape from the ideal shape. The optimization model is:

- minimize: mass, rms shape deviation;
- constraints: lower and upper bounds on design variables, and lower bounds on eigenfrequencies in stowed and orbit configuration.

Thus, different configurations and related system equations have to be observed in one problem statement.

A typical example of a high precision reflector for the **F**ar **I**nfrared and **S**ubmillimeter space **T**elescope FIRST is shown in Fig. 6.2/2. It has to be stowed for launch and is then accurately unfolded in orbit to a diameter of 8 m. The total shape error (including those under environmental loads) must not exceed 10 μm.

Because of the high ratio of strength and stiffness to mass and the very low coefficient of thermal expansion α_T, carbon fibre composites are extensively applied, i.e. the corresponding laminate and sandwich structures are thus part of the design variables. For fibre laminates or a honeycomb core sandwiched between these laminates, the mass m and the weighted sum of the different α_T, which occur because of anisotropic material properties, are taken as objectives and are to be minimized:

$$f_1(\mathbf{x}) = m\ (t_1, t_2, \ldots t_n)\ ,$$
$$f_2(\mathbf{x}) = \sum_j w_j\, \alpha_{Tj}^2\ (t_1, t_2, \ldots, t_n;\ \alpha_1, \alpha_2, \ldots, \alpha_n)\ . \qquad (6.2\text{-}1)$$

The constraints are: failure criteria, stiffnesses, lower/upper bounds on design variables;

The system equations are: laminate plate theory for plates consisting of stacked orthotropic plies.

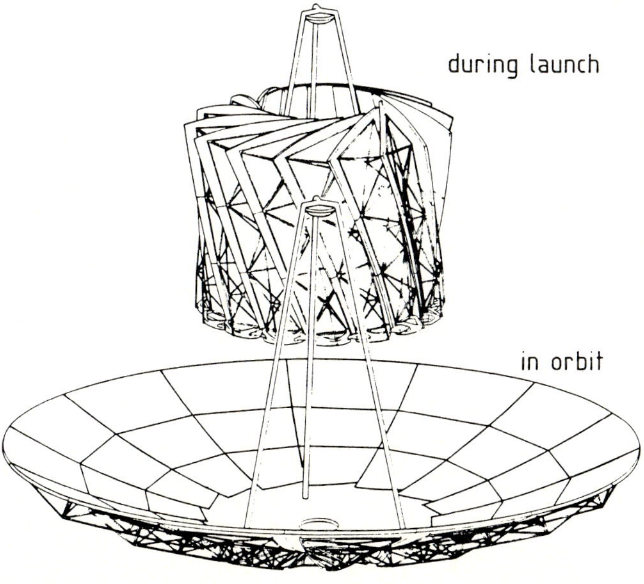

Fig. 6.2/2. High precision reflector FIRST (diameter ~ 8 m)

In (6.2-1) the design variables t_i and α_i are the thickness and fibre angle of the i-th lamina, respectively. The weighting factors w_j emphasize the importance of low α_{Tj} in specific directions. Since the α_{Tj} may become negative for carbon fibre laminates, it is important to use square (or absolute) values within the objective function vector. This design task is formulated as a multicriteria problem instead of constraining the α_T and minimizing mass because the selection of reasonable and feasible constraint levels for α_T might be problematic. On the one hand small α_T are desirable while, on the other hand, constraint levels which are too small could cause difficulties in finding feasible variables or could even generate an empty feasible set. Since in laminate plate theory fibre angles are introduced via trigonometric functions, the nonconvexity of the corresponding optimization problems is often more prevalent than in many other types of structural optimization in statics.

6.2.3 Considerations on Optimality Criteria

In Section 1.2 optimality conditions, similar to those of the Kuhn-Tucker conditions for scalar optimization, have been discussed for multicriteria problems. They may be used for the verification of optimality after a design optimization process has been applied, or they may serve as a base-line for constructing optimization strategies. In addition, some more general observations based on optimality criteria can be made for optimal solution sets and the corresponding properties for certain types of structures with mass and stiffness as objective functions. In order to make the discussion more explicit, the three bar truss of Fig. 6.2/3 is used for illustration. The cross-sectional areas x_1, x_2, and x_3 are to be determined so that the objectives of structural mass (volume V) and the work W_j of the applied forces (as a measure of compliance) are minimized while stress constraints are satisfied, i.e.

$$\text{minimize } (V(\mathbf{x}),\ W_1(\mathbf{x}),\ W_2(\mathbf{x}))^T \tag{6.2-2}$$

$$\text{subject to} \quad \sigma_{ij}(\mathbf{x})/\sigma_i^u - 1 \le 0\ , \qquad i = 1,2,3; \qquad j = 1,2\ ,$$

$$x_i - x_i^u \le 0\ ,$$

$$\text{where} \quad W_j = \frac{1}{2} F_j^T u_j(\mathbf{x}) \tag{6.2-3}$$

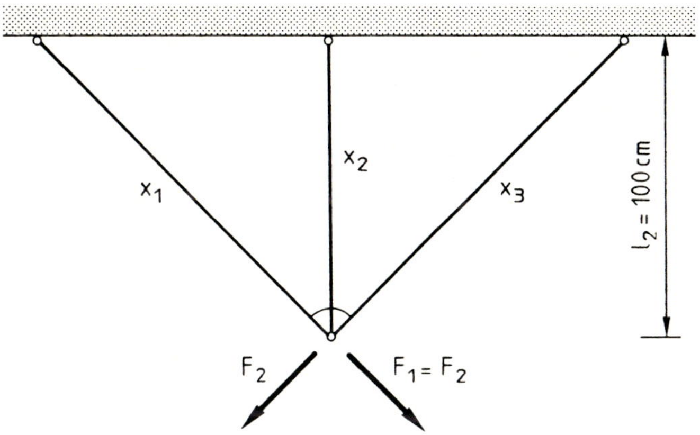

Fig. 6.2/3. Three bar truss with the loads F_1 and F_2
$\sigma_{adm} = 2 \cdot 10^2$ N/mm^2, $E = 7.5 \cdot 10^4$ N/mm^2

is the work of the applied force in the j-th loading case and i is the number of bars. The symmetry of the truss and its loading shall be reflected in the preference order, i.e. $f_2(\mathbf{x}) = f_3(\mathbf{x})$. Fig. 6.2/4 shows the relevant part of the transformation of the design space into the criterion space. The strong line represents the Pareto-optimal or functional-efficient solutions with the numbers being the weighting factors w used to transform the multicriteria problem into a scalar problem via

$$p[\mathbf{f}(\mathbf{x})] = w \cdot V + 0.5\,(1-w)\,(W_1 + W_2) \quad . \qquad (6.2\text{-}4)$$

Although the order of magnitude of the objectives is identical, it is interesting to note that only small differences within the objectives occur for a range of $0.6 < w \leq 1$ while for $w \leq < 0.4$ significant variations in f_1 are achieved. This shows that an a priori and fixed selection of weighting factors is questionable.

Since this problem is not strictly convex, different design variable vectors may exist for one Pareto-optimal point on the functional-efficient boundary (see Fig. 6.2/4). This, for example, applies to the following case:

$$\mathbf{f}(\mathbf{x}) = \begin{bmatrix} f_1(\mathbf{x}) \\ f_2(\mathbf{x}) \\ f_3(\mathbf{x}) \end{bmatrix} = \begin{bmatrix} 252 \\ 2050 \\ 2050 \end{bmatrix} \rightarrow \begin{array}{l} \mathbf{x}_1 = (0.7,\ 0.57,\ 0.7)^T \ , \\ \mathbf{x}_2 = (0.79,\ 0.3,\ 0.79)^T \ . \end{array}$$

It is also important to consider the hyperbolic behaviour of the major part of the functional-efficient boundary which is shown in Fig. 6.2/4.

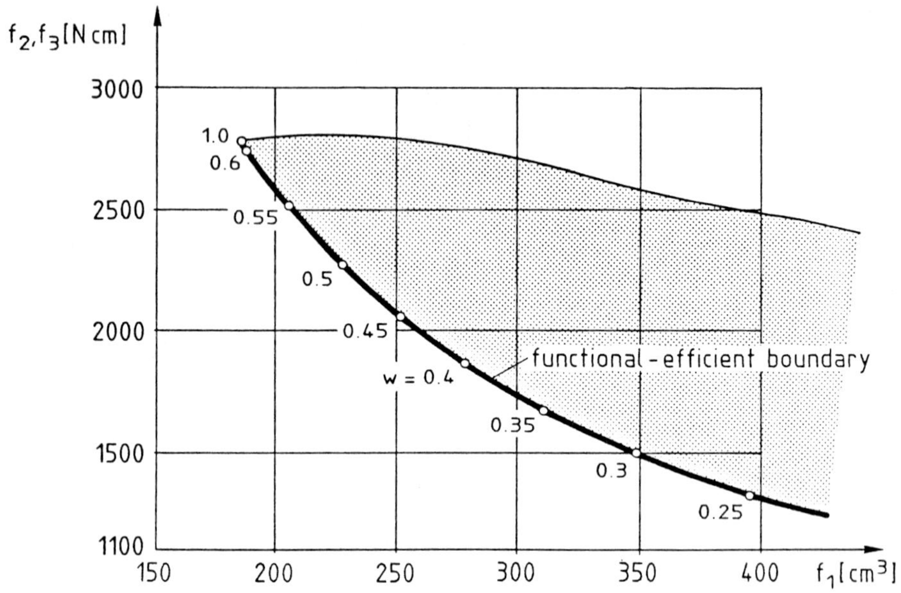

Fig. 6.2/4. Objective function space for a three bar truss

The multiplication of the design variables belonging to e.g. $\mathbf{f}(\mathbf{x}) = (252, 2050, 2050)^T$ by a factor of 1.4 yields the objective function values $\mathbf{f}(\mathbf{x}) = (350, 1465, 1465)^T$. These also prove to be Pareto-optimal. As shown by the optimality criteria in [3], the cross-sectional areas of any truss structure with Pareto-optimal objectives (6.2-2) can be mulitplied by a certain factor to generate another Pareto-optimal solution. This can be generalized for other types of structures, and the dynamic compliance may be included.

6.2.4 Solution Strategies

Solution strategies have to guarantee Pareto-optimality as well as to allow the introduction of the preference order while also satisfying practical requirements such as computational efficiency, ease of use etc. One way to classify the different strategies for multicriteria problems is to divide them into " a priori" and "ad hoc"-methods. In the first case, which forms a basis for the latter one, the user has to specify the preference order *before* a computation is started , whereas in the latter case the user is able to specify and to modify it *during* the computational process. The common feature of both cases is the mapping of the multicriteria problem into single criteria problems which may change in the case of "ad hoc"-methods.

Two very frequently applied a priori-methods are the objective weighting approach and the constraining of all but one of the objective functions. The approach of the type (1-11)

$$p[\mathbf{f}(\mathbf{x})] = \sum w_j f_j(\mathbf{x}) \tag{6.2-5}$$

is relatively simple, but a reasonable a priori-selection of w_j is often problematic. This mainly applies to objectives with a very different order of magnitude and a very different physical meaning. Moreover, in nonconvex cases this transformation is not unique. Constraining all but one of the objectives has to be used with care in nonconvex problems, and a specification of reasonable constraint levels might be difficult. Thus, the implementation of interactive and ad hoc-methods is highly desired. This concept was outlined in Chapter 3 and will be emphasized in the following as well.

In the case of the interactive approach of the preference objective, as originally published by Geoffrion (see Section 3.2.5) and Dyer [4], one has to determine beforehand to what extent an increase in one objective will be accepted in favour of the decrease of another objective. The decision maker compares two different objective function vectors proposed by the computer: one objective function is a baseline version whereas the other one varies the change Δf_j of objective f_j until a state of indifference to the baseline vector is reached. This means that

$$\frac{t_j}{t_i} \approx -\frac{\Delta f_i}{\Delta f_j} \; . \tag{6.2-6}$$

By varying the index j, all weighting factors can be determined up to a constant factor. A consistency check can be introduced by

$$\frac{t_j}{t_i} \overset{!}{\approx} \frac{t_j}{t_k} \cdot \frac{t_k}{t_i} \; . \tag{6.2-7}$$

When all weighting factors have been determined, a search direction or a direction of design improvement $\boldsymbol{\delta}$ can be determined by any technique that requires gradients only (e.g. methods of feasible direction). Establishment of the search direction by the computer is then followed by

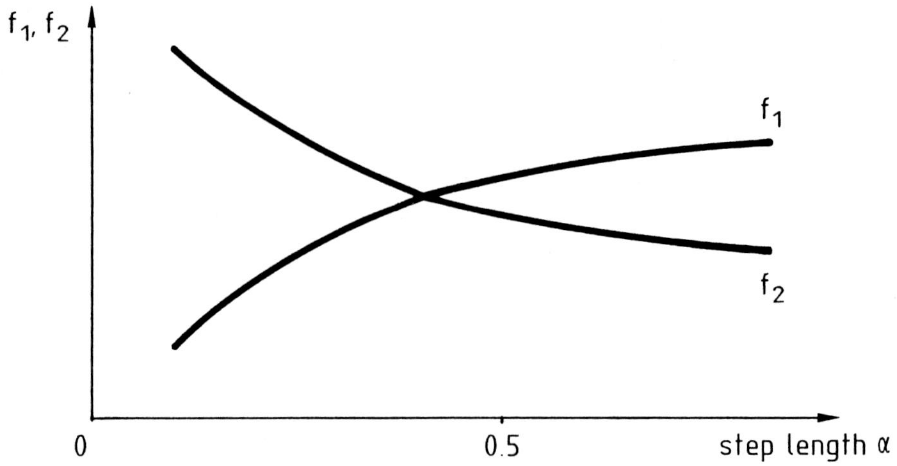

Fig. 6.2/5. Change of (hypothetical) objectives as a function of step length α

a one-dimensional search which determines a positive value $\alpha \leq \alpha_{max}$ so that for

$$\bar{\mathbf{x}} = \mathbf{x} + \bar{\alpha}\,\boldsymbol{\delta}$$

the objective function vector $\mathbf{f}(\bar{\mathbf{x}})$ is preferred against all other $\mathbf{f}$ for any (positive) $\alpha \leq \alpha_{max}$. In many cases α_{max} is determined by using appropriate interpolation functions for the constraints. Again, the step length $\bar{\alpha}$ can and should be determined interactively by assessing the behaviour of the different objective functions. Fig. 6.2/5 outlines a typical situation where objective 1 represents the mass and objective 2 the compliance function. If the decision maker is ready to accept a relatively high increase of the mass f_1 in order to decrease the compliance f_2, $\bar{\alpha}$ will be selected in the range of 0.6 to 0.8. In the opposite case $\bar{\alpha}$ would be chosen at about 0.2.

6.2.5 Examples of Spacecraft Structural Optimization

The following examples illustrate how the optimization of spacecraft structures is applied. These examples put emphasis on multicriteria problems and are furthermore related to the problem statements of Section 6.2.2.

l) Optimal design of fibre laminates

Fibre laminates, especially those based on carbon fibres are extensively applied to spacecraft structures because of their high stiffness and strength, their low specific mass, and their low α_T. These properties mostly depend on the laminate structure, and a reasonable constraint limit is often difficult to specify, especially for α_T. Thus, a multicriteria formulation of type (6.2-1) corresponds to a combined objective

$$p[\mathbf{f}(\mathbf{x})] = m(\mathbf{x}) + \sum_i w_i \alpha_{Ti}^2(\mathbf{x}) \ , \qquad (6.2\text{-}8)$$

where m is the mass and w_i stands for the weighting factors. Fig. 6.2/6 gives a first and simple example. It considers a stack of two lamina where the fibre angle and the lamina thickness are the design variables. In an extension not considered in detail here, the type of material (e.g. different glass (GFRP) and carbon-reinforced plastics (CFRP) has also been used as an additional variable. The objectives $f_1(\mathbf{x})$ = mass and $f_2(\mathbf{x}) = \alpha_{T1} + \alpha_{T2}$ are outlined in Fig. 6.2/7. Here, the fibre angle $\pm\,\alpha$ has been chosen as a parameter, and thicknesses are selected so that the failure criteria just take on critical values. It can be seen that minima are obtained for $\alpha = 0^o$ and $\alpha = 45^o$, respectively. Therefore, Pareto-optima are located between $0^o \le \alpha \le 45^o$. However, if the preference function $p[\mathbf{f}(\mathbf{x})]$ of (6.2-8) is used, the discrete minima of $p[\mathbf{f}(\mathbf{x})]$ are exactly at $\alpha = 0^o$ or $\alpha = 45^o$ depending on the weighting factor. For that reason the Pareto-optima other than at $\alpha = 0^o$ or $\alpha = 45^o$ cannot be computed numerically in this nonconvex

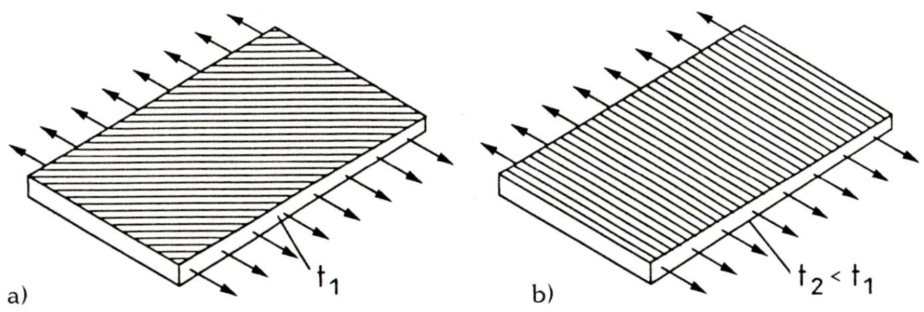

Fig. 6.2/6. Composite layer optimization including simultaneous material selection

a) Initial design with material GFRP

b) Optimal design with material CFRP (T800)

case whatever variation of w is used. This shows again that the transformation of multicriteria optimization problems into scalar optimization problems of type (6.2-5) have to be performed carefully.

2) Mass and dynamic stiffness optimization

Fig. 6.2/1 shows the sandwich cylinder of the main structure of a future multimirror X-Ray satellite XMM with a truss-type structure considered to be an alternative. Due to its height and its relatively large mirror masses of about 1 ton on top of the spacecraft, exceeding the specified, but nevertheless not very strict, lower bounds of the fundamental eigenfrequency poses a basic preliminary design driver. Thus, optimization computers run minimized structural mass while satisfying eigenfrequency constraints and lower bounds for the design variables which are the cross-sectional areas of truss and sandwich structures. For both versions mass could be improved by approximately 10% compared to each starting design, and the cylindrical sandwich shell structure is superior compared to a truss. Thus, a clear comparison between the two versions could be made on the basis of their individual optima.

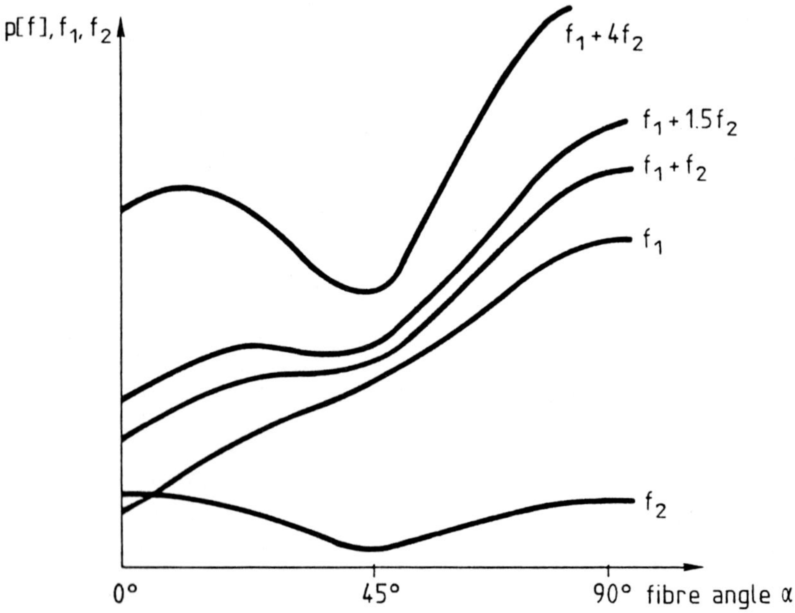

Fig. 6.2/7. Objectives $f_1(\mathbf{x})$ (weight) and $f_2(\mathbf{x}) = \sum_i |\alpha_T|_i$ as a function of the fiber angle α; preference objective $p[\mathbf{f}] = f_1 + w\, f_2$;

Since the constraint level of the eigenfrequency is not strictly fixed (10 Hz would be safe, but 6–8 Hz would still be possible from the point of view of a flight control), a parameter study of optimal mass vs frequency can be carried out as follows:

$$\text{minimize } p[\mathbf{f}(\mathbf{x})] = m(\mathbf{x}) + w\,\boldsymbol{\omega}(\mathbf{x}), \quad x_i^l \le x_i \le x_i^u , \qquad (6.2\text{-}9)$$

where $\boldsymbol{\omega}$ is the fundamental eigenfrequency. This is a simple linearly constrained nonlinear problem. A reasonable value for the weighting factor w can be derived from the relationship between w and the Lagrangian multipliers obtained by solving the Kuhn-Tucker equations for the frequency constrained problem discussed above. Nevertheless, it is still a problem to reasonably vary w within the anticipated frequency spectrum. This can be partly solved by adding frequency constraints to 6.2-9.

3) Optimization of a precision reflector

Precision or dimensionally stable spacecraft structures such as telescopes and antenna reflectors have to maintain their shape with high accuracy even under severe (mainly temperature) loads in order to provide the required electromagnetic performance. A typical example already mentioned is the far infrared observatory FIRST, as shown in Fig. 6.2/2, which is to be developed for ESA. Different disciplines (and design

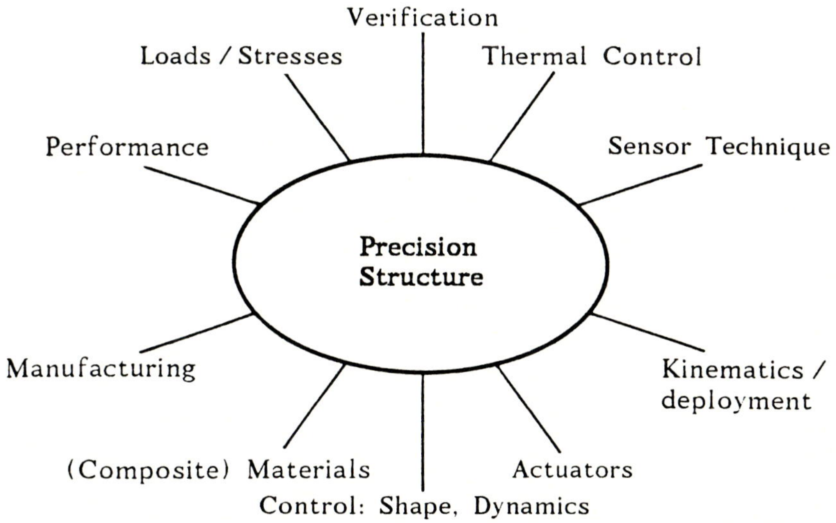

Fig. 6.2/8. Disciplines to be considered in precision structure design

objectives) are involved in the general task (Fig. 6.2/8). With respect to project scheduling and knowledge acquisition processes, a simultaneous consideration of all these aspects is neither feasible nor possible. The inherent iterative nature of the development process can be reflected by an appropriate decomposition into different subproblems.

The telescope FIRST is operating in the far infrared wavelength range and thus imposing severe shape and pointing accuracy requirements. The total rms surface deviation error is to be below 10 μm. The 8 m center fed Cassegrain antenna with quasi-rigid panels has to be folded for a launch by Ariane and has then to be deployed in orbit. The panels will consist of backup truss stiffened sandwich shells with fibre composite facesheet and honeycomb cores.

The iteration history for this dimensional stability optimization problem is presented in Fig. 6.2/9. Though the starting vector of the engineering design was already a very efficient one for this optimization, considerable improvements could still be achieved. Optimization can be favourably applied in such cases where one does not have a great deal of experience, in contrast to the classical weight minimization problem under stress constraints only.

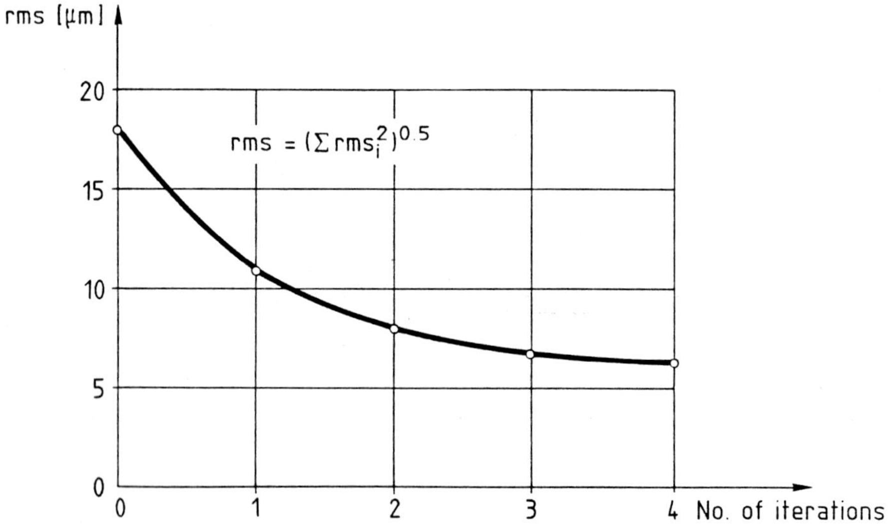

Fig. 6.2/9. Iteration history for shape deviation minimization of FIRST telescope

4) Shape optimization of axisymmetric shell structures

Axisymmetric shells are often applied in launcher and spacecraft structures, for example as a spacecraft central structure or launcher components. The shell structures often consist of a honeycomb core sandwiched between composite material facesheet, a stringer stiffened thin shell, or a corrugated shell. These structures should have a minimum weight and at the same time they should satisfy stiffness, strength, as well as global and local stability constraints. As an example the cylindrical and conical upper shell of a launcher providing the interface to the spacecraft is shown in Fig. 6.2/10. This sandwich shell structure SPELTRA with a diameter of about 5 m has to carry loads which generate forces on the bottom section of N = 730 kN and Q = 240 kN in the vertical and lateral direction, respectively, and a bending moment of M = 3800 kNm.

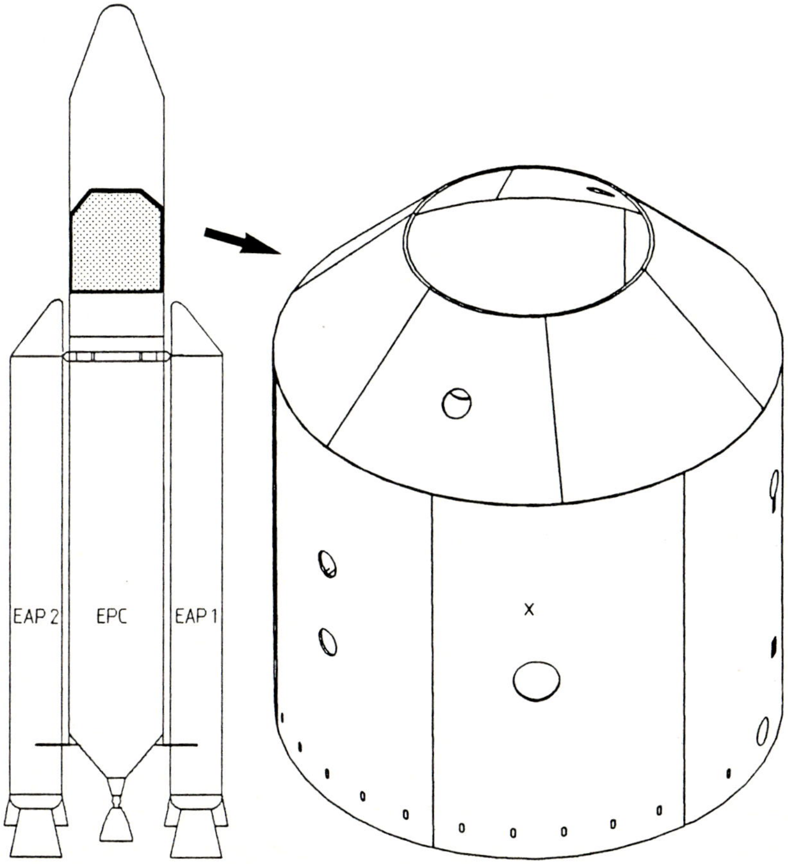

Fig. 6.2/10. Sandwich shell structure carrying the spacecraft in the upper part of a launcher (Ariane 5)

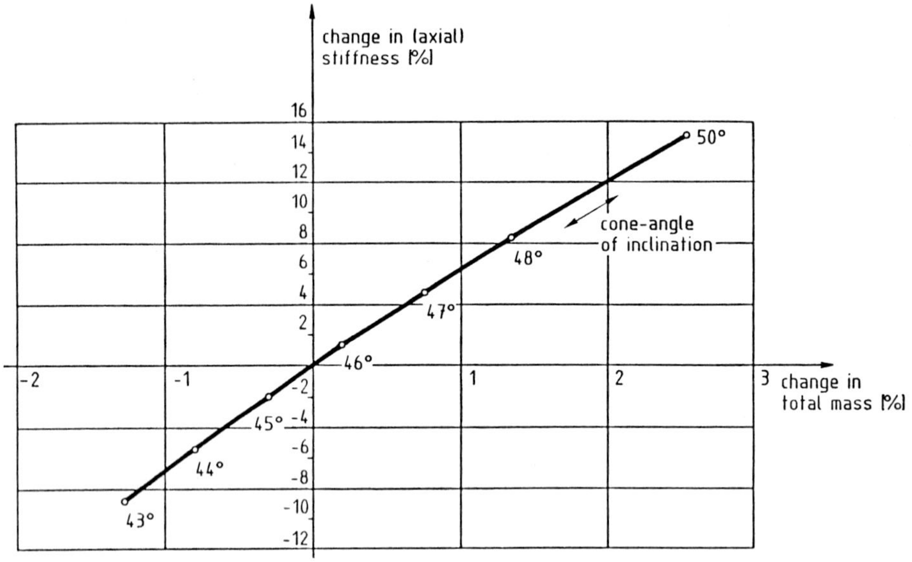

Fig. 6.2/11. Optimal trade-off between stiffness and mass of SPELTRA with a fiber layup and a conical shell inclination as design variables

In addition to design variables describing the sandwich structure or the positioning and stiffness of stringers, shape properties such as the shell contour between given interface coordinates can be used. This is done in order to minimize mass while at the same time maximizing the volume or stiffness or minimizing stress intensities at intersections of different shells. The multicriteria optimization results for mass and stiffness of the conical upper shell are shown in Fig. 6.2/11. The angle of inclination proved to be quite a sensitive parameter. Any deviation from the "linear" cone shape has actually no effect and is also not desirable from a manufacturing point of view.

6.2.6 Conclusion

The application of mathematical structural optimization proves to be worthwhile because it delivers optimal results for design problems which can be stated mathematically and because it can save development time and costs.

Many design optimization problems are multicriteria problems and should be considered as such. Most of the solution algorithms for these problems require a transformation into single criteria problems. This is a nontrivial step for many practical cases. Nevertheless, after one gets acquainted with the problem behaviour, the approaches can be applied to relatively large and complex design problems with considerable benefit in improved structural performance. On the other hand, the approaches cannot yet be considered as black box tools in engineering practice with its very different design tasks. Instead, flexibility in using different algorithmic approaches and problem statements is required. Strategies for a proper decomposition of large and complex problems into smaller subproblems and means to introduce engineering knowlede into the optimization process will lead to further progress.

References

[1] Specht, B.; Helwig, G.: OPOS, A General Computer Package for Optimization of Structures, Report MEB 67/88. Dornier System. Friedrichshafen, 1988

[2] Baier, H.: Structural Control and Optimization of Structures with Constraints in Dynamics. Proceedings 5th Int. Modal Analysis Conf. London, (1987) 903-908

[3] Baier, H.: Algorithmen zur Ermittlung und Charakterisierung Pareto-optimaler Lösungen bei Entwurfsaufgaben für Tragwerke. Zeitschrift für Angewandte Mathematik und Mechanik (ZAMM), Band 57 (1977) 318-320

[4] Dyer, J.S.: A Time-Sharing Computer Program for the Solution of the Multicriteria Problem. Management Science 19 (1973) 1379-1383

7 MACHINE TOOL DESIGN

7.1 Application of Multicriteria Optimization Methods to Machine Tool Structural Design

M. Yoshimura

7.1.1 Evaluative Factors of Machine Tools

Machine tools are the most fundamental and essential machines in industrial manufacturing shops. Performance of machine tools in terms of machine accuracy (high precision) and machine productivity (high working efficiency) greatly depends on the static rigidity and dynamic characteristics of the machine tools [1-3]. Therefore, until now the design optimization of machine-tool structures has been studied intensively [4-8]. Small static and vibrational displacements are mainly evaluated for realizing high machine accuracy. Far smaller displacements, than those existing at the state of machine failure (breakdown), at which the stress is usually known only approximately, must be considered for evaluating the performance of machine tools. That is, the design of machine tools is a displacement criterion design problem. The static rigidity and dynamic characteristics also depend on the design of the machine structures which are composed of structural members and machine elements as well as of joints connecting the structural members and machine elements.

Mutual interaction often occurs between many characteristic and evaluative factors of machine-tool dynamics, and competitive and cooperative relations exist between them. Hence, the relationships between design variables and evaluative factors are very complicated. If those complicated relations are not clarified before the formulation of the design optimization problem, the possibility of a poor convergence to a local optimum is very high, and it is often difficult to obtain a design solution with a great improvement of the product performance (or great reduction of the product manufacturing cost). Furthermore, practical machine-tool structures have an enormous number of design variables. It is not easy to use mathematical programming methods to determine all the design variables simultaneously.

Fig. 7.1/1 shows the framework of simulation models of a milling machine and a vertical boring and turning lathe for structural analysis. In machine tools, static and dynamic behaviour between the tool (point A) and the work (point B) at the cutting point have direct effects on machine performance such as machine accuracy and machine productivity. For evaluating the performance, the tool work relative receptance frequency response R (=d/F) of the relative displacement d (between the tool point A and the work point B as shown in Fig. 7.1/1) caused by the exciting force F at the cutting point, such as shown in Fig. 7.1/2, is calculated by structural analysis. The receptance is also called "dynamic compliance".

There are two types of vibrations: forced and self-excited ones. In the case of forced vibrations, the relative displacement d is caused by the harmonic force F at an exciting force source point, while in the case of self-excited vibration (regenerative chatter) d is caused by F excited between points A and B at the cutting point. Regenera-

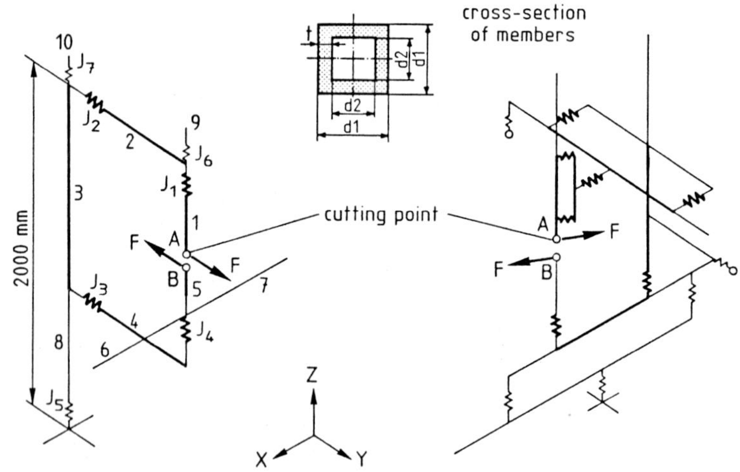

a) Milling machine

b) Single-column vertical boring and turning lathe

Fig. 7.1/1. Simulation models for the structural analysis of machine tools tive chatter is the most troublesome problem which must be evaluated at the design stage of machine tools because after construction of the machine tools it is very difficult to solve the problem.

The transfer function of the receptance frequency response, concerning the exciting force F in the cutting force direction, and the relative displacement d between the tool and the work at the cutting point at a given frequency ω, is expressed approximately, for vibrations with viscous damping, as follows:

$$R(\omega) = \frac{d}{F}(\omega) = \sum_{m=1}^{\infty}\left[\frac{\psi_m}{1-(\frac{\omega}{\omega_m})^2+2j(\frac{\omega}{\omega_m})\zeta_m}\right]. \qquad (7.1\text{-}1)$$

where j stands for the imaginary unit, and ω_m and ζ_m are the natural frequency and the damping ratio at the m-th natural mode. ψ_m is the modal flexibility between the exciting force and the displacement at the cutting point.

Since R(0) is equivalent to the static compliance ψ_s, substitution of zero for ω in the equation (7.1-1), establishes the following equation showing the relation between the modal flexibility ψ_m and the static compliance ψ_s [9]:

$$\psi_s = \sum_{m=1}^{\infty}\psi_m . \qquad (7.1\text{-}2)$$

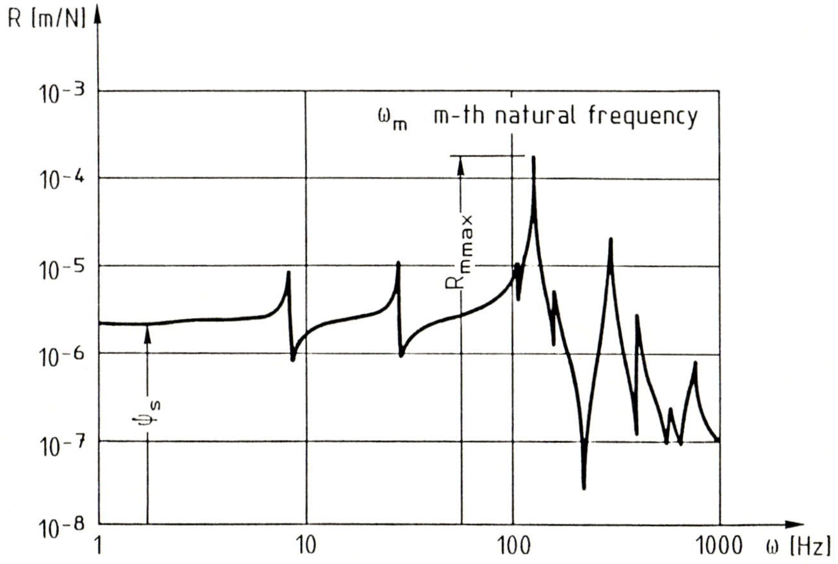

Fig. 7.1/2. Example of the receptance frequency response curve at the cutting point

That is to say, the total sum of the modal flexibility ψ_m over all natural modes is equal to the static compliance ψ_s. Furthermore, the following relation is obtained from the equation (7.1-2).

$$\sum_{m=1}^{\infty} (\psi_m / \psi_s) = 1 \ . \tag{7.1-3}$$

ψ_m / ψ_s stands for the contribution of the m-th natural mode to the static compliance and also serves as a basis for checking the balance of the flexibility among all natural modes. The damping ratio of a practical machine-tool structure is , for example, from 0.01 to 0.1. The outline of a receptance frequency response curve, as shown in Fig. 7.1/2, is determined by the values of modal flexibilities rather than by the values of damping ratios.

The modal flexibility ψ_m at the natural frequency ω_m, having the maximum receptance value $R_{m.max}$, has the maximum or almost maximum value among all modal flexibilities over the entire frequency range. Usually, the ratio ψ_m / ψ_s is 0.2 to 0.9 depending on the structural design. When the receptance value at the m-th natural frequency is greater than that at any natural mode, the magnitude of $R(\omega_m)$, i.e., $R_{m.max}$ can be approximated by the following equation:

$$|R(\omega_m)| = R_{m.max} \cong \frac{\psi_m}{2\zeta_m} \ . \tag{7.1-4}$$

The following evaluative factors are considered at the design stage of machine-tool structures:

(i) Static Compliance ψ_s

If the relative displacement between the tool and the work at the cutting point caused by the cutting force is large, sufficient machine accuracy cannot be obtained. The static compliance ψ_s is equal to the relative displacement between the tool and the work caused by a unit cutting force. It is obtained as the receptance value at the frequency ω of zero on the receptance frequency response such as shown in Fig. 7.1/2 and must be small.

(ii) Maximum Value of Receptance over the Entire Frequency Range $R_{m.max}$

If regenerative chatter occurs in an operational state of a machine tool, the vibrational displacement between the tool and the work increases greatly, and the surface roughness of the workpiece also gets worse. This results in extremely low machine accuracy and low machine productivity. The stability limit against regenerative chatter depends on the maximum value $R_{m.max}$ of the tool-work relative receptance R (the maximum real negative part is evaluated in a strict theoretical sense [2,10], but because it requires an unnecessarily complicated analysis, the absolute maximum value $R_{m.max}$ is usually used for practical application). In Fig. 7.1/2 the maximum receptance value $R_{m.max}$ occurs at the m-th natural mode. In order to increase the stability limit against regenerative chatter, $R_{m.max}$ must be decreased. As shown in Eq. (7.1-4), minimization of $R_{m.max}$ is realized by minimizing the modal flexibility ψ_m and maximizing the damping ratio ζ_m.

(iii) Total Weight of the Structural Members W_T

A machine-tool structure is composed of structural members such as columns and cross-rails, machine elements such as gears and motors, and joints which connect the structural members and the machine elements. It is usually necessary to minimize the total weight of the structural members and the machine elements to achieve savings of material cost and of operational power cost to ease the positioning control, etc.

(iv) Natural Frequency ω_m

In forced vibrations, natural frequencies around forced vibrational frequencies must be evaluated.

(v) Damping Ratio ζ_m

Greater damping ratios are preferable. Especially, the damping ratio at the natural mode having a great receptance value should be large.

Among the aforementioned evaluative factors, the value of the static compliance ψ_s depends on the design of structural members, indicated by the heavy lines, and joints, indicated by the heavy spring marks as shown in Fig. 7.1/1. These structural members and joints support the static cutting force at the cutting point. The path of heavy lines from point A to point B is called "the static force loop". (The structural members 1 through 5 and the joints J_1 through J_4 are the structural members and joints on the static force loop in Fig. 7.1/1 a)). The total weight of the structural members on the static force loop is denoted by W_s.

The design of the structural members, machine elements, and joints outside the static force loop has no effect on the static compliance ψ_s, but it has a great effect on the ratio ψ_m/ψ_s of modal flexibility ψ_m at each natural mode to the static compliance ψ_s. (The structural members 6 through 10 and the joints J_5 through J_7 are the structural members and joints outside the static force loop in Fig. 7.1/1 a)). When the ratio ψ_m/ψ_s is close to unity, the potential (strain) energy distribution at the static force application state is similar to that at the m-th natural mode. In this case, at the m-th natural mode, potential energy is almost never stored in the structural members, machine elements, and joints outside the static force loop.

When design changes for improving one characteristic lead to an improvement of another characteristic here, the two characteristics are said to be in a "cooperative" relationship. On the other hand, when design changes for improving one characteristic bring about degradation of another characteristic, the relationship between the two characteristics is described as a "competitive" one.

7.1.2 Competitive and Cooperative Relationships between Evaluative Factors

a) Pareto optimum solution set between two objectives minimizing the total weight and the static compliance

The static compliance at the points considered (such as the cutting point) and the total weight of the structural members having an influence on the static compliance are the most fundamental characteristics. For higher machine accuracy and productivity, it is required that both the total weight and the static compliance should be as small as possible. However, maximization of the static rigidity of a structural system (that means minimization of the potential energy stored at the member or minimization of the static compliance) and minimization of the weight of the structural system have a conflicting relationship in the Pareto-optimum solution set in the criterion space of the two objectives.

When n is the total number of the structural members on the static force loop, the total weight of the structural members on the static force loop W_s reads as follows:

$$W_s = \sum_{i=1}^{n} \rho_i l_i A_i g \qquad (7.1\text{-}5)$$

where l_i, A_i, ρ_i and g are the length, the cross-sectional area, the mass density of the i-th structural member and the gravitational acceleration, respectively.

When the static force F is applied between points A and B at the cutting point of a machine tool model (such as shown in Fig. 7.1/1), the static compliance ψ_s is calculated using the potential energy U_i stored at the i-th structural member as follows:

$$\psi_s = \frac{2}{F^2} \sum_{i=1}^{n} U_i \quad . \tag{7.1-6}$$

If the static external force F is assumed to be a unit force, then (7.1-6) becomes:

$$\psi_s = 2 \sum_{i=1}^{n} U_i \quad . \tag{7.1-7}$$

Now, we consider a case where the structural model of the static force loop is composed of structural members having square cross-sections. Design variables are the outside width and the inside width of the cross-section at the i-th structural member which are denoted $d1_i$ and $d2_i$, respectively. The constraints are written as follows:

$$g_{1i} = d1_i - d_{imax} \le 0 \ , \qquad i = 1,2,...,n \ , \tag{7.1-8}$$

$$g_{2i} = d_{imin} - d2_i \le 0 \ , \qquad i = 1,2,...,n \ , \tag{7.1-9}$$

$$g_{3i} = 2t_{imin} - (d1_i - d2_i) \le 0 \ , \qquad i = 1,2,...,n \ , \tag{7.1-10}$$

where d_{imax}, d_{imin}, and t_{imin} are the upper bound of the cross-sectional outside width, the lower bound of the cross-sectional inside width, and the lower bound of the wall thickness t of the i-th structural member, respectively.

Then, a multicriteria optimization problem minimizing both the total weight of the structural members on the static force loop W_s and the static compliance ψ_s is expressed according to (1-8):

$$\min_{x \in \mathbb{R}^n} \left(W_s, \psi_s \right)^T$$

subject to constraints:

$$g_{1i} \le 0, \qquad i = 1,2,...,n \ , \tag{7.1-11}$$

$$g_{2i} \le 0, \qquad i = 1,2,...,n \ , \tag{7.1-12}$$

$$g_{3i} \le 0, \qquad i = 1,2,...,n \ . \tag{7.1-13}$$

The multicriteria optimization problem considered here is most fundamental. The problem can easily be solved by the method of objective weighting. The multiple objectives, minimization of the total weight of the structural members on the static force loop W_s and the static compliance ψ_s are transformed into a scalar preference function $p[f(\mathbf{x})]$ by introducing a weighting factor $w(0<w<1)$ and using the maximum value W_{smax} of W_s and the maximum value ψ_{smax} of ψ_s as follows (see (1-11)):

$$p[f(\mathbf{x})] = w \frac{W_s}{W_{smax}} + (1-w)\frac{\psi_s}{\psi_{smax}} \quad . \tag{7.1-14}$$

The Lagrangian L is then formulated by applying multipliers β_{1i}, β_{2i}, and β_{3i} $(i = 1,2,\ldots,n)$ to constraint functions of Eqs. (7.1-11) through (7.1-13) as follows (see (1-5)):

$$L(x,\beta) = p[f(\mathbf{x})] + \Sigma\,(\beta_{1i}g_{1i} + \beta_{2i}g_{2i} + \beta_{3i}g_{3i}) \tag{7.1-15}$$

The Kuhn-Tucker necessary conditions for optimality (1-6) lead to the following equations [12]:

$$\frac{\partial L}{\partial d1_i} = \frac{2a_i\,d1_i}{W_{smax}}\,w_i - \frac{8e_i d1_i^3}{\psi_{smax}(d1_i^4 - d2_i^4)^2}(1-w_i) + \beta_{1i} - \beta_{3i} = 0 \,,$$

$$\frac{\partial L}{\partial d2_i} = -\frac{2a_i d2_i}{W_{smax}}\,w_i + \frac{8e_i d2_i^3}{\psi_{smax}(d1_i^4 - d2_i^4)^2}(1-w_i) - \beta_{2i} + \beta_{3i} = 0 \,;$$

$$\beta_{1i}(d1_i - d_{imax}) = 0 \,,$$

$$\beta_{2i}(d_{imin} - d2_i) = 0 \,,$$

$$\beta_{3i}[2t_{imin} - (d1_i - d2_i)] = 0 \,;$$

$$\beta_{1i} \geq 0 \,, \tag{7.1-16}$$

$$\beta_{2i} \geq 0 \,,$$

$$\beta_{3i} \geq 0 \,;$$

$$d1_i - d_{imax} \leq 0 \,,$$

$$d_{imin} - d2_i \leq 0 \,,$$

$$2t_{imin} - (d1_i - d2_i) \leq 0 \,;$$

$$a_i = l_i \rho_i g$$

$$e_i = \frac{2l_i^3}{E_i} S_i^2 + \frac{6l_i}{E_i} M_i^2 - \frac{6l_i^2}{E_i} S_i M_i \,,$$

where S_i and M_i are the shear force and the bending moment which act upon the i-th structural member, and E_i is Young's modulus of the i-th structural member.

By solving the system of equations (7.1-16), the design variables of all structural members on the static force loop are obtained corresponding to the value of the weighting factor w_i(i = 1,2,...,n) as shown in Fig. 7.1/3. The results in Fig. 7.1/3 are for a structural model in which each structural member has a square cross-sectional shape, and only shear forces and bending moments are applied to the structural member. Similar results can also be obtained for machine-tool spindle designs [13], more complicated configurations and other types of external forces.

The shaded region of Fig. 7.1/4 corresponds to the design space feasible for the two objective functions W_s and ψ_s, and the heavy solid line shows the Pareto-optimum solution set for minimization of both W_s and ψ_s. When increasing the weighting factor w from 0 to 1, the optimum design solution moves from point P_1 to point P_2 along the heavy solid line. Such figures as Fig. 7.1/4 are here referred to as "the weight-compliance diagram".

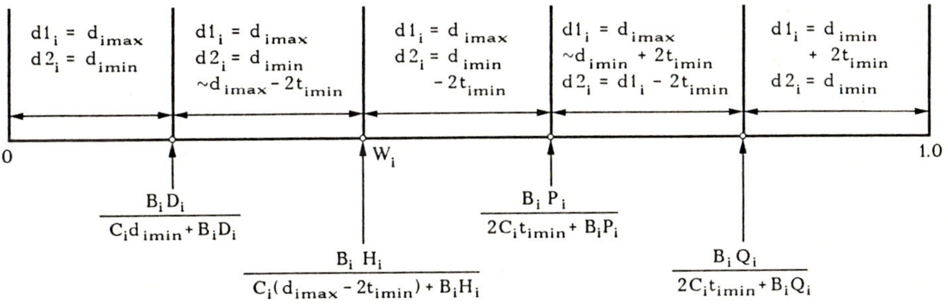

$$B_i = \frac{8e_i}{\psi_{smax}}, \quad D_i = \frac{d^3_{imin}}{(d^4_{imax} - d^4_{imin})^2}, \quad P_i = \frac{d^3_{imax} - (d_{imax} - 2t_{imin})^3}{(d^4_{imax} - (d_{imax} - 2t_{imin})^4)^2},$$

$$C_i = \frac{2a_i}{W_{smax}}, \quad H_i = \frac{(d_{imax} - 2t_{imin})^3}{(d^4_{imin} - (d_{imax} - 2t_{imin})^4)^2}, \quad Q_i = \frac{(d_{imin} + 2t_{imin})^3 - d^3_{imin}}{((d_{imin} + 2t_{imin})^4 - d^4_{imin})^2}.$$

Fig. 7.1/3. Distributions of the weighting factor w_i and the design variables on the Parteto-optimum solution set

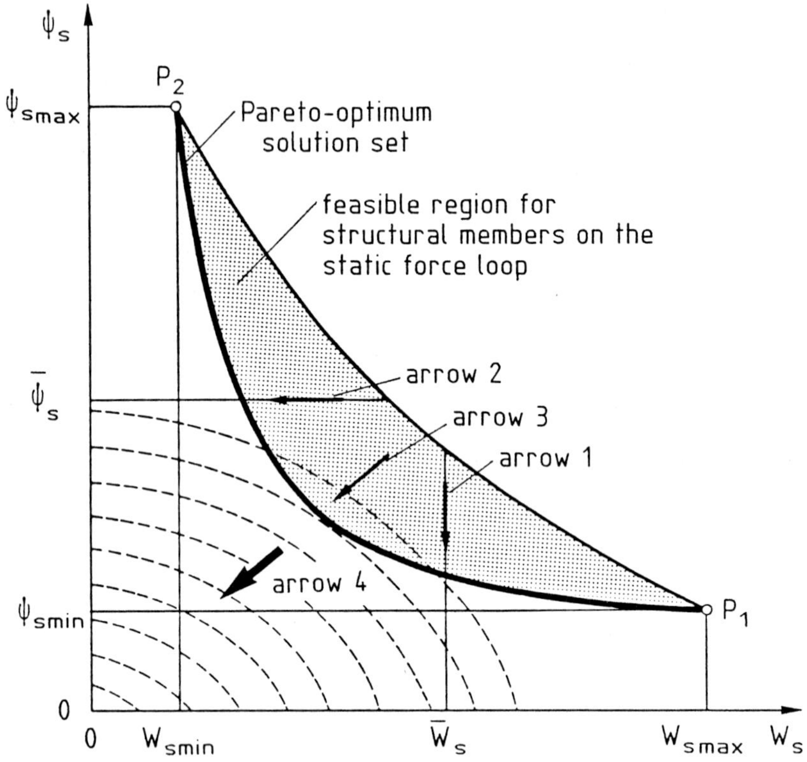

Fig. 7.1/4. Weight-compliance diagram

b) Relationship between the Pareto-optimum solution set and the dynamic characteristics

The Pareto-optimum or functional-efficient solution set, obtained in the foregoing section for a two-objective optimization problem minimizing W_s and ψ_s, is closely related to the optimization of vibrational characteristics and dynamical stability of machines.

Both, the smaller W_s and the smaller ψ_s (as shown by the arrow 3 in Fig. 7.1/4) lead to the smaller maximum receptance value $R_{m.max}$, the higher natural frequency having the greatest receptance value ω_m, and the larger damping ratio at the natural mode ζ_m. That means, the farther the design solution is away from the origin of the weight-compliance diagram, the greater $R_{m.max}$ becomes, the lower ω_m becomes, and the smaller ζ_m becomes. From such considerations it can be said that each broken curve in Fig. 7.1/4 expresses the contour line of a different constant value of $R_{m.max}$, ω_m, or ζ_m, although the configuration of the contour line depends on the characteristic under consideration and the unit scale of W_s and ψ_s. Arrow 4 in Fig. 7.1/4 indicates decreasing direction of $R_{m.max}$ or increasing direction of ω_m or ζ_m. The minimum (or maximum) point of the contour line is the origin of the weight-compliance diagram. Therefore, for the structural members on the static force loop, a design in a Pareto-optimum solution (a solution on the thick solid line in Fig. 7.1/4) located as near as possible to the origin brings about a greater reduction of $R_{m.max}$ and more significant increases of ω_m and ζ_m.

A point in the Pareto-optimum solution set of a two objectives optimization problem minimizing W_s and ψ_s is said to be "a contact point of static and dynamic characteristics" for realizing the most stable dynamics of the machine. A different point may be selected according to the dynamic characteristic considered. Some of the competitive relations between dynamic characteristics originate from this competitive relation between W_s and ψ_s in the functional-efficient solution set.

7.1.3 Design Optimization Procedures

The algorithm for the design optimization of minimizing the maximum receptance value (that is, maximizing stability against chatter vibration) consists roughly of the following three phases [11]:

Phase I: Obtaining the Pareto-optimum solution set of a multicriteria problem minimizing W_s and ψ_s

Step 1. A multicriteria design optimization problem minimizing both the total weight of the structural members on the static force loop W_s, and the static compliance ψ_s is solved, and the Pareto-optimum solution set is obtained. As can be understood easily, the minimum value of the static compliance ψ_s is obtained when the spring stiffness of each of the joints on the static force loop equals its own upper bound. Therefore, in this step each spring stiffness value of the joints on the static force loop is set equal to its own upper bound.

Step 2. Based on the evaluation of the relation between W_s and ψ_s in the Pareto-optimum set, the design variables of the structural members corresponding to the static compliance are determined when the static compliance ψ_s is defined.

Phase II: Minimization of the ratio of the maximum modal flexibility $\psi_{m.max}$ to the static compliance ψ_s

Step 3. The ratio ψ_m/ψ_s of the modal flexibility ψ_m at the m-th natural mode, having the maximum modal flexibility value to the static compliance ψ_s, is minimized by the design of the structural members, machine elements, and joints outside the static force loop. Such a decrease of the ratio ψ_m/ψ_s corresponds to an increase of the ratio U_J/U_T. U_J represents the potential energy stored at the joints outside the static force loop, while U_T represents the total potential energy of the complete system. Increasing the ratio U_J/U_T can lead to an increase of the damping ratio ζ_m since joints have a far greater vibrational energy dissipation ability than structural members. Therefore, minimization of the ratio ψ_m/ψ_s is cooperative with the maximization of the damping ratio ζ_m as shown on the bold printed curve in Fig. 7.1/5 (ζ_s is the damping ratio when ψ_m/ψ_s is unity). Thus, minimization of the ratio ψ_m/ψ_s results in minimization of the maximum modal flexibility and maximization of the damping ratio, both of which are required to minimize the maximum receptance value.

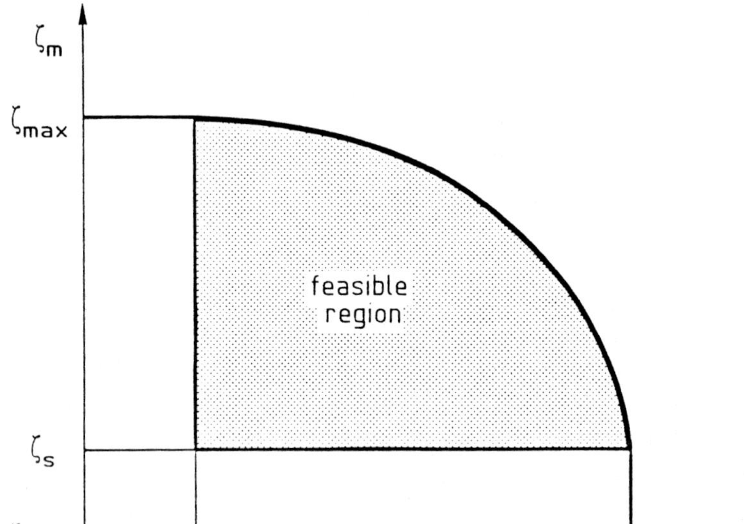

Fig. 7.1/5. Relation between the modal flexibility ratio and the damping ratio

Phase III: Maximization of the damping ratio

Step 4. The damping ratio ζ_m at the m-th natural mode is maximized by adjusting the damping coefficients of the joints on the static force loop and outside the static force loop.

Step 5. If the maximum receptance value $R_{m.max}$ is sufficiently small, the procedure is ended. Otherwise, the rigidity of joints on the static force loop is decreased within the upper bound $\Delta\psi_s^u$ of the incremental quantity $\Delta\psi_s$ of the static compliance ψ_s, and the damping ratio ζ_m is increased by adjusting the damping coefficients of the joints on the static force loop.

If the static compliance ψ_s is not clearly defined in Step 2, the temporary value of ψ_s is first given in the Pareto-optimum solution set. ψ_s is then moved in the Pareto-optimum solution set and is finally determined so that the minimum value of $R_{m.max}$ is obtained in Step 4.

As one can understand from the foregoing algorithm, all design variables are divided into three groups: (1) design variables of the structural members on the static force loop, (2) design variables of the structural members and joint stiffnesses outside the static force loop, and (3) the damping coefficients of all the joints. In Step 1 and 2, the structural model for optimization is composed only of strucural members and joints on the static force loop. At this time, a static structural analysis is conducted. In Step 3, only design variables of the structural members and joint stiffnesses outside the static force loop are determined. In this case, vibrational analysis is conducted for the structural model of the structural members, machine elements, and joints over the complete structural system, but damping factors are not included in the analysis. In Step 4, only the damping coefficients of the joints are design variables. Vibrational characteristics including damping effects are calculated for the complete structural system. The design decision problem in each procedure is far simpler and easier than that in usual design optimization methods in which all design variables are determined at the same time.

In the modelling of whole machine-tool structures the techniques of simplified structural models are used according to the principles of the design optimization methods in three phases: simplification, optimization and realization [7,8]. For such simplified models, the Pareto-optimum solution sets are obtained analytically.

7.1.4 Application Examples

A simplified milling-machine model shown in Fig. 7.1/1 a) is used for demonstrating the availability of the design optimization method. The machine structural model is composed of 10 box-type members having square cross-sections and 7 joints expressed by spring marks. The outside width d1 and the inside width d2 of the cross-section of each structural member as well as the equivalent spring stiffness, and the damping coefficients of each joint are said to be design variables.

a) Maximization of stability against chatter vibration

Phase I:

Step 1. Fig. 7.1/6 shows the criterion space for two objectives minimizing both the total weight of the structural members on the static force loop W_s, and the static compliance ψ_s. The relation between the total weight W_s and the static compliance ψ_s in the Pareto-optimum solution set is shown by the bold printed curve in Fig. 7.1/6. The shaded portion corresponds to the feasible design space. Fig. 7.1/7 shows the design variables of the structural members 2, 3, and 4 corresponding to the weighting factor. The values of d_{imax}, d_{imin}, and t_{imin} for the i-th structural member are written under the number of the structural member.

Step 2. From the Pareto-optimum solution set shown in Fig. 7.1/6 it is obvious that the static compliance ψ_s suddenly increases by decreasing the total weight W_s from point P_3 while the total weight W_s suddenly increases by decreasing the static compliance ψ_s . Since point P_3 is considered to be the most preferable point for minimizing both W_s and ψ_s , it was selected as the solution of the design variables of the structural members on the static force loop.

Phase II:

Step 3. In this example, only the design of the joints outside the static force loop was optimized in this step as the design of the structural members and machine elements outside the static force loop is often determined by the design and the functional requirements. Then, the ratio ψ_m/ψ_s of the modal flexibility ψ_m at the m-th natural mode having the maximum model flexibility value to the static compliance ψ_s was minimized by adjusting spring stiffnesses of the joints J_5, J_6, and

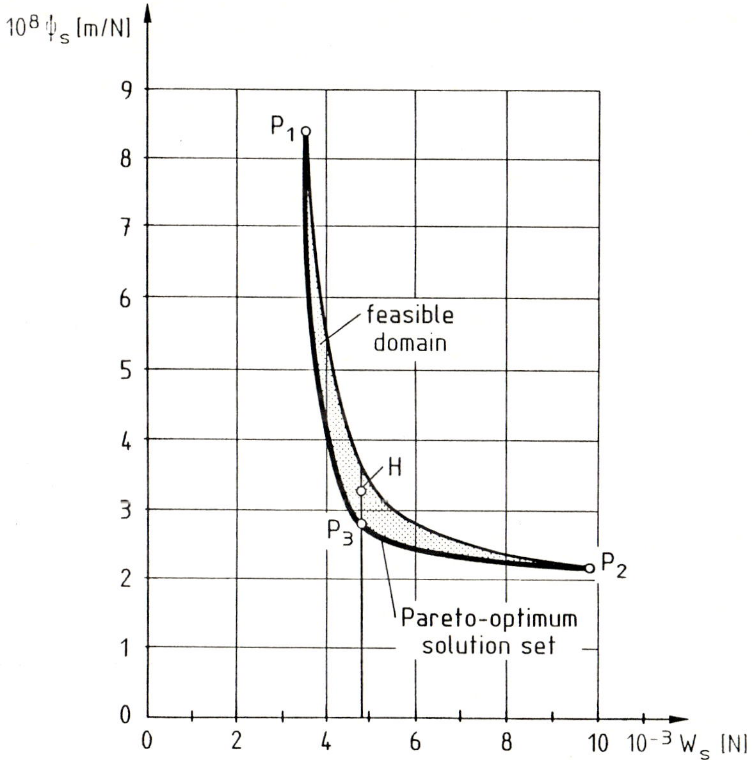

Fig. 7.1/6 Weight-compliance diagram for the milling-machine model

J_7 outside the static force loop. In this optimization, the simplex method [14], one of the direct search methods for nonlinear functions, was used.

Phase III:

Step 4. Maximization of the damping ratio corresponds to minimization of the maximum receptance value $R_{m.max}$. The maximum receptance value $R_{m.max}$ was minimized by adjusting the damping coefficients of all joints in the complete system. In this minimization, the simplex method was used as well. The solid line in Fig. 7.1/8 shows the receptance frequency response after completion of this step. The circle "o" indicates the maximum receptance.

Step 5. The procedure was stopped as the maximum receptance value was sufficiently small.

For comparison, the broken line in Fig. 7.1/8 shows the receptance frequency response when all design variables were optimized at the same time by the direct search method (an extended method was used with a suitable modification for dealing with constraints), using minimization of the maximum receptance value $R_{m.max}$ as the objective. The dotted line shows the receptance frequency response when the design variables at point H in Fig. 7.1/6 were selected in Step 2 and the rest of the procedures were conducted according to the algorithm described in Section 7.1.3. The receptance frequency response shown by the solid line has the minimum value of the maximum receptance and demonstrates the effectiveness of the design optimization method proposed here.

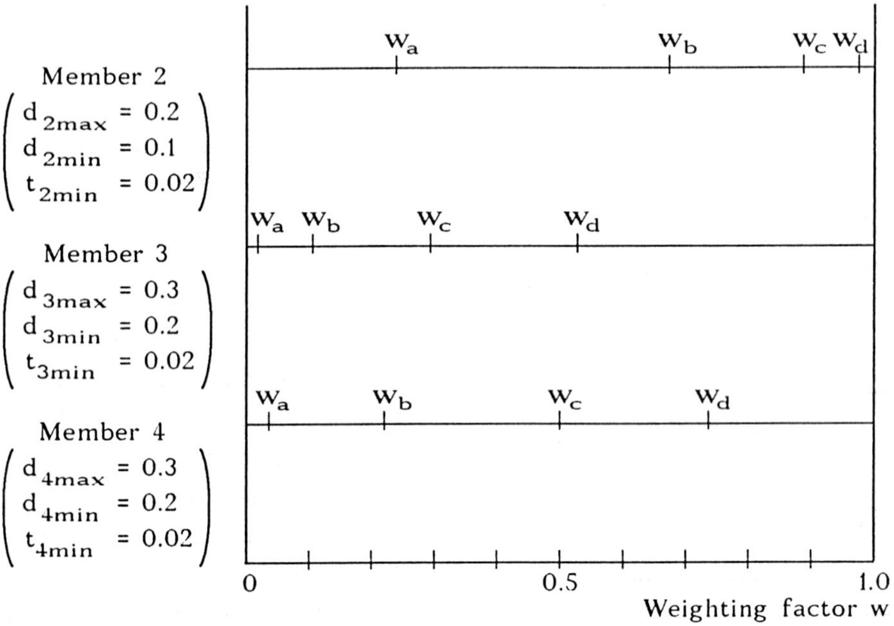

w	$d1_i$	$d2_i$
$0 \sim w_a$	d_{imax}	d_{imin}
$w_a \sim w_b$	d_{imax}	$d_{imin} \sim d_{imax} - 2t_{imin}$
$w_b \sim w_c$	d_{imax}	$d_{imax} - 2t_{imin}$
$w_c \sim w_d$	$d_{imax} \sim d_{imin} + 2t_{imin}$	$d1_i - 2t_{imin}$
$w_d \sim 1.0$	$d_{imin} + 2t_{imin}$	d_{imin}

Fig. 7.1/7. Design variables of the structural members 2,3, and 4 [m]

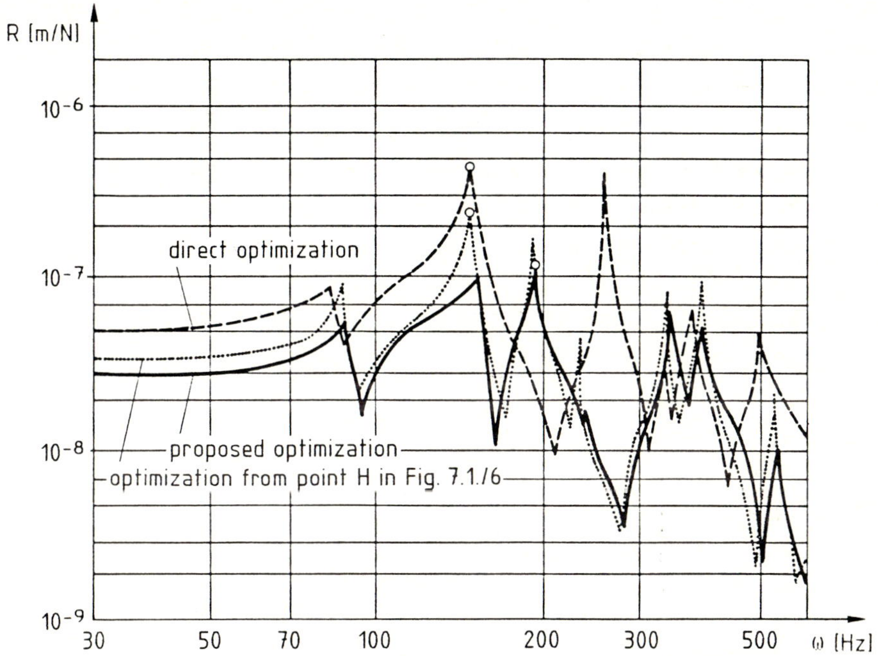

Fig. 7.1/8. Receptance frequency responses for different optimization procedures

b) Decision making for selection of structural material

The Pareto-optimum solution set can be regarded as a barrier for the further improvement of product performance. The application of new materials for improving the product performance and reducing the product manufacturing cost is an easy and hopeful strategy for breaking through this barrier. However, it is not always wise and sometimes even dangerous to anticipate the effectiveness of new materials when considering only the material qualities and the characteristics of material test pieces.

General decision making procedures for material choices are constructed using the relational diagram of the Pareto optimum solution sets concerning the total weight and the static compliance [15].

Fig. 7.1/9 shows an example of a weight-compliance diagram of the milling machine model shown in Fig. 7.1/1 a) in which all structural members on the static force loop are made of either alumina ceramics, epoxy resin concrete, cast iron, or mild steel. The Young's modulus E, mass density ρ and material damping ratio ζ_s of the materials are shown in Table 7.1/1. The lines in bold print of Fig. 7.1/9 show the Pareto-optimum solution sets of the two objectives: minimization of the total weight W_s and minimization of the static compliance ψ_s. The shaded regions of Fig. 7.1/9 correspond to the feasible design spaces. Each of the broken curves indicates the contour line of a constant maximum receptance value.

The figure shows that the feasible design solution located at the point of contact with the contour line which has the minimum receptance value corresponding to the solution which minimizes the maximum receptance value $R_{m \cdot max}$. From the receptance value at such a design point for each material an evaluation of the improvements brought about by using the new material can be conducted quantitatively . It can be seen in the example of Fig. 7.1/9 that alumina ceramics bring about minimization of the maximum receptance value.

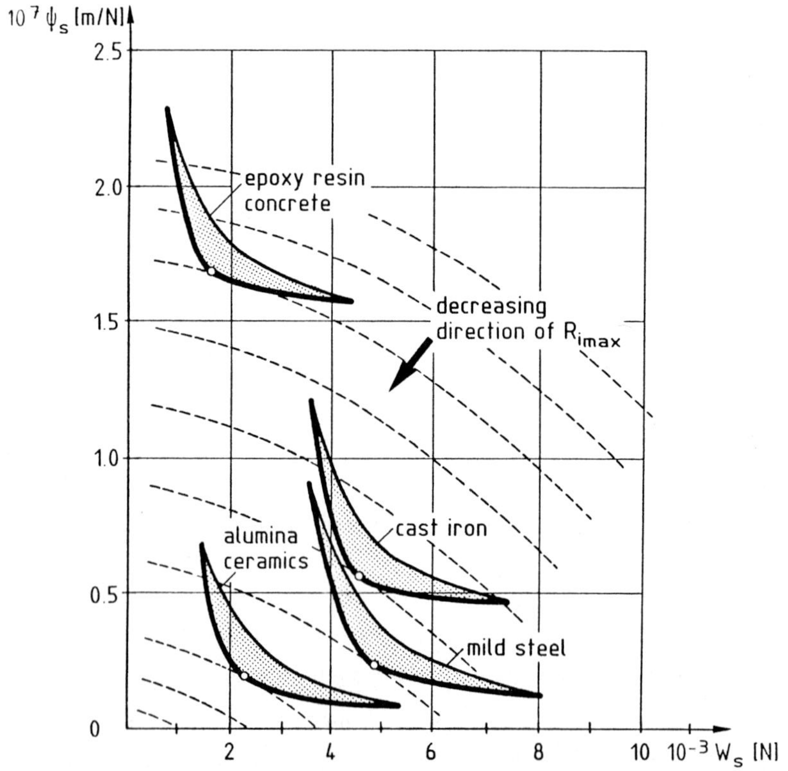

Fig. 7.1/9. Weight-compliance diagram of the milling-machine model for different member materials

Material	Young's modulus E [GPa]	Mass density ρ [$10^3 kg \cdot m^{-3}$]	Material damping ratio ζ_s
Mild steel	205	7.8	0.0000796
Cast iron	100	7.0	0.000159 ~0.000477
Epoxy resin concrete	33	2.5	0.00318
Alumina ceramics	280	3.4	0.0000955

Table 7.1/1. Material properties

These fundamental procedures can be applied to various purposes when choosing new materials. By using these methods, material selection and determination of design variables are simultaneously conducted and the advantages of using a new material can be evaluated quantitatively.

Decision making based on a comparison of material qualities alone may lead to a wrong conclusion. Decision making should be conducted by solving an optimization problem formulation by means of systematic analyses.

7.1.5 Conclusion

In this section multicriteria optimization techniques are used for obtaining optimum design solutions effectively and easily (or for making decisions of design choices). The methods described here can be applied not only to machine tools but also to other machines such as industrial robots, three dimensional measuring apparatus, etc. requiring high precision and high working efficiency.

Here, conflicting relationships between characteristics concerning the product performance are considered. Designers of machine products are required to design a product having a higher product performance and a lower product manufacturing cost. However, the two objectives of improving the product performance and lowering the product manufacturing cost generally have a conflicting relationship. This fact is very important for the "Integration of CAD and CAM (Computer Aided Design and Manufacturing)" which is nowadays frequently required. A multicriteria optimization technique can also be applied to obtain the integrated optimum design solutions of product design and process design [16].

References

[1] Tobias, S.A.: Machine Tool Vibration. London: Blackie, 1965

[2] Koenigsberger, F.; Tlusty, J.: Machine Tool Structures. Vol. 1. Oxford: Pergamon Press, 1970

[3] Weck, M.; Teipel,K.: Dynamisches Verhalten Spanender Werkzeugmaschinen. Berlin: Springer-Verlag, 1977

[4] Reddy, C.P.; Rao, S.S.: Automated Optimum Design of Machine-Tool Structures for Static Rigidity, Natural Frequencies and Regenerative Chatter Stability. ASME Journal of Engineering for Industry. Vol. 100, No.2 (1978) 137-146

[5] Rao, S.S.; Grandhi, R.V.: Optimum Design of Radial Drilling Machine Structure to Satisfy Static Rigidity and Natural Frequency Requirements. ASME Journal of Mechanisms, Transmissions, and Automation in Design. Vol. 105, No. 2 (1983) 236-241

[6] Reddy, V.R.; Sharan, A.M.: Design of Machine Tool Spindles Based on Transient Analysis. ASME Journal of Mechanisms, Transmissions, and Automation in Design. Vol. 107, No. 3 (1985) 346-352

[7] Yoshimura, M.; Hamada, T.; Yura, K.; Hitomi, K.: Design Optimization of Machine-Tool Structures With Respect to Dynamic Characteristics. ASME Journal of Mechanisms, Transmissions, and Automation in Design. Vol. 105, No. 1 (1983) 88-96

[8] Yoshimura, M.; Takeuchi, Y.; Hitomi, K.: Design Optimization of Machine-Tool Structures Considering Manufacturing Cost, Accuracy, and Productivity. ASME Journal of Mechanisms, Transmissions, and Automation in Design. Vol. 106, No. 4 (1984) 531-537

[9] Yoshimura, M.: Evaluation of Forced and Self-Excited Vibrations at the Design Stage of Machine-Tool Structures. ASME Journal of Mechanisms, Transmissions, and Automation in Design. Vol.108, No. 3 (1986) 323-329

[10] Merritt, H.E.: Theory of Self-Excited Machine-Tool Chatter. ASME Journal of Engineering for Industry. Vol. 87, No. 4 (1965) 447-454

[11] Yoshimura, M.: Design Optimization of Machine-Tool Dynamics Based on Clarification of Competitive-Cooperative Relationships Between Characteristics. ASME Journal of Mechanisms, Transmissions, and Automation in Design. Vol.109, No. 1 (1987) 143-150

[12] Kuhn, H.W.; Tucker, A.W.: Nonlinear Programming. Proceedings of Second Berkeley Symposium on Mathematical Statistics and Probability (1951) 481-492

[13] Yoshimura, M.; Hamada, T.; Yura, K.; Hitomi, K.: Multiobjective Design Optimization of Machine-Tool Spindles. ASME Journal of Mechanisms, Transmissions, and Automation in Design. Vol. 106, No. 1 (1984) 46-53

[14] Nelder, J.A.; Mead, R.: A Simplex Method for Function Minimization. Computer Journal. Vol. 7. (1965) 308-313

[15] Yoshimura, M.: Decision Making in the Choosing of New Materials from the Standpoint of Machine Structural Dynamics. ASME Journal of Mechanisms, Transmissions, and Automation in Design, Vol. 111, No. 1 (1989) 110-116

[16] Yoshimura, M.; Itani, K.; Hitomi, K.: Integrated Optimization of Machine Product Design and Process Design. International Journal of Production Research, Vol. 27, No. 8 (1989) 1241-1256

7.2 A Decomposition Method for Multicriteria Optimization and its Application to Machine Tool Spindle Design

J. Montusiewicz , A. Osyczka, J. Zamorski

7.2.1 Introduction

In order to solve multicriteria optimization design problems as formulated in Chapter 1, numerous computer procedures have been developed (see Chapters 2 and 3). Their efficiency, however, drastically decreases with the increase of the number of design variables. Even for single criterion problems, as shown by Himmelblau [4], the number of function evaluations grows expotentially with the number of elements in **x**. In addition to these difficulties, the use of conventional optimization methods for solving complex design problems is inconvenient for several reasons. A change in a part of the problem involves a recalculation of the whole model; the designer has little influence on the computer optimization process; finding the global minimum is difficult. Recently, great efforts have been made to develop decomposition methods for both single [1,5,6], and multiple criteria optimization problems [2,9,10]. These methods are used for decomposing a system into subsystems, each of which can be easily solved. Interactions between the subsystems considered as parametric and/or constraint coordinations are taken into account while solving the optimization model. The idea of decomposition combined with a multicriteria approach to problem-solving resulted in a new method called a Four-Stage Multicriteria Optimization Strategy employed for dealing with complex problems of optimum design. The strategy supported by CAMOS, a Computer Aided Multicriteria Optimization System [7], is used for designing a machine tool spindle system with hydrostatic bearings. This complex optimization problem illustrates the possibilities and advantages of the strategy.

7.2.2 Four-Stage Multicriteria Optimization Strategy

A complex optimum design problem belongs to a class of optimization models consisting of subsystems. Models like those can be decomposed into subsystems to be analysed separately. In the process of multicriteria optimization these subsystems will be called local tasks. The complete design problem in the multicriteria optimization process, called the global task, will be solved after the analyses of the local tasks. Obviously, the global task is correlated with local tasks which in turn can also be correlated with each other. In the multicriteria optimization model the description of this correlation will be called parametric and constraint coordination.

Figure 7.2/1 outlines the main idea of the proposed Four-Stage Multicriteria Optimization (MO) Strategy. In this strategy the vector of design variables **x** will be divided into three vectors as follows:

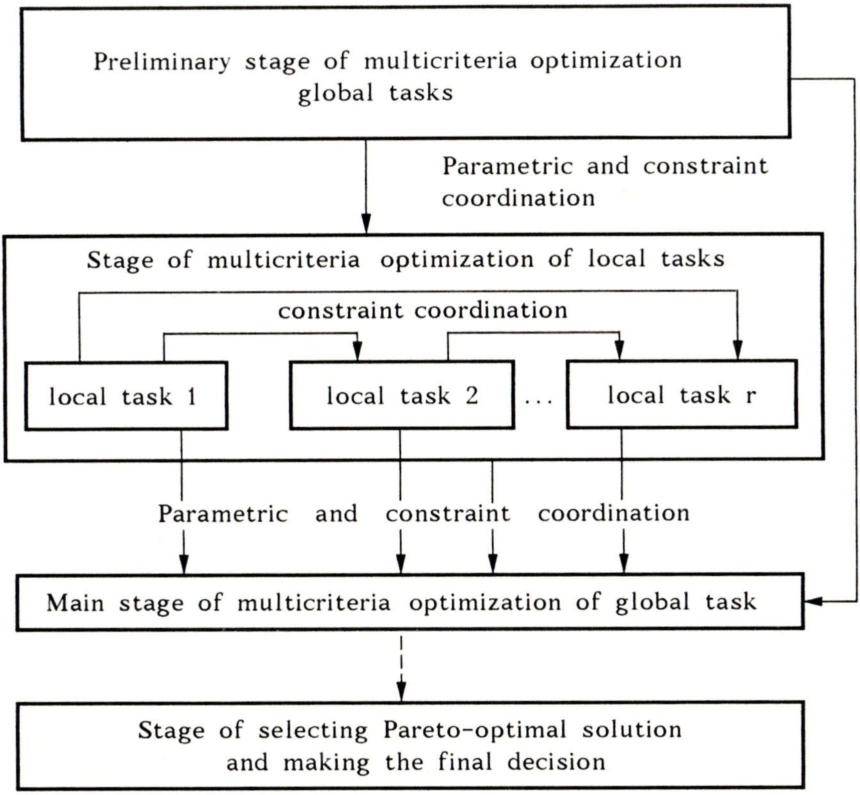

Fig. 7.2/1. Four-stage multicriteria optimization strategy

$\mathbf{x} = (\mathbf{u},\mathbf{v},\mathbf{z})^T$ (7.2-1)

$\mathbf{u} = (x_1,\ldots,x_k)^T$ - vector of global task design variables,

$\mathbf{v} = (x_{k+1},\ldots,x_l)^T$ - vector of global task design variables which act as coordinate variables between the global task and the local tasks,

$\mathbf{z} = (x_{l+1},\ldots,x_n)^T$ - vector of the design variables of all local tasks.

The space and subspaces of design variables will be denoted as:

X' - space of the vector $\mathbf{x}$,

U' - subspace of the vector $\mathbf{u}$ of the global task,

V' - subspace of the coordinate variables of the vector $\mathbf{v}$,

Z' - subspace of the vector $\mathbf{z}$ of local tasks,

where

$$X' = U' \times V' \times Z' \ . \tag{7.2-2}$$

Let $\mathbf{g}^u(\mathbf{u},\mathbf{v}) \le \mathbf{0}$ and $\mathbf{h}^u(\mathbf{u},\mathbf{v}) = \mathbf{0}$ define the constraints of the global task whereas $\mathbf{g}^l(\mathbf{v},\mathbf{z}) \le \mathbf{0}$ and $\mathbf{h}^l(\mathbf{v},\mathbf{z}) = \mathbf{0}$ define the constraints of the local tasks. Let U, V, Z_v be the sets of feasible solutions of the vectors $\mathbf{u}$, $\mathbf{v}$, $\mathbf{z}$, respectively. These sets are defined as follows:

$$U = \{\mathbf{u}\colon\ \mathbf{g}^u(\mathbf{u},\mathbf{v}) \le \mathbf{0},\ \mathbf{h}^u(\mathbf{u},\mathbf{v}) = \mathbf{0}\} \ , \tag{7.2-3a}$$

$$V = \{\mathbf{v}\colon\ \mathbf{g}^u(\mathbf{u},\mathbf{v}) \le \mathbf{0},\ \mathbf{h}^u(\mathbf{u},\mathbf{v}) = \mathbf{0}\} \ , \tag{7.2-3b}$$

$$Z_v = \{\mathbf{z}\colon\ \mathbf{g}^l(\mathbf{v},\mathbf{z}) \le \mathbf{0},\ \mathbf{h}^l(\mathbf{v},\mathbf{z}) = \mathbf{0}\} \ . \tag{7.2-3c}$$

Stage 1:

Let $\mathbf{x}^w = (\mathbf{u},\mathbf{v})^T$ be the vector of design variables defined in the subspace F' which is the subspace of the preliminary stage of MO of the global task where

$$F' = U' \times V' \ . \tag{7.2-4a}$$

The set of feasible solutions $F \subset F'$ is defined as

$$F = U \times V \ . \tag{7.2-4b}$$

The vector function of the global task is

$$\mathbf{f}^w(\mathbf{x}^w) = (f_1^w(\mathbf{x}^w),\ldots,f_k^w(\mathbf{x}^w))^T \ , \tag{7.2-5}$$

where $f_j(\mathbf{x}^w)$ is the j-th objective function considered at the preliminary stage.

The multicriteria optimization problem can be formulated here as follows. Find the vector $\mathbf{x}^{*w}$ such that

$$p[\mathbf{f}(\mathbf{x}^{*w})] = \min_{\mathbf{x}^w \in F} p[\mathbf{f}(\mathbf{x}^w)] \ . \tag{7.2-6}$$

Note that optimal values of both **u** and **v** are to be found at this stage.

Between the preliminary stage of MO of the global task and the stage of local tasks two types of coordination may occur: parametric coordination and constraint coordination. Parametric coordination takes place when objective functions of local tasks are extremized in relation to some design variables described by the vector **z**, while the remaining design variables which also describe individual local tasks and which are denoted by vector **v** are constant and treated as parametric coordination variables. These variables are design variables at the preliminary stage of the global task. The constraint coordination takes place when the values of some design variables of the global task influence the constraints of the local tasks.

Stage 2:

At this stage several local tasks may appear. These tasks can be considered separately, and thus the vector **z** will be divided into separate vectors $\mathbf{z}^i$ each of which refers to the i-th local task where $i = 1,\dots,r$. The set of feasible solutions Z_v can be separated as follows

$$Z_v = Z_v^1 \times Z_v^2, \dots, \times Z_v^r \ , \tag{7.2-7a}$$

where Z_v^i is the set of feasible solutions for the i-th local task defined as:

$$Z_v^i = \{z^i : \mathbf{g}^i(\mathbf{v},\mathbf{z}^i) \le \mathbf{0},\ \mathbf{h}^i(\mathbf{v},\mathbf{z}^i) = \mathbf{0}\} \ ; \quad i = 1,..,r \ . \tag{7.2-7b}$$

The subsets Z_v^i are still dependent on the vector of coordination variables **v**, but each of the subsets describes the set of feasible solutions for the vector $\mathbf{z}^i$. The vector of design variables for the i-th local task is

$$\mathbf{x}^{li} = (\mathbf{v},\mathbf{z}^i)^T \ , \tag{7.2-8}$$

and the vector of objective functions for the i-th local task is

$$\mathbf{f}^i(\mathbf{v},\mathbf{z}^i) = (f_1^i(\mathbf{v},\mathbf{z}^i),\ldots,f_{m_i}^i(\mathbf{v},\mathbf{z}^i))^T, \tag{7.2-9}$$

where $f_j^i(\mathbf{v},\mathbf{z}^i)$ is the j-th component of the i-th local task vector function, $j = 1,\ldots,m_i$ and $i = 1,\ldots,r$.

The MO-problem for a local task is to find the vector $\mathbf{z}^{*i} \in Z_v^i$ such that

$$p[\mathbf{f}^i(\mathbf{v},\mathbf{z}^{*i})] = \min_{\substack{\mathbf{z}^i \in Z_v \\ \mathbf{v} \in V}} p[\mathbf{f}^i(\mathbf{v},\mathbf{z}^i)]\ . \tag{7.2-10}$$

Local tasks may correlate forming a certain hierarchical system [6, 9]. This means that in a set of local tasks one leading task and one or many tasks correlated with the leading task can be distinguished. It may also happen that the task correlated with some leading task becomes the leading task for one or several other local tasks. The first situation in which one task influences another local task, i.e. certain design variables of the leading task act as parameters or form constraints for the task corelated with it, is called the constraint coordination of the first order. The second situation in which the correlated task is the leading task for another local task, is called the constraint coordination of the second order. Similarly, one can specify the coordination of the third order and so on.

Assume that the first local task is the leading task of the first order for the second local task. Then, the vector of the design variables of the leading task $\mathbf{x}^{11} = (\mathbf{v},\mathbf{z}^1)^T$ is such that the vector $\mathbf{z}^1$ has two separate parts as

$$\mathbf{z}^1 = (\mathbf{z}^{12},\mathbf{v}^{12})^T \tag{7.2-11}$$

where $\mathbf{z}^{12} = (z_1^1,\ldots,z_{S1}^1)^T$ and $\mathbf{v}^{12} = (z_{S1+1}^1,\ldots,z_{t1}^1)^T$.

The vector $\mathbf{v}^{12}$ is called the constraint coordination vector between the first and the second local tasks. In this case the set of feasible solutions for the second local task is

$$Z_v^2 = \{\mathbf{z}^2\colon\ \mathbf{g}^2(\mathbf{v},\mathbf{z}^2,\mathbf{v}^{12}) \le \mathbf{0},\ \mathbf{h}^2(\mathbf{v},\mathbf{z}^2,\mathbf{v}^{12}) = \mathbf{0}\}\ . \tag{7.2-12}$$

Note that the constraints which define the set of feasible solutions depend on the vector $\mathbf{v}^{12}$ (compare (7.2-7b) and (7.2-12)).

If the first local task is the leading task for the p-th local task, then the vector $\mathbf{z}^1$ is

$$\mathbf{z}^1 = (\mathbf{z}^{1p},\mathbf{v}^{1p})^T \tag{7.2-13a}$$

where $\mathbf{z}^{1p} = (z_1^1,\ldots,z_{R1}^1)^T$ and $\mathbf{v}^{1p} = (z_{R1+1}^1,\ldots,z_{t1}^1)^T$.

The set of feasible solutions for the p-th local task is

$$Z_v^p = \{\mathbf{z}^p\colon \mathbf{g}^p(\mathbf{v},\mathbf{z}^p,\mathbf{v}^{1p}) \le \mathbf{0} \ , \ \mathbf{h}^p(\mathbf{v},\mathbf{z}^p,\mathbf{v}^{1p}) = \mathbf{0}\} \quad . \tag{7.2-13b}$$

Finally, if the p-th local task is the leading task for the q-th task, the vector $\mathbf{z}^p$ is

$$\mathbf{z}^p = (\mathbf{z}^{pq},\mathbf{v}^{pq})^T \tag{7.2-14a}$$

where $\mathbf{z}^{pq} = (z_1^p,\ldots,z_{Sp}^p)^T$ and $\mathbf{v}^{pq} = (z_{Sp+1}^p,\ldots,z_{tp}^p)^T$.

The set of feasible solutions for the q-th task is

$$Z_v^q = \{\mathbf{z}^q\colon \mathbf{g}^q(\mathbf{v},\mathbf{z}^q,\mathbf{v}^{pq}) \le 0 \ , \ \mathbf{h}^q(\mathbf{v},\mathbf{z}^p,\mathbf{v}^{pq}) = 0\} \quad . \tag{7.2-14b}$$

Stage 3:

Here, the global task enters the main stage of MO in which the coordination vectors obtained at the stage of local tasks and at the preliminary stage of the global task are incorporated into the optimization process. Between local tasks and the global task constraint coordination takes place. The coordination vectors are $\mathbf{v}^i$ for $i = 1,\ldots,r$. Moreover, parametric coordination expressed by the vectors $\mathbf{v}$ and $\mathbf{q}$ also takes place. The vector $\mathbf{v}$ expresses the parametric coordination between the preliminary stage of MO of the global task and the main stage of this optimization. The vector $\mathbf{q}$ expresses the parametric coordination between the local tasks and the global task of the main stage. The vector $\mathbf{q}$ consists of quantities which at the stage of MO of local tasks were decision variables and of the quantities which are the functions of the vectors $\mathbf{v}$ and $\mathbf{z}$. Note that certain quantities which describe the global task can be estimated by the designer at the preliminary stage. These quantities are evaluated at the stage of MO of local tasks, and they become parameters for the main stage of MO of the global task.

At the main stage of MO of the global task the vector of the design variables is $\mathbf{u} = (x_1,...,x_k)^T$ and the set of feasible solutions has the form

$$\hat{U} = \{\mathbf{u}: \mathbf{g}^u(\mathbf{u},\mathbf{v},\mathbf{v}^1,\mathbf{v}^2,...,\mathbf{v}^r,\mathbf{q}) \leq \mathbf{0} \ , \ \mathbf{h}^u(\mathbf{u},\mathbf{v},\mathbf{v}^1,\mathbf{v}^2,...,\mathbf{v}^r,\mathbf{q}) = \mathbf{0}\}. \quad (7.2\text{-}15)$$

Obviously, in (7.2-15) some vectors $\mathbf{v}^i$ can be zero vectors.

The vector of the objective functions can be written as:

$$\mathbf{f}^k(\mathbf{u},\mathbf{v},\mathbf{q}) = (f_1^k(\mathbf{u},\mathbf{v},\mathbf{q}),...,f_p^k(\mathbf{u},\mathbf{v},\mathbf{q}))^T \quad . \quad (7.2\text{-}16)$$

The MO-problem at this stage is to find the vector $\mathbf{u}^* \in \hat{U}$ such that

$$p[\mathbf{f}^k(\mathbf{u}^*,\mathbf{v},\mathbf{q})] = \min_{\mathbf{u} \in \hat{U}} p[\mathbf{f}^k(\mathbf{u},\mathbf{v},\mathbf{q})] \quad . \quad (7.2\text{-}17)$$

Stage 4:

The optimization strategy of the previous stages results in obtaining a certain set of solutions which are optimal with respect to the assumed criteria for the global and the local tasks. For the final evaluation of these solutions a combined vector function $\mathbf{f}(\mathbf{x})$ is created. The vector $\mathbf{f}(\mathbf{x})$, where $\mathbf{x} = (\mathbf{u},\mathbf{v},\mathbf{z})^T$, has the form

$$\mathbf{f}(\mathbf{x}) = (\mathbf{f}^k(\mathbf{u},\mathbf{v},\mathbf{q}), \mathbf{f}^l(\mathbf{v},\mathbf{z}))^T \quad (7.2\text{-}18)$$

where $\mathbf{f}^k(\mathbf{u},\mathbf{v},\mathbf{q})$ is the vector of the objective functions of the main stage of the global task (see (7.2-17)) and

$\mathbf{f}^l(\mathbf{v},\mathbf{z})$ is the vector of objective functions of local tasks

which has the form $\quad \mathbf{f}^l(\mathbf{v},\mathbf{z}) = (f_1^l(\mathbf{v},\mathbf{z}),...,f_S^l(\mathbf{v},\mathbf{z}))^T \quad . \quad (7.2\text{-}19)$

The vector function $\mathbf{f}^1(\mathbf{v},\mathbf{z})$ should only contain these objective functions of the local tasks which are important for the whole optimization model. Some of the objective functions of the local tasks given in (7.2-9) can be of local importance only and thus in order to operate in a reasonable size of the objective space these functions may be neglected at this stage. If some objective functions are additive, then the i-th component of the vector $\mathbf{f}^1(\mathbf{v},\mathbf{z})$ has the form

$$f_i^1(\mathbf{v},\mathbf{z}) = \sum_{j=1}^{r} f_i^j(\mathbf{v},\mathbf{z}^j) \quad . \tag{7.2-20}$$

At this stage the set of solutions X is finite and determined, and thus the optimization problem is to select the set of Pareto-optimal solutions X^* from the set of X. This can be done by using the algorithm presented in [8]. The choice of the solution from the set X^* is the designer's final decision.

In MO strategy described here the weighting min-max method has been used for generating representative subsets of Pareto-optimal solutions at each of the first three stages. The procedure of generating the solutions can be carried out automatically. However, it is better if the designer has access to the decision making process at each stage of the procedure. In the latter case the optimization process is arranged in an interactive way in which the designer has influence on the solutions generated by the optimization system. The idea of such an approach is presented in Fig. 7.2/2. The results of the designer's decisions at each stage are stored, e.g. in direct access files and then taken into consideration in the optimization process at the stages to follow. A tree graph in Fig. 7.2/2a represents this generation and selection process. The number of solutions which are evaluated at the final stage depends on the number of solutions generated at each stage and on the number of solutions passed on to the remaining stages. At each stage the designer makes heuristic decisions. The loops denoted by broken lines in Fig. 7.2/2b shows the possibilities of returning to any node of the tree graph in order to obtain another set of solutions which he expects to be valuable for further considerations.

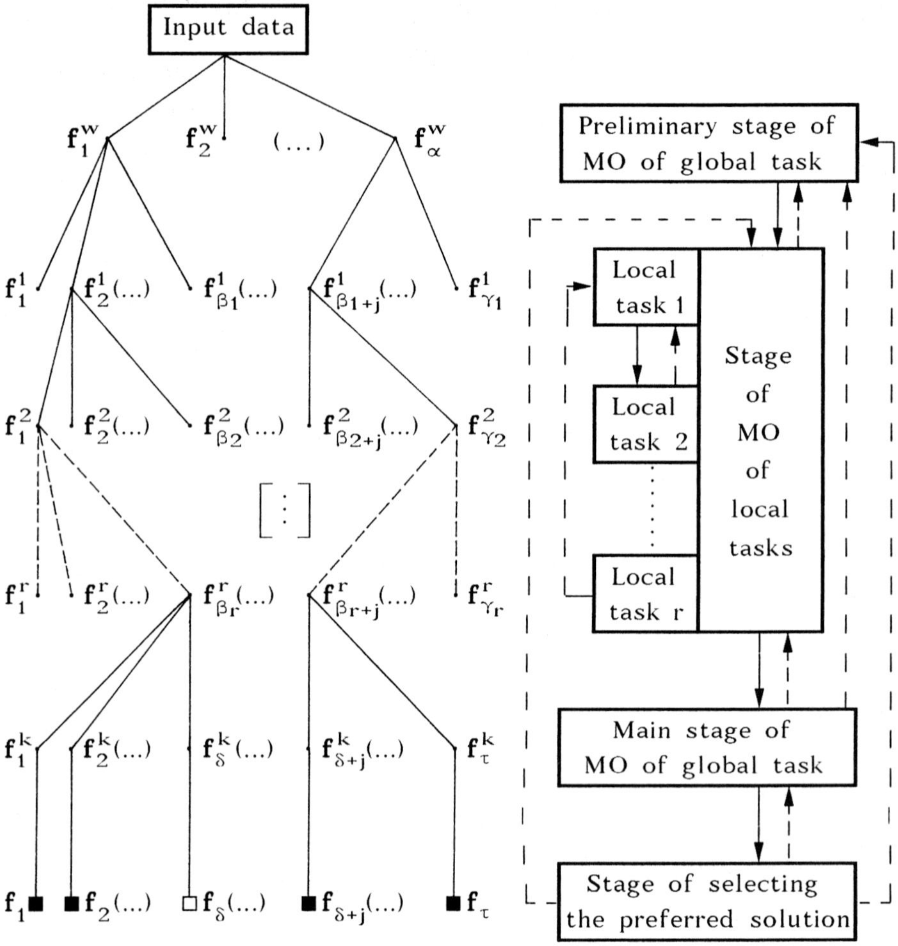

a) b)

----->	reflexive relation to the earlier stage based on the designer's decision,
■	Pareto-optimal solution ,
□	non-Pareto-optimal solution ,
$\mathbf{f}^{w}_{\alpha}$	vector of objective functions at the preliminary stage (7.2-5) ,
$\mathbf{f}^{1}_{\beta 1}, \mathbf{f}^{2}_{\beta 2}, \mathbf{f}^{r}_{\beta r}$	vectors of objective functions of the local tasks (7.2-9),
$\mathbf{f}^{k}_{\delta}$	vector of objective functions at the main stage (7.2-16),
$\mathbf{f}_{\delta}$	vector of objective functions of the global task (7.2-18).

Fig. 7.2/2. Process of generating solutions by means of the four-stage MO strategy
a) tree graph , b) possible loops

7.2.3 Application of the Strategy to the Design of a Machine Tool Spindle System with Hydrostatic Bearings

The Four-Stage Multicriteria Optimization Strategy discussed in Section 7.2.2 is applied to solving a complex problem which occurs in the design of a machine tool spindle supported by hydrostatic bearings. A typical example of a lathe structure with a machine tool spindle system presented in Fig. 7.2/3 is considered. The optimization model of this system covers only its static behaviour.

It is assumed that the spindle system contains the following subsystems: the front hydrostatic journal bearing A, the end hydrostatic journal bearing B, the hydrostatic thrust bearing of A, the spindle S, and the nose C of the spindle. Since the spindle nose has to be chosen from the catalogue before starting the design process the designer has to decide on the type and dimensions of this nose. In this way the dimensions of the spindle nose become constant parameters of the system.

Stage 1:

a) Vector of Design Variables

$$\mathbf{x}^{w} = (\mathbf{u},\mathbf{v})^{T}, \quad (7.2\text{-}21)$$
$$\mathbf{u} = (x_1,x_2,x_3)^{T},$$
$$\mathbf{v} = (x_4,x_5,x_6)^{T},$$

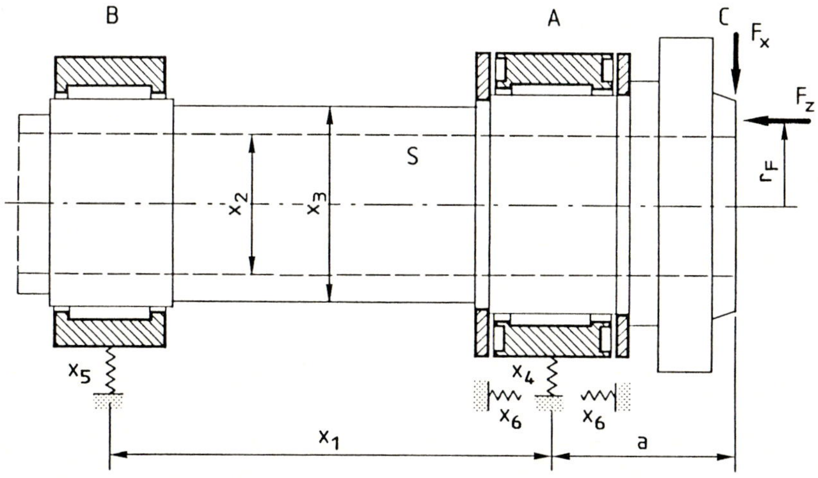

Fig. 7.2/3. Scheme of the spindle system

where

x_1 - distance between supports [mm],
x_2 - inner diameter of the spindle [mm],
x_3 - outer diameter of the spindle [mm],
x_4, x_5, x_6 - initial static stiffnesses of the journal bearings A and B, and the thrust bearing, respectively.

Here, the vector $\mathbf{v}$ is the parametric coordination vector whereas the constraint coordination vector has the form

$$\mathbf{v}^w = (x_2, x_3)^T \quad . \tag{7.2-22}$$

b) Vector of objective functions

$$\mathbf{f}^w(\mathbf{x}^w) = \left[f_1^w(\mathbf{x}^w),\ f_2^w(\mathbf{x}^w),\ f_3^w(\mathbf{x}^w),\ f_4^w(\mathbf{x}^w)\right]^T , \tag{7.2-23}$$

where
$f_1^w(\mathbf{x}^w)$ - radial stiffness of the spindle system,
$f_2^w(\mathbf{x}^w)$ - mass of the spindle with the spindle nose,
$f_3^w(\mathbf{x}^w)$ - radial runout of the spindle nose,
$f_4^w(\mathbf{x}^w)$ - axial stiffness of the spindle system.

The first and the fourth objective function are to be maximized whereas the second and the third one are to be minimized.
The first objective function is

$$f_1^w(\mathbf{x}^w) = \frac{F_x}{d_x} \quad [\mathrm{N}/\mu\mathrm{m}], \tag{7.2-24a}$$

where F_x radial force,
d_x radial deflection calculated using formula (7.2-12) from [3].

The second objective function is

$$f_2^w(\mathbf{x}^w) = 10^{-6} \rho \sum_j A_j\, l_j \quad [\mathrm{kg}], \tag{7.2-24b}$$

where A_j cross-section area of the j-th part of the spindle [mm^2] ,
l_j length of the j-th part of the spindle [mm] ,
ρ mass density [g/cm^3] .

The third objective function is

$$f_3^w(\mathbf{x}^w) = \Delta_{Ar} + \frac{a}{x_1}(\Delta_{Ar} + \Delta_{Br}) \quad [\mu\mathrm{m}] , \tag{7.2-24c}$$

where Δ_{Ar} and Δ_{Br} are the assumed radial runouts of the front and the end bearings, respectively.

The fourth objective function is

$$f_4^w(\mathbf{x}^w) = \frac{F_z}{d_z} \qquad [N/\mu m] , \tag{7.2-24d}$$

where F_z - axial force,
d_z - axial deflection calculated using formula (18) from [3].

c) Constraints

The constraints imposed on the vector $\mathbf{x}^w$ describe the upper and the lower bounds of design variables and the required proportions between the spindle stiffness and the bearing stiffnesses.

It should be mentioned here that in the optimization model considered at this stage some quantities which are used for calculations of the objective functions are estimated "a priori". These quantities, for example dimensions of the bearings, will be evaluated at the second stage, and they are introduced to the global task at the third stage by means of the parametric coordination vector **q**.

Stage 2:

At this stage three separate local tasks are distinguished:

1. optimization of the thrust bearing,
2. optimization of the front journal bearing,
3. optimization of the end journal bearing.

These tasks are hierarchically ordered in such a way that the optimization of the thrust bearing is the leading task of the first order for the optimization of the front journal bearing, and the optimization of the front journal bearing is the leading task of the second order for the optimization of the end journal bearing.

Task 1:

This task consists of optimizing the thrust hydrostatic journal bearing shown schematically in Fig. 7.2/4.

Vector of design variables

$$\mathbf{x}^{l1} = (\mathbf{v}, \mathbf{z}^1)^T , \tag{7.2-25}$$

where z_1^1 - supply pressure for fluid [MPa],

z_2^1 - orifice diameter [mm],

z_3^1 - depth of the pocket [mm],

z_4^1 - pocket land length [mm],

z_5^1 - initial fluid film thickness [μm],

z_6^1 - outer diameter of the bearing [mm],

z_7^1 - inner diameter of the bearing [mm].

Vector of objective functions

$$\mathbf{f}^1(\mathbf{x}^{l1}) = (f_1^1(\mathbf{x}^{l1}),\ f_2^1(\mathbf{x}^{l1}))^T \tag{7.2-26a}$$

where $f_1^1(\mathbf{x}^{l1})$ - total power loss in the bearing [kW],

$f_2^1(\mathbf{x}^{l1})$ - size of the bearing expressed by

$$f_2^1(\mathbf{x}^{l1}) = 0.25\pi[(z_6^1)^2 - (z_7^1)^2]\ [\mathrm{mm}^2]. \tag{7.2-26b}$$

Constraints

Constraints $g_j^{l1}(\mathbf{v},\mathbf{z}^1,\mathbf{v}^w) \le 0$ for $j = 1,\ldots,14$ are side constraints of the design variables. The 15th constraint defines the load capacity of the bearing

$$g_{15}^{l1}(\mathbf{v},\mathbf{z}^1,\mathbf{v}^w) \equiv P_{min} - P \le 0, \tag{7.2-27}$$

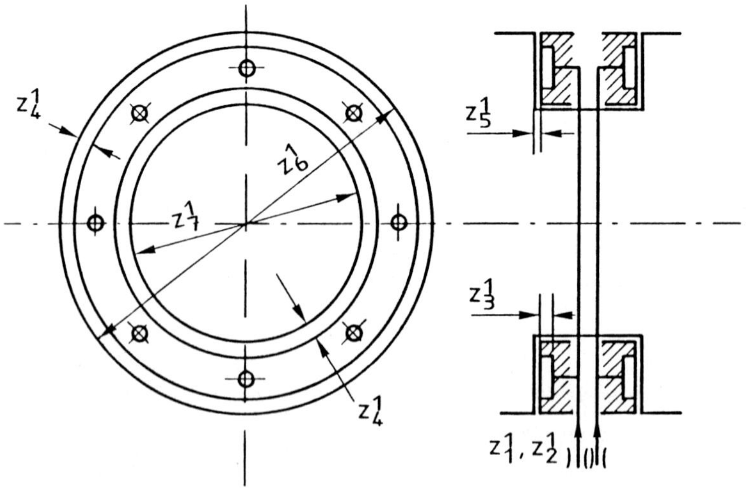

Fig. 7.2/4. Scheme of the hydrostatic thrust bearing

where P_{min} - minimal load capacity of the thrust bearing,
P - calculated load capacity of the thrust bearing.

Coordination

The constraint coordination vector for task 2 is $\mathbf{v}^{12} = (z_7^1)$.

Task 2:

This task consists of optimizing the front hydrostatic journal bearing shown in Fig. 7.2/5.

Vector of design variables

$$\mathbf{x}^{12} = (\mathbf{v}, \mathbf{z}^2)^T , \qquad (7.2\text{-}28)$$

where z_1^2 - orifice diameter [m] ,
z_2^2 - supply pressure of the fluid [MPa] ,
z_3^2 - angle of the circumferential land length [rad] ,
z_4^2 - angle of the circumferential pocket length [rad] ,
z_5^2 - depth of the pocket [mm] ,
z_6^2 - axial pocket land length [mm] ,
z_7^2 - radial clearance [μm] ,
z_8^2 - length of the bearing [mm] ,
z_9^2 - diameter of the bearing [mm] .

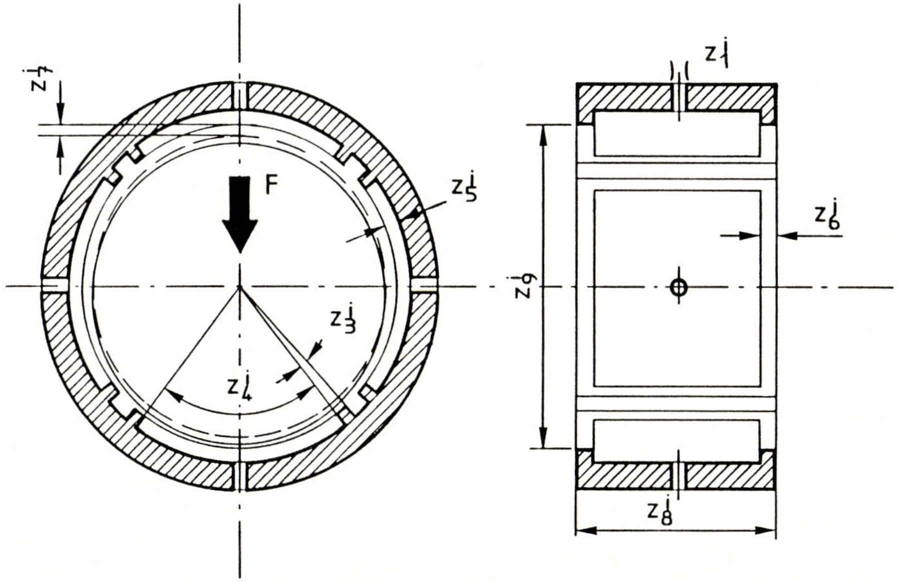

Fig. 7.2/5. Scheme of the hydrostatic journal bearing

Vector of objective functions

$$\mathbf{f}^2(\mathbf{x}^{l2}) = (f_1^2(\mathbf{x}^{l2}),\ f_2^2(\mathbf{x}^{l2}))^T, \tag{7.2-29a}$$

where $f_1^2(\mathbf{x}^{l2})$ total power loss in the bearing [kW],
$f_2^2(\mathbf{x}^{l2})$ size of the bearing expressed by

$$f_2^2(\mathbf{x}^{l2}) = \pi z_9^2 \cdot z_8^2 \quad [\mathrm{mm}^2]\ . \tag{7.2-29b}$$

Constraints

The constraints $g_j^2(\mathbf{v},\mathbf{z}^2,\mathbf{v}^w,\mathbf{v}^{12}) \le 0$ for $j = 1,...,18$ are the side constraints of the design variables. The 19th constraint defines the load capacity of the journal bearing and has a similar form as given in (7.2-27).

Note that the diameter of the front journal bearing z_9^2 has to be equal to the diameter of the thrust bearing z_7^1. This is expressed by the constraint coordination vector $\mathbf{v}^{12} = (z_7^1)$. Thus, the constraints imposed on z_9^2 have the form

$$g_{17}^2(\cdot) \equiv z_9^2 - z_7^1 \le 0 \quad ,$$

$$g_{18}^2(\cdot) \equiv z_7^1 - z_9^2 \le 0 \quad .$$

Coordination

The constraint coordination vector for task 3 is $\mathbf{v}^{23} = (z_9^2)$.

Task 3:

This task consists of optimizing the end journal hydrostatic bearing. The optimization model is identical with the model for Task 2. The vector of the design variables is $\mathbf{x}^{l3} = (\mathbf{v},\mathbf{z}^3)$. The elements of vector $\mathbf{z}^3$ have the same meaning as those of vector $\mathbf{z}^2$. Similarly, the vector of the objective function $\mathbf{f}^3(\mathbf{x}^{l3})$ has the same elements as the vector of the objective function $\mathbf{f}^2(\mathbf{x}^{l2})$. Also, the constraints are the same as in Task 2. The only difference is that the constraint coordination vector is $\mathbf{v}^{23}$ instead of $\mathbf{v}^{12}$.

Stage 3:

At this stage the vector of the design variables is $\mathbf{u} = (x_1,x_2,x_3)^T$ with elements as in formula (7.2-21). The objective function vector is

$$\mathbf{f}^k(\mathbf{u},\mathbf{v},\mathbf{q}) = [(f_1^k(\mathbf{u},\mathbf{v},\mathbf{q}),\ f_2^k(\mathbf{u},\mathbf{v},\mathbf{q}),\ f_3^k(\mathbf{u},\mathbf{v},\mathbf{q})]^T \ , \tag{7.2-30}$$

where $f_1^k(\cdot)$, $f_2^k(\cdot)$, $f_3^k(\cdot)$ have the same meaning as functions $f_1^w(\cdot)$, $f_2^w(\cdot)$, $f_3^w(\cdot)$ in formula (7.2-23).

The parametric coordination vector $\mathbf{q}$ represents the following quantities evaluated at the stage of the local tasks: the diameter and the length of the thrust bearing, the diameters and the lengths of the front, and the end journal bearings.

The constraint coordination vectors from the stage of local tasks are $\mathbf{v}^1 = \mathbf{0}$, $\mathbf{v}^2 = \mathbf{0}$, $\mathbf{v}^3 = (z_9^3)^T$.

The constraints imposed on decision variables are the side constraints, the same as those in the preliminary stage of MO of the global task.

Stage 4:

At this stage all solutions generated at previous stages are evaluated by taking into consideration the combined vector function of the form

$$\mathbf{f}(\mathbf{x}) = (\mathbf{f}^k(\mathbf{u},\mathbf{v},\mathbf{q}),\ \mathbf{f}^l(\mathbf{v},\mathbf{z}))^T , \tag{7.2-31a}$$

where

$$\mathbf{f}^k(\mathbf{u},\mathbf{v},\mathbf{q}) = (f_1^k(\mathbf{u},\mathbf{v},\mathbf{q}),\ f_2^k(\mathbf{u},\mathbf{v},\mathbf{q}),\ f_3^k(\mathbf{u},\mathbf{v},\mathbf{q}))^T , \tag{7.2-31b}$$

$$f^l(\mathbf{v},\mathbf{z}) = \sum_{i=1}^{3} f_1^i(\mathbf{v},\mathbf{z}^i) . \tag{7.2-31c}$$

with
- $f_1^k(\cdot)$ - radial stiffness of the spindle system,
- $f_2^k(\cdot)$ - mass of the spindle system,
- $f_3^k(\cdot)$ - radial runout of the spindle nose,
- $f^l(\cdot)$ - total power loss in the bearings.

From the set of all solutions a set of Pareto-optimal solutions is to be selected using for example the algorithm presented in [8]. This set is then evaluated by the designer who makes the final decision on the choice of the preferred solution.

Fig. 7.2/6 summarizes the main points of the Four-Stage Strategy. On the basis of this strategy a computer program for the optimum design of the machine tool spindle system has been developed. All stages of the strategy are realized by means of a dialogue with a computer. After the final stage of the strategy a graphical illustration of the optimum design is screened or plotted. The computer program is coded in FORTRAN and implemented on an IBM PC/XT/AT.

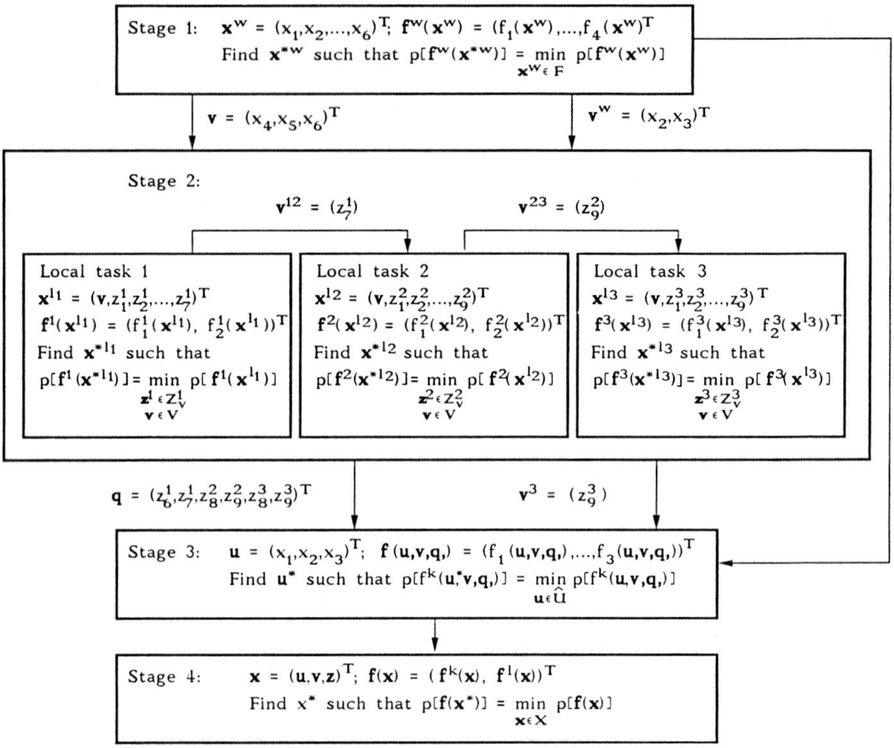

Fig. 7.2/6. Four-stage multicriteria optimization strategy for machine tool spindle system design

7.2.4 Numerical Example

Consider the problem of designing a machine tool spindle system based on the example presented in [11] for which the basic data are

- radial force $F_x = 2000$ N,
- axial force $F_z = 500$ N,
- radius of the acting forces $r_F = 50$ mm,
- assumed radial run-outs of the bearings $\Delta_{Ar} = \Delta_{Br} = 2$ µm,
- assumed bounds on the basic design variables $600 \leq x_1 \leq 1000$; $x_2 \geq 45$ [all in mm], $x_4 \geq 200$; $x_5 \geq 200$; $x_6 \geq 400$ [all in N/µm] .

The preliminary stage of MO of the global task has been carried out for two spindle noses No. 6 and No. 8 of the "Camlock" type. The weighting min-max method (see Section 1.2.2) has been used for generating the solutions. Note that the number of generated solutions at each stage depends on the designer. After generating optimal solutions at each stage, the choice is made as to which of them are to be considered in further calculations. This is shown in Fig. 7.2/7 where the nodes which have no other continuations are disregarded. In this example, 21 solutions have been passed to the stage of final selection of Pareto-optimal solutions. From this set, 19 were Pareto-optimal. The min-max optimal solution is:

$$\mathbf{f}(\mathbf{x}) = (75.3,\ 20.5,\ 2.84,\ 0.358)^T,$$
$$\mathbf{x} = (600.0,\ 82.7,\ 56.1)^T,$$
$$\mathbf{z}^1 = (1.55,\ 0.31,\ 5.0,\ 2.4,\ 30.0,\ 130.0,\ 93.0)^T,$$
$$\mathbf{z}^2 = (0.12,\ 2.0,\ 0.115,\ 1.112,\ 5.0,\ 3.6,\ 43.5,\ 82.8,\ 93.0)^T,$$
$$\mathbf{z}^3 = (0.11,\ 2.0,\ 0.115,\ 1.176,\ 4.1,\ 4.6,\ 40.8,\ 82.8,\ 84.7)^T.$$

Another solution from the set of Pareto-optimal solutions which is better considering the stiffness of the spindle system but at the same time worse considering the mass of the spindle, and the total power loss in the bearings yields the results:

$$\mathbf{f}(\mathbf{x}) = (84.2,\ 24.6,\ 2.82,\ 0.425)^T,$$
$$\mathbf{x} = (600,\ 81.2,\ 45.0)^T,$$
$$\mathbf{z}^1 = (1.92,\ 0.31,\ 3.8,\ 3.1,\ 32.3,\ 128.8,\ 94.7)^T,$$
$$\mathbf{z}^2 = (0.12,\ 1.9,\ 0.100,\ 1.207,\ 4.8,\ 2.5,\ 45.9,\ 82.8,\ 94.7)^T,$$
$$\mathbf{z}^3 = (0.12,\ 1.9,\ 0.087,\ 1.217,\ 4.8,\ 2.5,\ 38.7,\ 79.7,\ 83.0)^T.$$

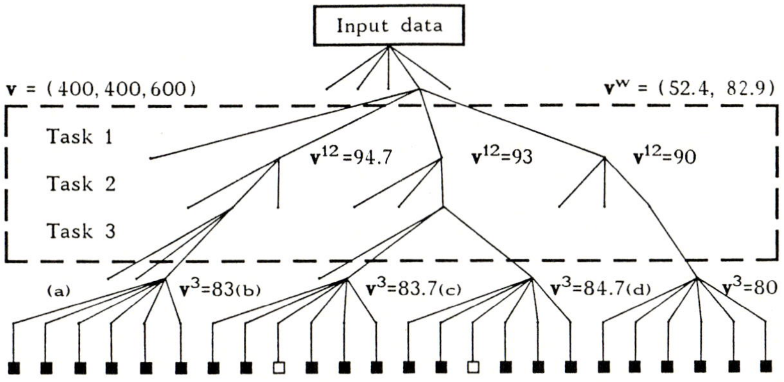

$\mathbf{q}_{(a)}$ = (94.7, 14, 94.7, 82.8, 83, 79.7) $\mathbf{q}_{(c)}$ = (93, 16, 93, 82.9, 84.7, 82.8)

$\mathbf{q}_{(b)}$ = (93, 16, 93, 82.9, 83.7, 82.5) $\mathbf{q}_{(d)}$ = (93, 16, 93, 87.8, 80, 88.4)

Fig. 7.2/7. Graph of generated solutions at all stages of the strategy

A computer graphics employed in the optimization program allows the designer to have a simplified design of the spindle system for any of the solutions which helps him to make the right decision. An example of the graphical representation of the solution for the min-max optimum is shown in Fig. 7.2/8.

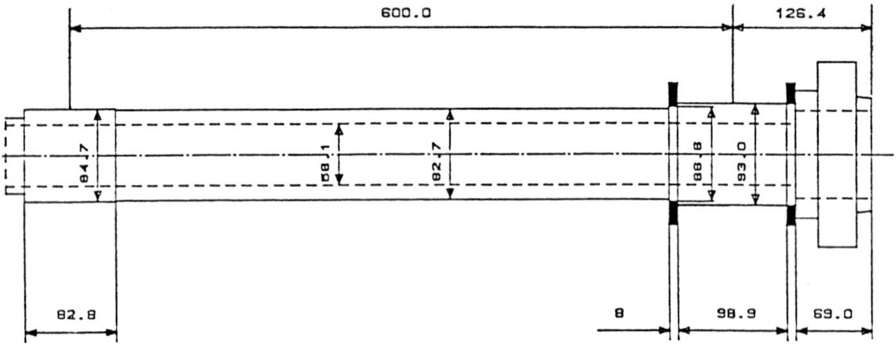

a)

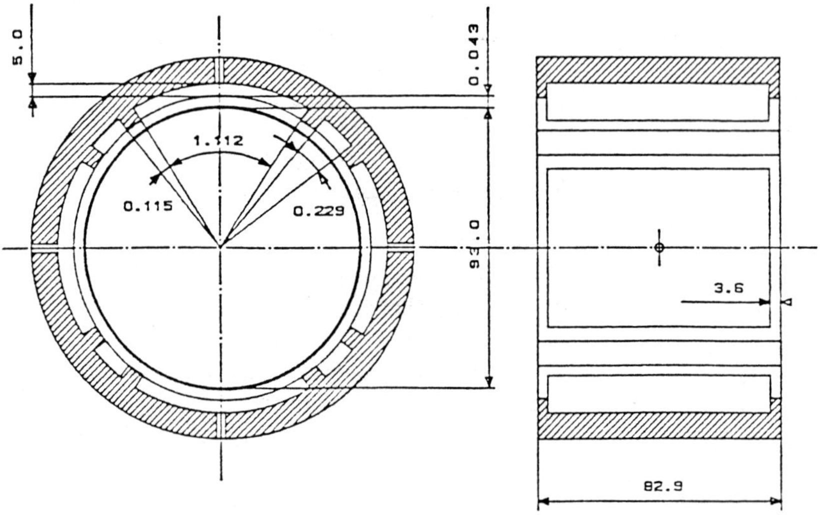

b)

Fig. 7.2/8. Computer graphic output for an optimal solution of a machine-tool spindle
a) Spindle
b) Journal bearing "A"

7.2.5 Conclusion

The paper shows that a decomposition approach to multicriterion optimization problems provides the designer with an effective tool for solving complex design tasks. The proposed Four-Stage Multicriteria Strategy improves the efficiency of optimization algorithms, and it is one of the most cited reasons to resort to multistage optimization. Moreover, the strategy provides the possibility of optimizing individual subsystems without the need to recalculate the complete system.

The strategy is made to conform better to the design process itself, especially for the system for which the division between subsystems is evident. It has a heuristic character, and thus heuristic rules used by the designer can be implemented to the optimization process. A knowledge-based approach may be used in the future to formalize these rules, and thus further steps in the direction of automation of design can be taken.

The optimization strategy is applied to solving the real life problem of designing a machine tool spindle system with hydrostatic bearings. A computer aided optimum design software package is developed for this application. Numerical results of the presented example show the advantages of this software regarding the ease of its application as well as the usefulness of the strategy to the designer.

References

[1] Barthelemy, J.-F.,M: Engineering Application of Heuristic Multilevel Optimization Methods. In: Eschenauer, H.; Thierauf, G.: "Discretization Methods and Structural Optimization - Procedures and Applications." Berlin, Heidelberg, New York, London, Paris, Tokyo: Springer-Verlag (1989) 24-31

[2] Bremicker, M.; Eschenauer, H.; Post, P.: On a Decomposition Technique within the Scope of a Design Optimization Process. In: Eschenauer, H.; Thierauf, G.: "Discretization Methods and Structural Optimization - Procedures and Applications." Berlin, Heidelberg, New York, London, Paris, Tokyo: Springer (1989) 87-94

[3] Giebner, E.: Bearing Arrangement Design for Machine Tool Spindles, Ball Bearing Journal 216 (1983) 1-13

[4] Himmelblau, D.M.: Applied Nonlinear Programming. New York: McGraw-Hill Book Co. 1972

[5] Johnson, R.C.; Benson, R.C.: A Basic Two-Stage Decomposition Strategy for Design Optimization. Journal of Mechanisms, Transmissions, and Automation in Design, Vol. 106 (1984) 380-386

[6] Johnson, R.C.; Benson, R.C.: A Multistage Decomposition Strategy for Design Optimization. Journal of Mechanisms, Transmissions, and Automation in Design, Vol. 106 (1984) 387-393

[7] Osyczka, A.: Computer Aided Multicriterion Optimization System (CAMOS). In: Eschenauer, H.; Thierauf, G.: "Discretization Methods and Structural Optimization - Procedures and Applications". Berlin, Heidelberg, New York, London, Paris, Tokyo: Springer-Verlag (1989) 263-270

[8] Osyczka, A.: Multicriterion Optimization in Engineering with FORTRAN Programs. Chichester: John Wiley & Sons 1984

[9] Reimers, U.: A Method for Solving the Decentralized Hierarchical Multiple Objective Decision Making Problem. Manuskripte aus dem Institut für Betriebswirtschaftslehre der Universität Kiel, No. 154, 1984

[10] Shimizu, K.; Aiyoshi,E.: Hierarchical Multi-Objective Decision Systems for General Resource Allocation Problems. Journal of Optimization Theory and Application, Vol. 35, No. 4, 1981

[11] Szimanowicz, M.A.; Kundraszow, L.W.: Sistiema Program dlia Rasczeta Nagruzocznych Charakteristik Szpindielia, Ustanowliennogo w Gidrostaticzeskich Oporach, Stanki i Intstrumienty, No. 3 (1985) 9-12

[12] Waleckx, J.: Best Compromise Stiffness/Speed for Precision Machine Tool Spindles, Ball Bearing Journal, 208 (1981) 11-21

8 METAL FORMING AND CAST METAL TECHNOLOGY

8.1 Optimal Die Design for Symmetric Strip Drawing

W. Stadler

8.1.1 Introduction and Review

Traditionally, there appear to be two optimal die designs, the conical die and the perfect die. The former is in general use because of the ease with which it may be manufactured; the latter is not used at all because it is considered to be too costly to manufacture. In essence, the conical die is a "throw-away" die. The advent of extremely hard materials, however, makes it worthwhile to construct long-lasting optimal dies.

In most plasticity books including discussions on drawing or extrusion, it is mentioned that the conical die minimizes the required axial drawing force. A formal treatment of this problem is included here to introduce the control theoretic methods subsequently used in some multicriteria problems. The "perfect die" was presented by Richmond and Davenpeck [1] in 1962. Assuming strip drawing without friction and without hardening, the equations of flow of a rigid-perfectly plastic material are hyperbolic, and the two families of maximum shear stress surfaces are characteristics designated as α- and β-lines. With U and V as the components of the velocity vector in the α and β directions, respectively, they obtain the streamlines for diagonal flow based on the assumption U=V. Streamlines which exhibit the desired amount of contraction may then be used as the corresponding die profile. In a later paper, Richmond and Morrison [2] extended their results to axisymmetric wire drawing and pointed out that their procedure did not yield a unique solution. With this in mind they ultimately select the shortest die from among those that are feasible. Subsequent discussions of this type of die design by Hill [3,4] provide a setting of this approach within general properties possessed by diagonal flows; he also points out that the resultant forming process minimizes the work of deformation. This clearly follows from the fact that all of the work goes into the homogeneous contraction of the material; that is, no excess work is being done. There are some additional comments on this type of die design in Sortais and Kobayashi [5]. A summary of perfect die design appears in the plasticity text by Johnson and Mellor [6].

The optimality of this type of die is intuitive rather than formal. It is akin to the optimality implied by fully stressed design in structural optimization. One thing does stand out clearly in this approach: the dependence of the die shape on the plastic behaviour postulated within the governing equations for the deformation process, a dependence which is often suppressed in optimal structural design.

An optimal design more in keeping with a formal approach is given by Thomsen [7,8] in a report and a related paper. Based on crack propagation arguments from fracture mechanics and a desire to extend the tool life, he considers the interaction of the plastic deformation process with an elastic die. In particular, he minimizes the maximum tensile stress subject to a stress constraint involving the von Mises stress. Among other problems he deals with the optimal lamination for a die of given shape and with the design of a suitable shape for the homogeneous die.

Overall, it is generally accepted that the die may be considered to be rigid when compared to the intended plastic deformation of the formed material. This viewpoint will be taken in the following discussion. The relevant equilibrium and failure models as well as the criteria are introduced in the next section. Subsequently, these form the basis for some single- and multicriteria optimization problems.

8.1.2 The Mathematical Model

In summary, the model is based on a steady state of flow of an ideally plastic material through dies for which the friction between die and material is negligible. Although this is the general setting, some results including friction are also presented.

a) The governing differential equation

The classic work by Hoffman and Sachs [9] suffices as desk reference for the following derivations and discussions. In essence, the models given there for conical dies are modified to allow for a nonlinear variation of the die profile. Because of space limitations, the presentation will focus on strip drawing without backpull.

A summation of forces on a segment of a strip of unit width yields the quasi static equilibrium condition (Fig. 8.1/1).

$$\frac{d}{dz}(\sigma(z)y(z)) + p\frac{dy(z)}{dz} - \mu p = 0 \quad . \tag{8.1-1}$$

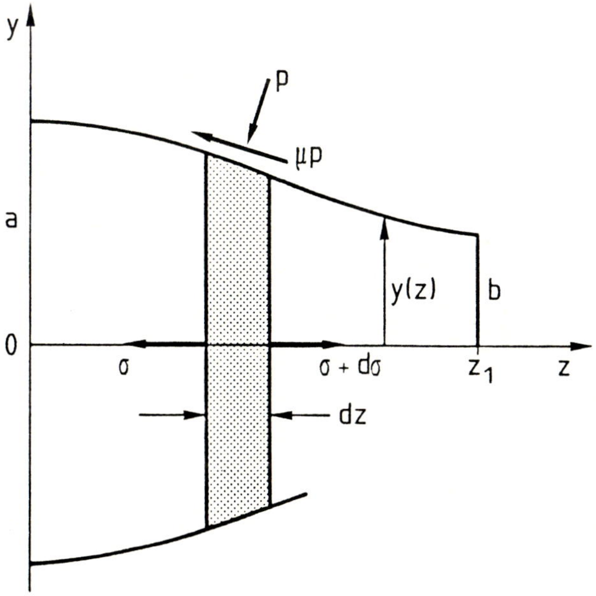

Fig. 8.1/1. Equilibrium

Here, σ denotes the stress in the z-direction, p is the normal pressure between die and material, μ is the dynamic friction coefficient, and $2y(z)$ is the diameter at z. The die length (which may be unspecified) is denoted by z_1, and the material flow is assumed to be from left to right. with the drawing force applied at $z = z_1$, the righ end of the die.

A suitable yield condition is given by the von Mises-Hencky condition of plasticity

$$(\sigma_1 - \sigma_2)^2 + (\sigma_2 - \sigma_3)^2 + (\sigma_3 - \sigma_1)^2 = 2\sigma_0^2 \, , \tag{8.1-2}$$

where the σ_i (i=1,...,3) are the principal stresses and where σ_0 is some uniaxial yield stress. The assumption of plane strain and the use of Saint Venant's stress-strain relation eventually yield

$$(\sigma_1 - \sigma_2) = \frac{2}{\sqrt{3}} \sigma_0 \quad . \tag{8.1-3a}$$

The further assumption that the principal stresses may be identified with $\sigma_1 = \sigma$ and $\sigma_2 = -p$ then implies

$$\sigma + p = \frac{2}{\sqrt{3}} \sigma_0 \quad . \tag{8.1-3b}$$

Consequently, the governing equation for strip drawing becomes

$$\frac{d}{dz} (\sigma(z)y(z)) + (\frac{2}{\sqrt{3}} \sigma_0 - \sigma(z)) (\frac{dy(z)}{dz} - \mu) = 0 \quad . \tag{8.1-4}$$

b) The total axial resistance

The backpull is taken to be zero; that is, the force at the die entry is zero. The stress vector at any z is (see Fig. 8.1/2)

$$\tau = \frac{p}{\sqrt{1 + (y')^2}} [(y' - \mu) \hat{i} + (-1 - \mu y') \hat{j}] \quad . \tag{8.1-5}$$

The total axial resistance is thus given by

$$R = 2\int_0^{z_1} (\frac{2}{\sqrt{3}} \sigma_0 - \sigma(z)) (\frac{dy(z)}{dz} - \mu) \, dz \quad . \tag{8.1-6}$$

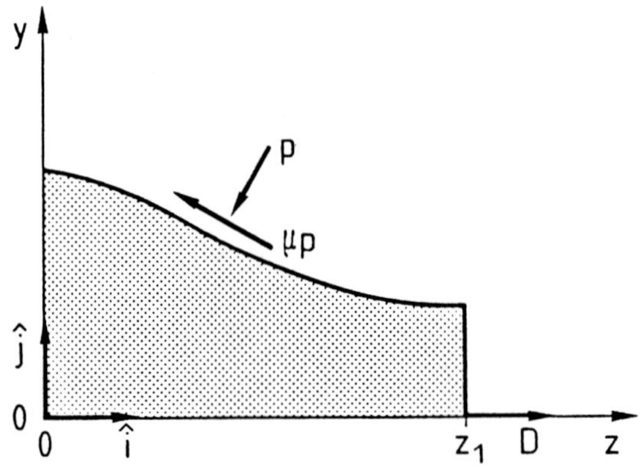

Fig. 8.1/2. Axial resistance

c) The total contraction

Rather than specifying the total amount of contraction, it may be taken as a criterion with

$$C = \int_0^{z_1} \frac{dy}{dz}(z)\, dz \quad . \tag{8.1-7}$$

d) Total work of plastic deformation

The derivation of the expression for the total plastic work follows that of [10]. The work per unit volume is

$$w = \int_0^t \dot{w} dt = \frac{2}{\sqrt{3}} \sigma_0 \int_0^t \sqrt{\frac{1}{2} \dot{\varepsilon}_{ij} \dot{\varepsilon}_{ij}}\, dt \ , \tag{8.1-8a}$$

where the $\dot{\varepsilon}_{ij}$ are the strain rates. The first step in integrating this expression is the assumption of proportional straining; that is, the ratios of the components of strain remain constant with respect to time (they may still depend on the location within the body). One has

$$\dot{\varepsilon}_{11} : \dot{\varepsilon}_{22} : \dot{\varepsilon}_{33} : \dot{\varepsilon}_{12} : \dot{\varepsilon}_{23} : \dot{\varepsilon}_{31} = \varepsilon_{11} : \varepsilon_{22} : \varepsilon_{33} : \varepsilon_{12} : \varepsilon_{23} : \varepsilon_{31} =$$
$$= 1 : a_{22} : a_{33} : a_{12} : a_{23} : a_{31} \quad . \tag{8.1-9}$$

As a consequence, the time integration may be carried out to obtain

$$w = \frac{2}{\sqrt{3}} \sigma_0 \sqrt{\frac{1}{2} \varepsilon_{ij} \varepsilon_{ij}} \quad . \tag{8.1-8b}$$

An integration over the material volume yields the total work as

$$W = \frac{2}{\sqrt{3}} \sigma_0 \int_V \sqrt{\frac{1}{2} \varepsilon_{ij} \varepsilon_{ij}}\, dV \quad . \tag{8.1-10a}$$

When principal axes of strain are used, this becomes

$$W = \frac{2}{\sqrt{3}} \sigma_0 \int_V \sqrt{\varepsilon_1^2 + \varepsilon_1 \varepsilon_2 + \varepsilon_2^2}\, dV \quad , \tag{8.1.-10b}$$

and for plane strain, this reduces to

$$W = \frac{2}{\sqrt{3}} \sigma_0 \int_V |\varepsilon_1| dV \quad . \tag{8.1-10c}$$

For the case of strip drawing, as considered here, the logarithmic principal strain in the direction normal to the axis of drawing is

$$|\varepsilon_1| = \ln \frac{a}{y(z)} , \tag{8.1-11}$$

where a defines the size of the die entry at z = 0.
The work integral then is

$$W = \frac{2}{\sqrt{3}} \sigma_o \int_0^{z_1} y(z) \ln \frac{a}{y(z)} dz \quad . \tag{8.1-10d}$$

e) The surface area of the die

The total surface of the strip of unit width will be approximated with

$$S = 2 \int_0^{z_1} \sqrt{1 + \left(\frac{dy(z)}{dz}\right)^2}\, dz \approx 2 \int_0^{z_1} \left[1 + \frac{1}{2}\left(\frac{dy(z)}{dz}\right)^2\right] dz \quad . \tag{8.1-12}$$

These are the criteria to be considered along with the equilibrium condition, with the die profile as the design variable. The first problem is a single criterion control problem for the determination of the die profile yielding a minimum of the drawing force. The remaining problems are bicriteria problems.

8.1.3 The Ubiquitous Conical Die

All of the following problems are formulated either as single criterion or bicriteria optimal control problems. For the general control theoretic aspects and notation, the reader is referred to a standard text such as Lee and Markus [11]; for the multicriteria optimal control problem and methods, a recent volume by the author [12] is recommended (see also Section 1.2.3). In order to make the general results of optimal control more easily accessible, it is useful to introduce standard notation. With $x_1 = y$, $x_2 = \sigma$, $\frac{dy}{dz} = u$, and independent variable t, the governing differential equation takes on the standard state space form

$$\dot{x}_1 = u ,$$

$$\dot{x}_2 = -k \frac{u}{x_1} , \qquad k = \frac{2}{\sqrt{3}} \sigma_o , \tag{8.1-13}$$

with boundary conditions to be specified in the individual problems. Note that $\mu = 0$ is assumed for the remainder. The same notation is introduced for the criteria.

1) The minimum drawing force

Clearly, minimizing the resistance above also minimizes the drawing force D. The present simple example is included to clarify some of the optimality statements in the literature. In the problem formulation, the die length t_1 is left open and the amount of contraction from $x_1(0) = a$ to $x_1(t_1) = b$ is specified. The pressure p is taken to be constant and the allowable slope is taken to be bounded by $-c \le u \le 0$. The expression for R (8.1-6) may then be written as

$$R = 2p[y(z_1) - y(0)] - 2p\mu \int_0^{z_1} dz \,. \qquad (8.1\text{-}14a)$$

Since constants do not enter the optimization process, one is left with minimizing

$$R = - \int_0^{z_1} dz \,. \qquad (8.1\text{-}14b)$$

This leads to the following formal problem statement.

Problem. Minimize

$$f_1\,(u(\cdot)) = - \int_0^{t_1} dt \qquad (8.1\text{-}15a)$$

subject to

$$\dot{x}_1 = u,\ x_1(0) = a,\ x_1(t_1) = b \,, \qquad (8.1\text{-}15b)$$

with control constraint set

$F = \{\text{Lebesgue measurable } u(\cdot) : -c \le u \le 0\}$

with t_1 not specified.

In essence, this is a time optimization problem. The Hamiltonian for the problem is

$$H(x_1, u, \lambda_0, \lambda_1) = \lambda_0(-1) + \lambda_1 u \qquad (8.1\text{-}16a)$$

with $\lambda_0 \le 0$ and with adjoint equation

$$\dot{\lambda}_1 = - \frac{\partial H}{\partial x_1} = 0 \,. \qquad (8.1\text{-}16b)$$

The use of the transversality conditions leaves $\lambda_1(0)$ and $\lambda_1(t_1)$ arbitrary and one thus has $\lambda_1(t_1) = \lambda_1 = \text{const}$. The necessary conditions for optimal control require

$$\sup_{u \in F} H(x_1^*, u, \lambda_O, \lambda_1^*) = 0 . \tag{8.1-17}$$

The abnormal problem with $\lambda_O = 0$ implies $\lambda_1 u = 0$. Since $u(t) = 0$ does not satisfy the boundary conditions and $\lambda_1 = 0$ leads to $(\lambda_O \ \lambda_1) \equiv 0$, a contradiction of necessary conditions, it follows that the abnormal problem has no solution. Thus, $\lambda_O = -1$ may be used in the remainder. Now, depending on the sign of λ_1, one has $u = -c$ or $u = 0$. The only possible choice turns out to be $\lambda_1 < 0$, resulting in

$$1 + \lambda_1 c = 0 \rightarrow \lambda_1^* = -\frac{1}{c} . \tag{8.1-18}$$

The integration of the state equation yields

$$x_1^*(t) = -ct + a . \tag{8.1-19a}$$

The impositon of the boundary conditon at $t = t_1$ then yields

$$t_1^* = \frac{a - b}{c} . \tag{8.1-19b}$$

Here, the extremal solution $u^*(t) = -c$ is also optimal. Thus, the conical die with maximum possible slope and length just sufficient to attain the desired contraction is the optimal die when the pressure is constant and Coulomb friction is included.

Remark 1: Note carefully that once the maximum slope was specified, the optimal die length was also fixed. Furthermore, this problem is essentially geometrical, since the physical description of the formed material has been omitted as a constraint in the problem. A proper formulation is given within the subsequent bicriteria problems.

Remark 2: It is well known that the maximum contraction within the present model is attained when the drawing stress $x_2(t_1)$ becomes equal to the uniaxial critical stress σ_O. The result of a use of the governing differential equation is

$$\sigma_O = \frac{2}{\sqrt{3}} \sigma_O \ln \frac{a}{b} \rightarrow b = .4206a \tag{8.1-20}$$

or a 58% reduction. Naturally, it will be assumed that any of the following contraction requirements or results do not exceed this limit.

2) The natural shape of the die

Recall that a natural structural shape is the result of the simultaneous minimization of the mass and the strain energy of a loaded structure subject to the equilibrium of the structure [12]. In the present context, there is no retrievable stored energy, since all of the strain is permanent. It thus becomes logical to define the natural die as one which minimizes both the work of deformation and the total mass of the material segment involved in the forming process subject to the governing equation for the plastic deformation. With the total contraction and the length of the die specified, one then has the following multicriteria optimal control problem (see Section 1.2.3).

Problem: Obtain Pareto-optimal solution(s) $u^*(\cdot) \in F$ for the criteria

$$f_1(u(\cdot)) = \int_0^{t_1} x_1(t) \ln \frac{a}{x_1(t)} \, dt \quad \text{and} \tag{8.1-21a}$$

$$f_2(u(\cdot)) = \int_0^{t_1} x_1(t) \, dt$$

subject to

$$\dot{x}_1 = u, \quad x_1(0) = a, \; x_1(t_1) = b,$$

$$\dot{x}_2 = -k \frac{u}{x_1}, \quad x_2(0) = 0, \; x_2(t_1) \text{ arbitrary}, \tag{8.1-21b}$$

and $u(\cdot) \in F = \{u(\cdot) \text{ Lebesgue measurable: } -c \le u \le 0\}$.

The H-function is

$$H = -c_1 x_1 \ln \frac{a}{x_1} - c_2 x_1 + s(t) u, \tag{8.1-22}$$

where

$$s(t) = \lambda_1 - \lambda_2 \frac{k}{x_1} \tag{8.1-23}$$

is called the switching function. Here, and henceforth, the argument of H will be omitted. The adjoint equations are

$$\dot{\lambda}_1(t) = -\frac{\partial H}{\partial x_1} = c_1\left(-1 + \ln\frac{a}{x_1}\right) + c_2 - \lambda_2 \frac{k}{x_1^2} u\,,$$
$$\dot{\lambda}_2(t) = -\frac{\partial H}{\partial x_2} = 0 \tag{8.1-24}$$

with $\lambda_1(0)$ and $\lambda_1(t_1)$ arbitrary, and with $\lambda_2(t_1) = 0$ as implied by a use of the necessary transversality conditions.

Using the maximum principle for multicriteria control problems requires

$$\sup_{u \in F} H = \text{const.}$$

since t_1 is prescribed. Clearly, the choice of $u(\cdot)$ on the boundary of F. satisfying the condition, depends on the sign of the switching function $s(\cdot)$. If it can be shown that $s(\cdot)$ is monotonic, then there is at most one switch in control. Only the possibility $s(t) < 0$ will be considered here.

Since $\lambda_2(t_1) = 0$ implies $\lambda_2(t) \equiv 0$, the derivative of the switching function is

$$\dot{s}(t) = \dot{\lambda}_1(t) = c_1\left(-1 + \ln\frac{a}{x_1}\right) + c_2\;. \tag{8.1-25}$$

An integration of the state equations provides

$$x_2(t) = -k \ln\frac{x_1(t)}{a} = k \ln\frac{a}{x_1(t)} \tag{8.1-26a}$$

so that s (t) may also be written in the form

$$\dot{s}(t) = c_1\left(-1 + \frac{\sqrt{3}}{2}\frac{x_2(t)}{\sigma_0}\right) + c_2\,. \tag{8.1-26b}$$

In accordance with Remark 2, the investigation is again restricted to $x_2 \le \sigma_0$. With this in mind, the constraint

$$0 < \frac{c_2}{c_1} < 1 - \frac{\sqrt{3}}{2} = 0.134$$

then implies $\dot{s}(t) < 0$, so that there is at most one switch in control.

It can be shown that the assumption $s(0) < 0$, corresponding to a switch in control from $u(t) = -c$ to $u(t) = 0$ leads to a contradiction; this leaves $s(0) < 0$, and a switch from $u(t) = 0$ to $u(t) = -c$ as the remaining possibility.

With this control, the state is given by

$$x_1(t) = \begin{cases} a, & 0 \le t \le t_0 , \\ c(t_0 - t) + a , & t_0 < t \le t_1, \end{cases} \tag{8.1-27a}$$

$$x_2(t) = -k \begin{cases} 0, & 0 \le t \le t_0, \\ \ln\left[\frac{c}{a}(t_0 - t) + 1\right], & t_0 < t \le t_1 , \end{cases} \tag{8.1-27b}$$

where $t_0 = t - \frac{1}{c}(a - b)$. It remains to show $s(0) > 0$. For $0 \le t \le t_0$ an integration of $\dot{s}(t)$ together with $s(t_0) = 0$ yield

$$s(t) = (c_1 - c_2)(t_0 - t) \quad ,$$

so that $s(0) = (c_1 - c_2)\, t_0 > 0$ in view of the restriction $c_2 < 0.134c_1$. Hence, the control

$$u^*(t) = \begin{cases} 0, & 0 \le t \le t_0 , \\ -c, & t_0 < t \le t_1 \end{cases} \tag{8.1-28}$$

is an extremal control and the expressions for $x_1(t)$ and $x_2(t)$ above are the corresponding extremal states. The resulting conical die is shown in Fig. 8.1/3.

Remark 3: It is of interest to note that a die with lesser slope connecting the points $(0,a)$ and (t_1,b), a perfectly feasible choice, is not extremal. The solution above exists as long as $t_1 > \frac{1}{c}(a - b)$.

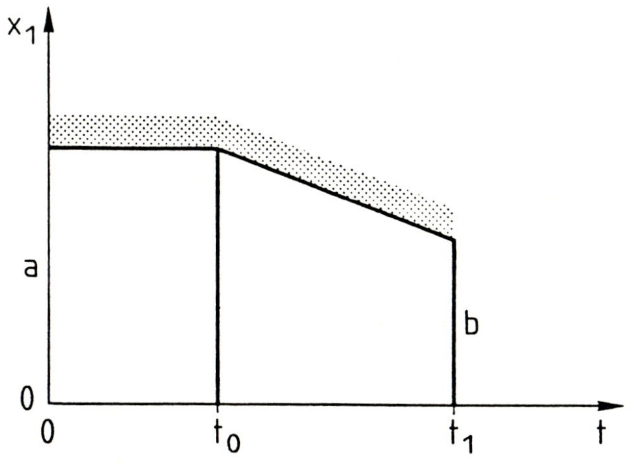

Fig. 8.1/3. The natural die

Remark 4: Note that besides the physical restriction $x_2(t) \leq \sigma_0$ the solution is independent of the choice of c_1 and c_2. That is, there exists a unique extremum which simultaneously "minimizes" both of the criteria. If sufficient conditions were also to be satisfied, this would be termed an utopian or ideal solution in the parlance of multicriteria optimization.

3) Maximum contraction and minimum drawing force

Clearly, the maximum contraction is obtained when

$$C = \int_0^{z_1} \frac{dy}{dz}(z)dz = y(z_1) - y(0) \tag{8.1-29}$$

is minimized, and minimizing the resistance will again minimize the drawing force. With specified die length and initial opening and with no backpull, one has the following nonlinear multicriteria control problem.

Problem: Obtain Pareto-optimal controls $u^*(\cdot) \in F$ for the criteria

$$f_1(u(\cdot)) = \int_0^{t_1} u(t)dt \quad \text{and} \quad f_2(u(\cdot)) = \int_0^{t_1} (k - x_2(t))\, u(t)dt \tag{8.1-30a}$$

subject to

$$\begin{aligned} \dot{x}_1 &= u, & x_1(0) &= a, \quad x_1(t_1) \text{ arb.,} \\ \dot{x}_2 &= -k\frac{u}{x_1}, & x_2(0) &= 0, \quad x_2(t_1) \text{ arb.,} \end{aligned} \tag{8.1-30b}$$

and $u(\cdot) \in F = \{u(\cdot) \text{ L-measurable: } -c \leq u \leq 0\}$.

The H-function for this problem is

$$H = s(t)u, \tag{8.1-31}$$

where

$$s(t) = -c_1 - c_2(k - x_2) + \lambda_1 - \lambda_2 \frac{k}{x_1} \tag{8.1-32}$$

is the switching function. The adjoint equations are

$$\begin{aligned} \dot{\lambda}_1 &= -\lambda_2 \frac{k}{x_1^2} u, \\ \dot{\lambda}_2 &= -c_2 u \end{aligned} \tag{8.1-33}$$

with $\lambda_1(t_1) = \lambda_2(t_1) = 0$ as implied by the transversality conditions. Again, a use of the multicriteria maximum principle for optimal control requires

$$\sup_{u \in F} H = \text{const.}$$

The "time" rate of change of the switching function is

$$\dot{s}(t) = c_2 \dot{x}_2 + \dot{\lambda}_1 - \dot{\lambda}_2 \frac{k}{x_1} + \lambda_2 \frac{k}{x_1^2} \dot{x}_1 =$$

$$= - c_2 k \frac{u}{x_1} - \lambda_2 \frac{k}{x_1^2} u + c_2 u \frac{k}{x_1} + \lambda_2 \frac{k}{x_1^2} u \equiv 0. \tag{8.1-34a}$$

Consequently, $s(t) \equiv$ const, and there is no switch in optimal control, Furthermore,

$$s(t_1) = -c_1 + c_2(x_2(t_1) - k) < 0 , \tag{8.1-34b}$$

when Remark 2 is taken into consideration. The extremal control is

$$u^*(t) = -c , \tag{8.1-35}$$

yielding a conical die, as long as $x_1(t_1) \geq 0.4206a$. The corresponding extremal state is given by

$$x_1^*(t) = -ct + a ,$$
$$x_2^*(t) = -k \ln \frac{a - ct}{a} . \tag{8.1-36}$$

4) Maximum contraction and minimum surface area

From a physical point of view, minimizing the surface area is an indirect way of taking into consideration the effect of friction. The main reason for including this problem, however, was to provide a clearer indication of the manner in which multicriteria optimization provides for the trade-off between the individual criteria. The die length and the initial opening of the die are taken to be specified. The only constraint imposed upon the boundary slope is that it remains finite. The result is the following unconstrained, nonlinear, multicriteria optimal control problem.

Problem: Obtain Pareto-optimal controls $u^*(\cdot) \in F$ for the criteria

$$f_1(u(\cdot)) = \int_0^{t_1} u(t)dt \quad \text{and} \quad f_2(u(\cdot)) = \frac{1}{2}\int_0^{t} u^2(t)dt \tag{8.1-37a}$$

subject to

$$\begin{aligned} \dot{x}_1 &= u, & x_1(0) = a,\ x_1(t_1) \text{ arbitrary,} \\ \dot{x}_2 &= -k\frac{u}{x_1}, & x_2(0) = 0,\ x_2(t_1) \text{ arbitrary} \end{aligned} \tag{8.1-37b}$$

with $u(\cdot) \in F = \{u(\cdot) \text{ L-measurable: } |u| < \infty\}$.

The H-function for this problem is

$$H = -c_1 u - c_2 \frac{1}{2} u^2 + \lambda_1 u - \lambda_2 k \frac{u}{x_1} . \tag{8.1-38a}$$

The corresponding adjoint equations are given by

$$\begin{aligned} \dot{\lambda}_1 &= -\lambda_2 k \frac{u}{x_1}, \\ \dot{\lambda}_2 &= 0 \end{aligned} \tag{8.1-39}$$

with $\lambda_1(t_1) = \lambda_2(t_1) = 0$ as implied by the transversality conditions. Integration of the adjoint equations thus results in $\lambda_1(t) = \lambda_2(t) \equiv 0$, so that

$$H = -c_1 u - c_2 \frac{1}{2} u^2. \tag{8.1-38b}$$

The maximum principle requires

$$\sup_{u \in F} H = \text{const.}$$

A necessary condition for a supremum of H with respect to u is

$$\frac{\partial H}{\partial u} = 0 = -c_1 - c_2 u , \tag{8.1-39a}$$

and an extremal control is thus given by

$$u^*(t) = -\frac{c_1}{c_2} . \tag{8.1-39b}$$

The corresponding extremal state is

$$\begin{aligned} x^*(t) &= a - \frac{c_1}{c_2} t , \\ x_2^*(t) &= -k \ln(1 - \frac{c_1}{c_2 a} t) . \end{aligned} \tag{8.1-40}$$

Obviously, the extremal die is again a cone. Increasing c_1 (emphasizing the contraction) decreases the die boundary slope; increasing c_2 "flattens" the cone. As before, these adjustments in the ratio c_1/c_2 are limited to the physically feasible ones, as implied in Remark 2. Here, the ratio is thus limited by

$$\frac{c_1}{c_2} \leq 0.5794 \frac{a}{t_1} .$$

That is, physical considerations limit the extent of the Pareto set.

8.1.4 Conclusion

Based on necessary conditions for multicriteria optimal control [11], one may draw the following conclusions. For the strip drawing of an ideally plastic material when friction is negligible, the conical die is an extremal solution for the more obvious criteria arising in such a drawing process. From a mathematical point of view, this is essentially due to the linearity of H in the control $u(\cdot)$ in conjunction with the assumed boundedness of the control. In every case, the behaviour of the switching function played a crucial role. However, the conical die was a solution in the last problem as well, one which was nonlinear in the control $u(\cdot)$.

The present model is the simplest one describing the plastic behaviour of continua. Still, it may not be far off the mark in the description of the well-lubricated steady-state drawing process. Eventually, the investigation should obviously be extended to include friction and should also be based on a more sophisticated mathematical description of the plastic deformation process.

References

[1] Richmond, O.; Davenpeck, M.L.: A Die Profile for Maximum Efficiency in Strip Drawing. Proceedings of the Fourth U.S. National Congress of Applied Mechanics held at University of California, Berkeley, California, June 18-21, (1962) 1053-1057

[2] Richmond, O.; Morrison, H.L.: Streamlined Wire Drawing Dies of Minimum Length. Journal of the Mechanics and Physics of Solids. Vol. 15, (1967) 195-203

[3] Hill, R.: A Remark on Diagonal Streaming in Plane Plastic Strain. Journal of the Mechanics and Physics of Solids. Vol. 14, (1966) 245-248

[4] Hill, R.: Ideal Forming Operations for Perfectly Plastic Solids. Journal of the Mechanics and Physics of Solids. Vol. 15, (1967) 223-227

[5] Sortais, H.C.; Kobayashi, S.: An Optimum Die Profile for Axisymmetric Extrusion. International Journal of Machine Tool Design and Research. Vol.8 (1967) 61

[6] Johnson, W.; Mellor, P.B.: Engineering Plasticity. London. Van Nostrand Reinhold Co., 1973

[7] Thomsen, S.G.: Numerical Solution of the Extrusion Problem with Respect to Vectoroptimization (in Danish). DCAMM Report No. S40. The Danish Center for Applied Mathematics and Mechanics, The Technical University of Denmark, Lyngby, Denmark, July 1987

[8] Thomsen, S.G.: Optimization of Dies for Axisymmetric Cold Forward Extrusion. Engineering Optimization, Vol. 11 (1987) 13-20

[9] Hoffmann, O.; Sachs, G.: Introduction to the Theory of Plasticity for Engineers. New York: McGraw-Hill Book Co. 1953

[10] Avitzur, B.: Metal Forming: Processes and Analysis. Huntington, NY: Robert E. Krieger Publishing Co. 1979

[11] Lee, E.B.; Markus, L: Foundations of Optimal Control Theory. New York: John Wiley & Sons, Inc. 1967

[12] Stadler, W. (ed.): Multicriteria Optimization in Engineering and in the Sciences. New York: Plenum Press 1988

8.2 Optimal Layouts of Chilled Cast-Iron High Quality Rollers

M. Bremicker, H. Eschenauer, H.-W. Wodtke

8.2.1 Introduction

Chilled rollers are employed in various branches of industry such as iron steel industry, plastics processing industry, paper industry, rubber industry, food industry, and in machine and plant techniques. The selection of the roller design and its material depends on the specification requirements. A drawing of a "thermal roller", particularly used in the paper manufacturing industry, is shown in Fig. 8.2/1. In many cases the material used is chilled iron. The cross-section of the central part of the roller (the bold drawn part in Fig. 8.2/1) shows the working layer constituting the exterior of the roller and the barrel core. Whereas the exterior is made of white martensite solidified cast-iron, the basic material of the barrel core is a special gray cast-iron. The latter is a rather soft material with good ductility and damping properities suitable for dynamically loaded machine parts. On the other hand, the

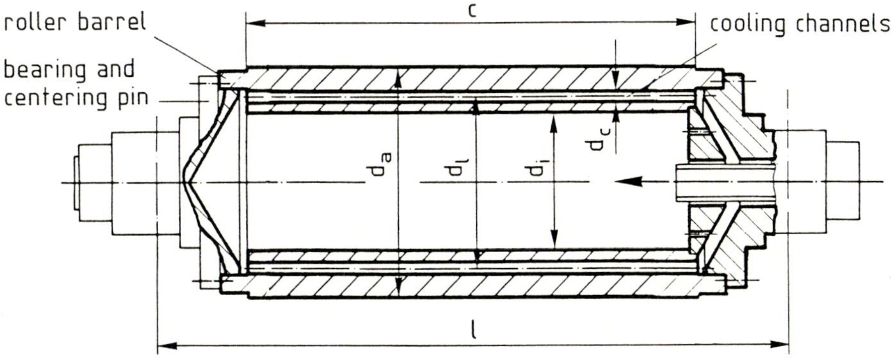

Fig. 8.2/1. Scheme of a thermal roller

Dimensions: d_i = 530mm, d_a = 860mm,
d_l = 640mm, d_c = 30mm,
c = 4090mm, l = 5710mm.

white solid cast-iron is a very hard, wear-resistant but rarely ductile material. It has a high degree of compressive strength, wear resistance, and retention of hardness. Therefore, this material is also used for rollers which are exposed to high temperatures. Experience shows that the hardening depth should not be more than 8–10% of the barrel diameter. As shown in Fig. 8.2/2, three radial zones can be distinguished:

1. zone with approximately constant hardness,
2. transition zone,
3. barrel core with lower hardness.

During the operation of the rollers the materials are subjected to temperature changes and thereby to the very dangerous thermal stresses. Here, the temperature distribution inside the rollers is important. Unalloyed roller materials are relatively unaffected by changes of temperature; highly alloyed, wear resistant roller materials require either adequate cooling or heating which can be achieved by boreholes for coolants or heating liquids (oil), as shown in Fig. 8.2/1.

An important task is the calculation of stresses at critical points of the heterogeneous roller barrel such as the rims of the boreholes or the martensite layers. This contribution deals with an optimization model whose aims should include weight minimization and at the same time minimization of stresses as well as observance of numerous constraints. The task is to find an optimal roller barrel layout for this multicriteria optimization problem. The investigations are restricted to the central part of the roller barrel, i.e. to the state of plane strain.

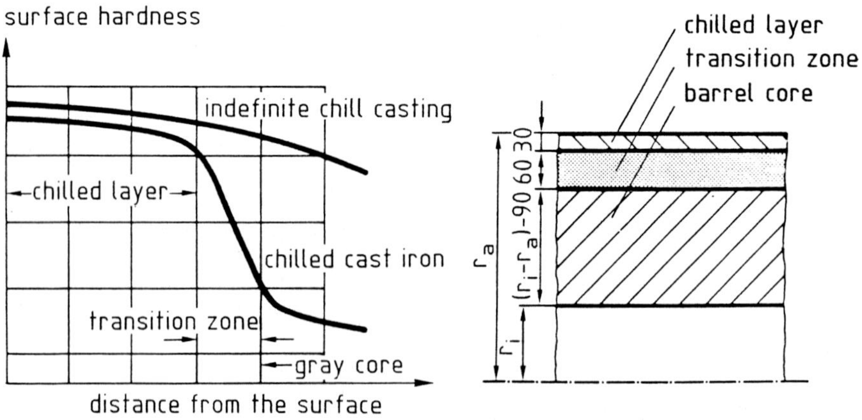

Fig. 8.2/2. Layered structure of a chilled iron roller

8.2.2 Definition of the Optimization Problem

According to Chapter 2 of this book, it is essential to establish and to define the optimization model in order to solve an optimization problem successfully. Therefore, design objectives and constraints are to be formulated by means of mathematical functions which depend on the design variables. The design problem itself is described with regard to Section 2.4.1:

a) Objectives

The following objectives ought to be fulfilled:

- minimum roller weight $f_1(\mathbf{x})$,
- minimum stresses at the critical points (borehole rims and martensite layer) $f_2(\mathbf{x})$,
- minimum roller barrel deformation $f_3(\mathbf{x})$.

Because of the contradictory nature of the objectives, an optimal compromise solution has to be found.

The first objective, the *roller weight*, can be determined from the relation

$$f_1(\mathbf{x}) := W(\mathbf{x}) = \sum_{i=1}^{k_s} \rho_i V_i(\mathbf{x}) - n_B \rho_B V_B(\mathbf{x}) \longrightarrow \text{minimum} \qquad (8.2\text{-}1)$$

with k_s number of material layers considered in the discrete model (see Section 8.2.3),

ρ_i density of the material layer i,

V_i volume of the material layer i,

n_B number of peripheral boreholes,

ρ_B density of the material layer in which the peripheral boreholes are placed,

V_B volume of the peripheral boreholes.

As far as the second objective *stress minimization* is concerned, two critical points must be considered in the case of a heterogeneous roller barrel:

- the borehole area,
- the martensite layer.

The stresses are calculated by using the structural analysis model described in Section 8.2.3. As the model calculates only a global stress distribution and neglects the stress concentrations at the borehole rims, these have to be determined additionally by means of stress concentration factors. According to Neuber [5], the stress concentration factor of a borehole in a disk, subject to compressive or tensile loads in one direction, is $k_f = 3$, which will be used here for calculation purposes. In order to ascertain whether the stress concentrations are underrated due to this approximation, a parallel stress computation of the initial design and the optimal design is carried out using the finite element method. Hence it follows that the agreement between both of the results is absolutely sufficient. Thus, two alternative formulations for $f_2(\mathbf{x})$ are to be considered.

$$f_2^M(\mathbf{x}) := \{ \max_{\Omega_M} \sigma_p \} \longrightarrow \text{minimum} \qquad (8.2\text{-}2a)$$

and/or

$$f_2^B(\mathbf{x}) := \{ \max_{\Omega_B} k_f \cdot \sigma_{pgl} \} \longrightarrow \text{minimum} \qquad (8.2\text{-}2b)$$

with Ω_M martensite layer area,
Ω_B borehole area,
k_f stress concentration factor for borehole (k_f=3),
σ_p reference stress,
σ_{pgl} global principal stress in the borehole area.

Two different formulations are used for the reference stresses:

- Hypothesis of maximum strain energy / Principal stresses by von Mises

This theory is valid for most materials subject to dynamic loads but is also often used for static loads:

$$\sigma_p = \sqrt{\frac{1}{2}\left((\sigma_{rr}-\sigma_{\varphi\varphi})^2 + (\sigma_{\varphi\varphi}-\sigma_{zz})^2 + (\sigma_{zz}-\sigma_{rr})^2 + 6\,(\tau_{r\varphi}^2 + \tau_{\varphi z}^2 + \tau_{zr}^2)\right)} \qquad (8.2\text{-}3a)$$

with σ, τ normal and shear stresses in radial (r), tangential (φ) and longitudinal (z) directions.

- Hypothesis of maximum principal stresses

This hypothesis is generally applied to brittle materials such as cast-iron:

$$\sigma_p = \max\{\sigma_1, \sigma_2, \sigma_3\}\,, \qquad (8.2\text{-}3b)$$

with $\sigma_{1,2,3}$ principal stresses.

Both reference stresses are evaluated at every step in order to use the larger one for the objective formulation.

In order to determine the third objective function *stiffness of the roller barrel*, the maximum deformation at its middle part is used. Thus, this objective function reads

$$f_3(\mathbf{x}) = \max w_{diff} = \max \{ w_{tot} - w_{fl} \} \longrightarrow \text{minimum} \qquad (8.2\text{-}4)$$

with w_{tot} total deformation relating to the support points,
w_{fl} deformation at the ends of the roller barrel relating to the support points (see Fig. 8.2/1).

For calculating the deformations w_{tot} and w_{fl}, Bernoulli's beam theory is used. In the case of heterogeneous roller barrels, the various elastic moduli of the individual layers are taken into account. However, it must be pointed out that with regard to the length-diameter ratio of about 5, the shear effect should be considered. In the case of a homogeneous roller barrel, calculations using the beam theory regarding the shear deformation effects lead to an increase of w_{diff} of about 10%.

b) Constraints

Three different types of constraints are taken into consideration:

- stress constraints for the entire structure,
- geometrical constraints (secondary structural conditions),
- design variable bounds.

All constraints listed above can be formulated as inequality constraints which for numerical reasons are expressed in a standardized form (2-4):

$$g_j(\mathbf{x}) = \begin{cases} \dfrac{\tilde{g}_j(\mathbf{x})}{g_{j\,fea}} - 1 \le 0\,, & \text{if } \tilde{g}_j \overset{!}{\le} g_{j\,fea}, \\[2ex] 1 - \dfrac{\tilde{g}_j(\mathbf{x})}{g_{j\,fea}} \le 0\,, & \text{if } \tilde{g}_j \overset{!}{\ge} g_{j\,fea} \end{cases} \qquad (8.2\text{-}5)$$

with the limited functions $\tilde{g}_j$ and the bound values $g_{j\,fea}$.

c) Design variables

The design variables of the heterogeneous roller barrel are shown in Fig. 8.2/3.

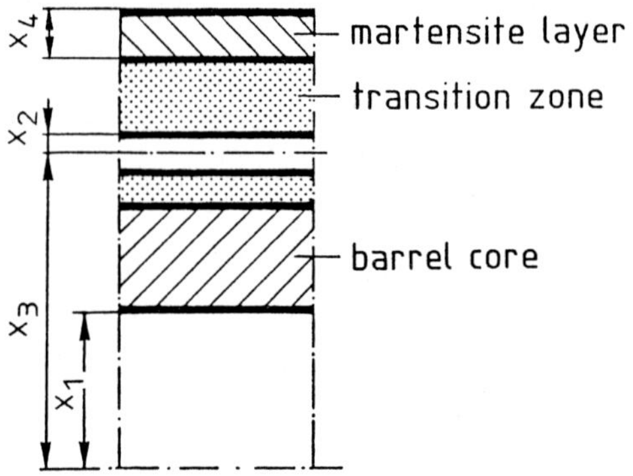

Fig. 8.2/3. Design variables of the heterogeneous roller barrel

By means of a preprocessor the design variables are transformed into analysis variables required by the structural analysis program. Thus, the following steps have to be carried out for every design variable x_i (i = 1,...,4):

x_1 inner radius of the roller, assignment of the radius value,

x_2 bolt circle radius of the circumferential oil boreholes:
- linear interpolation of interface radii and material properties in the transition zone (10 layers);
- calculation of temperatures at the layer bounds; starting from the heat-flow equation the inner temperature of each layer ${}^{k}\Theta_i$ can be calculated as follows:

$$ {}^{k+1}\Theta_i = {}^{k}\Theta_i - \frac{\Phi \ln\left(\frac{{}^{k+1}r_i}{{}^{k}r_i}\right)}{2\pi c\, {}^{k}\lambda} \tag{8.2-6}$$

with
k layer index,
c roller barrel length,
${}^{k}r_i$ inner radius of the k-th layer,
${}^{k}\lambda$ thermal conductivity of the k-th layer,
Φ heat flow through the multilayer tube.

${}^{1}\Theta_i$ and Φ are given values. The thermal conductivity λ of the borehole layer is multiplied by the circumferential portion of the boreholes;

		Heterogeneous		
		Barrel core	Transition zone	Martensite Layer
E	N/mm^2	$1.10 \cdot 10^5$	linear interpolat.	$1.8 \cdot 10^5$
α	1/K	$1.15 \cdot 10^{-5}$	linear interpolat.	$8.8 \cdot 10^{-6}$
λ	W/mK	50.	linear interpolat.	20.
ρ	kg/m^3	$7.7 \cdot 10^3$	$7.7 \cdot 10^3$	$7.7 \cdot 10^3$
ν	—	0.29	0.29	0.29

Table 8.2/1. Material properties of the heterogeneous roller barrel

- determination of the layer of the transition zone or the barrel core belonging to the borehole circle and, corresponding to that, definition of the tangential stress-free layer as well,

x_3 radius of the oil boreholes:
- inclusion of thickness and the circumferential portion of tangential stress-free layer (see [4]),

x_4 thickness of the martensite layer:
- assignment of the corresponding material radius,
- calculation of the thickness of the transition zone ($=3x_4$),
- determination of the interface temperatures according to (8.2-6).

d) Material properties

The material properties of the heterogeneous cylinder made of chilled cast-iron are compiled in Table 8.2/1.

8.2.3 Structural Analysis

By limiting the analysis to the middle part of a roller, a simplified structural analysis approach was developed in order to reduce the

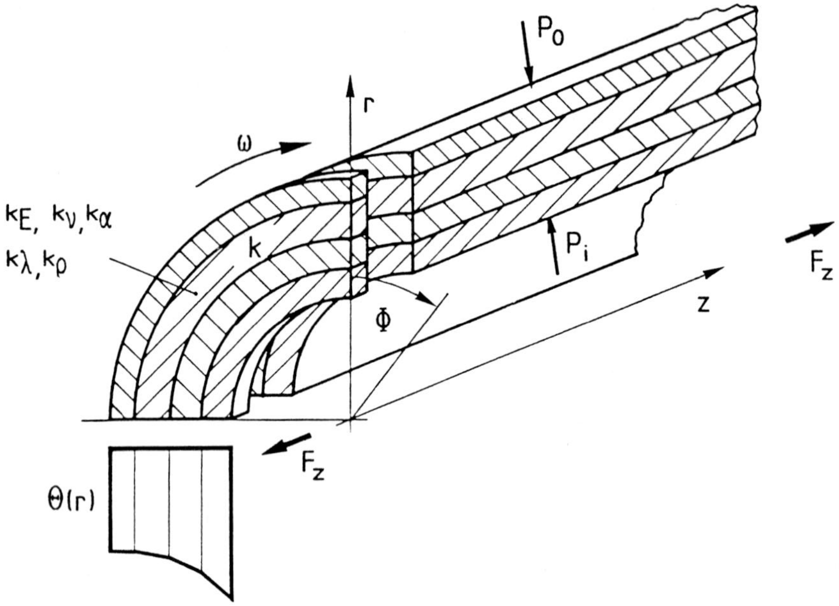

Fig. 8.2/4. Structural model of the middle part of the roller

computing time of the optimization process. Fig. 8.2/4 shows a part of the cylinder which consists of several layers and which is subject to the loads: inner and outer pressure p_i, p_a, axial force F_z, temperature distribution Θ^r and angle velocity ω. As all loads are axisymmetric, the governing basic equations of the thick-walled cylinder (equilibrium conditions, strain-displacement equations and material law) can be simplified. Solving them for the displacements yields a heterogeneous Euler's differential equation of the second order for every layer k:

$$\frac{d^2u}{dr^2} + \frac{1}{r}\frac{du}{dr} - \frac{u}{r^2} = \frac{1+\nu}{1-\nu}\,\alpha\,\frac{d\Theta}{dr} - \frac{(1+\nu)(1-2\nu)}{(1-\nu)\,E}\,\rho\,\omega^2 r\,. \qquad (8.2\text{-}7)$$

The solution of (8.2-7) is

$$u(r) = D_1(r)\,p_i - D_2(r)\,p_a + D_3(r,\Theta) + D_4(r,\omega) - \varepsilon_{zz} r \nu \qquad (8.2\text{-}8)$$

with

$$D_1(r) = \frac{r_i^2}{\tilde{E}\,\Delta(r^2)}\left(r + \frac{r_a^2}{r(1-2\nu)} \right),$$

$$D_2(r) = \frac{r_a^2}{\tilde{E}\,\Delta(r^2)} \left(r + \frac{r_i^2}{r(1-2\nu)} \right),$$

$$D_3(r,\Theta) = \frac{E\alpha T(r_a)}{\tilde{E}(1-\nu)\,\Delta(r^2)} \left(r + \frac{r_i^2}{r(1-2\nu)} \right) + \frac{1+\nu}{1-\nu}\,\alpha\,\frac{1}{r}\,T(r)\,,$$

$$D_4(r,\omega) = \frac{\rho\,\omega^2}{8\,\tilde{E}(1-\nu)} \left((3-2\nu)\,\frac{r\,\Delta(r^4)}{\Delta(r^2)} + (3-2\nu)\,\frac{r_a^2 r_i^2}{r(1-2\nu)} - r^3 \right),$$

$$T(r) = \int_{r_i}^{r} \Theta(\bar{r})\,\bar{r}\;d\bar{r}\,, \qquad \tilde{E} = \frac{E}{(1+\nu)(1-2\nu)}\,,$$

$$\Delta(r^4) = r_a^4 - r_i^4\,, \qquad \Delta(r^2) = r_a^2 - r_i^2\,,$$

and the material properties

- E Young's modulus ,
- ν Poisson's ratio ,
- α thermal expansion coefficient,
- λ thermal conductivity,
- ρ mass density .

In the following, all quantities of a k-th layer are denoted by an index k. Quantities without an index are valid for the complete structure.

We now consider the complete cylinder consisting of n layers. Starting with the equilibrium conditions corresponding to Fig. 8.2/5, we have for each layer

$$^{k}p_a = {}^{k+1}p_i\,, \tag{8.2-9a}$$

and with the given inner and outer pressure loads $p_i = {}^{1}p_i$ and $p_a = {}^{n}p_a$ we obtain n-1 unknown stresses $^{k}p_i$, k=2,...,n which can be determined from the compatibility conditions of each layer surface:

$$^{k}u(^{k}r_a) = {}^{k+1}u(^{k+1}r_i) - {}^{k,k+1}\Delta u\,, \tag{8.2-9b}$$

where the interference $^{k,k+1}\Delta u$ can be considered optionally.

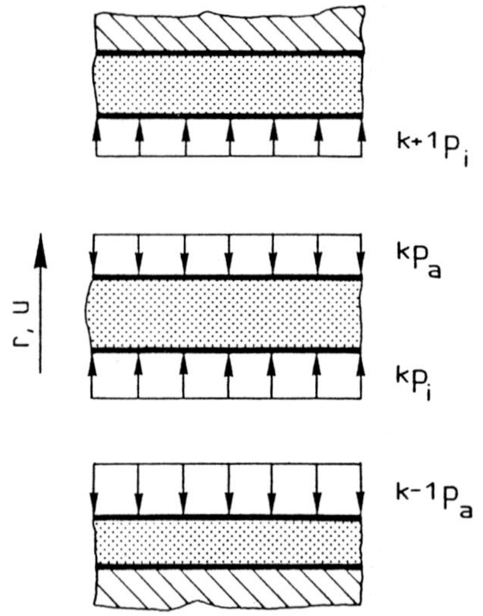

Fig. 8.2/5. Free body diagram of individual layers

Eqs. (8.2-9a,b) together with (8.2-8) yield:

$$\Big\{ {}^{k}p_i \, {}^{k}D_1(r) + {}^{k+1}p_i - \left[{}^{k+1}D_1(r) - {}^{k}D_2(r) \right] + {}^{k+2}p_i \, {}^{k+1}D_2(r) + $$
$$+ \, \varepsilon_{zz} \, r \left[{}^{k+1}\nu - {}^{k}\nu \right] \Big\} \Big|_{{}^{k+1}r_i} = \tag{8.2-10}$$
$$= \Big\{ {}^{k+1}D_3(r,\Theta) - {}^{k}D_3(r,\Theta) + {}^{k+1}D_4(r,\omega) - {}^{k}D_4(r,\omega) - {}^{k,k+1}\Delta u \Big\} \Big|_{{}^{k+1}r_i} .$$

In (8.2-10), the axial strain ε_{zz} is still unknown. It can be determined from the global axial equilibrium condition. According to the principle of de Saint-Venant, the stresses and displacements at a certain distance from the end of the cylinder are not affected by the way the pressure resultant and the axial force F_z are applied to the structure. Thus, it can be written:

$$2\pi \int_{r_i}^{r_a} \sigma_{zz}(r) \, r \, dr = 2\pi \sum_{k=1}^{n} \int_{{}^{k}r_i}^{{}^{k}r_a} {}^{k}\sigma_{zz}(r) \, r \, dr = \tag{8.2-11}$$
$$= F_z + \pi (p_i r_i^2 - p_a r_a^2) .$$

The determination of the axial tension of the k-th layer from the basic solution and the integration of (8.2-11) yield an equation for defining the axial strain ε_{zz}:

$$\varepsilon_{zz} \frac{1}{2} \sum_{k=1}^{n} {}^{k}E \; {}^{k}\Delta(r^2) + \sum_{k=1}^{n} {}^{k}\nu \left({}^{h}r_i^2 \; {}^{k}p_i - {}^{k+1}r_i^2 \; {}^{k+1}p_i \right) = \tag{8.2-12}$$

$$= \sum_{k=1}^{n} {}^{k}E \; {}^{k}\alpha_T \; {}^{k}T({}^{k}r_a) - 2 \sum_{k=1}^{n} {}^{k}\Omega \; {}^{k}\Delta(r^4) \; {}^{k}\nu + \frac{1}{2} \left(p_i r_i^2 - p_a r_a^2 \right) + \frac{1}{2\pi} F_z \; .$$

Establishing (8.2-10) for every layer and adding (8.2-12) provide a linear system of equations for the determination of the inner pressures ${}^{k}p_i$ and the axial strain ε_{zz} [4,7]. After solving these equations, stresses and deformations can be calculated for any r.

If the flow of tangential stresses in axially symmetric structures is disturbed by holes, spokes or shovels, arranged in regular distances, very small tangential stress occurs in such zones. In order to be able to consider its influence on the global stress distribution, we introduce a layer which is completely free of tangential stresses [4]. Here, the straps between the boreholes which transmit the radial stresses are assumed to be arranged so closely that they can be considered to be a homogeneous layer with reduced radial stiffness. Radial displacements and stresses of a layer without tangential stiffness can be determined from the basic equations and included into the equation system of the complete structure in a straightforward manner. However, it must be emphasized that the stress peaks at the borehole boundaries cannot be determined by this approach. For the calculation of a structure with several layers, including those without tangential stress, the analysis program FIZYL has been developed. It is used for the optimization and turns out to be more efficient than the usual FE-methods. Fig. 8.2/6 shows the results of an analytical calculation using FIZYL and a FE-calculation for a hollow cylinder consisting of two layers with different ${}^{k}\alpha$ subject to a thermal load.

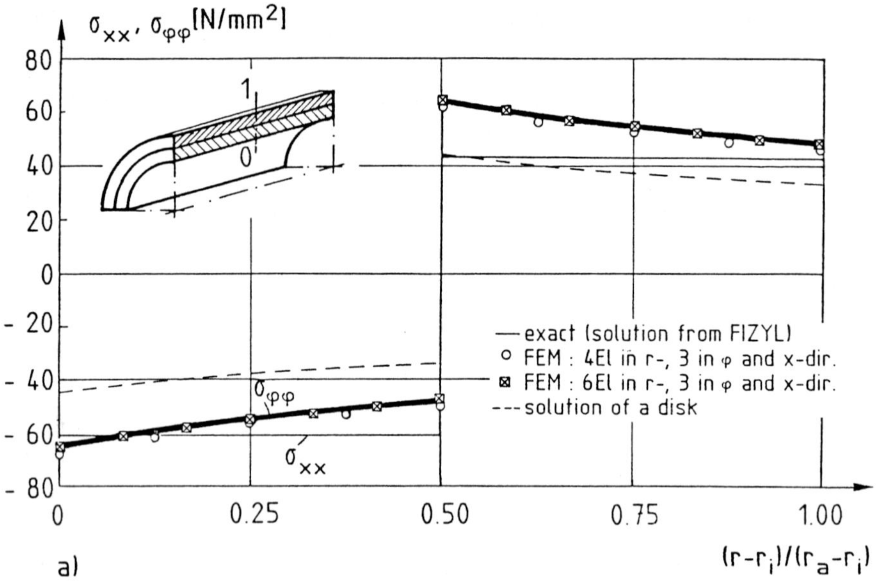

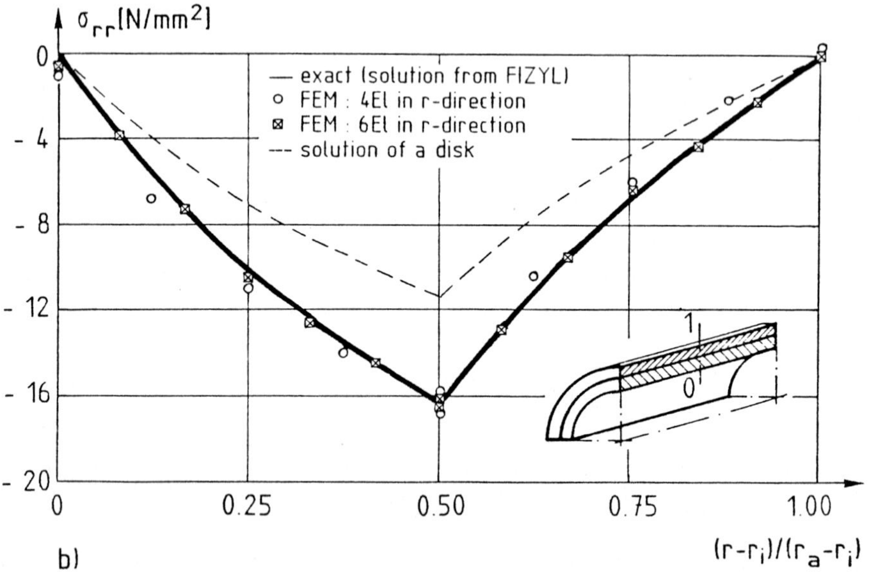

Fig. 8.2/6. Stresses of a two-layered hollow cylinder under thermal load

a) circumferential and axial stresses b) radial stresses

8.2.4 Realization of an Optimization Loop

This section describes how the design model and the calculation model are realized in a software system. A large part of the program module is taken from the optimization procedure SAPOP (see Chapter 2). Starting with a vector $\mathbf{x}_0$ to be given by the user, the design variables are transformed into analysis variables by means of a preprocessor (design model). These are given to the structural analysis model (here: the program FIZYL) which yields several state quantities. The postprocessor (evaluation model) uses these quantities for calculating the objective functions and the constraints which are employed by the optimization algorithm to determine improved design variables. Then the procedure starts again. For the given optimization problem, this procedure is illustrated in more detail in Fig. 8.2/7.

In order to solve the multicriteria, nonlinear and constrained optimization problem, the algorithms SEQLI and GREGA were applied in an augmented and modified form in order to compare their efficiency (see Section 2.5). Thus, it is possible to check the optimal solutions obtained which is very important for the verification of solutions of nonlinear optimization problems.

8.2.5 Optimization Results

a) Functional-efficient solutions

Fig. 8.2/8 shows the functional-efficient solution set. Additionally, the initial design and the "minimal deformation"-roller without boreholes is shown. Furthermore, calculations for a homogeneous roller barrel (graphite cast-iron) are carried out.

Due to the increased Young's modulus of the martensite layer, smaller deformations occur in the case of heterogeneous rollers. Fig. 8.2/9 shows that for small deformations the optimal weights of the heterogeneous rollers are smaller than those of the homogeneous rollers whereas in the area of large deformations a contrary effect occurs.

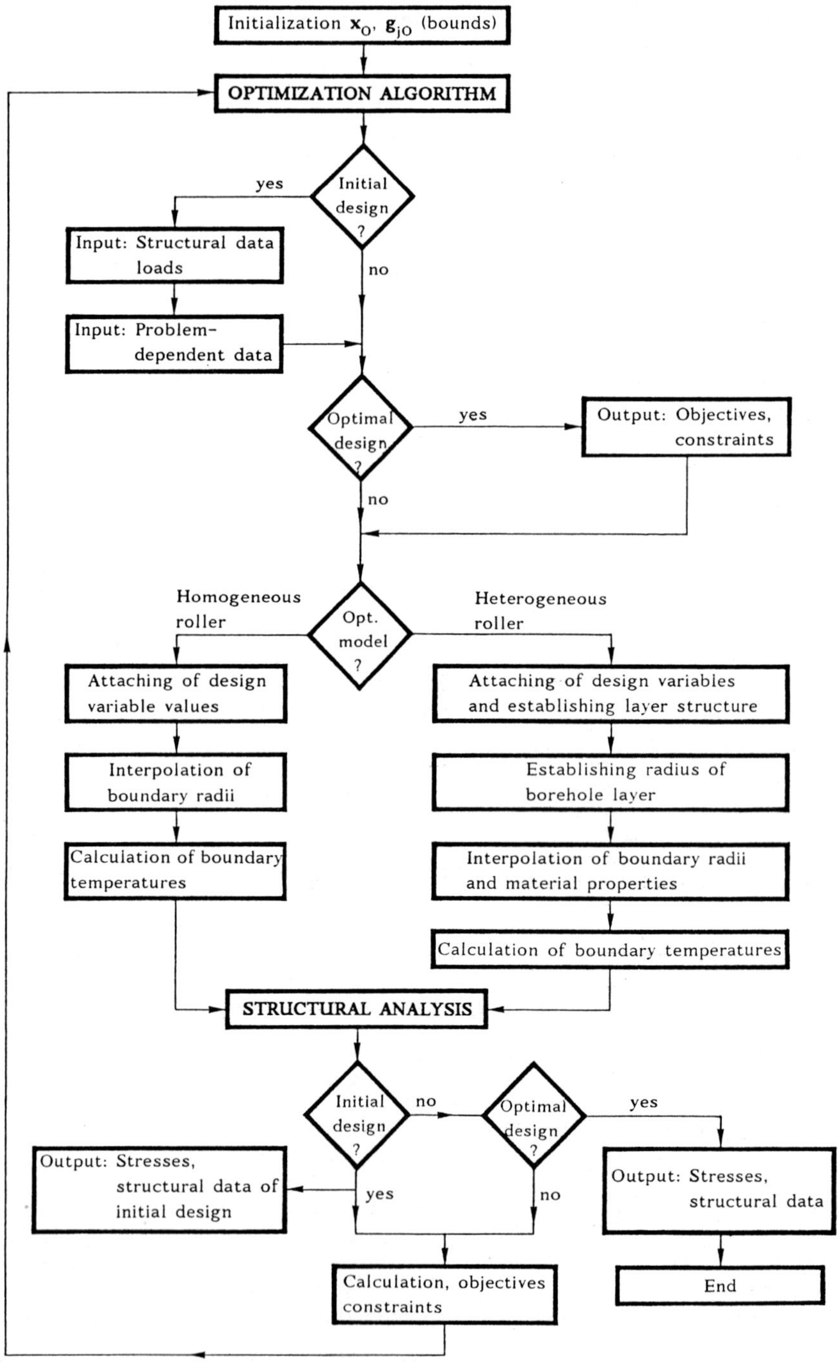

Fig. 8.2/7. Flow chart of the optimization loop

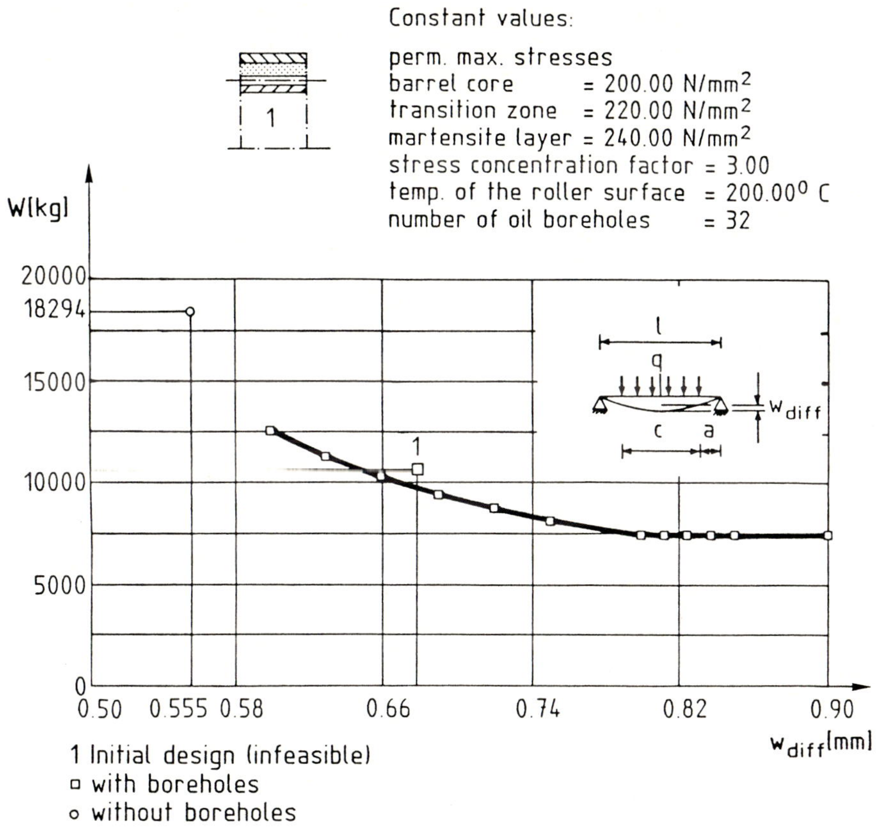

Fig. 8.2/8. Functional-efficient boundary after optimization of heterogeneous roller barrels

Thus, it can be concluded that a chilled cast-iron design is recommendable if special deformation demands have to be met. Compared to it, a graphite cast-iron design is more appropriate for lower stiffness demands. The fields in which the heterogeneous roller is more advantageous are predominant. However, it must be pointed out that these first investigations neglect surface hardness as well as production costs.

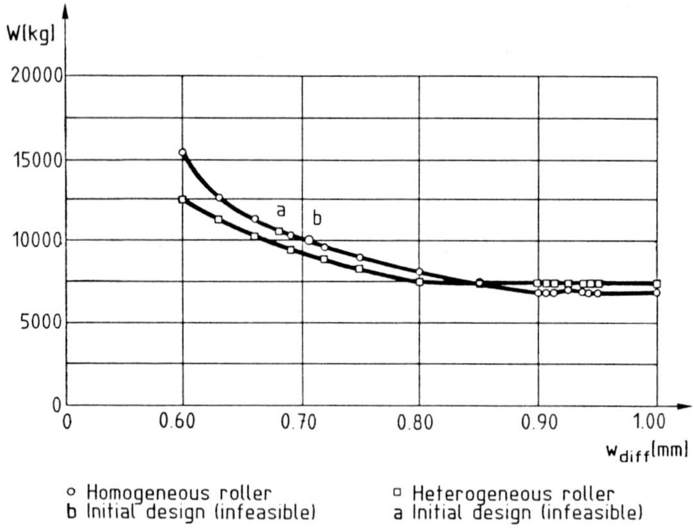

Fig. 8.2/9. Functional-efficient boundaries of roller barrels with comparable stress levels

b) Selected Solutions

Two solutions are now presented in more detail:

- deformation requirement of w_{diff} = 0.6135mm (Fig. 8.2/10a). This "strong" requirement can be met with considerably less weight than in the case of a homogeneous roller barrel. The reason for this is the higher YOUNG's modulus of the martensite layer.
- deformation requirement of w_{diff} = 0.9mm (Fig. 8.2/10b).

In the case of the optimal stress designs shown in Table 8.2/2 and Fig. 8.2/11, the maximum stress level in the borehole area and the martensite layer is directly minimized according to (8.2-5). The weight of the roller is limited to W = 13000kg. In order to be able to demonstrate the strong influence of the deformation requirement on the optimal solution, three computations with only slightly different admissable deformations are carried out. The influence on the stresses is evident: a slight relaxation of the deformation constraint leads to a relatively large reduction of the stress objective function values. Design No. 3 is the optimal stress design; it is not possible to reduce the stresses any further, not even by increasing the admissible level of deformation.

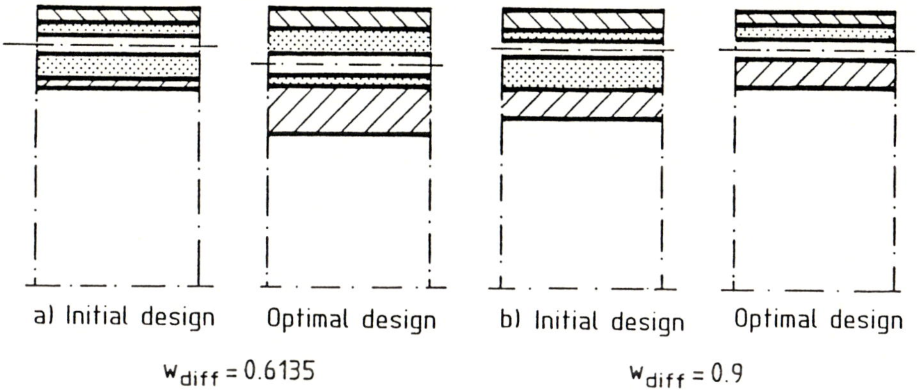

Fig. 8.2/10. Geometry of initial and optimal designs in relation to different deformation constraints

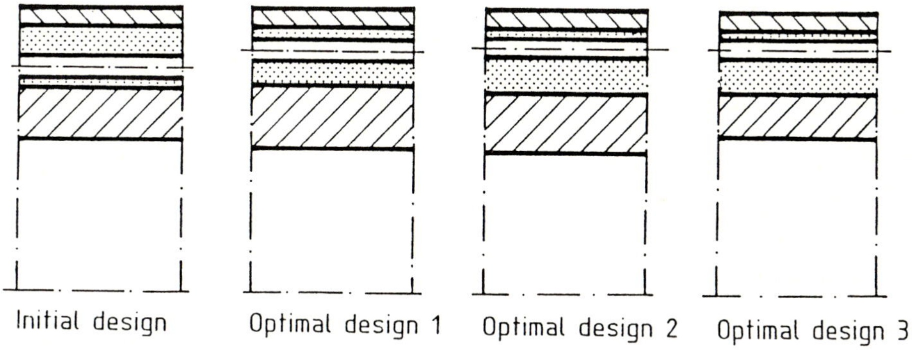

Fig. 8.2/11. Geometry of the optimal design in stress optimization including different deformation requirements (heterogeneous roller barrel)

8.2.6 Conclusion

This paper deals with a multicriteria design optimization problem of chilled cast-iron rollers. The design variables, the objectives, and the constraints required for the structural optimization of a special

		Initial design	Optimal design		
			1	2	3
x_1	mm	230.00	215.12	215.12	215.3
x_2	mm	15.00	15.00	15.00	15.00
x_3	mm	340.00	364.5	373.88	375.00
x_4	mm	30.00	30.00	30.00	25.27
f_1=min. weight	kg	12347.50	13000.00	13000.00	2994.00
f_2=min. w_{diff}	mm	0.5966	0.6100	0.6200	0.6302
f_3=min. stress	N/mm^2	286.66	172.34	141	129.1
heating temp.	^oC	279.48	255.97	247.33	246.32
w_{diff}	mm	0.6100	0.6100	0.6200	0.6300
weight	kg	13000.00	13000.00	13000.00	13000.00
stress concentr. factor		3.00	3.00	3.00	3.00
barrel length	mm	4090.00	4090.00	4090.00	4090.00

Table 8.2/2. Results of the stress minimization of the heterogeneous roller barrel

"flexitherm" roller barrel are defined, and a set of given specification values are incorporated. The design variables are represented by various geometrical dimensions of the roller barrel. The following objectives are taken into consideration:

minimum roller weight,

and maximum stiffness and/or minimum roller barrel deformation,

or minimum stresses at the critical points.

As these objectives are contradictory, the best possible compromise solutions can only be found by means of multicriteria optimization strategies. Furthermore, the observance of admissible stresses and other secondary design conditions must be guaranteed. Varying the demand levels of the individual objectives, the complete solution set of the optimization problem is determined as a functional-efficient or a Pareto-optimal solution set. Hence, the individual solutions can be selected with regard to the objectives.

As a result of the optimization calculations, the objectives of the roller barrel design can be considerably improved. For example, the maximum stresses in a homogeneous roller barrel can be reduced by about 25% while retaining the same weight and deformation values. In the case of a heterogeneous roller it is even possible to reduce the stress peaks by 60%.

As two different optimization algorithms (SEQLI2 [1,2] and GREGA [6] are used for separate optimization calculations, it is possible to confirm the results: both algorithms obtain the same solution, almost without exception, requiring an average of some 50 structural analyses including numerical sensitivity calculations. The total computing time using a MICROVAX II workstation computer is reduced to a few seconds only. The structural analysis results obtained by means of the structural analysis program FIZYL are confirmed by comparable finite element calculations. Further numerical results are given in [7,8].

Acknowledgement

The investigations were carried out within the scope of a research project supported by Walzen Irle Co., Netphen, West Germany. We would like to express our thanks for the excellent cooperation between the company and our research laboratory.

References

[1] Eschenauer, H.: Numerical and Experimental Investigations on Structural Optimization of Engineering Designs. DFG-Research Report, University of Siegen. Siegen: Bonn and Fries Publishing Co. 1985

[2] Eschenauer, H.; Post, P.U.; Bremicker, M.: Einsatz der Optimierungsprozedur SAPOP zur Auslegung von Bauteilkomponenten. Bauingenieur, 63, II (1988) 515-526

[3] Eschenauer, H.; Post, P.U.: Optimization Procedure SAPOP in Application on Optimal Layouts of Complex Structures. In: Rozvany, G.I.N.; Karihaloo, B.L. (eds.): Structural Optimization. Proceedings of the IUTAM-Symposium on Structural Optimization, Melbourne, Australia, 9-13 Feb. 1988. Dordrecht: Kluwer Academic Publishers (1988) 85-92

[4] Wodtke, H.W.: Aufstellen eines finiten Verfahrens zur Berechnung inhomogener, kreiszylindrischer Bauteile. Diploma Thesis, University of Siegen, Institute of Mechanics and Control Engineering 1987

[5] Neuber, H.: Kerbspannungslehre. Berlin, Heidelberg, New York, Tokyo: Springer Verlag, 3rd edition 1985

[6] Bremicker, M.: Entwicklung eines Optimierungsalgorithmus der generalisierten reduzierten Gradienten - Anwendung auf Beispiele der Strukturmechanik. Diploma Thesis, University of Siegen, Institute of Mechanics and Control Engineering 1986

[7] Eschenauer, H.; Bremicker, M.: Berechnung und Optimierung verschiedener Walzenkonstruktionen. Internal Reports 1-3 for the Research Project of Walzen Irle Co., University of Siegen, Research Laboratory for Applied Structural Optimization 1986

[8] Eschenauer, H.A.; Bremicker, M.: Optimization Modelling and Optimal Layouts of Heterogeneous, Thick-Walled, Chilled Cast-Iron Rollers. In: Rao, S.S. (ed.): Advances in Design Automation - 1988. Proceedings of the ASME Design Automation Conference, Kissimmee, USA. Sept. 25-28, 1988, ASME New York (1988) 189-194

8.3 A Mechanical Model for the Optimization and Simulation of the Metal Forming Process in Roller Levelling of Sheets

L. Henrich, K. Schiffner

8.3.1 Introduction

In order to automate manufacturing processes, mathematical and mechanical models for simulating their behaviour are becoming more and more important. These models help to establish control mechanisms for these processes. This paper presents a model for the metal forming processes in roller levelling of sheets.

The purchasers of sheet metal have strict requirements for the flatness and, additionally, a very low level of internal stresses is demanded. Divergence from flatness is caused by the manufacturing process in roll lines, by forming processes during handling and by transportation. Two different types of divergence occur [1]: Developable divergences which are constant along the sheet depth (Fig. 8.3/1a) and nondevelopable divergences which change along the sheet depth (Fig. 8.3/1b).

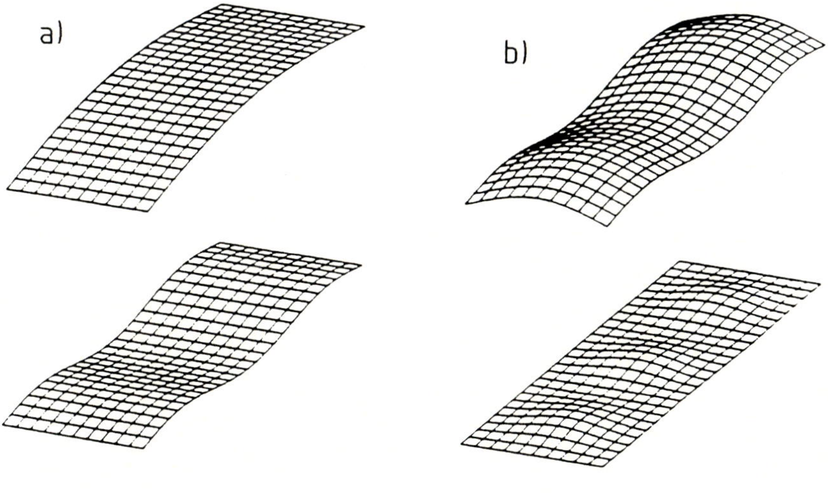

Fig. 8.3/1. Divergences of thin sheets
a) developable; b) nondevelopable

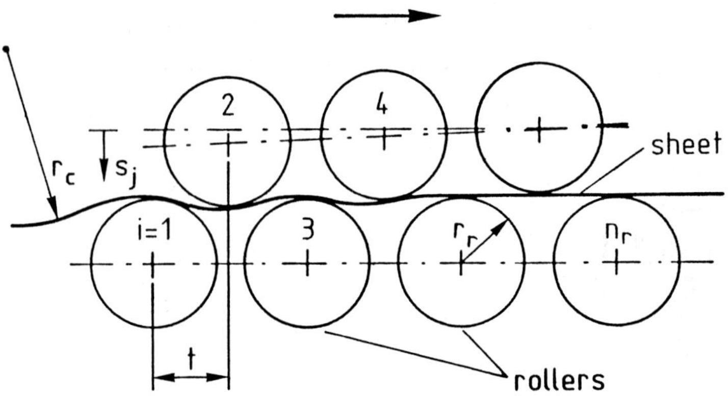

Fig. 8.3/2. Method of roller levelling

Both types of divergences are removed by a type of roller levelling where the sheet is bent in opposite directions by a certain number of rollers (Fig. 8.3/2). During this forming process, the sheet is loaded plastically and unloaded elastically several times.

In the removal of developable divergences, the adjustment parameters are the feed motions s_j of the upper rollers. The lower rollers are assumed to be on the same level. Most roller levellers admit a linear profile of feed motions of the rollers only. The linear profile is described by feed motions of the first and the last upper roller and is adjusted in such a way that the sheet curvatures $\varkappa$ decrease towards the end of the series of rollers ($\varkappa = \frac{1}{r_c}$). In order to optimize these process parameters, an optimization procedure is applied.

8.3.2 Mechanical Model

A general sketch of the mechanical model used for describing the forming process is shown in Fig. 8.3/3. The process can be approximated by a statically indeterminate, elastic-plastic beam deformation with displaced points of applied force. Each point of the beam cross-section will be subjected to a multiple bending process. In order to get a relation for the deformed structure of the sheet in the plant, a cubic spline S^i is used [2]:

$$S^i(x) = \sum_{k=0}^{3} a_{ik}(x - x_i)^k , \qquad i = 1, \ldots, (n_r - 1). \tag{8.3-1}$$

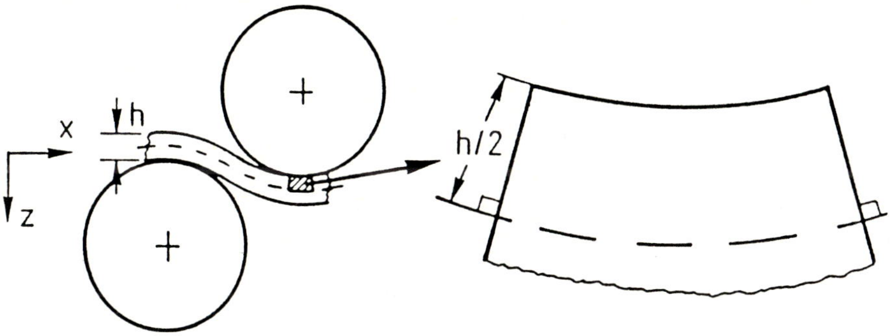

Fig. 8.3/3. Scheme of the mechanical model

The spline is fitted iteratively, assuming that only one contact point between the spline (sheet) and each circle (roller) exists. These n_r contact points are the spline points (n_r = number of rollers).

The cubic spline is the exact solution for the elastically deformed Bernoulli beam and is used here as a first approximation in the elastic-plastic forming. The stress-strain relation used for the investigations is shown in Fig. 8.3/4.

The maxima of the curvature appear at the spline points. The process is investigated in discrete steps with each step bounded by two points of the unloaded state (moment = 0). In Fig. 8.3/5, the known and unknown values for each discrete forming step are listed. The bending moment under the first and the last roller is zero so that $n_r - 2$ bends must be determined and counted by $i = 2,...,n_r - 1$.

With reference to Figs. 8.3/2 and 8.3/5, the following notations are used:

b	sheet depth,
c_h	hardening coefficient,
E	Young's modulus,
E_T	strain hardening modulus,
h	sheet thickness,
M^i	moment in loaded state,
M_u^i	elastic moment during unloading,
n_r	number of rollers,

r_c radius of curvature,
r_r radius of rollers,
S^i spline,
t pitch of rollers,
$\varepsilon^i(z)$ strain in loaded state,
$\varepsilon^i_{el}(z)$ elastic strain in unloaded state,
$\varepsilon^i_{pl}(z)$ plastic strain in unloaded state,
$\varepsilon^i_f(z)$ total strain in unloaded state,
$\varkappa^i$ sheet curvature at roller i in loaded state,
$\varkappa^i_b$ sheet curvature before roller i in unloaded state,
$\varkappa_{a\,(after)}$ residual sheet curvature after levelling,
$\varkappa_{b\,(before)}$ given sheet curvature before levelling (predeformation),
$\sigma^i(z)$ stress in loaded state,
$\sigma^i_{in}(z)$ internal stress in unloaded state,
$\sigma^i_u(z)$ elastic stress during unloading,
$\sigma^i_y(z)$ initial yield stress.

For a given curvature $\varkappa$ the stress-strain state $(\sigma(z), \varepsilon(z))$ and the resulting moment M^i are determined, taking into account the internal stresses $\sigma^i_{in}(z)$ and the strains $\varepsilon^i_f(z)$ caused by predeformations:

$$\varepsilon^i(z) = -[\varkappa^i z + \varepsilon^i_f(z)] \tag{8.3-2}$$

$$\sigma^i(z) = \begin{cases} E\,\varepsilon^i(z) + \sigma^i_{in}(z) & \text{for } \sigma^i \le \sigma^i_y\,, \\ \sigma^i_y(1-c_h) + c_h[E\,\varepsilon^i(z) + \sigma^i_{in}(z)] & \text{for } \sigma^i > \sigma^i_y\,, \end{cases} \tag{8.3-3}$$

$$M^i = 2b\int_0^{h/2} \sigma^i(z)\,z\,dz\,. \tag{8.3-4}$$

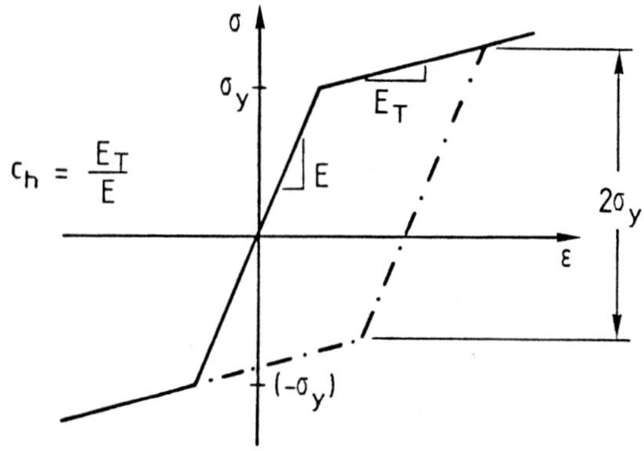

Fig. 8.3/4. Stress-strain curve with kinematic hardening

During the loading process, the yield stress increases locally due to kinematic hardening. This effect is taken into account by the discretization method in the z-direction. The parameters after unloading are determined using the following equations:

$$-M^i = M_u^i = -2b \int_0^{h/2} \sigma_u^i(z)\, z\, dz\,, \tag{8.3-5}$$

$$\sigma_{in}^i(z) = \sigma^i(z) - \sigma_u^i(z)\,, \tag{8.3-6}$$

$$\varepsilon_{el}^i(z) = \sigma_{in}^{i+1}(z)/E \tag{8.3-7}$$

The stresses $\sigma_u(z)$ connected with M_u are linear in z. There is always an elastic region around the neutral line $-z_{pl} \le z \le z_{pl}$ so that the remaining curvature after unloading $\varkappa_b^{i+1}$ can be calculated by (8.3-8). The total remaining strains after unloading $\varepsilon_f\ (z)$ consist of elastic parts $\varepsilon_{el}^{i+1}(z)$ and plastic parts $\varepsilon_{pl}^{i+1}(z)$ (Fig. 8.3/6):

$$\varkappa_b^{i+1} = \frac{-\varepsilon_{el}^{i+1}(z_{pl})}{z_{pl}} = \frac{-\varepsilon_f^{i+1}(z=h/2)}{h/2}\,, \tag{8.3-8}$$

$$\varepsilon_{pl}^{i+1}(z) = \varepsilon_f^{i+1}(z) - \varepsilon_{el}^{i+1}(z)\,. \tag{8.3-9}$$

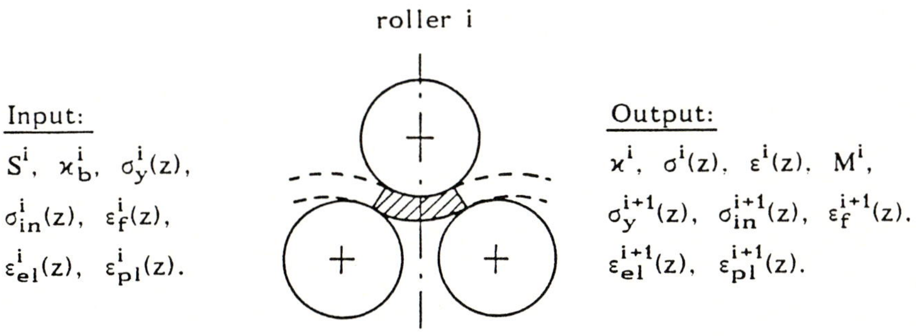

Fig 8.3/5. Discrete forming step of the levelling process

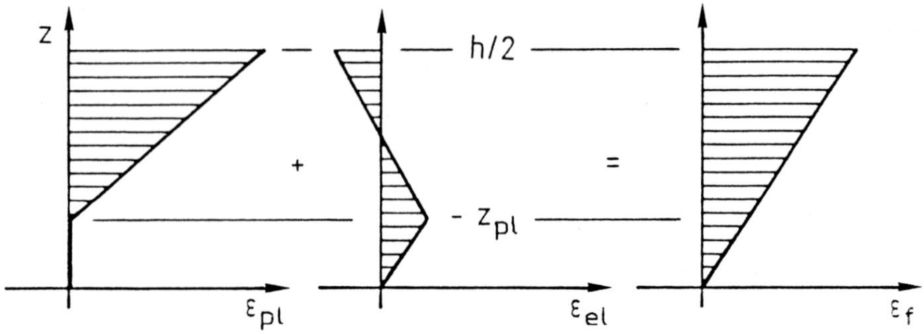

Fig. 8.3/6. Strain state after unloading

8.3.3 Formulation of the Optimization Problem

The sheet has predeformations $\varkappa_{bmin} \le \varkappa_b \le \varkappa_{bmax}$ before levelling, and it should have minimal deformations $\varkappa_a$ after levelling. Design variables are the feed motions of the rollers. The predeformations before levelling are represented by m curvatures $\varkappa_{bk}$:

$$\varkappa_{bk} = \varkappa_{bmin} + (k-1)(|\varkappa_{bmax}| - |\varkappa_{bmin}|)/(m-1), \quad k = 1,\dots,m. \qquad (8.3\text{-}10)$$

When the feed motions s_j are known, the residual deformation $\varkappa_{ak}$ can be calculated for each curvature $\varkappa_{bk}$ by using the mechanical model presented above. The aim of the optimization process is to get an adjustment for the roller leveller which results in minimal residual deformations $\varkappa_{ak}$ for all m predeformations $\varkappa_{bk}$ before levelling. This can lead to a conflict of objectives, because an adjustment which minimizes the residual deformation for a particular predeformation can be unfavourable for other predeformations. Thus, the problem can be formulated as a multicriteria optimization problem. The m residual deformations $\varkappa_{ak}$ (k = 1,...,m) are the objective functions. The feed motions s_j(j = 1,...,n) are the design variables. The symbol j, in contrast to the symbol i, is used to count the upper rollers only. Using the standard formulation of a multicriteria optimization problem (see [3] and chapter 1), the problem reads as follows:

$$\min \{\mathbf{f}(\mathbf{x}) : \mathbf{h}(\mathbf{x}) = \mathbf{0},\ \mathbf{g}(\mathbf{x}) \le \mathbf{0}\} \qquad (8.3\text{-}11)$$
$$\text{with} \quad \{\mathbf{x} \in \mathbb{R}^n,\ \mathbf{h}(\mathbf{x}) = \mathbf{0},\ \mathbf{g}(\mathbf{x}) \le \mathbf{0}\},$$

where: $\mathbf{f}^T(\mathbf{x}) = (\varkappa_{a1}, \ldots, \varkappa_{ak}, \ldots, \varkappa_{am})$ m objective functions,
$\mathbf{x}^T = (s_1, \ldots, s_j, \ldots, s_n)$ n design variables,
$\mathbf{h}$: q equality constraints, system equations,
$\mathbf{g}$: p inequality constraints.

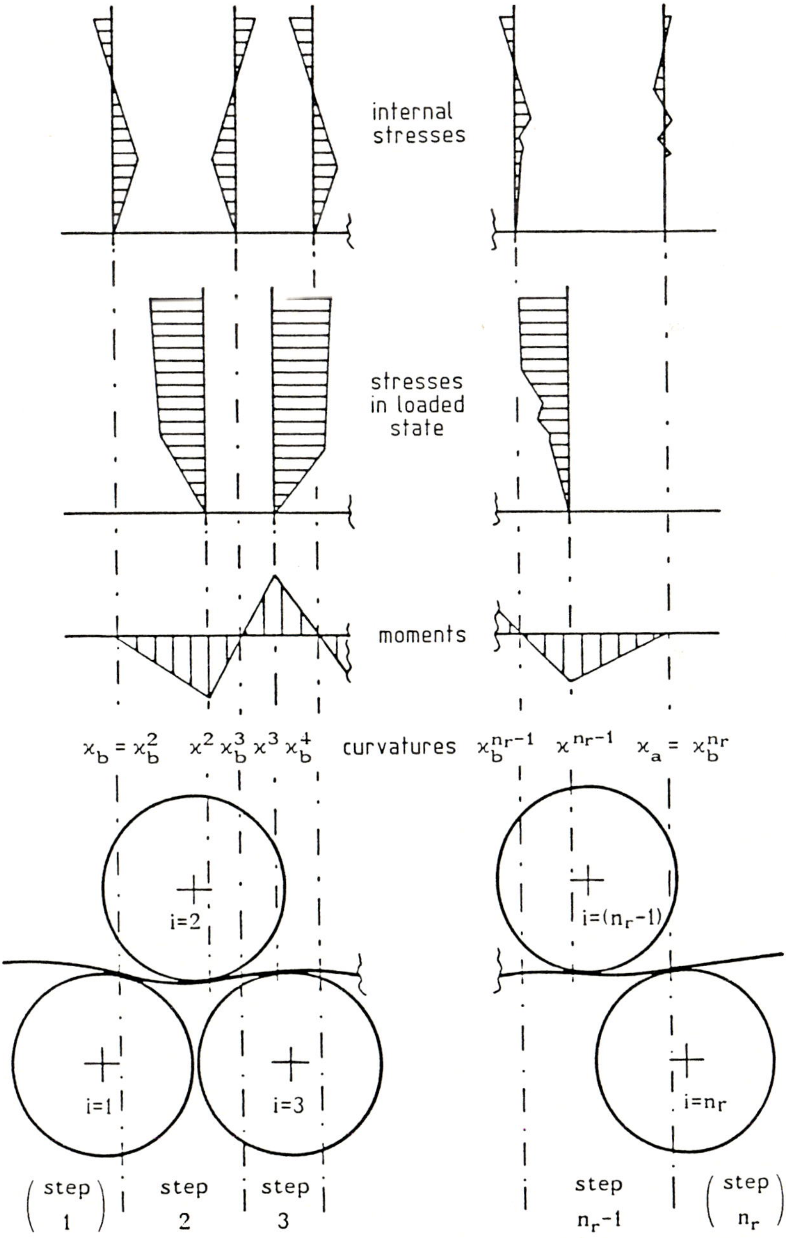

Fig. 8.3/7. Stress state, bending moment and curvature during levelling

The inequality constraints prove the validity of the mechanical model by checking if the contact force takes the form of pressure at each roller. Additionally, suitable upper and lower bounds for the design variables are to be considered. For numerical treatment, the multicriteria optimization problem described above is reduced to a scalar substitute problem by means of the relations in Section 1.2.2:

$$\min \{ p[\mathbf{f}(\mathbf{x})] : \mathbf{h}(\mathbf{x}) = \mathbf{0},\ \mathbf{g}(\mathbf{x}) \leq \mathbf{0} \}. \tag{8.3-12}$$

The preference function $p[\mathbf{f}(\mathbf{x})]$ is applied as the distance function (1-13b)

$$p[\mathbf{f}(\mathbf{x})] = \left(\sum_{k=1}^{m} \left(\frac{\varkappa_{ak}}{\varkappa_{Ok}} \right)^2 \right)^{1/2} \tag{8.3-13}$$

with the appropriate scaling factors $\varkappa_{Ok}$.

The optimization algorithm used here is a generalized reduced gradient method [4].

8.3.4 Results and Conclusion

In the following some results will be exemplified:

Material data				
Young's modulus	: E	=	210000	N/mm^2
initial yield stress	: σ_y	=	500	N/mm^2
hardening coefficient	: c_h	=	0.05	
Geometrical data of the sheet				
thickness	: h	=	2	mm
depth	: b	=	variable	
Geometrical data of the roller leveller				
number of rollers	: n_r	=	9	
radius of rollers	: r_r	=	25.5	mm
pitch of rollers	: t	=	26.0	mm

Table 8.3/1. Values for the reference configuration

The values in Table 8.3/1 describe a reference configuration for the calculations. Assuming a linear profile of the feed motion, the solution to the problem is reduced to the design variables s_1 and s_4.

Between these values a linear interpolation of the feed motions s_2 and s_3 is carried out. The predeformation region is represented by 10 curvatures $\varkappa_{bk}$ (8.3-10).

Fig. 8.3/7 shows the stress states and the curvatures of the sheet for the reference configuration during levelling. The predeformation is $\varkappa_b = 0.003\ \mathrm{mm}^{-1}$. It can be seen that the curvature increases at rollers 2 and 3. After being subjected to the third roller, the remaining curvature decreases until the value $\varkappa_a$ is obtained after the roller n_r-1. An increase of the curvature at rollers 2 and 3 allows different predeformations before levelling to be accomodated. Thus, after being subjected to the third roller, stress states and sheet curvatures $\varkappa^i$ are almost the same, independent of the predeformation $\varkappa_b$.

In order to verify the algorithm presented, the roller forces are compared with the values given by Zelikow [5] and Geleji [6]. The forces are determined for the reference configuration without hardening ($c_h = 0$), because the algorithms of Geleji and Zelikow do not consider hardening effects. Due to their assumptions, the values given by Geleji and Zelikow represent upper and lower bounds for the roller forces F_4 to F_{n_r}. Fig. 8.3/8 shows that the values calculated by the program RICHT, based upon the algorithm presented, are approximately identical with the values given by Geleji and Zelikow.

For a given range of predeformations, the program RICHT provides a range of roller forces. After being subjected to the third roller, differing predeformations are equalized so that a range of forces for the following rollers cannot be determined. A standardized interval of predeformations is applied to all further investigations taking into account that thick sheets normally have smaller predeformations than thin ones.

Below, optimal feed motions for different sheet thicknesses and yield stresses are presented and compared. In Fig. 8.3/9, the feed motion s_j for a roller is shown schematically.

Figures 8.3/10 and 8.3/11 show the feed motions for different sheet thicknesses and yield stresses. The investigations are limited to linear profiles of feed motions. It is not possible to compare the values of these figures with the values found in practical application, as the roller leveller is assumed to possess an infinite stiffness. The feed motion s_1

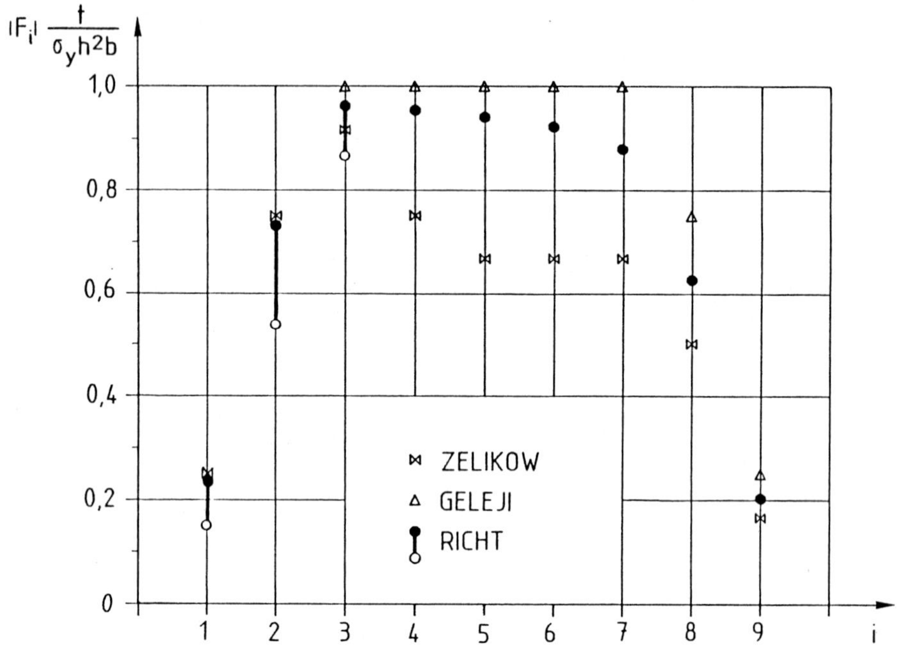

Fig. 8.3/8. Comparison of the forming forces F_i (scaled by the maximum force) for the reference configuration (c_h=0)
○ value for $\varkappa_b$ = -0.003 [1/mm]
● values for $\varkappa_b$ = 0 and 0.003 [1/mm]

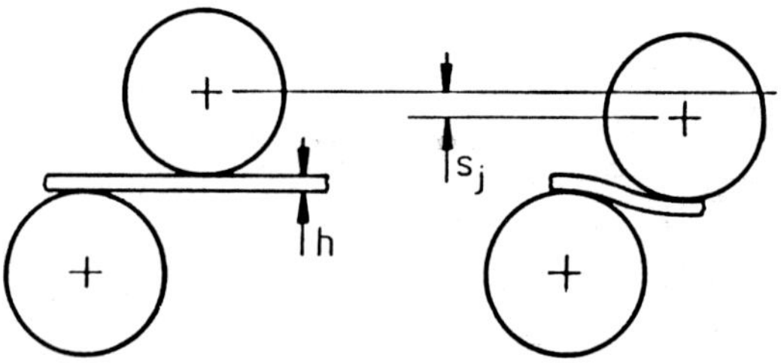

Fig. 8.3/9. Arrangement of the feed motion

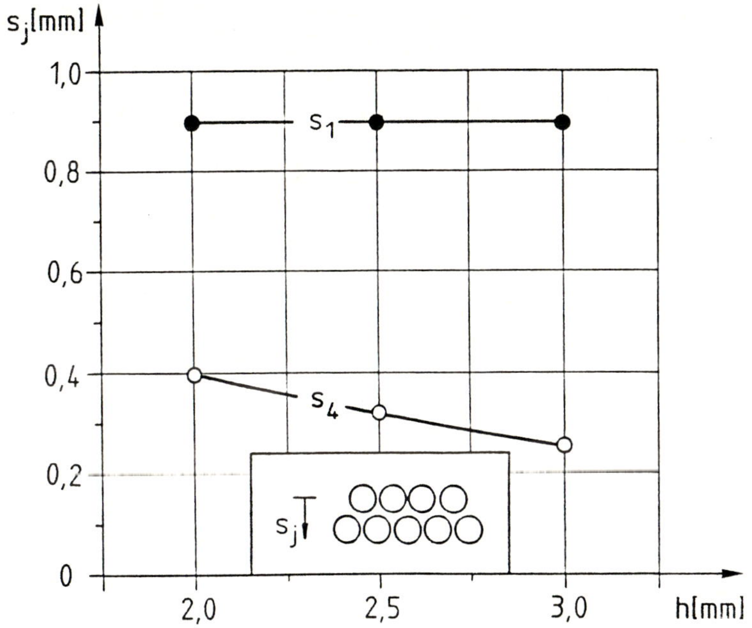

Fig. 8.3/10. Feed motions for different sheet thicknesses

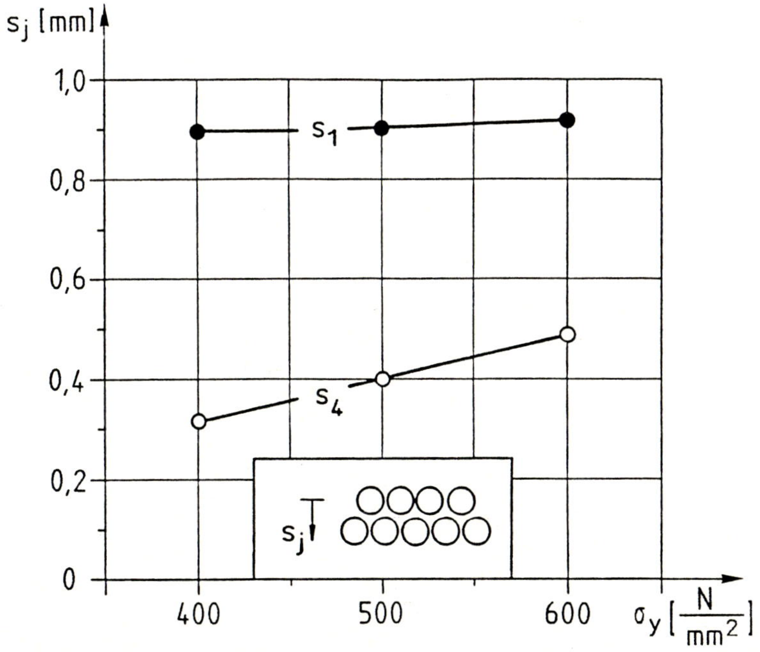

Fig. 8.3/11. Feed motions for different yield stresses

of the first upper roller is almost independent of the yield stress as well as of the sheet thickness. The feed motion s_4 of the last upper roller grows while increasing the yield stress σ_y and decreases while increasing the sheet thickness h. In general, the feed motions of the last upper roller is a function of the geometrical data of the roller and of the sheet data:

$$s_{j_r} = s_{j_r}(t, r_r, \sigma_y, h) , \qquad j_r = \frac{n_r - 1}{2} \; . \tag{8.3-14}$$

The results shown in Fig. 8.3/10 and 8.3/11 can be approximated by the following equation:

$$s_4 = \bar{s}_4 \left(\frac{\sigma_y}{\bar{\sigma}_y} \right) \left(\frac{\bar{h}}{h} \right) \tag{8.3-15}$$

with the values of the reference configuration:

$$\bar{s}_4 = 0{,}404 \text{ mm} , \quad \bar{\sigma}_y = 500 \text{ N/mm}^2 , \quad \bar{h} = 2 \text{ mm} .$$

In most cases, roller levellers meeting high accuracy requirements have got more rollers than the reference configuration. There are levellers with up to 23 rollers [7]. Calculations for configurations with

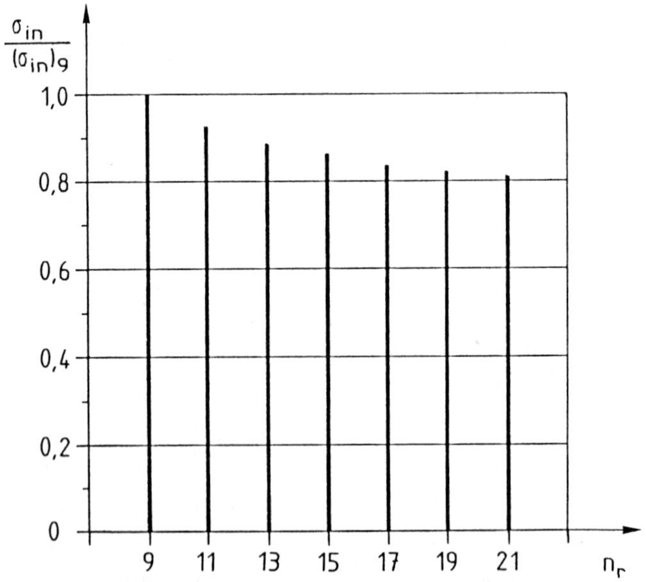

Fig. 8.3/12. Remaining internal stresses for a various number of rollers

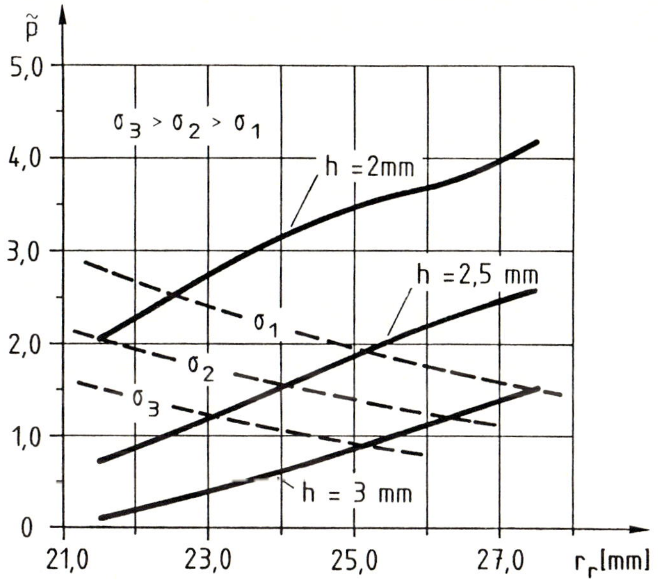

Fig. 8.3/13. Objective function values for different roller diameters

more rollers (Fig. 8.3/12) show that both the residual deformations and the residual internal stresses after levelling decrease. The values in Fig. 8.3/12 are scaled by the values of the reference configuration.

It could be shown as well that thick sheets are easier to level than thin sheets (Fig. 8.3/13). The ordinate values in Fig. 8.3/13 are scaled by the objective function value for $h = 3$ mm and $r_r = 25.5$ mm. It is obvious that the values for the objective function decrease with smaller roller diameters. Both effects are known from the practical application of levellers [7], but there are bounds for the roller diameters which result from the loads connected with levelling. In Fig. 8.3/13, lines of equal roller stresses are shown; thick sheets produce higher loads and need a greater roller diameter.

Thiele [8] suggests that the roller levellers should be modified in such a way that a nonlinear profile can be adjusted in order to achieve better results. Fig. 8.3/14 shows the results obtained by calculations with a nonlinear profile for the reference configuration (4 upper rollers = 4 design variables). The comparison of a linear and a nonlinear profile shows that the use of the latter one slightly increases the sensitivity of the objective function depending on the feed motions.

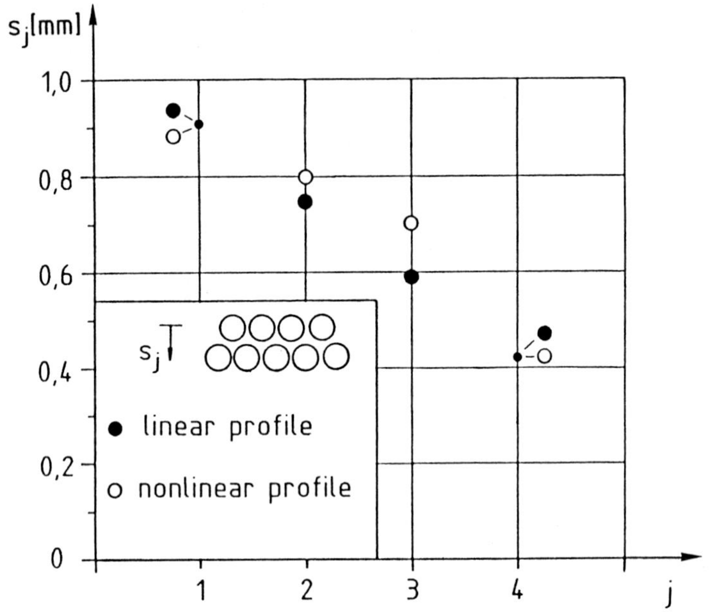

Fig. 8.3/14. Feed motions for a nonlinear and a linear profile

The results described here are based upon the geometrical data of the reference configuration given in Table 8.3/1. With the algorithm presented in this paper, optimal adjustments can also be calculated for other parameters, such as the number of rollers n_r, the radius of rollers r_r, and the pitch of rollers t.

References

[1] Kübert, M.; Schroeder, L.: Entstehung und Beseitigung von Planheitsfehlern bei Blechen und Bändern. Werkstatt und Betrieb 115, (1982) 477-479

[2] Schiffner, K.; Henrich, L.: A Model for Simulation and Optimization of the Metal Forming Process at Roller Levelling. In: Eschenauer, H.; Thierauf, G. (eds.): Discretization Methods and Structural Optimization - Procedures and Applications. Berlin: Springer Verlag (1989) 317-324

[3] Eschenauer, H.; Olhoff, N.: Optimization Methods in Structural Design. Mannheim: Bibl. Institut 1983

[4] Bremicker, M.: Dekompositionsstrategie in Anwendung auf Probleme der Gestaltsoptimierung, Dissertation. Universität-GH Siegen, VDI-Fortschrittsbericht. Reihe 1, Nr. 173. Düsseldorf: VDI-Verlag 1989

[5] Zelikow, A.I.: Hilfsmaschinen der Walzstraßen. Berlin: VEB Verlag Technik 1954

[6] Geleji, A.: Walzwerks-und Schmiedemaschinen. Berlin: VEB Verlag Technik 1961

[7] Kübert, M.; Schroeder, L.: Konstruktionsprinzipien von Walzenrichtmaschinen, Werkstatt und Betrieb 115 (1982) 559-562

[8] Thiele, H.: Möglichkeiten zur Beseitigung von Unplanheiten an kaltgewalzten Bändern, Metall 28 (1974) 968-975

9 CIVIL AND ARCHITECTURAL ENGINEERING

9.1 Multicriteria Optimization of Concrete Beams, Trusses, and Cable Structures

St. Jendo

9.1.1 Introduction

The optimization of civil engineering structures usually involves a number of requirements that should be met at the same time in order to obtain a useful design. In the case of single criteria optimization, one of the requirements is selected as the criterion of optimization while the remaining ones are met by including them into the constraints of optimization. However, while using this approach it is necessary to determine a priori the bounds which these requirements should fulfill a priori. Multicriteria (vector) optimization enables the designer to consider effectively all the different, mutually conflicting requirements inherent in the design problem. The criteria appearing most frequently in the optimization of structures are as follows [3]:

- *Minimum volume, weight or cost criterion*

According to these criteria, the optimal structure is the one with the minimum volume, weight or cost. It has been selected from the possible structures subjected to the given loading conditions and satisfying the given constraints. If the structure has been built of a number of different materials, it may be necessary to minimize each of them. For example, in the case of a prestressed concrete beam the minimum volumes of both the concrete and the prestressing cables may be required.

- *Minimum compliance or maximum structural stiffness*

For this criterion, among all possible structures subjected to the same loads and satisfying the given constraints, the optimal structure is that one whose internal work done by the loads due to the strains, has a minimum value. On the other hand, the work done by the loads on displacements in the equilibrium state is equal to the energy (potential) of elastic deformation and, thus, this energy also attains its minimum value.

- *Minimum displacements at selected points or regions of the structure*

In some cases, it is necessary to minimize changes in the shape of the structure under the action of different loading conditions. Such a requirement is involved, e.g., for the supporting structure of radio-telescopes or for some cable-suspended structures.

- *Maximum critical load*

It is obvious that buckling is important in very slender compressed elements. Therefore, they should be shaped in such a way as to obtain the maximum critical load. In some structures, local buckling is most important, and then it is necessary to ensure the maximum stiffness of particular structural elements.

- *Maximum natural frequency of free vibration*

Structures under dynamic loading conditions should be shaped in such a way that extreme values of the natural frequencies of free vibration can be obtained. In some cases it is sufficient to take into account only the first frequency of free vibration (unimodal optimization), but in others, analysis must be performed for the frequencies of higher vibration modes (multimodal optimization). This criterion can be modified in such a way that the maximum detuning of the vibrating system, i.e. achieving the maximum separation between the frequencies of natural and forced vibration can be obtained.

- *Maximum reliability or safety*

This criterion is always required, but it is especially important in aerospace engineering and in civil engineering structures built in seismic zones.

- *Maximum moment of inertia*

It may be required that the cross-sectional moment of inertia of bent or compressed structural elements should be maximized. The same requirement may hold for vibrating systems. For twisted elements it is necessary to maximize the polar moment of inertia of the cross-section. The criterion of maximum moment of inertia is consistent with many of the criteria mentioned above.

It should be emphasized that it is necessary to find the extremum of an objective function. The related constraints should be taken into account in order to satisfy any of the criteria mentioned above. In structural optimization the most important constraints involve:

- allowable stresses or the safety factors of the structure under all possible loading conditions,
- allowable displacements in a given structure,
- lower and upper bounds for design variables based on serviceability and technological reasons; e.g. on the minimum thickness of metal plates or maximum height of a bridge structure because of the requirements of a transportation system.

9.1.2 Optimization of a Prestressed Concrete I-beam

The optimum design problem for a prestressed concrete I-beam can be formulated as follows. Find the dimensions of the cross-sectional area of a prestressed concrete I-beam as well as the eccentricity of the prestressing force according to two criteria: minimum volumes of the concrete (A) and the prestressing steel (S). The extreme loading conditions represented by the bending moments M_1 and M_2 are considered. The vector of design variables $\mathbf{x} = (x_i)^T$, i=1,2,...,6, contains the following quantities: x_1 - beam depth, x_2 - flange width, x_3 - flange thickness, x_4 - web thickness, x_5 and x_6 - the values of the prestressing force and its eccentricity, respectively (Fig. 9.1/1).

The muticriteria optimization problem given in (1-2) is now expressed as

$$\min_{x \in \mathbb{R}^n} (A(\mathbf{x}),\ S(\mathbf{x}))^T, \tag{9.1-1}$$

subject to:

a) stress constraints in two loading conditions

$$x_5/A - x_5x_6x_1/2J + M_1\,x_1/2J \le \sigma_1^u, \tag{9.1-2a}$$

$$x_5/A + x_5x_6x_1/2J - M_1\,x_1/2J \ge \sigma_1^l, \tag{9.1-2b}$$

$$x_5/A - x_5x_6x_1/2J + M_2x_1/2J \geq \sigma_2^l, \qquad (9.1\text{-}2c)$$

$$x_5/A + x_5x_6x_1/2J - M_2x_1/2J \leq \sigma_2^u, \qquad (9.1\text{-}2d)$$

$$x_1 \leq x_1^u, \quad x_2 \leq x_2^u, \quad x_3 \geq x_3^l, \quad x_4 \geq x_4^l, \quad x_6 \leq x_1/2-d, \qquad (9.1\text{-}3)$$

where σ_i^l and σ_i^u, i=1,2, are lower and upper bounds of allowable stresses.

The components of the objective function vector can be written as follows:

$$A = 2x_2x_3 + x_4(x_1-2x_3), \qquad (9.1\text{-}4a)$$

$$S = x_5. \qquad (9.1\text{-}4b)$$

The curve segment ∂Y^*, given by

$$x_5 = (\sigma_1^u - M_1x_1^u/2J)/(1/A + (x_1^u/2-d)\, x_1^u/2J), \qquad (9.1\text{-}5)$$

contained between the two following curves

$$x_5 = (\sigma_2^u + M_2x_1^u/2J)/(1/A + (x_1^u/2-d)\, x_1^u/2J), \qquad (9.1\text{-}6a)$$

and

$$x_5 = (\sigma_1^l + M_2x_1^u/2J)/(1/A + (x_1^u/2-d)\, x_1^u/2J), \qquad (9.1\text{-}6b)$$

is the functional-efficient boundary of the set Y (efficient curve) (for details see [18,26]).

The preferable solution is selected on the basis of the utility function

$$U = -\Phi = -(A + kS), \qquad k = c_s\gamma / c_c\sigma_s^u \ (m^2/kN), \qquad (9.1\text{-}7)$$

where k is the ratio of the unit cost of the prestressing steel divided by the concrete unit cost. The following notation is used: c_c is the unit cost of 1 cubic meter of concrete; c_s is the unit cost of 1 kg of prestressing steel; γ is the bulk density of the prestressing steel, and σ_s^u is the allowable tension stress for the prestressing steel. The preferable solution occurs at one of the end points of the minimal curve or at the tangent point between the minimal curve and the indifference lines of the utility function $U = -\Phi = -(A+kS) = \text{const}$, which depends on the value of k (Fig. 9.1/1).

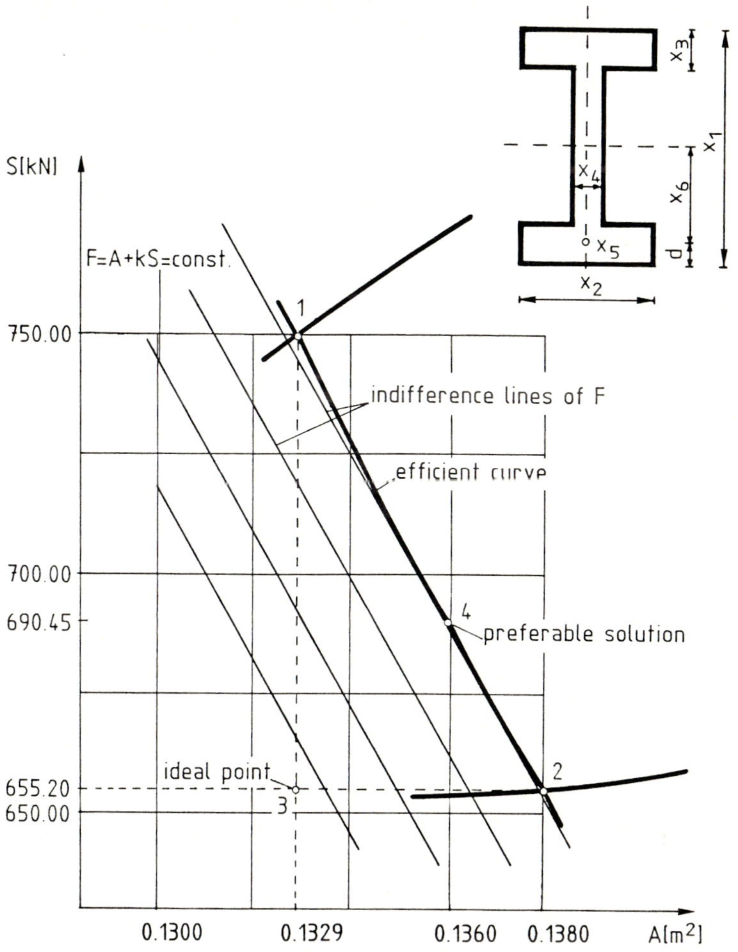

Fig. 9.1/1. Geometrical interpretation of the solution for the I-beam

Numerical example. The following data are used: $M_1 = 320$ kNm, $M_2 = 100$ kNm, $x_1^u = 0.70$ m, $x_2^u = 0.40$ m, $x_3^l = 0.10$ m, $x_4^l = 0.08$ m, $d = 0.07$ m, $\sigma_1^u = \sigma_2^u = 10000$ kN/m^2, $\sigma_1^l = \sigma_2^l = -500$ kN/m^2 and $k = 0.000054$ m^2/kN.

The solution is presented in Fig. 9.1/1 and includes the following information:

- *the coordinates of the end points of the efficient curve*

point No 1 ($A = 0.13286$ m^2, $S = 750.0$ kN);

point No 2 ($A = 0.13797$ m^2, $S = 655.2$ kN);

- *the ideal solution*

point No 3 ($\bar{A}$ = 0.13286 m^2, $\bar{S}$ = 655.2 kN);

- *the preferable solution*

point No 4 ($\tilde{A}$ = 0.13600 m^2, $\tilde{S}$ = 690.45 kN) and the corresponding compromise solution, i.e. the final vector of design variables $\tilde{x}_1$ = 0.70 m, $\tilde{x}_2$ = 0.40 m, $\tilde{x}_3$ = 0.125 m, $\tilde{x}_4$ = 0.08 m, $\tilde{x}_5$ = 690.45 kN, $\tilde{x}_6$ = 0.28 m;

- *the width of the efficient set*

ΔA = 0.00511 m^2, ΔS = 94.80 kN;

- *the sensitivity vector of the utility function* $U = -\Phi$ *at the preferable solution*

$\partial\Phi/\partial A = 1$, $\partial\Phi/\partial S = 0.000054$.

9.1.3 Optimization of Isostatic Trusses

The multicriteria optimization problem of isostatic steel truss structures can be formulated as follows. Find the cross-sectional areas of truss bars $\mathbf{x} = (x_i)^T$, i = 1,2,...,7, according to two criteria: minimum volume of truss (V) and minimum vertical displacement (δ_B) at point B (Fig. 9.1/2).

A multicriteria optimization problem (1-2) is now expressed as

$$\min_{\mathbf{x} \in \mathbb{R}^n} (V(\mathbf{x}), \delta_B(\mathbf{x}))^T, \tag{9.1-8}$$

subject to stress and side constraints, i.e.

$$X = \left\{\mathbf{x} \in \mathbb{R}^n : \sigma^l \le \sigma_i \le \sigma^u, \quad x_i \le x_i^u, \quad i = 1,\dots,7\right\}, \tag{9.1-9}$$

where σ^l and σ^u denote the allowable compression and tension stresses, respectively, and $\sigma_i = N_i/x_i$.

The components of the objective function vector for truss structures can be written in the following form:

$$V = \sum_{i=1}^{7} x_i l_i , \tag{9.1-10a}$$

$$\delta_B = \sum_{i=1}^{7} \partial/\partial F\left(N_i^2 \, l_i/(2Ex_i)\right). \tag{9.1-10b}$$

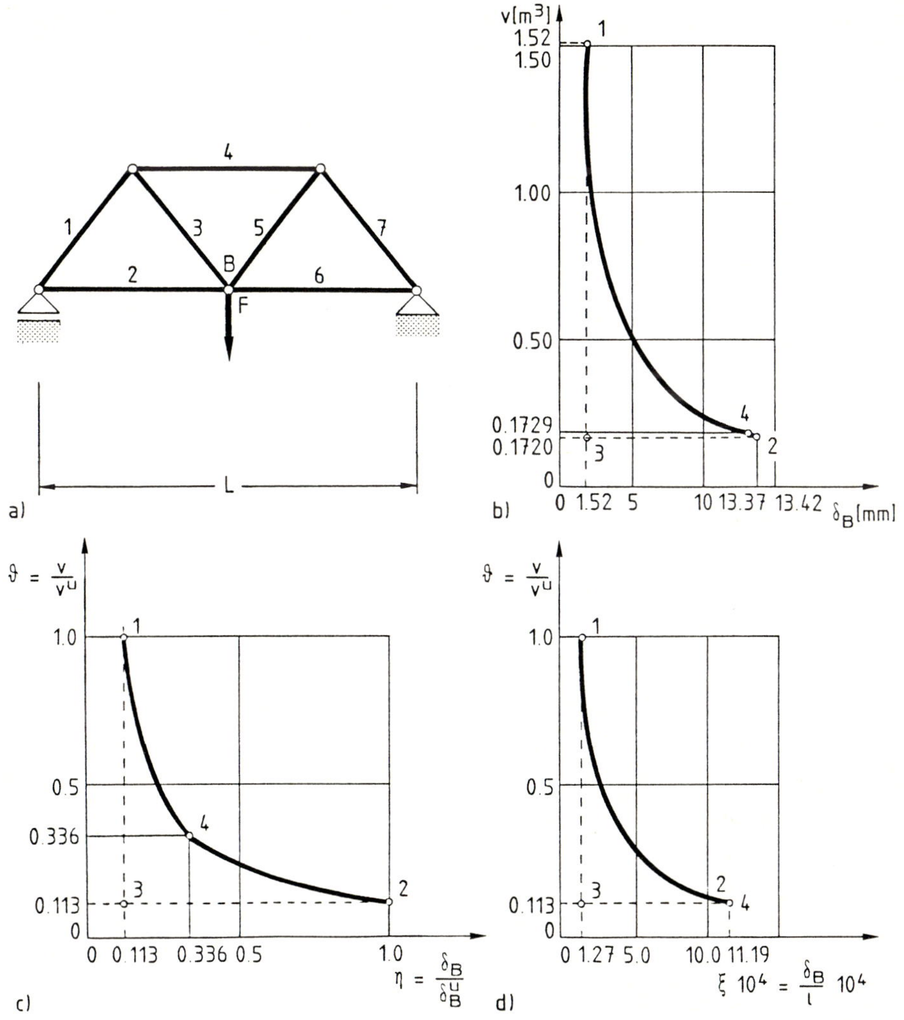

Fig. 9.1/2. Geometrical interpretation of the three solutions for isostatic truss using distance functions

The optimization problem is solved using dynamic programming methods [30]. The functional-efficient boundary ∂Y^* has the following form

$$V = F(5l_1+3l_2+3l_3+5l_4) \,/\, (32E\delta_B) , \tag{9.1-11}$$

and it is contained between V^l and V^u in the coordinate system (V,δ_B). The minimal feasible volume of truss V^l is found on the basis of allowable stresses in truss bars, i.e.

$$V^L = 2l_1\, 5F/(8\sigma^l) + 2l_2\, 3F/(8\sigma^u) + 2l_3\, 5F/(8\sigma^u) + l_4\, 3F/(4\sigma^l). \tag{9.1-12}$$

The maximal feasible volume of truss V^u is found on the basis of side constraints $x_i = x_i^u = x^u$, i.e.

$$V^u = (2l_1 + 2l_2 + 2l_3 + l_4)\, x^u \ . \tag{9.1-13}$$

Since there is no preference between the objective functions V and δ_B, the solution is found on the basis of the minimization of the distance function (1-13b). Three solutions based on different forms of the distance function are found.

- *Solution No. 1*

The distance function in dimensional form reads:

$$p[V,\delta_B] = \{(V-\overline{V})^2 + \mu^2(\delta_B - \overline{\delta}_B)^2\}^{1/2}, \qquad \mu = 1\ m^2 . \tag{9.1-14}$$

The ideal solution has the following coordinates

$$\overline{V} = F(5l_1/\sigma^l + 3l_2/\sigma^u + 5l_3/\sigma^u + 3l_4/\sigma^l)\ /\ 4\ , \tag{9.1-15a}$$

$$\overline{\delta}_B = F(5l_1 + 3l_2 + 5l_3 + 3l_4)^2 / \{32\ Ex^u(2l_1 + 2l_2 + 2l_3 + l_4)\}\ . \tag{9.1-15b}$$

Substituting (9.1-11) and (9.1-15) into (9.1-14) one gets

$$p[V,\delta_B] = \{\{V - F(5l_1/\sigma^l + 3l_2/\sigma^u + 5l_3/\sigma^u + 3l_4/\sigma^l)\ /\ 4\}^2 + \tag{9.1-16}$$
$$+\mu^2\{(F^2(5l_1+3l_2+5l_3+3l_4)^4/(1024\,E^2)\}\ \{1/V - 1/(x^u(2l_1+2l_2+2l_3+l_4))\}^2\}^{1/2}.$$

The necessary condition for an extremum gives the following equation:

$$V^4 - \{F(5l_1/\sigma^l + 3l_2/\sigma^u + 5l_3/\sigma^u + 3l_4/\sigma^l)\ /\ 4\}\ V^3 + \tag{9.1-17}$$
$$+\ \mu^2\{(F^2(5l_1+3l_2+5l_3+3l_4)^4\ /\ (1024\,E^2x^u(2l_1+2l_2+2l_3+l_4))\}V\ -$$
$$-\ \mu^2\{F^2(5l_1+3l_2+5l_3+3l_4)^4\ /\ (1024\,E^2)\}\ = \ 0\ ,$$

which allows us to determine the preferable solution $V^l \le \tilde{V} \le V^u$.

- *Solution No. 2*

Taking the distance function in dimensionless form one gets

$$p[V,\delta_B] = \{\{(V - \overline{V})\ /\ V^u\}^2 + \{(\delta_B - \overline{\delta}_B)\ /\ \delta_B^u\}^2\}^{1/2}\ . \tag{9.1-18}$$

Introducing the notation

$$\nu = V/V^u\ , \qquad \eta = \delta_B/\delta_B^u\ , \qquad \beta = V^l/V^u\ , \tag{9.1-19}$$

and using an analogous procedure to the one above, one gets

$$\nu^4 - \beta\nu^3 + \beta^2\nu - \beta^2 = 0 \, , \tag{9.1-20}$$

which allows us to determine the preferable sulution $\tilde{\nu}$.

	Solution No. 1	Solution No. 2	Solution No. 3
Equation of effective set	$V[m^3] = \frac{0.002311}{\delta_B[m]}$	$\nu = \frac{0.113137}{\eta}$	$\nu = \frac{0.0001267}{\xi}$
Coordinates of end points of effective set			
- point No. 1	$V = 1.52\ m^3$ $\delta_B = 0.00152\ m$	$\nu = 1$ /$V = 1.52\ m^3$/ $\eta = 0.113$ /$\delta_B = 0.00152\ m$/	$\nu = 1$ /$V = 1.52\ m^3$/ $\xi = 0.000127$ /$\delta_B = 0.00152\ m$/
- point No. 2	$V = 0.172\ m^3$ $\delta_B = 0.0134\ m$	$\nu = 0.113$ /$V = 0.172\ m^3$/ $\eta = 1$ /$\delta_B = 0.0134\ m$/	$\nu = 0.1132$ /$V = 0.172\ m^3$/ $\xi = 0.001119$ /$\delta_B = 0.134\ m$/
Ideal solution /point No. 3/	$\bar{V} = 0.172\ m^3$ $\bar{\delta}_B = 0.00152\ m$	$\bar{\nu} = 0.113$ /$\bar{V} = 0.172\ m^3$/ $\bar{\eta} = 0.113$ /$\bar{\delta}_B = 0.00152\ m$/	$\bar{\nu} = 0.113$ /$\bar{V} = 0.172\ m^3$/ $\bar{\xi} = 0.000127$ /$\bar{\delta}_B = 0.00152\ m$/
Preferable solution /point No. 4/	$\tilde{V} = 0.1729\ m^3$ $\tilde{\delta}_B = 0.0134\ m$	$\tilde{\nu} = 0.336$ /$\tilde{V} = 0.5107\ m^3$/ $\tilde{\eta} = 0.336$ /$\tilde{\delta}_B = 0.00453\ m$/	$\tilde{\nu} = 0.1132$ /$\tilde{V} = 0.172\ m^3$/ $\tilde{\xi} = 0.001119$ /$\tilde{\delta}_B = 0.0134\ m$/
Design variables $x_1[m^2]$ $x_2[m^2]$ $x_3[m^2]$ $x_4[m^2]$	 0.005034 0.003017 0.005026 0.006030	 0.014846 0.008907 0.014846 0.017814	 0.0050 0.0030 0.0050 0.0060
Width of effective set	$\Delta V = 1.348\ m^3$ $\Delta\delta_B = 0.0119\ m$	$\Delta\nu = 0.887$ /$\Delta V = 1.348\ m^3$/ $\Delta\eta = 0.887$ /$\Delta\delta_B = 0.0119\ m$/	$\Delta\nu = 0.887$ /$\Delta V = 1.348\ m^3$/ $\Delta\xi = 0.000993$ /$\Delta\delta_B = 0.0119\ m$/
Sensitivity vector	$\frac{\partial F}{\partial V} = 0.0757$ $\frac{\partial F}{\partial \delta_B} = 0.9971$	$\frac{\partial F}{\partial \nu} = 0.7071$ $\frac{\partial F}{\partial \eta} = 0.7071$	$\frac{\partial F}{\partial \nu} = 0$ $\frac{\partial F}{\partial \xi} = 1$

Table 9.1/1. List of solutions for multicriteria optimization of an isostatic truss

- *Solution No. 3*

Taking the distance function in another dimensionless form

$$p\{V,\delta_B\} = \{\{(V-\bar{V}) \;/\; V^u\}^2 + \{(\delta_B - \bar{\delta}_B) \;/\; L\}^2\}^{1/2} \tag{9.1-21}$$

and introducing a notation analogous to the one above

$$\nu = V/V^u, \; \xi = \delta_B/L, \quad \alpha = P(5l_1 + 3l_2 + 5l_3 + 3l_4) \;/\; (32\,E)\,, \tag{9.1-22}$$

yields

$$(V^u)^2 \nu^4 - V^u V^l \nu^3 + (\alpha^2/L^2)\,\nu - \alpha^2/L^2 = 0\,, \tag{9.1-23}$$

which allows us to determine the preferable solution $\tilde{\nu}$.

- *Numerical example*

The following data are taken: $F = 2000$ kN, $E = 200$ kN/mm^2, $\sigma^l = \sigma^u = 250$ MPa, $l_1 = l_3 = l_5 = l_7 = 5.00$ m, $l_2 = l_4 = l_6 = 6.00$ m, $x^u = 40000$ mm^2. Three solutions for an optimal truss are presented in Table 9.1/1 and shown in Figure 9.1/2. The functional-efficient sets and ideal solutions are independent of the form of the distance function. Unfortunately, the preferable solution and the sensitivity vector essentially depend on the form of the distance function.

9.1.4 Optimization of Cable Structures

1) Characteristics of cable structures

This section deals with an optimization problem of single-layer cable systems which are often used as carrying elements in electric power lines, hanging rope-ways or large-span roofing structures as shown in Fig. 9.1/3. Cable structures are substantially different from other types of structures because they are capable of assuming a variety of shapes under the action of different loadings. That is why static and dynamic analyses of cable structures are different from those commonly used for stiff structures displacing only slightly from an unloaded reference configuration. The first difference occurs because of the flexible shape of the structure so that the principle of structural rigidity cannot be applied here. The second difference consists of the fact that the principle of superposition is inapplicable. This follows from the geometric nonlinearity of cables which is caused by large changes in

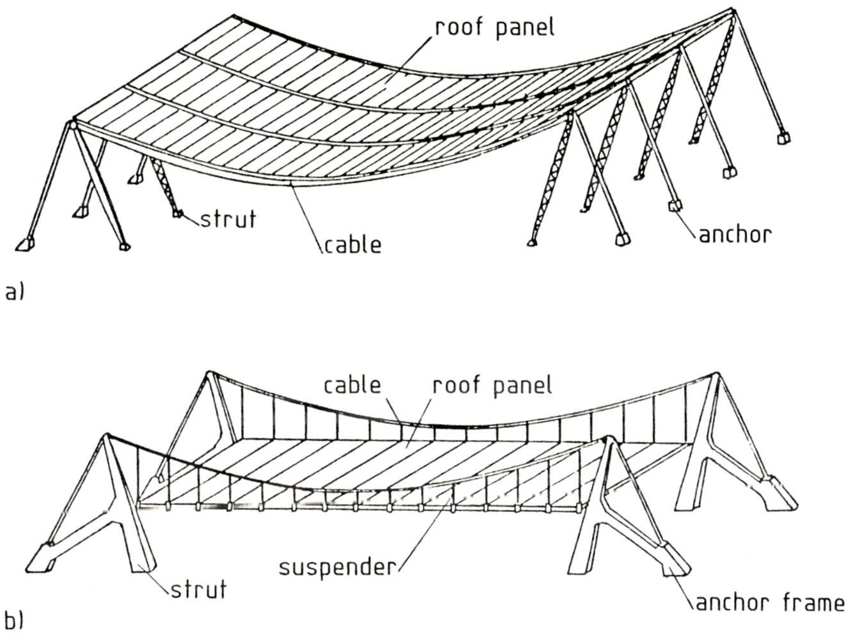

Fig. 9.1/3. Schematic layouts of single-layer cable systems

cable shapes due to varying loadings. In order to write the equilibrium and deformation equations, it is necessary to take into account all the loads acting on the structure which has no a priori determined shape.

2) Basic relationships for flat sag cables

A nomogram for determining the lowest frequency for symmetric in-plane free vibrations of the extensible cable is shown in Fig. 9.1/4. It can be seen that maximal values of the lowest natural frequencies occur for very flat sag cables; i.e. $\eta = f/l = 0.03 \div 0.05$ corresponds to the interval (20 - 200) of the parameter $\xi = ql/A$. The following notation is used: f, l, A and q denote sag, span, cross-section area and uniform load of cable, respectively. On the other hand, the optimal cable sags determined according to the minimum weight criterion always occur on the boundary of the feasible domain (Fig. 9.1/5) given by side, stress, and displacement constraints (see [4]). Comparing the results given above, one may conclude that a set of compromise solutions should be found using a multicriteria optimization approach.

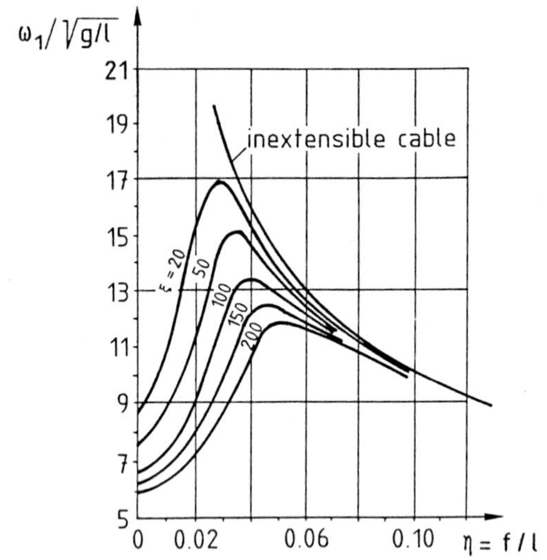

Fig. 9.1/4. Nomogram for determining the lowest frequency for symmetric in-plane free vibration

3) Formulation of the bicriteria optimization problem

The bicriteria optimization problem for single-layer cable systems with the assumptions of flat sags and Hooke's law for material behaviour can be formulated as follows. Find the cable sag $\eta = f/l$ and cross-section area (A) for a given cable span (l), static loading q(x), modulus of elasticity E, dynamic loading p(x,t) and allowable stress σ'' so that the weight of the single-layer cable system

$$W = \gamma A s = \gamma A l (1 + 8\eta^2 / 3) \tag{9.1-24}$$

attains a minimal value, and the lowest natural frequency of the in-plane free vibrations

$$\omega_1 = 2 a_1 \{gH_{st} / (ql^2)\}^{1/2} \tag{9.1-25}$$

attains a maximal value. Here H_{st} and g denote the cable force under static loading q and acceleration of gravity, respectively. This corresponds to the symmetric in-plane eigenmode

$$X_1(x) = C_1 \{1 - \cos a_1 (2x/l) - \tan a_1 \sin a_1 (2x/l)\} , \tag{9.1-26}$$

where C_1 is a constant and a_1 can be determined from the following transcendental equation

$$\tan a_1 - a_1 + a_1^3 \{H_{st}/(16 EA \eta^2)\} = 0 . \tag{9.1-27}$$

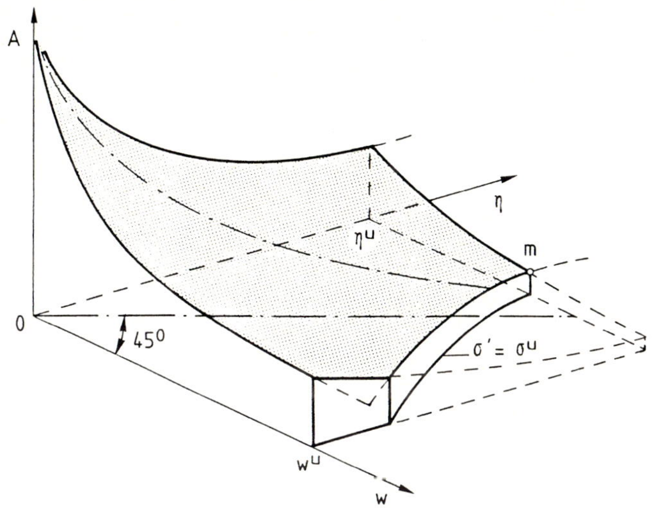

Fig. 9.1/5. Weight of cable versus sag (f) and displacement (w)

It is assumed here that the catenary can be replaced by a parabolic curve of the second order

$$y = 4f(x/l - x^2/l^2) \tag{9.1-28}$$

because of the flatness. In this case the approximate length of the cable is

$$s = l(1 + 8\eta^2/3) , \tag{9.1-29}$$

and γ is the bulk density of the cable material.

The bicriteria optimization problem given in (1-2) can now be expressed as

$$\min_{\eta, A} \ (W(\eta, A), \ \omega_1(\eta, A))^T \tag{9.1-30}$$

subject to the following system of inequality and equality constraints:

a) *side constraints*

$$0.01 \le \eta \le 0.1 , \qquad A > 0 ; \tag{9.1-31}$$

b) *static governing equation*

$$\eta = \{(ql/H_{st})^2 - 24 H_{st}/(EA)\}^{1/2}/8 ; \tag{9.1-32}$$

c) *stress constraint*

$$\sigma_{max} = (H_{st} + H_d) / (A \cos a_{max})$$
$$= (H_{st} + H_d)(1 + 16\eta^2) / A \le \sigma^u \,, \qquad (9.1\text{-}33)$$

where $H_d(t) = (EA/l) \int_0^l (\partial w / \partial x)\, y'(x)\, dx$, $y' = dy/dx$,

with $w(x,t) = \Phi(t)\, X_1(x)$;

d) *dynamic displacement constraint*

$$w_{max} = \Phi(t)\, X_1(x) \le w^u \,, \qquad (9.1\text{-}34)$$

where $\Phi(t)$ is determined by the dynamic magnification factor based on the steady state response only. The last two constraints arise from the dynamic response and can be calculated on the basis of a dynamic loading represented by

$$p(x,t) = p_0 \sin \bar{\omega} t \,. \qquad (9.1\text{-}35)$$

Substituting (9.1-32) by (9.1-24) gives the following cable weight function

$$W = \gamma A l \left\{ 1 + (q l / H_{st})^2 / 24 - H_{st} / (EA) \right\} . \qquad (9.1\text{-}36)$$

4) Solution of the optimization problem

In order to solve the bicriteria optimization problem it is necessary to determine the set of feasible solutions in the design space (A, η) and the set of functional-efficient solutions in the criteria space (ω_1, W). The sets of feasible and functional-efficient solutions should satisfy the constraints (9.1-31) to (9.1-34) as given above. In order to find the sets of feasible and efficient solutions, two problem formulations have been checked. In the first formulation we maximize the first natural frequency of free vibrations for a given cable weight $W = \text{constant}$ and take the constraints (9.1-31) to (9.1-34) into account. The second one deals with the minimization of the cable weight for a given frequency considering the same group of constraints (9.1-31) to (9.1-34). Both formulations give the same sets of feasible and functional-efficient solutions as shown in Figs. 9.1/6 and 9.1/7.

The optimization problems were solved by two different methods. The first one uses the method of Lagrange multipliers and requires the gradients of objective functions and constraints. The gradients were

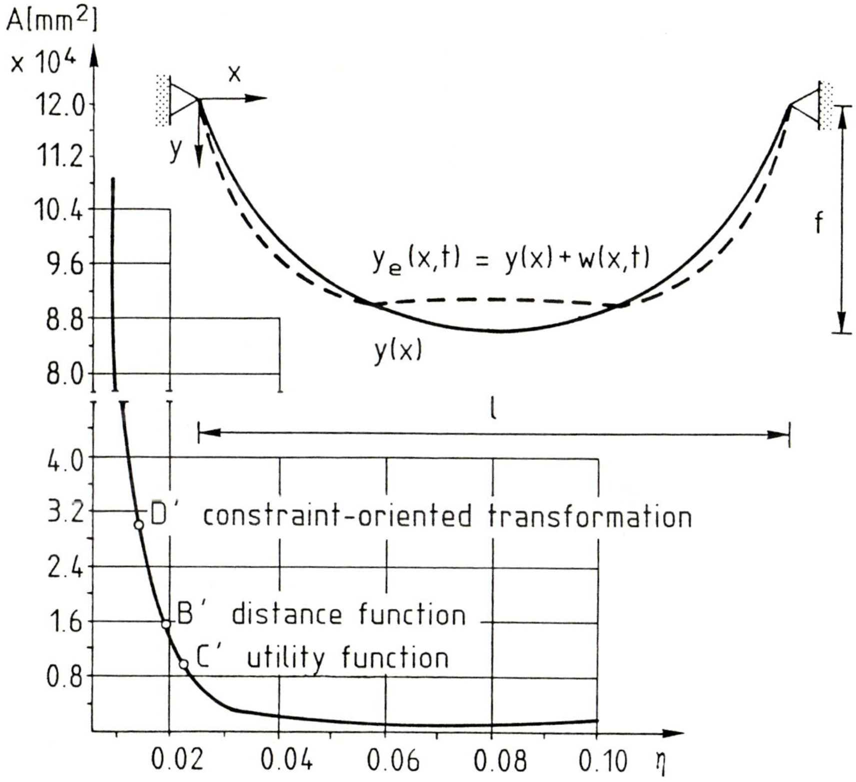

Fig. 9.1/6. Set of functional-efficient solutions for the cable

evaluated numerically. The second algorithm is based on the Rechenberg's evolution strategy which works with random numbers and can by itself determine the improvements of the objective function. This algorithm does not require any evaluation of gradients.

5) Choosing a preferable solution

The set of Pareto-optimal solutions shown in Fig. 9.1/7 contains a number of solutions one of which is to be chosen. For choosing a suitable solution from the set of efficient solutions some methods are available (see section 1.3.2).

Using the distance function to find a preferable solution we have $m=2$ and we choose $r=2$. The following data have been taken in the numerical example: $q=1$ kN/m, $E=200$ kN/mm^2, $\sigma^u=1.2$ kN/mm^2, $p=0.001$ kN/m, $\overline{\omega}=1$ rad/sec, $\delta=0.06$, where δ is the logarithmic decrement of damping. The objective functions are $f_1(x)=W(\eta, A)$ and $f_2(x)=\omega_1(\eta, A)$. The ideal point (solution) satisfying the constraints (9.1-31) to (9.1-34) was found

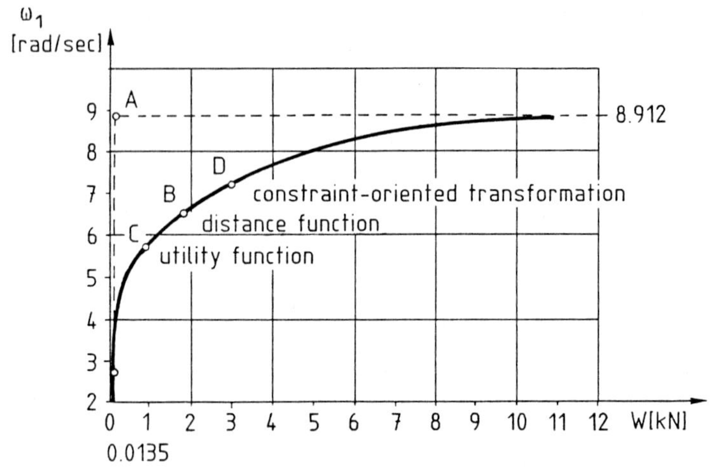

Fig. 9.1/7. Set of functional-efficient solutions for the cable

numerically and is shown in Fig. 9.1/7 as the point A with the coordinates $\overline{W}=0.0135$ kN, $\overline{\omega}_1=8.91$ rad/sec. It does not belong to the set of functional-efficient solutions. The suitable solution was found numerically by minimizing the following distance function

$$p\{W,\omega_1\} = \{(W-\overline{W})^2 - \mu^2(\omega_1-\overline{\omega}_1)^2\}^{1/2}, \tag{9.1-37}$$

where $\mu=1$ kN-sec/rad. The preferable solution obtained by this distance function is shown as point B($\tilde{W}^*=1.60$ kN, $\tilde{\omega}_1^*=6.46$ rad/sec) in Fig. 9.1/7 corresponding to the point B'($\eta=0.0185$, $A=15990$ mm^2) in Fig. 9.1/6.

Next, the following solution is determined by using the utility function method (objective weighting) with the weighting factors $w_1=0.5$ and $w_2=0.5$. In order to find the preferable solution by this method, we maximize the utility function

$$U(f) = \sum_{i=1}^{2} w_i f_i(\mathbf{x}) = -\{w_1 W(\eta,A) - w_2\,\omega_1(\eta,A)\} \tag{9.1-38}$$

subject to the constraints (9.1-31) to (9.1-34). The preferable solution found numerically is shown as the point C($\tilde{W}^*=0.97$ kN, $\tilde{\omega}_1^*=5.92$ rad/sec) in Fig. 9.1/7 corresponding to the point C'($\eta=0.023$, $A=9700$ mm^2) in Fig. 9.1/6.

The preferable solution can also be found by employing the method of constraint-oriented transformation (see section 1.3.2c)). As the main objective, the first natural frequency of free vibrations has been chosen, i.e.

$$\max_{\eta,\mathbf{A}} \omega_1 = 2a_1\{g H_{st}/(q\,l^2)\}^{1/2}, \tag{9.1-39}$$

subject to constraints (9.1-31) to (9.1-34) and the additional constraint concerning the cable weight

$$W^l \le W(\eta, A) \le W^u \ . \qquad (9.1\text{-}40)$$

The lower and upper bounds for the cable weight can be established on the basis of the allowable interval for cable sags which was taken as follows:

$$\eta^l \le \eta \le \eta^u \ , \qquad (9.1\text{-}41)$$

i.e. $W^l = W(\eta^u)$ and $W^u = W(\eta^l)$. We have taken $\eta^l = 0.015$ and $\eta^u = 0.1$. The preferable solution is shown as the point $D(W^* = 3.00$ kN, $\omega_1^* = 7.176$ rad/sec) in Fig. 9.1/7 corresponding to the point $D'(\eta = 0.015$, $A = 30000$ mm^2) in Fig. 9.1/6.

9.1.5 Further Applications of Multicriteria Optimization

Stadler [32,34] was the first to apply multicriteria optimization in structural mechanics. He considered the bicriteria optimization problem with weight and stored energy as criteria and obtained analytical solutions for some structural elements, calling the results "natural structural shapes". Leitmann [25] and Stadler [32-35] applied the control theory for solving some multicriteria optimization problems in linear viscoelasticity. Baier [1,2] studied multicriteria optimization of structures from a general point of view, taking weight and energy stored under separate loading conditions as optimization criteria. Truss optimization with weight and displacement criteria were considered by Koski [21, 22, 23], Koski and Silvennoinen [24]. Osyczka [27, 28] and Rao and Hati [29] considered some conflicting objectives in solving some mechanical optimization problems. Eschenauer [6-8] and Sattler [31], among others, applied the multicriteria optimization approach to spatial structures which support radio telescopes, assuming the following criteria: minimum weight and minimum displacements of the radio telescope reflector surface from its initial configuration under selected loading conditions. Also, Eschenauer [9,10] considered the multicriteria optimization of fiber-reinforced plates and disks, choosing weight and deformation as optimization criteria. Gerasimov, Potchman and Skalozub [13] considered multicriteria optimization of bar, plate and shell structures according to minimum weight or cost criteria. Carmichael [5] solved multicriteria optimization problems using the method of constraint-oriented transformations. Jendo and Marks [18,19] applied multicriteria optimization for solving some civil engineering structures like flexural elements and truss structures using weight and displacements or potential energy as optimization criteria. Jendo, Marks and Thierauf [20] considered multicriteria optimization of an isostatic truss using dynamic programming. Jendo [14-16] considered multicriteria optimization of single-layer cable systems choosing minimum weight and maximal lowest frequency of free vibrations as optimization criteria. Some aspects of these problems, including dynamic plasticity of cable systems, were also considered by Eschenauer and Jendo [11]. Eschenauer, Kneppe and Stenvers [12] compared deterministic and stochastic multicriteria optimization of beams and shell structures. Watkins and Morris [36] applied multicriteria optimization to laminated composites.

References

[1] Baier, H.: Über Algorithmen zur Ermittlung und Charakterisierung Pareto-optimaler Lösungen bei Entwurfsaufgaben elastischer Tragwerke. ZAMM, 57 (1977) 318-320

[2] Baier, H.: Mathematische Programmierung zur Optimierung von Tragwerken insbesondere bei mehrfachen Zielen. Dissertation D17. Darmstadt, 1978

[3] Brandt, A.M. (ed.): Criteria and Methods of Structural Optimization. The Hague, Boston, Lancaster: PWN-Polish Scientific Publishers (Warszawa) and Martinus Nijhoff Publishers 1984

[4] Brandt, A.M. (ed.): Foundations of Optimum Design in Civil Engineering. Dordrecht, Boston. Lancaster: PWN-Polish Scientific Publishers (Warszawa) and Murtinus Nijhoff Publishers 1989

[5] Carmichael, D.G.: Computation of Pareto Optima in Structural Design. Int. J. Num. Meth. Eng. 15 (1980) 925-934

[6] Eschenauer, H.: Eine Vektoroptimierungsaufgabe der Strukturmechanik. Karl-Marguerre-Gedächtniskolloquium. THD Schriftenreihe Wissenschaft und Technik Bd. 16 (ISBN 0172-2085) (1980) 89-101

[7] Eschenauer, H.: Anwendung der Vektoroptimierung bei räumlichen Tragstrukturen. Der Stahlbau 4 (1981) 110-115

[8] Eschenauer, H.: Vector Optimization in Structural Design and its Application on Antenna. Proc. of Euromech-Colloquium 164 on Optimization Methods in Structural Design. In: Eschenauer, H.; Olhoff, N. (eds.): Mannheim: Bibl. (1983) 146-155

[9] Eschenauer, H.: Optimierung ebener Flächentragwerke aus Verbundwerkstoff. Z. Flugwiss. Weltraumforsch. Vol. 8, 6, (1984) 367-378

[10] Eschenauer, H.: Numerical and Experimental Investigations on Structural Optimization of Engineering Design. A Research Project of the German Research Association (GRA) 1980-1984. Research Laboratory for Applied Structural Optimization. Institute of Mechanics and Control Engineering. University of Siegen, 1986

[11] Eschenauer, H.: Jendo, S.: Multiobjective Optimization of Cable Structures. ZAMM 66, 5 (1986) T342-343

[12] Eschenauer, H.: Kneppe, G.; Stenvers, K.H.: Deterministic and Stochastic Multiobjective Optimization of Beam and Shell Structures. J. Mech. Transm. and Autom. in Design. Vol 108 (1986) 31-37

[13] Gerasimov, E.N.; Potchman, Ju.M.; Skalozub, V.V.: Multicriteria Structural Optimization (in Russian) Vistcha Schola. Kiev 1985

[14] Jendo, S.: Some Problems of Multicriteria Optimization of Cable Structures. IFTR Reports 43. Warsaw, 1985

[15] Jendo, S.: Multicriteria Optimization of Cable Systems. Proc. of the IASS Symp. on Membrane Structures and Space Frames. Sept. 1986. Osaka, Japan, Elsevier. Amsterdam. Vol 2 (1986) 71-78

[16] Jendo, S.: Multicriteria Optimization of Single-layer Cable Systems. Arch. of Mech. 38, 3 (1986) 219-234

[17] Jendo, S.; Marks, W.: Multiobjective Structural Optimization. In: Lecture Notes in Control and Information Science. Vol. 84 (1986) 365-374. In: Thoma, M.; Wyner, A.: System Modelling and Optimization. Proc. of the 12th IFIP Conference. Budapest, Hungary Sept. 1986 ed. by Prekopa, A.; Szelezsan, J.; Strazicky, B.; Berlin: Springer Verlag

[18] Jendo, S.; Marks, W.: Formulation of Structural Optimization Problems (in Polish). Rozp. Inzyn. (Eng. Trans.) 34, 4 (1986) 457-481

[19] Jendo, S.; Marks, W.: Multicriteria Optimization in Civil Engineering Design. Proc. of the Int. Conf. on Optimization Techniques and Applications. Singapore: (1987) 998-1006

[20] Jendo, S.; Marks, W.; Thierauf, G.: Multicriteria Optimization in Optimum Structural Design. In: Sage, A.P. (ed.): Large Scale Systems. North Holland: Elsevier Science Publishers B.V. 9 (1985) 141-150

[21] Koski, J.: Truss Optimization with Vector Criterion. Tampere University of Technology. Tampere, Finland. Publications No. 6, 1979

[22] Koski, J.: Multicriteria Optimization in Optimum Structural Design. In: Atrek, E. et al. (eds.): New Directions in Optimum Structural Design. Chichester: John Wiley & Sons (1984) 483-503

[23] Koski, J.: Bicriterion Optimum Design Method for Elastic Trusses. Acta Polytechnica Scandinavica. Mech. Eng. Series No. 86. Helsinki, 1984

[24] Koski, J.; Silvennoinen, R.: Pareto Optima of Isostatic Trusses. Comp. Meth. Appl. Mech. Eng. 31, 3 (1982) 265-279

[25] Leitman, G.: Some Problems of Scalar and Vector-Valued Optimization in Linear Viscoelasticity. JOTA 23, 1 (1977) 93-99

[26] Marks, W. Optimum Design of Prestressed Bending Elements (in Polish), IFTR Reports 52. Warsaw, 1978

[27] Osyczka, A.: An Approach to Multicriterion Optimization Problems for Engineering Design. Comput. Meth. Appl. Mech. Eng. 15 (1978) 309-333

[28] Osyczka, A.: Multicriterion Optimization in Engineering. Chichester, England: Ellis Horwood Publishers 1984

[29] Rao, S.S.; Hati, S.K.: Game Theory Approach in Multi-Criteria Optimization of Function Generating Mechanisms. Trans. ASME, J. of Mech. Eng. 101 (1979) 398-405

[30] Save, M.; Warner, W.: (eds.): Structural Optimization. Vol. 2: Mathematical Programming. New York, London: Plenum Press 1989

[31] Sattler, H.J.: Ersatzprobleme für Vektoroptimierungsaufgaben und ihre Anwendung in der Strukturmechanik. Düsseldorf: VDI-Verlag, Series 1, No. 88, 1982

[32] Stadler, W.: Preference Optimality and Applications of Pareto Optimality. In: Leitmann, G.; Marzollo, A. (eds.): Multicriteria Decision Making. CISM Courses and Lectures No. 211. Wien, New York: Springer-Verlag 1975

[33] Stadler, W.: Natural Structural Shapes of Shallow Arches. J. Appl. Mech. Trans. ASME 44, 2 (1977) 291-298

[34] Stadler, W.: Natural Structural Shapes (the static case). Quart. J. Mech. Appl. Math. 31, 2 (1978) 196-217

[35] Stadler, W.: Preference Optimality in Multicriteria Control and Programming Problems. Nonlinear Anal. Theory, Methods and Applications 4, No. 1 (1980) 51-65

[36] Watkins, R.I.; Morris, A.J.: A Multicriteria Objective Function Optimization Scheme for Laminated Composites for Use in Multi-level Structural Optimization Schemes. Comp. Meths. in Appl. Mech. Eng. 60 (1987) 233-251

9.2 Design Decision Making Using Pareto-Optimal Dynamic Programming

J.S. Gero, M.A. Rosenman

9.2.1 The Stage-State Model of Optimization

The model

When designing using the systems approach, the given system is decomposed into a number of subsystems, and for each subsystem a set of feasible alternatives is selected by the designer. This system may be modelled as a stage-state model where the stages are the decision variables representing the subsystems, and the states are the feasible values of the decision variables representing the feasible alternatives to the subsystems. The optimization problem is to select a state at each stage so as to optimize a given objective function. There will exist relationships between states in one stage and states in other stages not necessarily adjacent (see Fig. 9.2/1). The notations

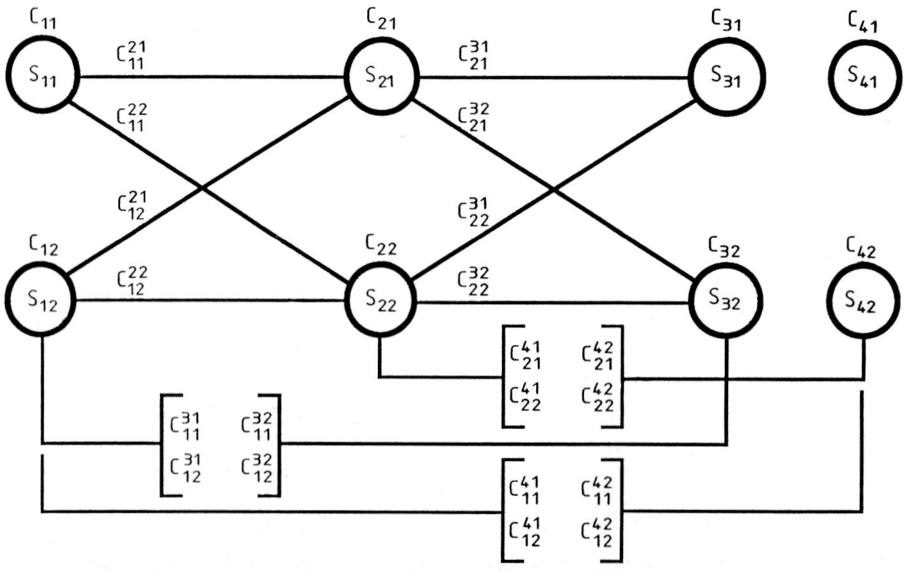

Fig. 9.2/1. Typical four-stage state model with nonlinear relationships

of Fig. 9.2/1 read as follows:

S_{nm} - the m-th state of the n-th stage,
C_{nm} - the "cost" associated with the m-th state of the n-th stage,
C_{ki}^{lj} - the "cost" associated with the arc between the i-th state, in the k-th stage and the j-th state in the l-th stage.

Single-criterion objective functions

A single-criterion objective function aggregates a set of objectives into a single vector such that values obtained may be ranked along the vector. Thus, the "best" solution is the one that ranks highest (maximization) or lowest (minimization), as the case may be. Where such an approach can be used, it greatly simplifies the procedure. For example, many problems can be reduced to one of cost minimization while other requirements are specified as constraints.

Multicriteria objective functions

Many design problems, however, cannot be reduced to a single-criterion objective. For example, in road design the problem may be to minimize costs while minimizing the impact to the environment (maximizing light input while minimizing heat flow through an exterior wall) [5]. If the various objectives conflict, then a multiobjective problem exists. Multiobjective functions may be multiattribute or multicriteria. In multiattribute functions the various attributes are commensurate and may be aggregated into a single function if desired, e.g. capital, overhead and labour costs may be aggregated into the single criterion, total costs whereas in multicriteria functions the various criteria are noncommensurate and cannot be aggregated.

There are basically two problems associated with multiobjective optimization problems. Firstly, the generation of the set of Pareto-optimal solutions or an approximation of it and secondly, the selection of a preferred solution. There are many techniques available, based on preference and nonpreference methods [2,6] for generating the set of Pareto-optimal solutions (see Chapter 1). The most commonly used method is the weighting method in which relative weights are assigned to each of the objectives and the objective function is aggregated into a single criterion. With one set of weights this method can be used to generate some of the Pareto solutions. A change in the relative weights is necessary to generate other solutions. However, arriving at the values of the weights which will produce a new set of Pareto solutions is not obvious. The NISE-method will compute all the weights to generate the entire convex part of the Pareto set, but it is suitable only for problems of up to three dimensions [1].

There is no accepted analytical method for selecting a preferred solution. The weighting method (and other implicit preference methods) may be used, but all these methods are based on a subjective selection of preferences which is usually meaningless. The preferred method is to present to designers a reduced set of Pareto solutions and let them make the final decision based on subjective tradeoffs. Reducing the Pareto set may be done using inverse goal programming [5] or by using clustering techniques [9].

9.2.2 Dynamic Programming

Conventional Dynamic Programming

Conventional Dynamic Programming (CDP) [8] is a stage-state optimization method. It is suitable for design problems for the following reasons:

- it can handle discrete and discontinuous variables as well as continuous functions,
- it can handle any type of objective function which is decomposable. It is especially useful when the objective function is irregular and given in a tabular form,
- it can handle constraints given explicitly or implicitly in an efficient manner,
- for a single-criterion problem it will always produce the global optimum solution,
- at one pass it will produce many near optimal solutions; thus, aiding in postoptimality analyses and helping designers in making a final selection.

Nonserial dynamic programming

While CDP is a powerful optimization method it relies on the problem being able to be formulated serially. However, in general, design problems are nonserial in nature. Therefore, a nonserial dynamic pro-

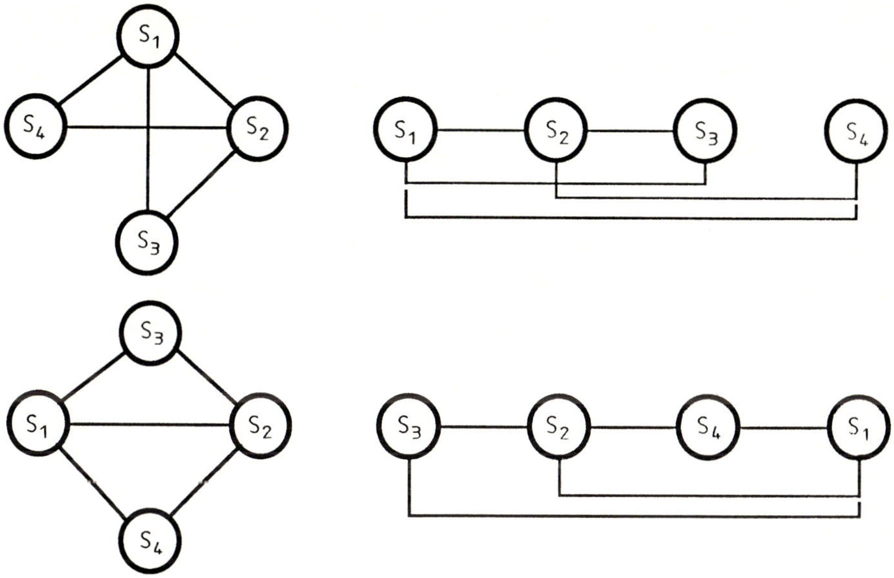

Fig. 9.2/2. Polygonal and linear graph theoretic representations of
a) Connections for a four-stage problem
b) Connections for a four-stage problem after reordering of stages to minimize nonseriality

gramming method using heuristics was developed to cope with realistic problems [7]. While the computational and storage requirements are larger than those for CDP, a large number of solutions is available for examination.

The stages are entered in an arbitrary order and, thus, a serial problem could be entered to appear nonserial. The methodology includes a procedure to reorder the stages to minimize the nonseriality. Fig. 9.2/2 shows the effect of applying this procedure to four-stage models.

Once the stages are ordered the optimization procedure may be carried out in a forward and/or reverse direction. Combining the results from both passes improves the accuracy of the solutions.

Multiobjective dynamic programming

In single-criterion dynamic programming, for each state at each stage an optimization is carried out and the optimal decision for that state selected. In multiobjective dynamic programming, the single-criterion optimization is replaced by a Pareto optimization, and the set of Pareto-optimal decisions for that state is selected. The weighting method may be used with some given set of weights to generate the Pareto suboptimal solutions. If the number of final Pareto solutions obtained is found to be insufficient, then further passes can be carried out with changed weight relationships, noting the problems discussed previously. Usually, however, a large enough number of solutions is obtained in one pass. Another possibility is to do without any preference function at all and to keep the various objectives disaggregated [8]. However, at present this technique has only been applied to serial dynamic progamming.

At any stage, the number of Pareto solutions may be too large for the program to handle. Similarly, the number of Pareto solutions available to designers at the final stage may be too large for them to evaluate. A method for reducing the number of Pareto solutions in a set, without stating any preference or ranking of the objectives, and applicable to dynamic programming is described in [9].

Postoptimality analysis

While the optimization procedure identifies the set of optimal solutions, designers may desire more information regarding the stability of the design decisions or information regarding other policies. Designers may wish to:

- carry out stability analyses of the design decision by determining the range within which the data associated with the alternatives and/or their connections may vary without the policy becoming suboptimal,
- select or "fix" particular alternatives and compare the results with the optimal solutions,
- if the weighting method is used, to carry out analyses on the bounds of the weights to determine which combination of weights causes other solutions to be the top ranked.

The large number of solutions generated by the dynamic programming procedure is used in carrying out these analyses with little additional increase in computation [11].

9.2.3 The Computer System - SID

SID (**S**ystem for **I**ntegrated **D**esign) is a computer system designed to:

- take in a decomposed stage-state representation of the system to be designed,
- take in a description of the performance objectives to be optimized,
- take in information regarding the performance of the various alternatives (states) and the relationship between them with respect to the various objectives,
- carry out an optimization and present an ordered set of optimal and nearly optimal solutions for single-criterion problems and an approximation to the Pareto set in the case of multicriteria problems using the weighting method; and
- carry out postoptimality analyses.

SID uses a command language for user interaction. Data may be input interactively or through a command file. Optimization is carried out using the heuristic nonserial dynamic programming algorithm with the weighting method. Either or both forward and reverse passes may be carried out. The number of decisions to be kept at each stage, the number of solutions to be ordered, and the number of solutions to be printed out are controlled by the user. Three types of postoptimality analyses may be carried out, namely: subsystem sensitivity analyses, connection sensitivity analyses, and analysis involving the "fixing" of alternatives.

A more complete description of SID is given in [10].

9.2.4 A Building Design Example Using SID

The problem

For a given building plan, various building components may be selected, thus generating the building attributes. Though many of the solutions will be feasible, we are only interested in the "better" ones, i.e. better in relation to a specified objective function. A building may be decomposed into subsystems in many ways, and the following example only illustrates one view applicable to a particular problem.

1. STRUCTURAL CROSS WALLS
1. In-situ RC
2. precast concrete panels
3. precast PC planks
4. In-situ concrete frame and concrete block infill
5. In-situ concrete frame and plasterboard infill
6. tunnel construction
7. steel STS and plasterboard infill
8. precast PC wall beam as STS

2. STRUCTURAL CROSS WALL FINISH
1. paint
2. render and paint
3. wallpaper
4. render and wallpaper

3. STRUCTURAL CROSS WALL ELECTRICAL
1. prelaid in precast elements
2. on-site distribution

4. NON-STRUCTURAL CROSS WALLS
1. none
2. concrete blocks
3. 1 1/2 hours fire-rated plasterboard

5. NON-STRUCTURAL CROSS WALL FINISH
1. none
2. paint
3. render and paint
4. wallpaper
5. render and wallpaper
6. sprayed texture finish

6. NON-STRUCTURAL CROSS WALL ELECTRICAL
1. none
2. on-site distribution

7. FLOOR
1. none
2. In-situ RC
3. metal deck and concrete topping
4. precast concrete floor panels
5. precast PC planks
6. precast concrete permanent formwork and concrete topping
7. patent beam and lightweight block infill and concrete topping

8. FLOOR ELECTRICAL
1. prelaid in precast elements
2. on-site distribution in structure
3. on-site distribution exposed

9. ROOM CEILING FINISH
1. paint
2. plaster and paint
3. sprayed vermiculite
4. plasterboard and paint
5. patent suspended

10. FACADE ELEMENT
1. precast concrete full-width balcony unit
2. precast concrete cupboard and half-width balcony unit
3. precast concrete sill unit
4. In-situ concrete full-width balcony unit
5. In-situ concrete cupboard and half-width balcony unit
6. In-situ concrete sill unit

11. GLAZED ELEMENT
1. full-width window wall
2. sliding doors only within precast element
3. sliding doors only on-site installation
4. full-width window within precast element
5. full-width window on-site installation

12. HVAC
1. perimeter ducting HV
2. perimeter ducting horizontal distribution
3. perimeter ducting LV horizontal distribution
4. corridor supply only

13. INTERNAL PARTITIONS
1. precast concrete wall panels
2. In-situ RC
3. precast PC planks
4. concrete block 100 mm
5. plasterboard wall construction

14. INTERNAL PARTITION FINISH
1. paint
2. render and paint
3. wallpaper
4. render and wallpaper
5. sprayed texture finish

15. INTERNAL PARTITION ELECTRICAL
1. prelaid in precast elements
2. on-site distribution

16. CORRIDOR PARTITION
1. precast concrete wall
2. In-situ RC
3. precast PC planks
4. concrete blocks 200 mm
5. 1 1/2 hours fire-rated plasterboard

17. CORRIDOR PARTITION FINISH
1. paint
2. render and paint
3. wallpaper
4. render and wallpaper
5. sprayed texture finish

18. DUCTWORK
1. prefabricated concrete duct and all plumbing
2. prefabricated steel frame and all plumbing
3. hole in floor and concrete block wall
4. hole in floor and precast concrete panel
5. hole in floor and precast plank
6. hole in floor fibre-cement

19. PLUMBING
1. none
2. conventional
3. prefabricated 'tree' and temp frame on-site erection

20. HALL AND CORRIDOR CEILING FINISH
1. paint
2. plaster and paint
3. sprayed vermiculite
4. plasterboard and paint
5. suspended plasterboard and paint
6. patent suspended

Fig. 9.2/3. Subsystems and alternatives for hotel unit

It is preferable that the scale of the subsystems, with respect to the objective function, is of the same order as otherwise one subsystem may have virtually no practical impact on the final solution no matter what alternative is chosen.

The example is that of a medium-size hotel. There are some major decisions to be made, namely: the method of construction (prefabricated, in-situ or a mixture), the standard of finish, and the provision of a balcony. There is also a fundamental decision to be made about the type of structure (cross walls, frame, staggered trusses or wall beams). The objective function is a multiattribute one in which total costs are a sum of the material, labour and overhead costs. There are costs associated with each component and costs associated with using a component in combination with other components.

The building is decomposed into 20 subsystems (decision variables) each of which has several alternatives (Fig. 9.2/3). There are approximately 5×10^{12} possible solutions, many of which are infeasible. The relation between the subsystems is shown in Fig. 9.2/4. The major task is to gather the cost information. Since we are seeking a comparison of design options, it does not matter if the cost data are outdated as long as the correct relativity is maintained.

Optimization procedure and results

A weighting of unity was selected for each of the attributes. Both a forward and a reverse pass optimization were carried out with 40 decisions kept at each stage, and the first 200 solutions based on the aggregate costs were ordered. Fig. 9.2/5 shows the optimal solution based on the aggregate cost.

This solution has a total cost of \$ 13.511, with material costs of \$ 9.053, labour costs \$ 3628, and overhead costs of \$ 830. The design decisions leading to this solution include precast concrete panels for cross walls and floors, rendered and painted wall finishes and steel-framed prefabricated plumbing units. The complete set of results included a listing of five solutions from the forward pass, five solutions from the reverse pass, and the first five solutions from the combined forward and reverse pass as well as a table and listing of the 58 Pareto-optimal solutions obtained. Fig. 9.2/6 shows the list of Pareto-optimal solutions while Fig. 9.2/7 shows Pareto-optimal solution No. 53 which is the solution of the lowest labour costs.

	1	2	3	4	5	6	7	8	9	10	11	12	13	14	15	16	17	18	19	20
1. Structural cross walls	•	■	■	■			■			■			■			■		■		
2. Structural cross walls finish		•																		
3. Struct. cross walls elect.work			•			■		■							■					
4. Non-struct. cross walls				•	■	■							■			■				
5. Non-struct. cross walls finish					•															
6. Non-struct. cross wall elect.work						•		■							■					
7. Floor							•	■	■	■			■			■		■		■
8. Floor electrical work								•	■						■					■
9. Room ceiling finish									•											
10. Facade element										•	■	■								
11. Glazed element											•	■								
12. HVAC												•								■
13. Internal partitions.													•	■	■	■				
14. Internal partition finish														•						
15. Internal partition elect.work															•					
16. Corridor partition.																•	■			
17. Corridor finish																	•			
18. Duct work																		•	■	
19. Plumbing																			•	
20. Hall & corridor ceiling finish																				•

Fig. 9.2/4. Binary association matrix for subsystems of hotel unit

COMBINED FORWARD AND BACKWARD PASS

SOLUTION NO. 1	TOT.COSTS	IN $	=	13511.00
	MAT.COSTS	IN $	=	9.053.00
	LAB.COSTS	IN $	=	3628.00
	OVER.COSTS	IN $	=	830.00

SUBSYSTEM	ALTERNATIVE
STRUCT. CROSS-WALLS	PRECAST CONC. PANELS
STRUCT. WALL FINISH	RENDER & PAINT
STRUCT. ELECTRICAL	PRELAID
NON-STRUCT. CR/WALLS	NONE
CROSS-WALL FINISH	NONE
CR/WALL ELECTRICAL	NONE
FLOOR	PRECAST CONC. PANELS
FLOOR ELECTRICAL	PRELAID
ROOM CEILING FINISH	PAINT ONLY
FACADE ELEMENT	P.C. CUPB/D & BALCONY
GLAZED ELEMENT	DOORS ONLY ON-SITE
HWAC	PERIMETER VERTICAL
INTERNAL PARTITIONS	PRECAST CONC. PANELS
INT. PART. FINISH	RENDER & PAINT
INT. PART. ELECTRICAL	PRELAID
CORRIDOR PARTITION	PLASTERBOARD 1 1/2 HRS
CORRIDOR FINISH	PAINT ONLY
DUCT	STEEL FRAME & PLUMBING
PLUMBING	NONE
HALL & CORRID. CEIL.	SPRAYED VERMICULITE

Fig. 9.2/5. Lowest aggregate cost solution

NONINF SOLN NO	ORDERED SOLN NO	TOT.COSTS $	MAT.COSTS $	LAB.COSTS $	OVER.COSTS $
1	1	13511.00	9053.00	3628.00	830.00
2	2	13516.00	9080.00	3666.00	770.00
3	3	13521.00	9033.00	3608.00	880.00
4	4	13526.00	9060.00	3646.00	820.00
5	6	13536.00	8998.00	3688.00	850.00
6	8	13541.00	9025.00	3726.00	790.00
7	10	13546.00	8978.00	3668.00	900.00
8	12	13551.00	9005.00	3706.00	840.00
9	13	13551.00	8963.00	3708.00	880.00
10	14	13551.00	8868.00	3773.00	910.00
11	15	13556.00	8990.00	3746.00	820.00
12	16	13556.00	8923.00	3723.00	910.00
13	18	13561.00	8943.00	3688.00	930.00
14	19	13561.00	9013.00	3643.00	905.00
15	20	13561.00	8848.00	3753.00	960.00
16	22	13566.00	8970.00	3726.00	870.00
17	23	13566.00	9040.00	3681.00	845.00
18	25	13566.00	8903.00	3703.00	960.00
19	28	13571.00	8703.00	3858.00	1010.00
20	32	13576.00	8908.00	3768.00	900.00
21	35	13581.00	8935.00	3806.00	840.00
22	37	13581.00	8683.00	3838.00	1060.00
23	40	13586.00	8888.00	3748.00	950.00
24	44	13591.00	8915.00	3786.00	890.00
25	45	13591.00	8778.00	3853.00	960.00
26	48	13596.00	8888.00	3713.00	995.00
27	49	13596.00	8833.00	3803.00	960.00
28	51	13596.00	9210.00	3646.00	740.00
29	54	13601.00	8915.00	3751.00	935.00
30	56	13601.00	8758.00	3833.00	1010.00
31	61	13606.00	8950.00	3761.00	895.00
32	63	13606.00	8888.00	3703.00	1015.00
33	64	13606.00	8813.00	3783.00	1010.00
34	66	13606.00	8883.00	3738.00	985.00
35	67	13606.00	9190.00	3626.00	790.00
36	71	13611.00	8915.00	3741.00	955.00
37	73	13611.00	8613.00	3938.00	1060.00
38	86	13621.00	8593.00	3918.00	1110.00
39	87	13621.00	8663.00	3873.00	1085.00
40	88	13621.00	8758.00	3808.00	1055.00
41	89	13621.00	9155.00	3706.00	760.00
42	101	13631.00	8643.00	3853.00	1135.00
43	102	13631.00	8803.00	3848.00	980.00
44	103	13631.00	8758.00	3798.00	1075.00
45	108	13636.00	8798.00	3793.00	1045.00
46	110	13636.00	8538.00	3943.00	1155.00
47	117	13641.00	8825.00	3831.00	985.00
48	118	13641.00	9208.00	3573.00	860.00
49	125	13646.00	9235.00	3611.00	800.00
50	126	13646.00	8798.00	3783.00	1065.00
51	129	13646.00	9170.00	3661.00	815.00
52	133	13651.00	8825.00	3821.00	1005.00
53	134	13651.00	9188.00	3553.00	910.00
54	141	13656.00	9215.00	3591.00	850.00
55	145	13656.00	8860.00	3876.00	920.00
56	166	13671.00	8138.00	4158.00	1375.00
57	179	13681.00	8118.00	4138.00	1425.00
58	200	13758.00	9288.00	3758.00	712.00

Fig. 9.2/6. List of Pareto solutions

NONINF SOLUTION NO. 53	TOT.COSTS	IN $	=	13651.00
	MAT.COSTS	IN $	=	9.188.00
	LAB.COSTS	IN $	=	3553.00
	OVER.COSTS	IN $	=	910.00

SUBSYSTEM	ALTERNATIVE
STRUCT. CROSS-WALLS	PRECAST CONC. PANELS
STRUCT. WALL FINISH	RENDER & PAINT
STRUCT. ELECTRICAL	PRELAID
NON-STRUCT. CR/WALLS	NONE
CROSS-WALL FINISH	NONE
CR/WALL ELECTRICAL	NONE
FLOOR	PRECAST CONC. PANELS
FLOOR ELECTRICAL	PRELAID
ROOM CEILING FINISH	PLASTER AND PAINT
FACADE ELEMENT	P.C. CUPB/D & BALCONY
GLAZED ELEMENT	DOORS ONLY ON-SITE
HWAC	PERIMETER VERTICAL
INTERNAL PARTITIONS	PRECAST CONC. PANELS
INT. PART. FINISH	RENDER & PAINT
INT. PART. ELECTRICAL	PRELAID
CORRIDOR PARTITION	PLASTERBOARD 1 1/2 HRS
CORRIDOR FINISH	PAINT ONLY
DUCT	HOLE & FIBRE CEMENT
PLUMBING	TREE & ON-SITE ERECTION
HALL & CORRID. CEIL.	SPRAYED VERMICULITE

Fig. 9.2/7. Pareto solution No. 53 - lowest labour costs

These results show that four main structural systems were listed, namely: precast concrete panels, precast concrete plank units, in-situ reinforced concrete, and steel-staggered truss. Also, there were different alternatives for all the subsystems except for subsystems: facade element, glazed element corridor partitions, and corridor partition finish.

While the results obtained from this optimization procedure are indicative of the set of decisions necessary for a good solution, additional information can be obtained to give aid in making a final selection by carrying out postoptimality analyses.

Postoptimality analyses

As stated previously, sixteen of the twenty subsystems display more than one alternative. The exceptions are the facade element (precast concrete cupboard unit and balcony), the glazed element (doors with on-site installation), the corridor partitions (plasterboard), and the corridor partition finish (paint). The stability of these four solutions suggest that we could accept them with confidence. For the remaining 16 subsystems, we can carry out sensitivity analyses. Fig. 9.2/8 shows the costs associated with switching from the optimal solution to another decision for each of the subsystems.

Subsystem	Difference in Costs of New Solution over Optimal Solution in $	Total Costs Associated with Optimal Alternative in $	Permissible Change (%)
1	40.00	3018.00	1.33
2	5.00	330.00	1.52
3	60.00	455.00	13.19
4	247.00	0.00	—
5	247.00	0.00	—
6	247.00	0.00	—
7	85.00	2615.00	3.25
8	247.00	280.00	8.21
9	10.00	330.00	3.03
12	247.00	2.950.00	8.37
13	25.00	775.00	3.23
14	25.00	240.00	10.42
15	25.00	410.00	6.10
18	50.00	1180.00	4.24
19	130.00	30.00	433.33
20	22.00	100.00	22.00

Fig. 9.2/8. Sensitivity analysis for subsystems drawn from the optimization results

Column 2 shows the difference in the total costs between the optimal solution obtained and the lowest cost solution with a different decision for that subsystem; column 2 shows the total costs attributable to the alternative solution taking into account the interrelationships between the decision; column 4 shows the percentage by which the costs for the optimal decision could increase before another solution becomes optimal. For example, if the costs of precast concrete panels would increase by more than 1.33% relative to the other costs, then the solution containing precast concrete planks will become optimal. On the other hand, the plumbing system costs can increase by 433% before another solution is better. This type of information tells us which cost information need to be accurate and which cost information is not sensitive to changes.

Postoptimality analyses can now be carried out to find other solutions for the subsystems facade element and corridor partitions. These remained stable in all the solutions obtained as precast concrete cupboard and balcony and plasterboard partitions, respectively. The new solution for the facade element is found in the set of 1200 nonordered final solutions obtained in the optimization procedure. This new solution (Fig. 9.2/9) has a new total cost of $ 13.916. In this solution, the new alternative for the facade element is precast concrete sill unit, and in addition the subsystem glazed element contains the new alternative of window with on-site fixing. It can be seen from Fig. 9.2/9 that the costs of the precast cupboard and balcony can increase by up to 38.4% before the alternative design becomes preferable. The new solution for corridor partitions is not found in any of the solution sets, and a new optimization is carried out with the alternative plasterboard set to infinity so that it will not appear. The new solution shows a change to precast concrete planks with a permissible change in total costs for the plasterboard partitions of 118%.

Finally, because we are interested in the tunnel formed system we can carry out another constrained optimization to find a solution with this alternative. This solution has a total cost of $ 13.651 and because of its low material costs is also a Pareto solution.

We have investigated the stability of solutions by looking at the sensitivity of cost information associated with given alternatives and also the costs of "fixed" alternatives. In addition, we wish to carry out analyses on the relativity between the various attributes of the objective function. We want answers to questions such as "what happens if

RESULTS OF POSTOPTIMALITY ANALYSIS

SOLUTIONS FOR SUBSYSTEM 10 NOT FOUND IN ORDERED SET OF 200 SOLUTIONS
DO YOU WANT TO SEARCH ON NON-ORDERED SET OF 1200 SOLUTIONS? (DEFAULTS TO NO-ELSE TYPE "YES OR Y)
?Y

SENSITIVITY ANALYSIS
SUBSYSTEM FACADE ELEMENT

SUBSYSTEM	ALTERNATIVE
STRUCT. CROSS-WALLS	STEEL S.T.S & PLAST/BD
STRUCT. WALL FINISH	PAINT ONLY
STRUCT. ELECTRICAL	ON-SITE DISTRIBUTION
NON-STRUCT. CR/WALLS	PLAST/BD 1 1/2 HRS
CROSS-WALL FINISH	PAINT ONLY
CR/WALL ELECTRICAL	ON-SITE DISTRIBUTION
FLOOR	PRECAST CONC.PANELS
FLOOR ELECTRICAL	ON-SITE EXPOSED
ROOM CEILING FINISH	PLAST/BD & PAINT
FACADE ELEMENT	P.C SILL UNIT
GLAZED ELEMENT	WINDOW ONLY ON-SITE
HVAC	CORRIDOR SUPPLY
INTERNAL PARTITIONS	PLASTERBOARD
INT.PART.FINISH	PAINT ONLY
INT. PART. ELECTRICAL	ON-SITE DISTRIBUTION
CORRIDOR PARTITION	PLASTERBOARD 1 1/2 HRS
CORRIDOR FINISH	PAINT ONLY
DUCT	HOLE & FIBRE CEMENT
PLUMBING	TREE & ON-SITE ERECTION
HALL & CORRID. CEIL.	SUSPEND. PLAST/BD & PAINT

TOT. COSTS OF NEW SOLUTION	IN $ =	13916.00
MAT.COSTS	IN $ =	9328.00
LAB.COSTS	IN $ =	3783.00
OVER.COSTS	IN $ =	805.00
DIFFERENCE BETWEEN OPT. SOLUTION	IN $ =	405.00
TOT. COSTS SUBSYSTEM 10	IN $ =	1055.00
PERMISSIBLE CHANGE IN TOT. COSTS ASSOCIATED WITH P.C CUPB/D & BALCONY	=	38.4%

Fig. 9.2/9. Results of the postoptimality analysis for subsystem facade element

material costs increase in relation to other costs?" or if "labour costs increase relatively?" or if "overhead costs increase relatively (perhaps due to rising interest rates)?" and "by how much can each of these costs rise relatively to the others before other solutions become optimal?" We can plot the set of 59 Pareto-optimal solutions obtained from the optimization and postoptimality procedures onto two-dimensional graphs for each combination of two attributes. Treating total costs as a fourth attribute, this gives us six graphs, three of which are shown in Figs. 9.2/10–9.2/12.

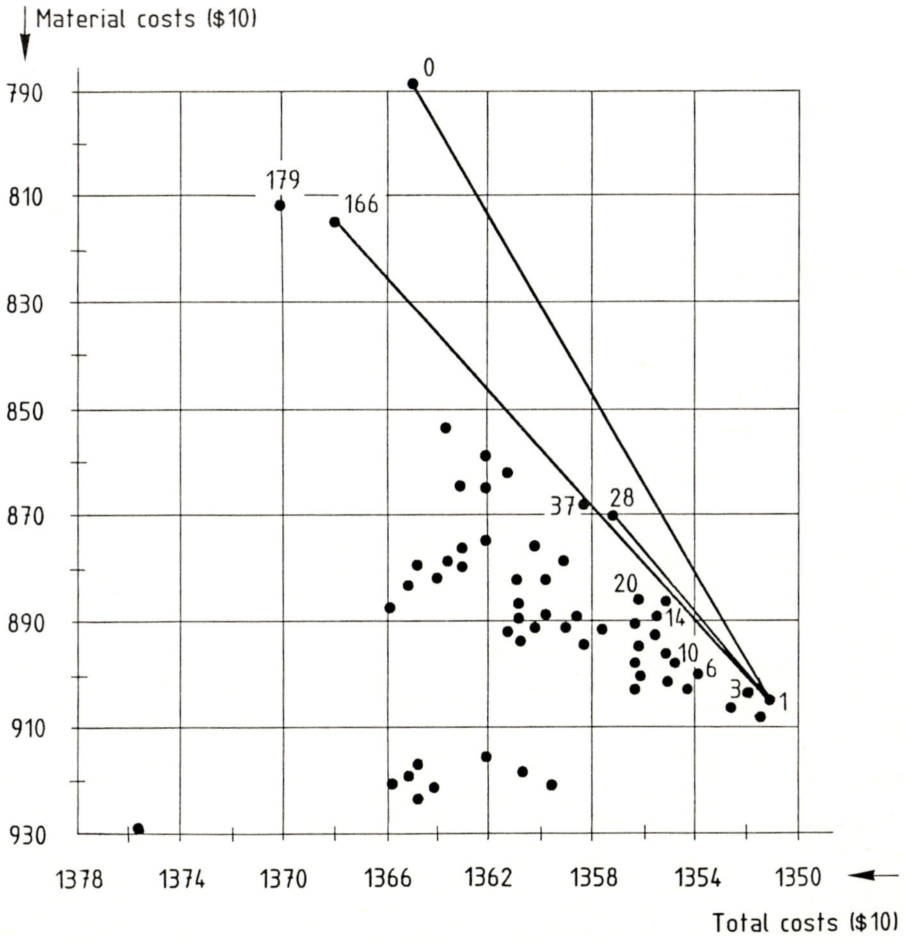

Fig. 9.2/10. Two-dimensional display of Pareto-optimal solutions: total and material costs

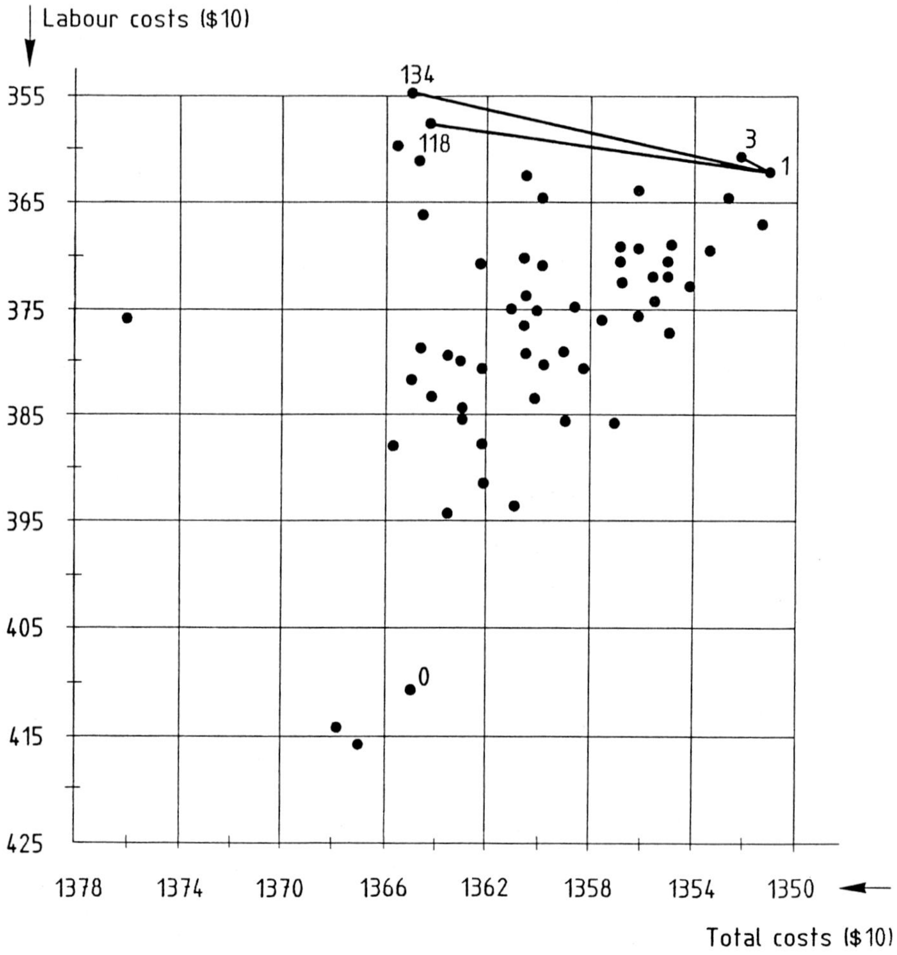

Fig. 9.2/11. Two-dimensional display of Pareto-optimal solutions: total and labour costs

The solutions are identified by their position in the ordered set of 200 solutions. Solution No. 0 refers to the tunnel formed alternative. For clarity's sake only those solutions deemed interesting were labelled. With unit weighting, solution No. 1 has been projected as the preferred solution. It is not, however, the solution with the lowest labour, material or overhead costs.

In Fig. 9.2/10, the graph of material costs against total costs, we can see that if material costs were to rise relative to the other costs,

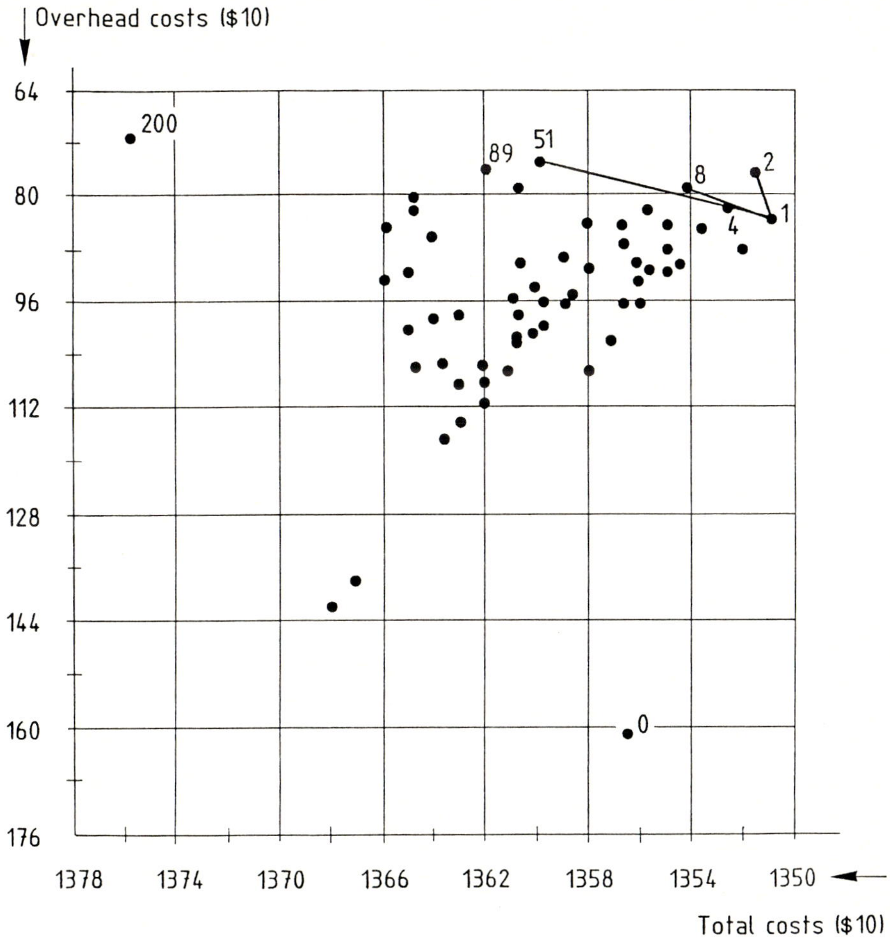

Fig. 9.2/12. Two-dimensional display of Pareto-optimal solutions: total and overhead costs

the first solution to replace solution No.1 as the preferred solution would be solution No. 0. This can be seen by drawing a line joining solutions 1 and 0. This line has a slope greater than that of other lines from solution No. 1 to other solutions (e.g. No. 166). The slope between the various solutions indicates the relative increase required in the weight of an attribute for a given solution to become the preferred solution. Solution No. 0 (tunnel formed structural system) has the lowest material costs and relatively high labour and overhead costs so that given a relative rise in material costs it would be the solution

ATTRIBUTE	SOLUTION NO.				
	0	2	3	51	134
MATERIAL COSTS	1.12	-	-	-	-
LABOUR COSTS	-	-	1.50	-	2.87
OVERHEAD COSTS	-	1.08	-	1.94	-

Fig. 9.2/13. Tabulation of critical weights with respect to solution No. 1

whose costs would increase the most slowly. We can see by the negative slope from solution No. 0 to the other solutions that any further relative increase in material costs would not cause the preferred solution to change from solution 0.

We can analyse Figs. 9.2/11 and 9.2/12 in a similar fashion. If labour costs were to increase relatively, then solution No. 3 would be the first solution to become preferred, followed by solution No. 134. However, because of the shallow slope between solutions 3 and 134, it would take a very large increase in the labour costs for solution No. 134 to be preferred over solution No. 3. If overhead costs were to increase relatively, then solution No. 2 would be the first solution to become preferred followed by the solution No. 51; but again this would have to be a very large increase. We can quantify the results indicated by the slopes of the lines [3] and display them as in Fig. 9.2/13.

9.2.5 Conclusion

While the above example showed an application using cost criteria, this in no way limits the generality of the system. Also, while the technique used for generating the Pareto set was the weighting method, the future versions of the system will use the nonpreference method developed by Rosenman and Gero [9].

The system SID is a tool to be used as an aid to designers in their selection of a design solution. No attempt is made to fix the designers' perception of the design problem as regards their goals, constraints, and their view of the structure of the problem. These considerations are still the designers' responsibilities.

The above building design example demonstrated that sufficient relevant solutions are generated in one pass of the dynamic programming procedure to give a good approximation to the Pareto set, thus offering designers sufficient choice in making a final selection. It was shown that a large number of possible solutions was first reduced to 59 Pareto-optimal solutions from which four solutions, namely Nos. 1,2,3, and 0 emerged as those most likely to be the most preferred solution depending on the particular conditions. The relevant information is displayed in an intelligible manner so that designers can either make a final decision or else perceive what extra information they require.

Much of this example dealt with the postoptimality analysis as the information obtained is as important, if not more important, than the results obtained from the optimization procedure alone. The ability to carry out various types of postoptimality analyses with the minimum of effort gives designers added confidence in their final decision.

References

[1] Balachandran, M; Gero, J.S.: The Noninferior Set Estimation Method for Three Objectives. Engineering Optimization 9(2) (1985) 77-88

[2] Cohon, J.L.: Multi-objective Programming and Planning. New York: Academic Press 1978

[3] Gero, J.S.: Radford, A.D.: Cameron, J.: Postoptimality Analysis for Multiattributive Objective Functions in Dynamic Programming. Engineering Optimization 4 (1979) 65-72

[4] Nemhauser, G.: An Introduction to Dynamic Programming. New York: Wiley 1966

[5] Radford, A.D.; Gero, J.S.: Multi-criteria Optimization in Architectural Design. In: Gero, J.S. (ed.): Design Optimization. New York: Academic Press (1985) 229-258

[6] Radford, A.D.; Gero, J.S.: Design by Optimization in Architecture, Building, and Construction. New York: Van Nostrand Reinhold 1988

[7] Rosenman, M.A.; Gero, J.S.: Heuristic Nonserial Programming for Large Problem. Engineering Optimization 4 (1980) 167-178

[8] Rosenman, M.A.; Gero, J.S.: Pareto-Optimal Serial Dynamic Programming. Engineering Optimization 6(4) (1983) 177-183

[9] Rosenman, M.A.; Gero, J.S.: Reducing the Pareto-Ooptimal Set in Multi-criteria Optimization (with applications to Pareto-optimal dynamic programming). Engineering Optimization 8(3) (1985a) 189-206

[10] Rosenman, M.A.; Gero, J.S.: A System for Integrated Optimal Design. In: Gero, J.S (ed.): Design Optimization. New York: Academic Press (1985b) 259-294

[11] Rosenman, M.A.; Radford, A.D.; Gero, J.S.: Postoptimality Analysis in Dynamic Programming. Engineering Optimization 4 (1980) 207-214

10 STRUCTURES MADE OF ADVANCED MATERIALS

10.1 Fibre-Reinforced Plate and Shell Structures under Various Loads - Proposals for their Optimization -

H. Eschenauer, W. Fuchs, P.U. Post

10.1.1 Introduction

As a result of continuous technical developments in materials and production methods, fibre-composite materials have acquired ever increasing fields of application over the last 20 to 30 years. A survey of ten fields in which fibre composite materials are used technically is presented, among others, in the conference proceedings [1,2]. Here, it becomes evident that fibre-reinforced plastics, due to their special physical, chemical, and mechanical characteristics, are employed not only in aircraft and spacecraft industry but also in civil and mechanical engineering.

The layout of fibre composite structures, compared to that of structures made of conventional homogeneous isotropic materials, is far more difficult because a fibre composite (laminate) consists of several unidirectional layers (UD-layers) arranged at different angles. Thus, it is a heterogeneous, anisotropic material. Fundamentally important publications describing anisotropic supporting structures have been written by Ambartsumyan [3,4] and Lekhnitskii [5]. Further publications can be found in the surveys of Ambartsumyan [6], Bert and Francis [7], and Habip [8].

The optimization of load-bearing structures made of fibre-composite materials started in the seventies following the presentation of accurate knowledge of the material behaviour of fibre-reinforced plastics. In their survey, Haftka and Prasad [9] mention a large number of references in which exclusively fibre composite panels or sandwich panels are optimized with respect to various objectives.

The use of vector optimization for load-bearing structures consisting of fibre composite materials is a topic taken up by very few authors. Adali [10] has investigated an asymmetrically composed panel by simultaneously minimizing the maximal deformation and maximizing the natural frequency and the buckling load. Eschenauer [11,12]. Eschenauer and Fuchs [13,14,15], and Eschenauer and Post [16,17] dealt with load-bearing structures under various loads (time-variant and time-invariant) using multicriteria optimization strategies regarding minimal deformation and minimal weight.

The following chapter provides a contribution to the structural analysis and optimization of fibre-composite load-bearing structures. For this purpose, the basic equations of the bending theory are used. In view of various objectives, the optimization is achieved by using vector optimization techniques, and optimization results for a spherical shell and a rectangular plate are presented. In order to enable the verification of the solutions of structural analysis and the results of optimization, these will be compared with the results of experimental investigations on a spherical shell.

10.1.2 Structural Modelling and Analysis

a) Basic equations of anisotropic composite structures under time-invariant loadings

The following bending theory equations are derived from the basic equations of heterogeneous anisotropic shell structures following [18]. This presupposes that the thickness of the structure t is small in comparison with other geometrical dimensions, and that the middle surface is divided into halves at each point. The structure is described by the Gaussian surface parameter ξ^α of the middle surface and by the coordinate ζ perpendicular to the middle surface. All Latin indices run from 1 to 3 and the Greek indices from 1 to 2. The direction perpendicular to the middle surface has the index 3 (Fig. 10.1/1).

Assuming that the deformations are small. the equilibrium conditions for the undeformed element read as follows:

$$\begin{aligned} N^{\alpha\beta}|_\alpha - Q^\alpha b^\beta_\alpha + p^\beta &= 0 \ , \\ Q^\alpha|_\alpha + N^{\alpha\beta} b_{\alpha\beta} + p &= 0 \ , \\ M^{\alpha\beta}|_\alpha - Q^\beta &= 0 \ , \end{aligned} \qquad (10.1\text{-}1)$$

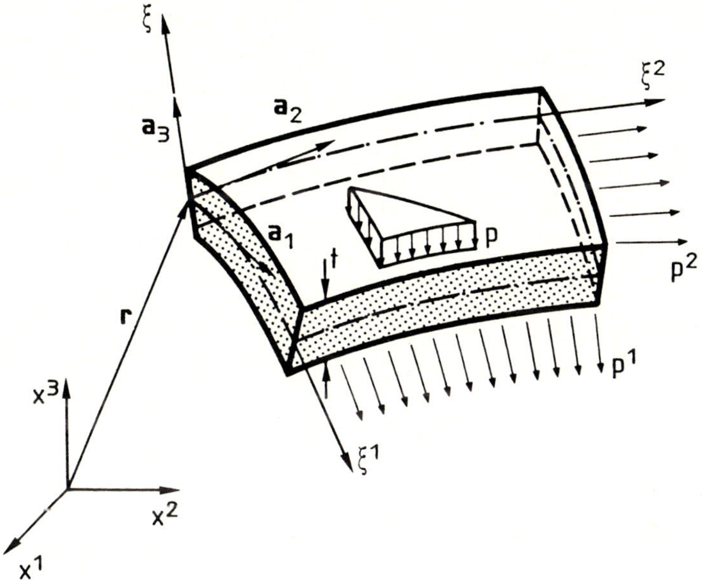

Fig. 10.1/1 Notations of a shell structure

where

$N^{\alpha\beta}$, $M^{\alpha\beta}$	- force and moment tensor,
Q^{α}	- shear force vector,
p^{β}, p	- tangential and normal surface loads,
$b_{\alpha\beta}$, b_{α}^{β}	- covariant and mixed curvature tensor,
$\mid$	- covariant derivative.

In order to ensure the symmetry of the stress tensor, not only the conditions of equilibrium but also the following conditions of symmetry must be fulfilled:

$$\varepsilon_{\alpha\beta} \, (N^{\alpha\beta} + M^{\rho\alpha} b_{\rho}^{\beta}) = 0 \quad , \qquad (10.1\text{-}2)$$
$$\varepsilon_{\alpha\beta} \, M^{\alpha\beta} = 0 \ ,$$

where $\varepsilon_{\alpha\beta}$ - permutation tensor [19].

For the selection of the displacement function

$$\mathbf{V} = (v_{\alpha} + \zeta\psi_{\alpha}) \, \mathbf{a}^{\alpha} + w\mathbf{a}^{3} \qquad (10.1\text{-}3)$$

with

v_α - displacement in the middle surface,
w - displacement perpendicular to the middle surface,
ψ_α - cross-section slopes,
$\mathbf{a}^i$ - contravariant basis of the middle surface,

it is assumed that the cross-sections perpendicular to the middle surface are plane after deformation, and that the displacement w is constant over the thickness. Thus, the strain-displacement equations of the shell structure including shear deformations yield

$$\begin{aligned} \alpha_{\alpha\beta} &= \frac{1}{2}\,(v_\alpha|_\beta + v_\beta|_\alpha - 2b_{\alpha\beta}w) \;, \\ \beta_{\alpha\beta} &= \frac{1}{2}\,(\psi_\alpha|_\beta + \psi_\beta|_\alpha - b^\rho_\alpha\, v_\rho|_\beta - b^\rho_\beta\, v_\rho|_\alpha + 2b_{\alpha\rho}\, b^\rho_\beta\, w) \;, \\ \gamma_\alpha &= \psi_\alpha + w|_\alpha + b^\rho_\alpha\, v_\rho \;, \end{aligned} \qquad (10.1\text{-}4)$$

where

$\alpha_{\alpha\beta}$ - strains of the middle surface,
$\beta_{\alpha\beta}$ - curvature of the middle surface,
γ_α - shearing strains perpendicular to the middle surface (note: $\gamma_\alpha = 2\gamma_{\alpha 3}$).

As a fibre-composite element consists of n_s single layers, each of them rotated at an angle α_k (Fig. 10.1/2). the material properties are discontinuous across the laminate thickness. This means, the material is heterogeneous and anisotropic. In contrast to homogeneous materials with constant material characteristics across the thickness, it is thus essential to indicate the material law for each of the n_s single layers. The material equations of a fibre-composite element have to satisfy the following conditions:

- linear elasticity of the material,
- no internal stresses,
- plane stress state,
- continuous displacements between the single layers of the laminate, i.e. ${}_k\alpha_{\alpha\beta} = \alpha_{\alpha\beta}$, ${}_k\beta_{\alpha\beta} = \beta_{\alpha\beta}$, ${}_k\gamma_\alpha = \gamma_\alpha$

and can be derived as follows [15]:

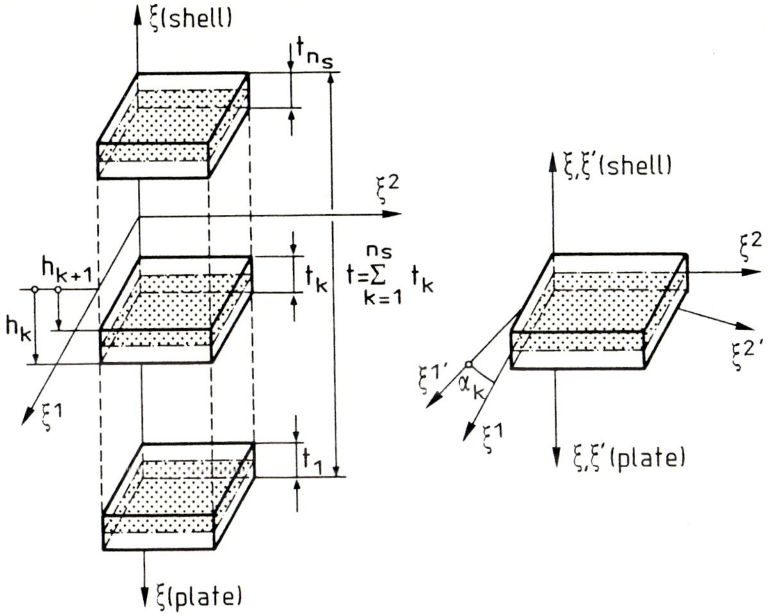

Fig. 10.1/2. Laminate of a multilayer composite

$$
\begin{aligned}
N^{\alpha\beta} &= A^{\alpha\beta\gamma\delta}\alpha_{\gamma\delta} + B^{\alpha\beta\gamma\delta}\beta_{\gamma\delta} - b^{\beta}_{\rho}\, M^{\alpha\rho} - N^{\alpha\beta}_{\Theta}, \\
M^{\alpha\beta} &= B^{\alpha\beta\gamma\delta}\alpha_{\gamma\delta} + K^{\alpha\beta\gamma\delta}\beta_{\gamma\delta} - M^{\alpha\beta}_{\Theta}, \\
Q &= S^{\alpha 3\gamma 3}\gamma_{\gamma},
\end{aligned}
\tag{10.1-5}
$$

where

$$A^{\alpha\beta\gamma\delta} = \sum_{k=1}^{n_s} \int_{h_k}^{h_{k+1}} \left[{}^{o}_{k}H^{\alpha\beta\gamma\delta} + ({}^{1}_{k}H^{\alpha\beta\gamma\delta} - 2H\, {}^{o}_{k}H^{\alpha\beta\gamma\delta})\, \zeta \right] d\zeta = \sum_{k=1}^{n_s} {}_{k}A^{\alpha\beta\gamma\delta},$$

$$B^{\alpha\beta\gamma\delta} = \sum_{k=1}^{n_s} \int_{h_k}^{h_{k+1}} \left[{}^{o}_{k}H^{\alpha\beta\gamma\delta} + ({}^{1}_{k}H^{\alpha\beta\gamma\delta} - 2H\, {}^{o}_{k}H^{\alpha\beta\gamma\delta})\, \zeta \right] \zeta\, d\zeta = \sum_{k=1}^{n_s} {}_{k}B^{\alpha\beta\gamma\delta},$$

$$K^{\alpha\beta\gamma\delta} = \sum_{k=1}^{n_s} \int_{h_k}^{h_{k+1}} \left[{}^{o}_{k}H^{\alpha\beta\gamma\delta} + ({}^{1}_{k}H^{\alpha\beta\gamma\delta} - 2H\, {}^{o}_{k}H^{\alpha\beta\gamma\delta})\, \zeta \right] \zeta^2 d = \sum_{k=1}^{n_s} {}_{k}K^{\alpha\beta\gamma\delta},$$

$$S^{\alpha 3\gamma 3} = k_{(\alpha)}\, k_{(\gamma)} \sum_{k=1}^{n_s} \int_{h_k}^{h_{k+1}} \left[{}^{o}_{k}C^{\alpha 3\gamma 3} + ({}^{1}_{k}C^{\alpha 3\gamma 3} - 2H\, {}^{o}_{k}C^{\alpha 3\gamma 3})\, \zeta \right] d\zeta,$$

$$N^{\alpha\beta}_{\Theta} = \sum_{k=1}^{n_s} \left({}_{k}A^{\alpha\beta\gamma\delta}\, {}^{o}_{k}\Theta_{\gamma\delta} + {}_{k}B^{\alpha\beta\gamma\delta}\, {}^{1}_{k}\Theta_{\gamma\delta} \right),$$

$$M_{\Theta}^{\alpha\beta} = \sum_{k=1}^{n_s} \left({}_k B^{\alpha\beta\gamma\delta} \, {}_k^o\Theta_{\gamma\delta} + {}_k K^{\alpha\beta\gamma\delta} \, {}_k^1\Theta_{\gamma\delta} \right) ,$$

$${}_k^o H^{\alpha\beta\gamma\delta} = {}_k H^{\alpha\beta\gamma\delta} ,$$

$${}_k^1 H^{\alpha\beta\gamma\delta} = b_\mu^\alpha \, {}_k H^{\mu\beta\gamma\delta} + b_\lambda^\beta \, {}_k H^{\alpha\lambda\gamma\delta} + b_\rho^\gamma \, {}_k H^{\alpha\beta\rho\delta} + b_\nu^\delta \, {}_k H^{\alpha\beta\gamma\nu} ,$$

$${}_k^o C^{\alpha 3\gamma 3} = {}_k C^{\alpha 3\gamma 3} ,$$

$${}_k^1 C^{\alpha 3\gamma 3} = b_\mu^\alpha \, {}_k C^{\mu 3\gamma 3} + b_\rho^\gamma \, {}_k C^{\alpha 3\rho 3} ,$$

$${}_k^o\Theta_{\gamma\delta} = {}_k\alpha_{T\gamma\delta} \, {}^o\Theta ,$$

$${}_k^1\Theta_{\gamma\delta} = {}_k\alpha_{T\gamma\delta} \, {}^1\Theta - \left(b_\gamma^\rho \, {}_k\alpha_{T\rho\delta} + b_\delta^\nu \, {}_k\alpha_{T\gamma\nu} \right) {}^o\Theta ,$$

$${}_k H^{*\alpha\beta\gamma\delta} = {}_k C^{*\alpha\beta\gamma\delta} - \frac{{}_k C^{*\alpha\beta 33} \, {}_k C^{*33\gamma\delta}}{{}_k C^{*3333}}$$

with

- ${}_k C^{ijkl}$ – elasticity tensor of the k-th layer,
- ${}_k\alpha_{T\gamma\delta}$ – tensor of the thermal expansion coefficients of the k-th layer,
- ${}^o\Theta$, ${}^1\Theta$ – constant and linear temperature distribution across the thickness of the structure,
- H – mean curvature.

The tensors $A^{\alpha\beta\gamma\delta}$, $B^{\alpha\beta\gamma\delta}$, and $K^{\alpha\beta\gamma\delta}$ include the strain, coupling, and bending stiffnesses of the structure. The shear stiffnesses perpendicular to the midsurface are included in the tensor $S^{\alpha 3\gamma 3}$. In the case of fibre-composite structures. $N_{\Theta}^{\alpha\beta}$ represents typical normal and shear forces and $M_{\Theta}^{\alpha\beta}$ represents moments resulting from temperature distribution.

In the third equation of (10.1-5) it is assumed that shear strains perpendicular to the midsurface are constant throughout the thickness of the laminate. As the shear stiffnesses perpendicular to the midsurface are practically the same for individual layers, it is suggested that the shear deformations of the single layers differ only slightly. Consequently, the cross-sections of the laminate remain plane. On account of these assumptions, the shear correction factors $k_{(\alpha)}$ are introduced in order to retain the material equations.

Fig. 10.1/2 shows that the local $\xi^{\alpha'}\zeta$-coordinate system of the k-th single layer need not coincide with the global ξ^{α} ζ-coordinate system. The combination of the local and the global coordinate systems follows from the transformation tensor [19]:

$$_k\left(\beta^{i}_{i'}\right) = \begin{bmatrix} \cos\alpha_k & \sin\alpha_k & 0 \\ -\sin\alpha_k & \cos\alpha_k & 0 \\ 0 & 0 & 1 \end{bmatrix} , \tag{10.1-6a}$$

where α_k is the fibre angle of the k-th layer. With the inverse transformation tensor

$$_k\left(\beta^{i'}_{i}\right) = {}_k\left(\beta^{-1}\right)^{i}_{i'} \tag{10.1-6b}$$

and the equation (10.1-6a), one obtains the elasticity tensor $_kC^{ijkl}$ and the following thermal tensor $_k\Theta_{mn}$ of the k-th layer, which are applicable to the global ξ^{α}, ζ-coordinate system

$$_kC^{ijmn} = {}_k\beta^{i}_{i'}\ {}_k\beta^{j}_{j'}\ {}_k\beta^{m}_{m'}\ {}_k\beta^{n}_{n'}\ {}_kC^{i'j'm'n'} , \tag{10.1-7a}$$

$$_k\Theta_{mn} = {}_k\beta^{m'}_{m}\ {}_k\beta^{n'}_{n}\ {}_k\Theta_{m'n'} . \tag{10.1-7b}$$

b) Augmented basic equations for a composite plate structure with regard to time-variant material behaviour and large deformations

The optimal layout of the fibre-composite components requires the consideration of the time-variant material behaviour if high shape accuracy within a given time period is demanded, and if particular deformation bounds have to be observed. When components are also exposed to natural climatic conditions, time and local variant hygrothermal loads have to be considered because of long-term diffusion processes. The investigations also have to take into account possible initial deflections [17].

Without deriving the plate equations in detail, we use the corresponding shell equations of the preceding section neglecting all members with curvature and completing terms with initial and large deflections as well. The equilibrium conditions of the plate then read (see 10.1-1):

$$N^{\alpha\beta}|_{\beta} + p^{\alpha} = 0 , \tag{10.1-8a}$$

$$\underline{N^{\alpha\beta}\hat{w}_{\alpha\beta}} + Q^{\alpha}|_{\alpha} + p - \underline{p^{\alpha}\hat{w}_{,\alpha}} = 0 \,, \tag{10.1-8b}$$

$$M^{\alpha\beta}|_{\beta} - Q^{\alpha} + m^{\alpha} = 0 \,. \tag{10.1-8c}$$

The underlined terms are the nonlinear deflection-dependent parts of the equilibrium condition which is perpendicular to the midsurface of the in-plane plate. They produce a coupling of the plate forces $N^{\alpha\beta}$ with the transverse forces Q^{α} and the moments $M^{\alpha\beta}$. The strains of the middle surface, its curvature and transversal shear strain can be established in analogy to (10.1-4):

$$\alpha_{\alpha\beta} = \frac{1}{2}\left[{}^{o}\hat{v}_{\alpha}|_{\beta} + {}^{o}\hat{v}_{\beta}|_{\alpha} + \hat{w}_{,\alpha}\hat{w}_{,\beta} - \tilde{w}_{,\alpha}\tilde{w}_{,\beta}\right], \tag{10.1-9a}$$

$$\beta_{\alpha\beta} = \frac{1}{2}\left[\hat{\psi}_{\alpha}|_{\beta} - \tilde{\psi}_{\alpha}|_{\beta} + \hat{\psi}_{\beta}|_{\alpha} - \tilde{\psi}_{\beta}|_{\alpha}\right], \tag{10.1-9b}$$

$$\gamma_{\alpha 3} = \hat{\psi}_{\alpha} + \hat{w}_{,\alpha} \,. \tag{10.1-9c}$$

Here, the terms marked with ^ belong to the total deformation state and those with ~ to the initial deformation state.

Linear viscoelastic material behaviour can be assumed for fibre composites with strains smaller than 0.5%. Such behaviour is described in Boltzman's superposition integral:

$$\sigma^{kl}(t) = \int_{-\infty}^{t} C^{klij}(t-\tau)\, d\gamma_{ij}(\tau) - \int_{-\infty}^{t} \beta_{T}{}^{kl}(t-\tau)\, d\theta(\tau) - \int_{-\infty}^{t} \beta_{C}{}^{kl}(t-\tau)\, dc(\tau) \,, \tag{10.1-10}$$

where C^{klij} is a creep function tensor, $\beta_T{}^{kl}$ a thermal stress tensor, $\beta_C{}^{kl}$ a hygrothermal stress tensor, θ - temperature, c - moisture concentration.

The evaluation of the superposition integral results from a superposition of single solutions for constant load steps

$$\sigma^{kl}(t) = \sum_{p} C^{ijkl}(t-t_s)\, {}^{s}\gamma_{ij}\, H(t-t_s) - \sum_{q} \beta_{T}{}^{kl}(t-t_q)\, {}^{q}\theta H(t-t_q) - \sum_{r} \beta_{C}{}^{kl}(t-t_r)\, {}^{r}c H(t-t_r) \tag{10.1-11}$$

with a unit step function $H(t-t_i)$.

For a laminate consisting of unidirectional single layers, the resultant force-strain equations are obtained after transforming the material law (10.1-11) of each layer into the global coordinate system, introducing the kinematics of the classical laminate theory, and integrating over the laminate thickness.

$$N^{\alpha\beta}(t) = \sum_p \left[A^{\alpha\beta\nu\mu}(\tau)\ {}^p\alpha_{\nu\mu} + B^{\alpha\beta\nu\mu}(\tau)\ {}^p\beta_{\nu\mu}\right] H(\tau)\Big|_{\tau=t-t_p} - N_\Theta^{\alpha\beta} - N_c^{\alpha\beta},$$

$$M^{\alpha\beta}(t) = \sum_p \left[B^{\alpha\beta\nu\mu}(\tau)\ {}^p\alpha_{\nu\mu} + K^{\alpha\beta\nu\mu}(\tau)\ {}^p\beta_{\nu\mu}\right] H(\tau)\Big|_{\tau=t-t_p} - M_\Theta^{\alpha\beta} - M_c^{\alpha\beta},$$

$$Q^{\alpha 3}(t) = \sum_p \left[S^{\alpha 3\nu 3}(\tau)\ {}^p\alpha_{\nu 3}\right] H(\tau)\Big|_{\tau=t-t_p} . \tag{10.1-12}$$

Thermal and moisture distributions have to be calculated for determining the appropriate resultant forces $N_\Theta^{\alpha\beta}$ and $M_\Theta^{\alpha\beta}$, or $N_c^{\alpha\beta}$ and $M_c^{\alpha\beta}$. In this case, a stationary temperature profile, linearly variable through the laminate thickness, shall be considered. The hygrothermal distribution in thickness direction results from solving the Fick's one-dimensional diffusion equation

$$\frac{\partial c}{\partial t} = \frac{\partial}{\partial z}\left\{D_z \frac{\partial c}{\partial z}\right\} , \tag{10.1-13}$$

where D_z is the diffusion coefficient in the z-direction. The solution for the unidirectional diffusion in a plate with thickness t and the boundary conditions $c(t=0,\ z=t/2) = c_o$, $c(t=0,\ z=t/2) = 0$, $c(t\rightarrow\infty,\ z=\pm t/2) = c_o$ reads as follows:

$$c(z,t) = c_o\left[1 - \frac{\pi}{2}\sum_{n=0}^{\infty}\frac{(-1)^n}{2n+1}\exp\left\{-\frac{(2n+1)^2\pi^2}{4h^2}D_z t\right\} \times \cos\left\{\frac{2n+1}{2h}\pi\left(z+\frac{h}{2}\right)\right\}\right] . \tag{10.1-14}$$

10.1.3 Objective Functions - Design Variables - Constraints

In the following cases, the two main objectives, namely the minimal weight $f_1(\mathbf{x})$ and the minimal deformations or stresses $f_2(\mathbf{x})$ shall be required for fibre-reinforced supporting structures. The deformation variables v_α, w, ψ_α and the resultant forces $N^{\alpha\beta}$, $M^{\alpha\beta}$, Q^α necessary for determining the objective function f_2 are obtained from a structural analysis by using equations (10.1-1) and (10.1-5). The calculation of the weight of the composite structures depends on the thickness t_k, the fibre volume fraction φ_{Fk}, the fibre density ρ_{Fk}, and the matrix density of the single layers ρ_{Mk} as well as on the surface A:

$$f_1 \mathrel{\hat{=}} W = g\sum_{k=1}^{n_s} t_k[\rho_{Fk}\varphi_{Fk} + \rho_{Mk}(1-\varphi_{Fk})] \int_A dA. \qquad (10.1\text{-}15)$$

The design variables have to describe the layout uniquely, i.e. the geometrical quantities and the physical properties of the material must be defined as accurately as possible. The most important design variables for fibre composite structures are:

- layer thicknesses t_k and laminate thickness t,
- fibre angles α_k,
- fibre volume fraction φ_{Fk},
- material properties of fibre and matrix, e.g. elasticity moduli, thermal expansion coefficients etc.

The elastic failure loads are the decisive constraints for fibre-composite structures such as bonding break and fibre break criteria, the buckling loads, and the fracture loads in the case of interlaminate shear stresses. In the case of sandwich components, the core must also be checked for shear failure.

As the buckling loads are not taken into account, small deformations were assumed for the derivation of the basic equations. The breaking failure due to interlaminar shear stresses was also neglected because an interlaminar break generally occurs on free edges and thus leads to debonding problems. This investigation deals only with such structures which have a layout causing debonding problems on the edges.

In order to comply with the above mentioned assumptions, the following constraints are to be taken into consideration [20]:

- Bonding Break Failure (BBF)

$$g_{1k}(\mathbf{x}) = \left(\frac{{}_k\sigma_{1'1'}}{\sigma_{1'1'HB}}\right)^2 + \left(\frac{{}_k\sigma_{2'2'}}{\sigma_{2'2'B}}\right)^2 + \left(\frac{{}_k\tau_{1'2'}}{\tau_{1'2'B}}\right)^2 - 1 \le 0 , \qquad (10.1\text{-}16)$$

- Fiber Break Failure (FBF)

$$g_{2k}(\mathbf{x}) = \left(\frac{{}_k\sigma_{1'1'}}{\sigma_{1'1'B}}\right)^2 - 1 \le 0 . \qquad (10.1\text{-}17)$$

10.1.4 Examples of Application

a) Spherical shell - optimum deformation layout

The optimal layout of a spherical shell (Fig. 10.1/3a) is to be found with regard to deformation and strain conditions. It is clamped at $\varphi = 90^o$ and has a constant wall thickness t. The acting load is a constant internal pressure $p_o = 10$ bar.

The "optimization objective" is to determine a shell with a minimal weight W and a minimal displacement max |w|. The corresponding objective function components then read as follows:

$$f_1(\mathbf{x}) \mathrel{\widehat{=}} W, \qquad f_2(\mathbf{x}) \mathrel{\widehat{=}} \max |w| . \tag{10.1-18}$$

The maximum deformation max |w| is obtained from the structural analysis program (LSAP **La**minated **S**hell **A**nalysis **P**rogram) developed in [15]. The weight W of the shell can be calculated from (10.1-15) as follows:

$$W = 2\pi a^2 g \sum_{k=1}^{n_S} t_k[\rho_{Fk}\varphi_{Fk} + \rho_{Mk}(1 - \varphi_{Fk})]. \tag{10.1-19}$$

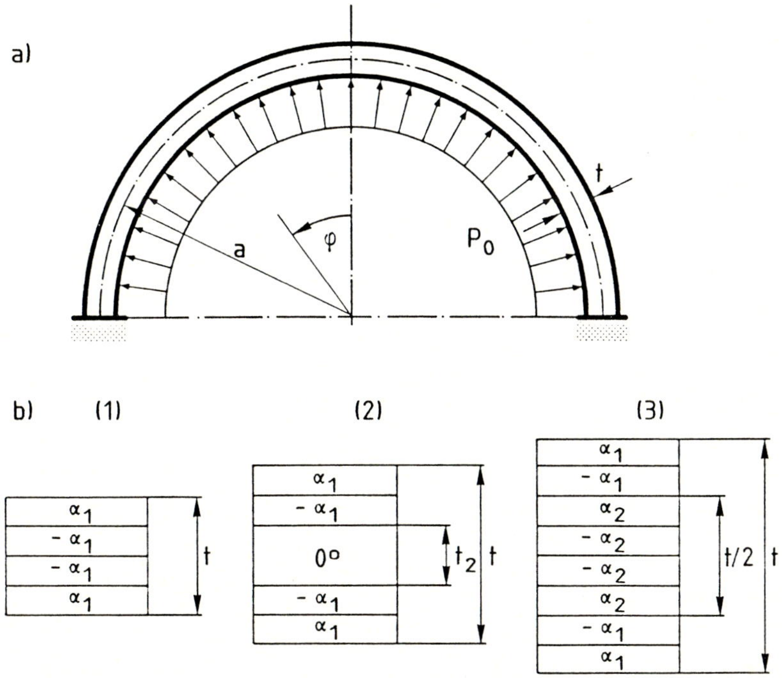

Fig. 10.1/3. Various spherical shells under constant internal pressure
a) Sketch of a spherical shell
b) Types of laminates

For the optimization of the spherical shell under internal pressure load only those types of composites presented in Fig. 10.1/3b) are taken into account.

The fibre angles α_1 and α_2, the laminate thickness t, and in the case of the composite structure (2) the thickness t_2 of the 0°-layer are selected as design variables. The variables are bounded by

$$0^\circ \le \alpha_1, \alpha_2 \le 90^\circ \ , \quad 0.2 \text{ mm} \le t \le 5 \text{ mm} \ , \quad t_2 \le t \ . \tag{10.1-20}$$

Using the min-max approach, the multi-criteria optimization problem can be reduced to the following scalar substitute problem (see 1-8):

$$p[f(\mathbf{x})] = \beta \ \bigcap \ \begin{aligned} & w_1 \frac{\max |w|}{\max |w_0|} - \beta \le 0 \ , \\ & w_2 \frac{W}{W_0} - \beta \le 0 \ , \end{aligned} \tag{10.1-21}$$

where w_0 is the deformation and W_0 the weight of the starting design.

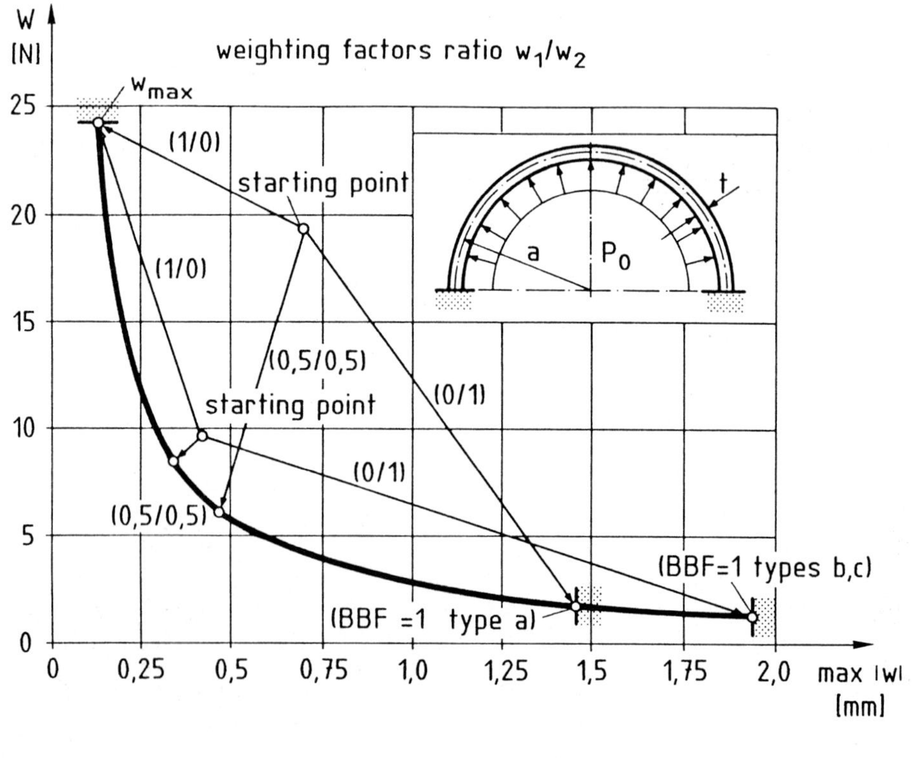

Fig. 10.1/4. Functional-efficient boundary of a spherical shell under constant pressure

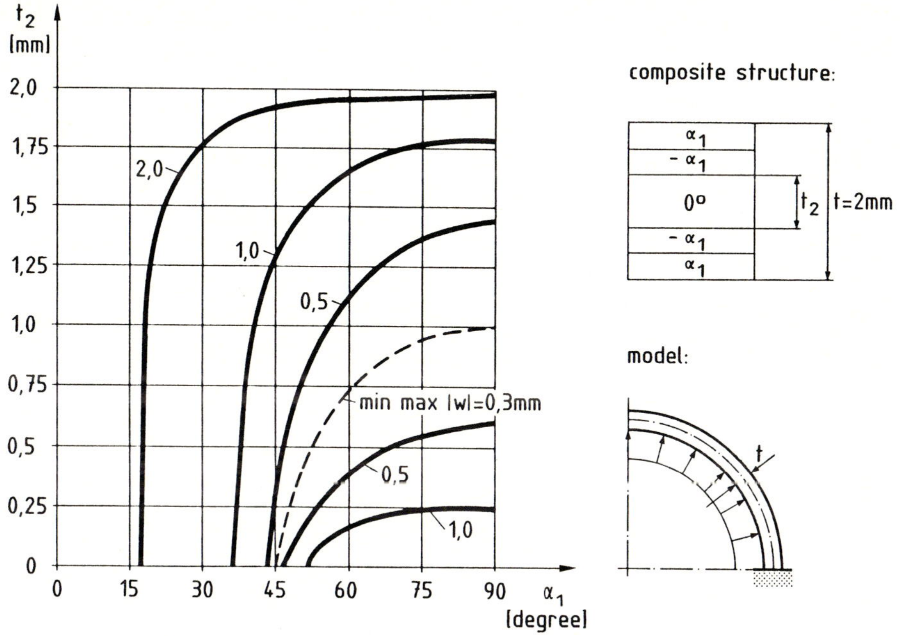

Fig. 10.1/5. Maximum deformation max |w| of type (2) performed by varying the layer thickness t_2 and the fiber angle α_1

In the deformation optimization of a shell under internal pressure load, all three designs provide more or less the same functional-efficient boundary (Fig. 10.1/4). For type (1), a global optimum exists at $\alpha_1^* = 45^\circ$. Nevertheless, whether or not the functional-efficient solutions are obtained, depends on the exact observance of the fiber angle $\alpha_1^* = 45^\circ$. Even small angle deviations $|\Delta\alpha_1| > 1^\circ$ lead to a considerable w-deformation.

In the case of composite type (1), the bonding break criterium of the weight minimization occurs considerably earlier than in the cases (2) and (3). At a fiber angle $\alpha_1^* = 45^\circ$, critical shear stresses occur in the individual layers next to the clamped edge. Composite type (2) is marked by the inclusion of the 0^0-layer thickness t_2 as an additional design variable. The layer thickness ration t_2/t can thus be made variable. In comparison to type (1), the composite type (2) with its three design variables α_1, t_2, and t shows a poor convergence behaviour compared to type (1). The calculated optimal design variables for a shell under internal pressure, lie within the angle range of 45° to 90° and in a wall thickness range of $0 \le t_2 \le t/2$. This can be explained by the contour plans shown in Fig. 10.1/5. Here, the maximal w-deformation depending on the design variables α_1 and t_2 is presented for t = 2 mm = const. and a crest structure for min max |w| = 0.3 mm can be easily noticed. Therefore, the design variables α_1 and t_2 are dependent on each other. For further optimization, the number of design variables can be reduced by putting $t_2 = t/2$. The functional-efficient boundary is now uniquely computable. If the convergence behaviour is sufficient, an optimal fibre layout of $\alpha_1 = 90°$ is achieved, and all the angles up to $\alpha_1 = 85°$ provide similar values. This is due to the flat deformation curve in the region of the global optimum which is clearly shown by the dotted line in Fig. 10.1/5.

The theoretical results have been proved valid by performing experimental investigations [15] of nonoptimal and optimal shells made of CFC (see Fig. 10.1/6). In the case of the nonoptimal shell (model 1), the reinforcing fibers run exclusively in the longitudinal direction whereas in the optimal shell (model 2) the longitudinal and lateral directions, corresponding to the optimization results, are reinforced for the same proportions.

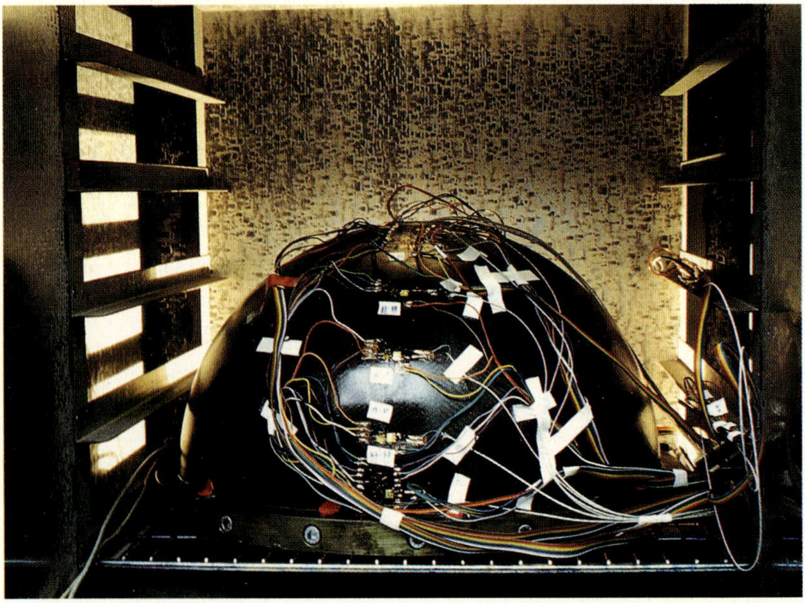

Fig. 10.1/6. Composite shell for strain gauge measurements

The deformations w (Fig. 10.1/7) show that despite the internal pressure acting on the spherical region the deformation of model 1 is negative. This is due to the high stiffness and the corresponding low strain ability in longitudinal direction whereby the shell attempts to retain its original meridian length. The deformation curve of model 2 shows that the theoretical maximum values (dotted line) for the optimal shell is well maintained.

In addition to deformation measurements, the strain measurements were carried out by using strain gauges (measuring grid arrangement 0°/45°/90°). It appears that besides the strains in meridional and lateral directions, additional shear strains occur in the shell. The reason for this is that the given fiber angles deviate from the nominal values. In Fig. 10.1/8, the measured shear strains $\gamma_{\varphi\vartheta}$ for model 2 are compared with the theoretically determined shearing strains. It was determined with the help of theoretical calculations that the laterally arranged fibers only deviate from their nominal values by approximately $\Delta\alpha = \pm\ 10°$.

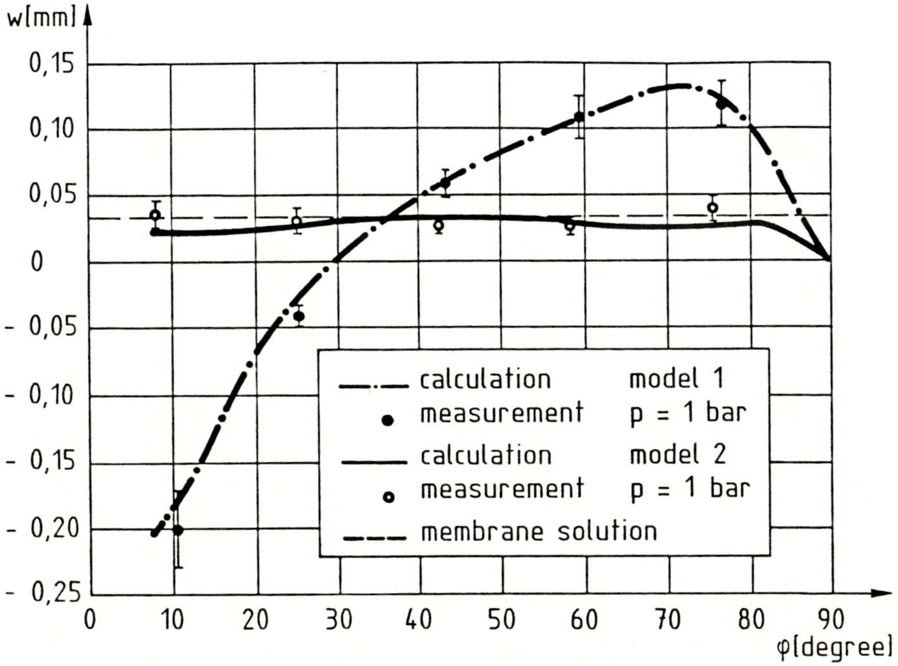

Fig. 10.1/7. Numerical and experimental displacements w of a spherical shell under internal pressure

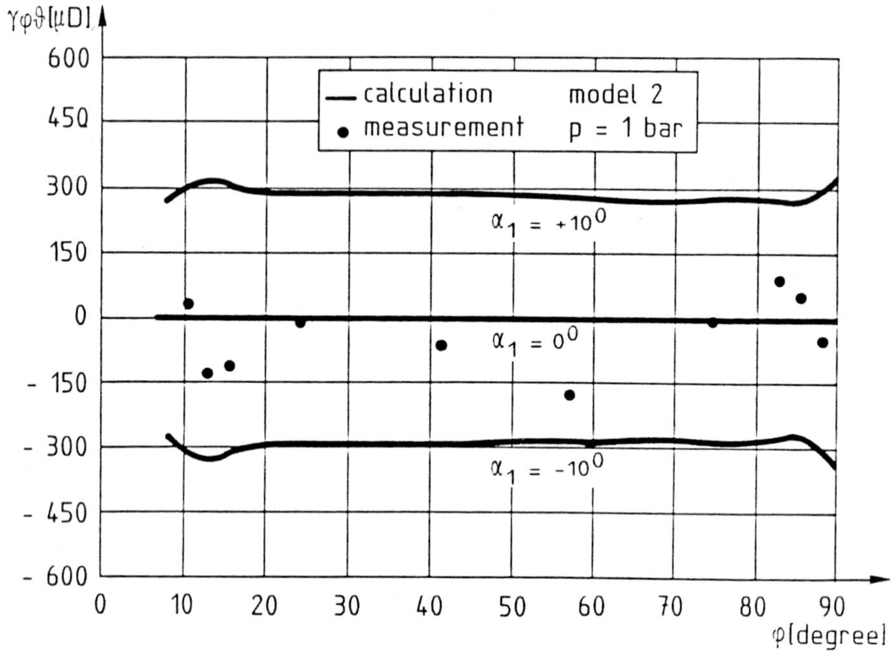

Fig. 10.1/8. Numerical and experimental shear strains of model 2 at $\zeta = +\frac{t}{2}$

b) Optimal plates with time-variant material behaviour

A viscoelastic plate is to be designed in such a way that during a certain time period objectives like weight $W(\mathbf{x})$ and deformation $w_{max}(\mathbf{x},t)$ are minimized. The constraints have to be fulfilled in the given time interval. Considering (1-8), the MO-problem marked by an objective conflict reads:

$$\text{"Min"}\left\{\mathbf{f}(\mathbf{x}) = \begin{bmatrix} f_1(\mathbf{x}) \\ f_2(\mathbf{x}) \end{bmatrix} = \begin{bmatrix} W(\mathbf{x}) \\ w_{max}(\mathbf{x},t) \end{bmatrix} \middle| \mathbf{g}(\mathbf{x},t) \leq 0;\ \mathbf{x} \in \mathbb{R}^n;\ \mathbf{g} \in \mathbb{R}^p;\ t \leq \bar{t}\right\}. \tag{10.1-22}$$

The MO-problem is transformed into a scalar substitute problem by means of a constraint-oriented transformation (see (1-14)):

$$\text{Min}\left\{ p[\mathbf{f}(\mathbf{x},t)] = W(\mathbf{x}) \,\middle|\, g_{p+1}(\mathbf{x},t) = w_{max}(\mathbf{x},t) - \bar{w} \leq 0; \right.$$
$$\left. \mathbf{g}(\mathbf{x},t) \leq 0,\ \mathbf{x} \in \mathbb{R}^n;\ g \in \mathbb{R}^{p+1};\ t \leq \bar{t} \right\}. \tag{10.1-23}$$

The vector of the design variables **x** contains the following component elements to be varied: fiber angle α_k, layer thickness t_k, values describing the component geometry etc. The constraint vector **g** consists of failure criteria, stability conditions, geometry and variable constraints. For a time $\overline{t}$ and for different values $\overline{w}$, the solution of the optimization problem (10.1-23) provides one point of the functional-efficient solution set of the MO-problem (10.1-22). If $\overline{t}$ is varied as well, then further points are established and summed up to a functional-efficient surface. The example of application deals with a simply supported CFC-plate under time-variant surface load and with constant thermal and hygrothermal distributions. The plate consists of two angle-ply laminates, i.e. they are arranged symmetrically about the middle surface. Fiber angle and layer thickness of the laminate are varied. The constraints are the fiber breaking and the bonding break failure criteria of the fiber [20]. The functional-efficient surfaces of Fig. 10.1/9 result from the solutions of plate optimization with and without thermal and hygrothermal load. The front boundary of the functional-efficient surface is represented by functional-efficient solutions for the time-independent case ($\overline{t}=0$). For $\overline{t} > 0$, the weight

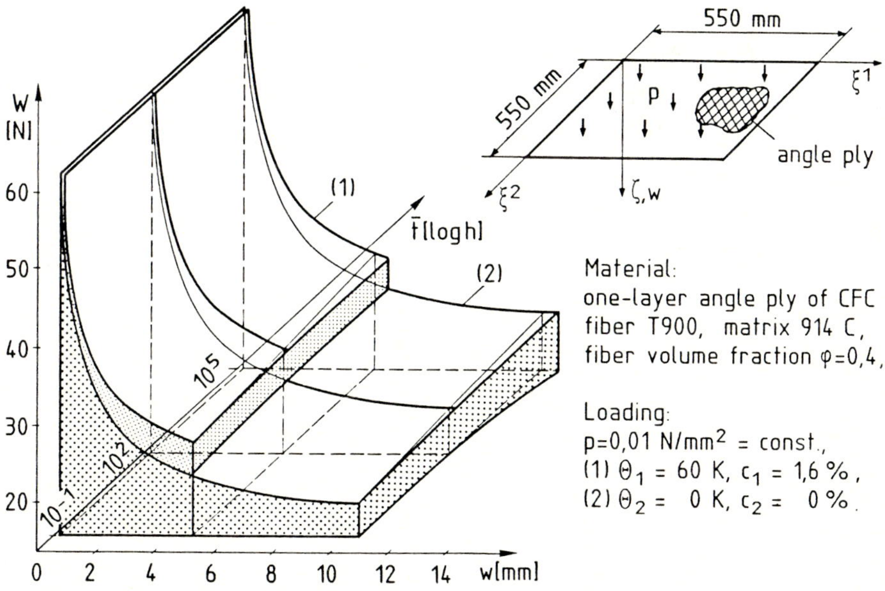

Fig. 10.1/9. Functional-efficient solutions of a simply supported plate

values belonging to the same deformation value increase. The higher optimal weights for thermal and hygrothermal loads are caused by additional loads as well as by changes of the material properties and the fracture behaviour [16,17].

10.1.5 Conclusion

This chapter contributes to the optimization of shell and plate structures made of Carbon Fibre Composite (CFC) material. With the requirements to minimize the weight of the structures and the maximal deformations, the optimization model includes two objective functions. Therefore, a multicriteria optimization problem is given which can be transformed into a scalar substitute problem by using either the constraint-oriented method (trade-off method) or the min-max method. First of all, for the determination of the maximum deformation as well as of the ultimate stresses, the basic equations of the structures have to be solved. Layer thickness and fiber angles were chosen as design variables. Among the essential constraints, failure criteria such as bonding break failure, fibrebreak failure, and especially shear failure of the core of the sandwich plate were used. As the elasticity tensor depends on the trigonometrical functions of the fiber angles, the optimization problem leads to a highly nonlinear and complex task for both structures.

In section 10.1.4, the results of the procedure concerning the optimal deformation layout of a spherical shell and a plate with time-variant material behaviour are discussed. Based on the results of the latter investigations, it can be concluded that for an optimal layout of fiber composite components, the time-variant material behaviour and the varying mechanical properties due to viscoelasticity and humid environment influences must be observed if the specification requirements shall be met.

A further problem of the optimal layout of thermally loaded plates is treated in Sections 10.2. and 10.3.

References

[1] Lenos, E.M.; Olpinger, D.W.; Burke, J.J. (eds.): Fibrous Composites in Structural Design. Proc. Conf. in Structural Design, San Diego, CA, 1978. New York: Plenum Press 1980

[2] N.N.: Konstruieren mit Verbund- und Hybridwerkstoffen. VDI-Berichte 563, Tagungsband der VDI-Gesellschaft Entwicklung, Konstruktion, Vertrieb. Fellbach 1985

[3] Ambartsumyan, S.A.: Theory of Anisotropic Shells. NASA Technical Translation, NASA TTF-118, 1964

[4] Ambartsumyan, S.A.: Theory of Anisotropic Plates. In: Ashton, J.E. (ed.) Progress in Materials Science Series, Vol. II. Stanford/CT: Technomic 1970

[5] Lekhnitskii, S.G.: Theory of Elasticity of an Anisotropic Body. San Francisco: Holden Day 1963

[6] Ambartsumyan, S.A.: Contributions to the Theory of Anisotropic Layered Shells. Appl. Mech. Rev. 15 (1962) 245-249

[7] Bert, C.W.; Francis, P.H.: Composite Material Mechanics: Structural Mechanics. AIAA J. 12 (1974) 1173-1186

[8] Habip, L.M.: A Survey of Modern Developments in the Analysis of Sandwich Structures. Appl. Mech. Rev. 18 (1965) 93-98

[9] Haftka, R.T.; Prasad, B.: Optimum Structural Design with Plate Bending Elements - A Survey. AIAA J. 19 (1981) 517-522

[10] Adali, S.: Multi-objective Design of an Antisymmetric Angle-ply Laminate by Nonlinear Programming. J. of Mechanisms, Transmission and Automation in Design, Trans. ASME 105 (1983) 214-219

[11] Eschenauer, H.: Optimierung ebener Flächentragwerke aus Verbundwerkstoff. Z. Flugwiss. Weltraumforsch. 8 (1984) 367-378

[12] Eschenauer, H.: Rechnerische und experimentelle Untersuchungen zur Strukturoptimierung von Bauweisen. DFG Forschungsbericht. Inst. Mechanik und Regelungstechnik, Universität-GH Siegen, 1985 (German and English version), Bonn+Fries Verlag Siegen, 1986

[13] Eschenauer, H.A.; Fuchs,W.: Optimum Design einer punktförmig belasteten Sandwichplatte mit orthotropen CFK-Deckschichten. Z. Angew. Math. Mech. 65 (1985) T 305 - T 306

[14] Eschenauer, H.A.; Fuchs, W.: Fiber-Reinforced Sandwich Plates under Static Loads - Proposals for their Optimization. Trans. ASME 108 (1986) 152-158

[15] Fuchs, W.: Strukturanalyse und Optimierung anisotroper Schalen aus Faserverbundwerkstoff. Dissertation Universität-GH Siegen, 1986

[16] Eschenauer, H.; Post, P.: Zur Optimierung von Bauteilen aus linear-viskoelastischem Verbundwerkstoff, Z. Angew. Math. Mech. ZAMM 68 (1988) T 176 - T 178

[17] Post, P.: Optimierung von Verbundbauweisen unter Berücksichtigung des zeitabhängigen Materialverhaltens. Dissertation Universität-GH Siegen. VDI-Fortschrittsbericht, Reihe 1, Nr. 172, Düsseldorf: VDI-Verlag 1989

[18] Schnell, W.; Eschenauer, H.: Elastizitätstheorie 2 - Schalen. Mannheim, Wien, Zürich: BI Wissenschaftsverlag 1984

[19] Eschenauer, H.; Schnell, W.: Elastizitätstheorie 1 - Grundlagen, Scheiben, Platten. Mannheim, Wien, Zürich: BI Wissenschaftsverlag 1986 (2.Aufl.)

[20] Puck, A.: Festigkeitsberechnungen an Glasfaser-Kunststoff-Laminaten bei zusammengesetzter Beanspruchung, Kunststoffe 59 (1969) 780-787

10.2 Multicriteria Thermoelastic Design of Antisymmetric Angle-Ply Laminates

S. Adali, K.J. Duffy

10.2.1 Introduction

Fibre-reinforced composite structures are often used in engineering applications where they are exposed to adverse environmental conditions. They may, for example, be required to operate in various temperature fields which may cause bending or buckling. The structure studied in this article is a simply supported antisymmetric angle-ply laminate subject to different thermal loads during its service life. Such structures may fail by buckling in a uniform temperature field [1] and undergo bending in a linear temperature field [2].

This observation indicates that the optimal design problem for an antisymmetric angle-ply laminate operating in different temperature fields should, by necessity, be posed as a vector optimization problem as it is the case in the present study. The objectives of the design optimization are the maximization of the buckling temperature and the minimization of the maximum thermal deflection. The design variables are the fibre orientations and layer thicknesses of individual laminae. The concept of Pareto-optimality is employed in determining compromise solutions which are presented in graphical and tabular form. Methods of objective weighting and min-max formulations are used in obtaining the Pareto-optimal solutions (see Section 1.2.2). The minimization of the objective functions is achieved by the method of downhill simplex which requires only function evaluations.

Previous works on the optimal design of thermoelastic laminates involve only single-objective designs. Optimal fibre orientations for maximum buckling temperatures are given in [1] and [3] for nonhybrid and hybrid antisymmetric angle-ply laminates. Optimal thermoelastic designs for symmetric laminates are obtained in [4,5]. Multicriteria design of laminates subject to mechanical loads is studied in [6,7,8,9]. For a review of the multicriteria structural design, the reader is referred to [8,9,10].

10.2.2 Thermoelastic Buckling and Bending Equations

A simply supported rectangular plate of the length a, width b, and thickness t is considered in the x, y, and z-directions. The plate itself is composed of an even number of orthotropic layers the fibres of which are oriented antisymmetrically with respect to the middle surface. The laminate has symmetry in layer thicknesses and antisymmetry in fibre orientations. Plates with these characteristics are commonly known as antisymmetric angle-ply laminates. During its service life, the plate may be subject to different thermal loads which lead either to a loss of stability by buckling or to bending. Starting from the basic equations in Section 10.1.2, a system of three coupled equations governing the thermal buckling of an antisymmetric angle-ply laminate is obtained (see [1]:

$$A_{11}u_{,xx} + A_{66}u_{,yy} + (A_{12} + A_{66})v_{,xy} - 3B_{16}w_{,xyy} - B_{26}w_{,yyy}$$
$$= N^{\theta}_{xx,x} + N^{\theta}_{xy,y} , \qquad (10.2\text{-}1a)$$

$$(A_{12} + A_{66})u_{,xy} + A_{66}v_{,xx} + A_{22}v_{,yy} - B_{16}w_{,xxx} - 3B_{26}w_{,xyy}$$
$$= N^{\theta}_{yy,y} + N^{\theta}_{xy,y} , \qquad (10.2\text{-}1b)$$

$$K_{11}w_{,xxxx} + 2(K_{12} + 2K_{66})w_{,xxyy} + K_{22}w_{,yyyy} - 3B_{16}u_{,xxy}$$
$$- B_{26}u_{,yyy} - B_{16}v_{,xxx} - 3B_{26}v_{,xyy} + M^{\theta}_{xx,xx} + 2M^{\theta}_{xy,xy} + M^{\theta}_{yy,yy}$$
$$= N_{xx}w_{,xx} + 2N_{xy}w_{,xy} + N_{yy}w_{,yy} , \qquad (10.2\text{-}1c)$$

where u, v, and w represent the displacement components in the x, y, z-directions, respectively, and a subscript after a comma stands for differentiation with respect to that variable. In general, similar notations as in Section 10.1 are used. The pure bending equations are obtained by setting the right hand side of (10.2-1c) equal to zero [2]. The laminate stiffnesses A_{ij}, B_{ij}, and K_{ij} are computed from

$$(A_{ij}, B_{ij}, \text{and } K_{ij}) = \int_{-t/2}^{t/2} (1, z, z^2)\,\overline{Q}_{ijk}dz , \qquad (10.2\text{-}2)$$

where $\overline{Q}_{ijk}$ is the plane stress reduced stiffness component of the k-th layer which can be computed as a function of the fibre orientation α_k using standard transformation relations (10.1-6a) [11].

The thermal forces and moment resultants appearing in equations (10.2-1) result from a change of $\Theta(x, y, z)$ in the temperature distribution of the laminate and are given by

$$\mathbf{N} = \int_{-t/2}^{t/2} \mathbf{Q}_k \, \boldsymbol{\alpha}_{Tk} \, \Delta T \, dz \, , \tag{10.2-3a}$$

$$\mathbf{M} = \int_{-t/2}^{t/2} \mathbf{Q}_k \, \boldsymbol{\alpha}_{Tk} \, \Delta T \, dz \, , \tag{10.2-3b}$$

where $\mathbf{N} = (N^{\Theta}_{xx} \, N^{\Theta}_{yy} \, N^{\Theta}_{xy})^T$, $\mathbf{M} = (M^{\Theta}_{xx} \, M^{\Theta}_{yy} \, M^{\Theta}_{xy})^T$,

$$\boldsymbol{\alpha}_{Tk} = (\alpha_{Tx} \, \alpha_{Ty} \, 2\alpha_{Txy})^T$$

and

$$\mathbf{Q}_k = \begin{bmatrix} {}_k\overline{Q}_{11} & {}_k\overline{Q}_{12} & {}_k\overline{Q}_{16} \\ {}_k\overline{Q}_{12} & {}_k\overline{Q}_{22} & {}_k\overline{Q}_{26} \\ {}_k\overline{Q}_{16} & {}_k\overline{Q}_{26} & {}_k\overline{Q}_{66} \end{bmatrix} \tag{10.2-4}$$

In equations (10.2-3), $\boldsymbol{\alpha}_{Tk}$ is the vector of thermal expansion coefficients for the k-th layer, the elements of which are given by

$$\alpha_{Tx} = \alpha_{T1}c^2 + \alpha_{T2}s^2, \; \alpha_{Ty} = \alpha_{T1}s^2 + \alpha_{T2}c^2, \; \alpha_{Txy} = (\alpha_{T1} - \alpha_{T2})sc, \tag{10.2-5}$$

where $c = \cos\alpha_k$, $s = \sin\alpha_k$, and α_{T1}, α_{T2} are thermal expansion coefficients in the fibre and transverse directions. The boundary conditions for the simply supported plate are given by

$$w = 0, \quad u = 0, \quad M_{xx} = 0, \quad N_{xy} = 0 \qquad \text{for } x = 0,a \, , \tag{10.2-6a}$$

$$w = 0, \quad v = 0, \quad M_{yy} = 0, \quad N_{xy} = 0 \qquad \text{for } y = 0,b \, . \tag{10.2-6b}$$

10.2.3 Multicriteria Design Problem

The design variables of the optimization problem are α_k and t_k, $k = 1,2,\dots,n_s$, where t_k stands for the layer thickness of the k-th lamina. The antisymmetry of the laminate requires that $\alpha_k = -\alpha_{n_s-k+1}$ and $t_k = t_{n_s-k+1}$ for all k. Moreover, the antisymmetric laminates are commonly constructed with fibres oriented alternatively at positive and negative angles, imposing the constraint

$$0 \le (-1)^{k+1} \alpha_k \le \frac{\pi}{2} \qquad \forall k. \tag{10.2-7}$$

The constant thickness constraint is given by

$$\sum_{k=1}^{n_s} t_k = t \ , \quad t_k \ge 0 \qquad \forall k. \tag{10.2-8}$$

In view of (10.2-7) and (10.2-8), the total number of design variables is $n_s - 1$ of which $n_s/2$ are fibre orientations and $(n_s/2) - 1$ are layer thicknesses.

The objectives of the design problem are to maximize the buckling temperature $\Theta_b(\alpha_k, t_k)$ under a temperature change of Θ_1 and to minimize the maximum transverse deflection $w_{max}(\alpha_k, t_k)$ under a temperature change of Θ_2. The temperature loadings Θ_1 and Θ_2 act on the plate at different times leading to buckling or bending, respectively. The buckling temperature Θ_b is calculated from

$$\Theta_b(\alpha_k, t_k) = \min_{m,n} \Theta\,(m,\ n,\ \Theta_1\,;\ \alpha_k, t_k) \tag{10.2-9}$$

where $\Theta(m,\ n,\ \Theta_1\ ;\ \alpha_k, t_k)$ is the buckling temperature and m and n are the numbers of half waves in the x and y directions, respectively. The maximum deflection $w_{max}(\alpha_k,\ t_k)$ is given by

$$w_{max}(\alpha_k, t_k) = \max_{x,y} |w\,(x,\ y,\ \Theta_2;\ \alpha_k, t_k)| \tag{10.2-10}$$

where $0 < x < a\ ,\ 0 < y < b$.

The design problem consists of maximizing Θ_b and minimizing w_{max} over the design variables α_k and t_k. The multicriteria problem can be expressed as:

$$\min_{\alpha_k, t_k} [-\Theta_b\,(\alpha_k, t_k),\ \ w_{max}\,(\alpha_k, t_k)\,]^T \tag{10.2-11}$$

subject to

$$0 \le (-1)^{k+1}\alpha_k \le \frac{\pi}{2} \quad \text{and} \quad \sum_{k=1}^{n_s} t_k = t,\ t_k \ge 0\ \forall k\ , \tag{10.2-12}$$

$$\text{where } k = 1,2,\ldots,n_s,\quad \alpha_k = -\alpha_{n_s-k+1} \quad \text{and} \quad t_k = t_{n_s-k+1}\ . \tag{10.2-12}$$

10.2.4 Solution Method

The solution of the multicriteria optimization problem involves the application of two distinct computational procedures as discussed in Section 1.1. One of them is the structural analysis to obtain $\theta_b\ (\alpha_k,\ t_k)$ and $w_{max}\ (\alpha_k,\ t_k)$ and the other one is the application of an optimization strategy to obtain the optimal solutions.

a) Solution of the buckling problem

We first consider the solution of the buckling problem with a uniform temperature rise of $\theta_1 = \theta$ constant. From equation (10.2-3) follows that $N^{\theta}_{xy} = 0$, $M^{\theta}_{xx} = M^{\theta}_{yy} = 0$ for θ = constant because of the antisymmetric nature of the laminate. Similarly, N^{θ}_{xx}, N^{θ}_{yy} and M^{θ}_{xy} are computed to be constants from equations (10.2-3)-(10.2-5). Substituting the values of the force and the moment resultants into (10.2-1) and noting that $N_{xx} = -N^{\theta}_{xx}$, $N_{yy} = -N^{\theta}_{yy}$, we find that the right hand sides of equations (10.2-1a) and (10.2-1b) are zero and that the right hand side of equation (10.2-1c) is $-N^{\theta}_{xx}w_{,xx} - N^{\theta}_{yy}w_{,yy}$. Hence, a uniform temperature rise θ_1 leads to buckling but not to bending in the case of antisymmetric angle-ply laminates.

The solutions for equations (10.2-1) and (10.6) are obtained by taking the displacement components

$$u_{mn} = A_{mn}\sin\alpha_m x\ \cos\beta_n y\ , \tag{10.2-13a}$$

$$v_{mn} = B_{mn}\cos\alpha_m x\ \sin\beta_n y\ , \tag{10.2-13b}$$

$$w_{mn} = C_{mn}\sin\alpha_m x\ \sin\beta_n y\ , \tag{10.2-13c}$$

where $\alpha_m = m\pi/a$ and $\beta_n = n\pi/b$. Substitution of (10.2-13) for (10.2-1) leads to a homogeneous system of equations. A nontrivial solution of this system is obtained by setting its coefficient matrix equal to zero from which the buckling temperatures $\Theta(m,n,\Theta_1;\ \alpha_k,t_k)$ are computed. The smallest buckling temperature $\Theta_b(\alpha_k,t_k)$ is determined from (10.2-9). For a more detailed discussion of the buckling analysis the reader is referred to [1].

b) Solution of the bending problem

Next, we consider the solution of the bending problem with a linear temperature variation of $\Theta_2 = z/t$. The deformation of the plate prior to buckling is obtained by solving the equations (10.2-1) without the nonlinear terms at the right-hand side of (10.2-1c)[2]. From equation (10.2-3) follows that $N^{\Theta}_{xx} = N^{\Theta}_{yy} = 0$, $M^{\Theta}_{xy} = 0$ for $\Theta = z/t$. N^{Θ}_{xy}, M^{Θ}_{xx} , and M^{Θ}_{yy} are expressed in terms of $\sin \alpha_m x \sin \beta_n y$ as a double Fourier series. Similarly, the displacement components are approximated in terms of u_{mn}, v_{mn}, and w_{mn} given by the equation (10.2-13). Substitution of these terms into (10.2-1) leads to a system of equations, the solution of which yields the transverse deflection $w(x,\ y,\ \Theta;\ \alpha_k,t_k)$. A detailed discussion of the bending analysis for antisymmetric angle-ply laminates is given in [2]. The maximum deflection $w_{max}(\alpha_k,t_k)$ is determined from (10.2-10) by searching for the maximum value of w over the plate domain $0 < x < a$, $0 < y < b$.

c) Pareto-optimal solutions

One can observe that the objective function (10.2-11) is evaluated by solving minimum and maximum problems for the buckling temperature and the maximum deflection, respectively. The multicriteria optimization problem stated in (10.2-11) and (10.2-12) requires the minimization of the objective function vector over the design variables α_k and t_k and leads to a saddle type problem. In obtaining the solution of this problem, we employ the concept of Pareto-optimality as expressed in Definition 1.4 (Chapter 1). In Chapter 1, several strategies for obtaining the Pareto-optimal solutions have been described. Here, the method of objective weighting is applied. From (1.11) and (10.2-11) the preference function $p[\Theta_b, w_{max}]$ is determined to be

$$p[\Theta_b, w_{max}] := w(-\Theta_b) + (1-w)\, w_{max} , \tag{10.2-14}$$

where $0 \le w \le 1$ is the weighting factor.

It can be noticed that the units of Θ_b (temperature) and w_{max} (length) differ. In consequence, there is a large difference between the numerical values of Θ_b and w_{max} which may lead to difficulties in obtaining the numerical solution of the problem. In order to overcome this difficulty, the objective functions are normalized and consequently the weighting factors reflect the relative importance of objectives more closely as suggested in [12]. The following normalized values have been chosen: the buckling temperature and the maximum deflection at $\alpha_k = (-1)^{k+1} \pi/4$, $t_k = 1/n_s$ for $k = 1,2,\ldots,n_s$ which are denoted as Θ_{45} and w_{45}. Thus,

$$\Theta_{45} = \min_{m,n} T(m,\ n,\ \Theta_1;\ (-1)^{k+1}\ \pi/4\ ,\ 1/n_s) , \tag{10.2-15a}$$

$$w_{45} = \max_{x,y} \mid w(m,\ n,\ \Theta_2\ ;\ (-1)^{k+1}\ \pi/4\ ,\ 1/n_s) . \tag{10.2-15b}$$

The normalized preference function used in the numerical computations is given by

$$p[\mathbf{f}(\alpha_k, t_k)] = w\ f_1(\alpha_k, t_k) + (1-w)\ f_2(\alpha_k, t_k) \tag{10.2-16}$$

with

$$\begin{aligned} \mathbf{f} &= (f_1, f_2)^T , \\ f_1(\alpha_k, t_k) &= -\Theta_b(\alpha_k, t_k)/\Theta_{45} , \\ f_2(\alpha_k, t_k) &= w_{max}(\alpha_k, t_k)/w_{45} , \end{aligned} \tag{10.2-17}$$

where Θ_b and w_{max} are defined by (10.2-9) and (10.2-10), Θ_{45} and w_{45} are defined by (10.2-15a) and (10.2-15b), respectively.

d) Min-max formulation

We determine the design at which the maximum of the relative deviations from the respective individual minima are minimized (see (1-19) and [11] for details). The preference function of the scalar substitute problem for the min-max formulation is given by

$$p[\mathbf{f}(\alpha_k, t_k)] := \max_{j=1,4} [z_j(\alpha_k, t_k)] \tag{10.2-18}$$

with

$$z_i(\alpha_k, t_k) = | f_i(\alpha_k, t_k) - \overline{f}_i | / \overline{f}_i \qquad (10.2\text{-}19a)$$

$$z_{i+2}(\alpha_k, t_k) = | f_i(\alpha_k, t_k) - \overline{f}_i | / f_i(\alpha_k, t_k) \qquad (10.2\text{-}19b)$$

where i = 1,2 and a bar denotes the minimum of the corresponding objective function. In equation (10.2-19), f_i's are defined in (10.2-17).

10.2.5 Numerical Results

Pareto-optimal thermoelastic designs of the laminated plates are obtained by solving the problem

$$\min_{\alpha_k, t_k} \; p[\mathbf{f}(\alpha_k, t_k)] \qquad (10.2\text{-}20)$$

where α_k and t_k satisfy (10.2-12) and $\mathbf{f}$ is defined in equation (10.2-17). In equation (10.2-20), $p[\mathbf{f}(\alpha_k, t_k)]$ is defined by (10.2-16) for Pareto-optimal designs and by (10.2-18) for min-max optimum designs. The numerical minimization is achieved by using the downhill simplex method as given in [13].

Design I (α_1)		Design II (α_1, α_2)	Design III ($\alpha_1, \alpha_2, t_1/t$)
Design for minimum w_{max} (w=0)			
α, t	24.1°	22.6° / -37.3°	25.4° / -25.7° / 0.146
w_{max}/w_{45}	0.548	0.506	0.433
θ_b/θ_{45}	1.568	1.621	1.838
Design for maximum θ_b (w=1)			
α, t	27.4°	25.7° / -32.8°	30.4° / -24.5° / 0.136
θ_b/θ_{45}	1.641	1.669	1.904
w_{max}/w_{45}	0.576	0.537	0.451

Table 10.2/1. Single-objective optimal designs with $n_s = 4$, $b/a = 2$

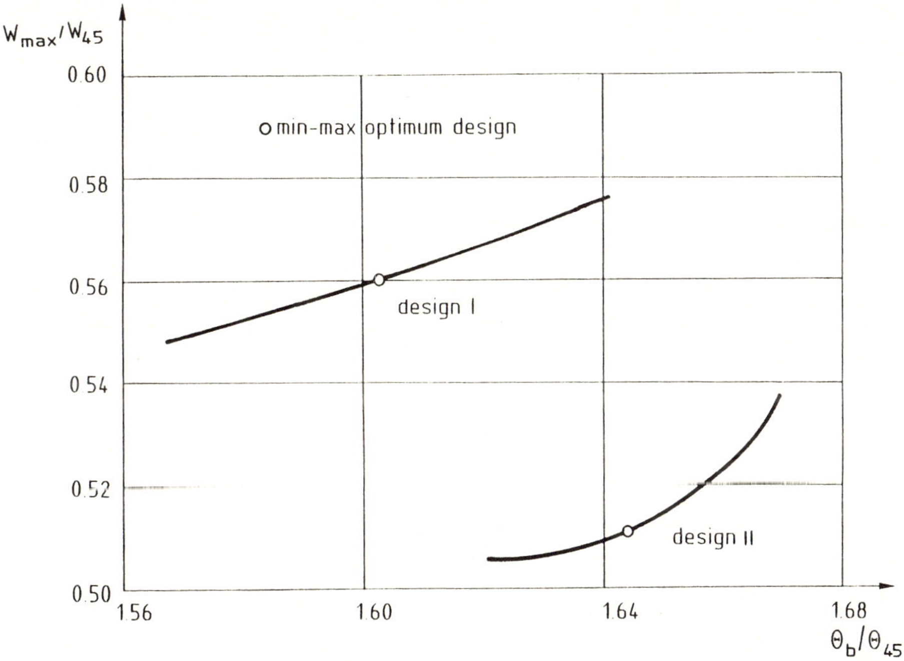

Fig. 10.2/1. Optimal trade-off curves and min-max optimum point for design I (α_1) and design II (α_1,α_2) with n_s = 4, b/a = 2

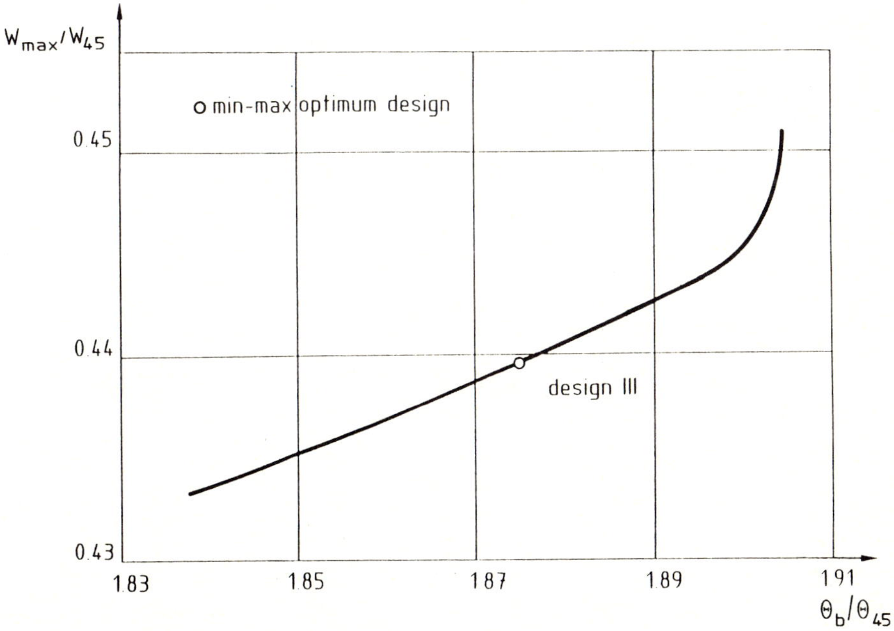

Fig. 10.2/2. Optimal trade-off curves and min-max optimum point for design III (α_1, α_2, t_1) with n_s = 4, b/a = 2

Numerical results are given for four-layered plates constructed of a graphite/epoxy material (T300/5208) which has the material constants E_1 = 181 GPa, E_2 = 10.3 GPa, G_{12} = 7.17 GPam $\nu 12$ = 0.28, α_{T1} = 0.02 10^{-6} K^{-1}, α_{T2} = 30.3 10^{-6} K^{-1} [14]. Optimal designs are given by taking the design variables as

i) Design I $\alpha_1/-\alpha_1/\alpha_1/-\alpha_1$, t_k = 0.25, k = 1,2,3,4
Design II $\alpha_1/\alpha_2/-\alpha_2/-\alpha_1$, t_k = 0.25, k = 1,2,3,4
Design III $\alpha_1/\alpha_2/-\alpha_2/-\alpha_1$, $t_1 = t_4$, $t_2 = t_3 = t - 2t_1$

The results for single-objective designs are given in Table 10.2/1. Note that the corresponding results at $\alpha = \pi/4$, i.e. θ_{45} and w_{45} provide a benchmark for the comparison of optimal designs against a standard plate. Thus, the quantities θ_b/θ_{45} and w_{max}/w_{45} indicate the efficiency of the designs. Table 10.2/1 shows that including the layer thickness t_1 among the design variables considerably improves the optimal design.

Design I (α_1)		Design II (α_1,α_2)	Design III (α_1,α_2,t_1/H)
		w = 0.1	
α,t	24.1°	22.9° / -35.9°	26.1° / -25.1°, 0.149
θ_b/θ_{45}	1.568	1.630	1.846
w_{max}/w_{45}	0.548	0.506	0.435
		w = 0.25	
α,t	24.3°	23.0° / -34.9°	28.5° / -26.2°, 0.145
θ_b/θ_{45}	1.573	1.634	1.902
w_{max}/w_{45}	0.550	0.506	0.446
		w = 0.5	
α,t	27.4°	25.3° / -34.5°	30.0° / -24.8°, 0.140
θ_b,θ_{45}	1.641	1.667	1.904
w_{max}/w_{45}	0.576	0.531	0.450
		Min-max optimum	
α,t	25.7°	23.7° / -34.4°	26.5° / 27.1°, 0.145
θ_b/θ_{45}	1.604	1.644	1.875
w_{max}/w_{45}	0.561	0.513	0.440

Table 10.2/2. Pareto-optimal designs with n_s = 4, b/a = 2

The numerical results for multicriteria designs are given in the form of optimal trade-off curves which show the Pareto-optimal solution set. Figure 10.2/1 shows the curves of w_{max}/w_{45} plotted against θ_b/θ_{45} for Designs I and II with b/a = 2. The min-max point for each case is also shown on the corresponding curve. The end points of each curve correspond to single-objective optimal designs.

Figure 10.2/2 shows the w_{max}/w_{45} vs. θ_b/θ_{45} curve for Design III with b/a = 2 . Design variables and the values of the objective functions for various values of the weighting factor w and min-max optimum designs are given for Designs I, II, and III in Table 10.2/2.

10.2.6 Conclusion

Pareto-optimal fibre orientations and layer thicknesses were determined for simply supported antisymmetric angle-ply laminates subject to uniform and linear temperature fields. The design objectives were specified as the maximization of the buckling temperature and the minimization of the maximum deflection. Methods of weighting objectives and min-max formulation were used to obtain the Pareto-optimal solutions. The objective functions are defined in comparison to the corresponding values of a laminate with equal layer thicknesses and with ply angles alternating at $\pi/4$. Failure criteria as constraints were not taken into consideration in this example.

Numerical results indicated that optimization improves the efficiency of a design considerably. The optimal trade-off curves were given and the location of the min-max optimum solution was determined.

References

[1] Tauchert, T.R.: Thermal Buckling of Thick Antisymmetric Angle-ply Laminates. Journal of Thermal Stresses, 10 (1987) 113-124

[2] Wu, C.H.; Tauchert, T.R.: Thermoelastic Analysis of Laminated Plates. 2 : Antisymmetric Cross-ply and Angle-ply Laminates. Journal of Thermal Stresses, 3 (1980) 365-378

[3] Adali, S.; Duffy, K.J.: Optimal Design of Antisymmetric Hybrid Laminates Against Thermal Buckling. Journal of Thermal Stresses, 13, (1990), to appear

[4] Schmit, Jr., L.A.; Tashkandi, M.A.: Influence of Temperature Change on Optimum Laminate Design. AIAA Journal, 15,9 (1977) 1238-1241

[5] Tauchert, T.R.; Adibhatla, S.: Optimum Thermoelastic Design of Laminated Plates. Journal of Thermal Stresses, 8 (1985) 11-24

[6] Adali, S.: Multi-objective Design of an Antisymmetric Angle-ply Laminate by Nonlinear Programming. ASME Journal of Mechanisms, Transmissions and Automation in Design, 105 (1983) 214-219

[7] Adali, S.: Design Sensitivity Analysis of an Antisymmetric Angle-ply Laminate. Enginering Optimization, 7 (1983) 69-83

[8] Eschenauer, H.: Optimierung ebener Flächentragwerke aus Verbundwerkstoff. Z. Flugwiss. Weltraumforsch. 8 (1984) 367-378

[9] Fuchs, W.: Strukturanalyse und Optimierung anisotroper Schalen aus Faserverbundwerkstoff. Dissertation Universität-GH Siegen 1986

[10] Adali, S.: Optimal Design of Vibrating Structures by Mathematical Programming. In: Levary, R.R. (ed.): Engineering Design - Better Results Through Operations Research Methods. North-Holland New York, Amsterdam, London: Elsevier Science Publishing Company 1988

[11] Jones, R.M.: Mechanics of Composite Materials. New York: McGraw Hill 1975

[12] Osyczka, A.: Multi-criterion Optimization in Engineering. New York, Chichester, Brisbane, Toronto: John Wiley 1984

[13] Press, W.H.; Flannery, B.P.; Teukolsky, S.A.; Vetterling, W.T.: Numerical Recipes. Cambridge: Cambridge University Press 1986

[14] Tsai, S.W.; Hahn, H.T.: Introduction to Composite Materials. Westport, Connecticut: Technomic Publishing Company 1980

10.3 Multicriteria Optimization and Advanced Materials in Telescope Design

K.-H. Stenvers

10.3.1 Motivation and Problem Formulation

The telescope is the basic instrument used in modern astronomy. Astronomers such as Kopernikus and Kepler developed their theories from simple observations with primitive instruments. The invention of Newton's reflector telescope in 1668 increased the interest in celestial research as, due to the small aperture of the human eye, most phenomena could only be observed with such an instrument. The construction of the first large telescope by Herschel in 1780 clearly showed that astronomical research depends on the efficiency of such instruments. In recent decades these instruments have become larger, more powerful and technically more complex, reflecting the state of technological development in industrialized nations. It is therefore not surprising that high technology is required in telescope construction and that technological capabilities are futher enhanced because of the very high demands of astronomy.

The increasing demands on the accuracy of a telescope depend on disclosing shorter and even unexplored wavelengths (< 1 mm) of cosmic radiation. The deviation of the total reflector surface from the ideal parabolic shape should be no more than 1/20 of the wavelength under investigation. This means that for the submillimeter range (approx. 0.3 mm) the deviation may not amount to more than 17 μm. Even in today's most advanced 30-meter-millimeter-wave Radio Telescope (MRT) in Spain [1], already mentioned in Section 1.3.1, the deviations arising in the steel structure under unfavourable temperature conditions (i.e. one-sided sun radiation) could only be controlled by insulating completely the exterior of the entire telescope as well as by using a very complicated and thus expensive temperature control system for the inner steel structure. For these reasons the telescope structure of the new 10-meter-Submillimeter Radio Telescope (SMT) for the Max-Planck

Institute in Bonn was developed using **C**arbon **F**iber-**R**einforced **P**lastic (CFRP). A special aspect in the application of CFRP is that components can be designed with very low coefficients of thermal expansion as well as a high degree of stiffness at low weight. In order to meet the specified accuracy requirements as far as possible the SMT-development was carried out by special *mathematical optimization procedures.* The material composition, various components, and also the entire structure were designed by a computer. This section treats the methods as well as some of the results of the structural optimization procedure.

10.3.2 Program System for Structural Analysis and Optimization

In Chapter 2 the basic structure of the optimization procedure SAPOP was introduced. In its essential points ("three columns") the following procedure is very similar to SAPOP. The basis of optimization is a "reasonable" structural analysis. The Finite-Element Method (FEM) has proved to be an adequate method for the solution of problems of structural mechanics. Excellent software with pre- and postprocessors is available for this task. It is quite obvious that optimization modules should be added to the existing programs. A more difficult task is the extensive and secure data handling.

Within the scope of developing telescopes, a program system for optimization problems (AOP, i.e. **A**nalysis and **O**ptimization **P**rogram) [2] was installed and developed by Krupp Industrietechnik. The program consists of four selfcontained units (Fig. 10.3/1) which are described in more detail below.

a) Structural analysis

The transformation of a given design into a mathematical-mechanical model is an important step on the way to define the problem-specific program parts more precisely. For AOP analyses, the FEM programs SAP IV [3] and ASKA [4] are available. ASKA has a very extensive element library which makes high level elements available to the user. The individual elements provide different problem-specific characteristics. The integrated program module ASKAMESH is available to the data input so that this time-consuming and expensive process can be considerably reduced. The ASKAVIEW postprocessor is used for the data output which simplifies and improves the interpretation of results.

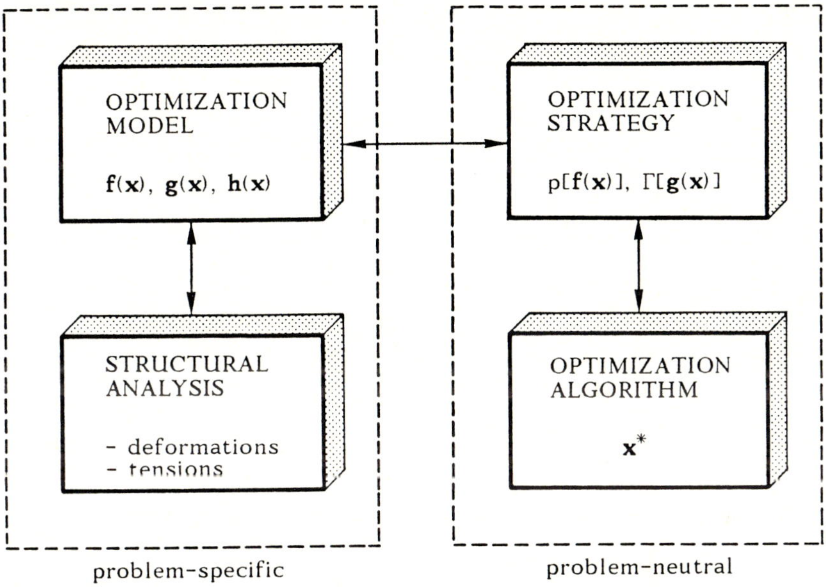

Fig. 10.3/1. AOP (**A**nalysis and **O**ptimization **P**rogram) program structure

b) Optimization model

An additional and very important step in optimization is the development of an optimization model. This includes the determination of design variables as well as objective and constraint functions.

c) Optimization strategy

This part of the program contains all transformation rules necessary for the transformation of a vector optimization problem into a scalar substitute problem. This includes the preference function (1-13b):

$$p\,[\mathbf{f}(\mathbf{x})] = \left\{ \sum_{i=1}^{m} w_i \left(\frac{f_i(\mathbf{x})}{\overline{f}_i}\right)^2 \right\}^{1/2} \qquad (10.3\text{-}1)$$

with $\sum w_i = 1,\ w_i \geq 0,\ \overline{f}_i \neq 0$.

Here, the decision maker has to indicate the objective function values $\overline{f}_1,\ldots,\overline{f}_m$ (specification values) to be attained as well as a quadratic penalty-function [5]:

$$\Gamma_j\,[g_j(\mathbf{x})] = \begin{cases} \dfrac{1}{g_j(\mathbf{x})} & \text{if } g_j \geq g_o, \\ \dfrac{1}{g_o}\left[\left(\dfrac{g_j(\mathbf{x})}{g_o}\right)^2 - 3\,\dfrac{g_j(\mathbf{x})}{g_o} + 3\right] & \text{if } g_j < g_o \end{cases}$$

(10.3-2)

with $g_o = C_o r^m$
$1/2 \geq m > 1/3$,
C_o = const.,
r = penalty parameter.

The extended substitute objective function

$$\Phi(\mathbf{x},r) = p[\mathbf{f}(\mathbf{x})] + r \sum_{j=1}^{q} \Gamma_j[g_j(\mathbf{x})] \tag{10.3-3}$$

contains p and Γ. The so-called multicriteria vector optimization task can be solved by considering constraints indirectly. This can be advantageous for certain problems. Direct strategies for processing the constraints are contained in the selected optimization methods.

d) Optimization algorithm

A method of zero order [6], a method of the quasi-Newton class [7], and an approximation method [2] have been chosen from the well-known methods as they prove to be efficient in other application examples (e.g. [8]). The latter method has been used for the subsequent optimization task and will therefore be explained in more detail: Within the approximation process the nonlinear starting problem is replaced by a subproblem requiring less calculation time and is solved using a strategy for nonlinear problems. This also allows the use of generally nonlinear subproblems. The k-th iteration step (k=1,2,...) is:

$$\min \left\{\hat{\Phi}^{(k)}(\mathbf{x},r) \mid g_{q+1}^{(k)}(\mathbf{x}) \geq 0 , \quad \mathbf{x}^l \leq \mathbf{x} \leq \mathbf{x}^u \right\} . \tag{10.3-4}$$

The solution of the subproblem k-l is the starting point of the problem. The function

$$\hat{\Phi}^{(k)}(\mathbf{x},r) = p[\hat{\mathbf{f}}(\mathbf{x})] + r \sum_{j=1}^{q} \Gamma_j[\hat{g}_j(\mathbf{x})] \tag{10.3-5}$$

contains the approximated objective function components f_i and the inequality constraints g_j, $\hat{f}_i$ and $\hat{g}_j$ are linear, quadratic or also generally nonlinear functions (e.g. approximation with spline-functions). With

$$g_{q+1}^{(k)}(\mathbf{x}) = R^{(k)} - | \mathbf{x}-\mathbf{x}^{(k)} | \geq 0 \tag{10.3-6}$$

a hypersphere with the radius $R^{(k)}$, limits on the permissible design space (move limits), are introduced. Its initial radius $R^{(O)}$ is a diagonal going through the design space which is limited by the upper and lower bounds of the design variables.

The reduction of the radius $R^{(k)}$ is achieved by multiplication by Fibonacci numbers, thus ensuring a sufficiently large overlapping of successive iterations. From any starting point $\mathbf{x}_O$, every optimal point $\mathbf{x}^*$ can be reached. On the other hand, after only 3-8 iterations (depending on the choice of the function approximation and the optimization task) the permissible design space is limited to a size for which further improvements of the objective functions are of little importance in practical application.

10.3.3 Application to a Submillimeter-Telescope

Under normal environmental conditions, a 10-meter-SMT is used for astronomical purposes in the submillimeter range without limitations. For designing purposes the following specifications have to be considered:

1. Reflector surface accuracy

Total deviation	≤ 17 μm rms,
- telescope structure	≤ 10 μm rms,
- panels	≤ 10 μm rms,
- measurement and adjustment	≤ 10 μm rms;

2. Positioning and tracking accuracy

- positioning error	≤ 1 arcsec,
- tracking error	≤ 1 arcsec;

3. Eigenfrequency

- total structure	≥ 5 Hz,
- reflector	≥ 10 Hz.

The mean quadratic deviation σ (rms) from a paraboloid with a best-fit surface is used as the measure for the reflector surface accuracy:

$$\sigma = \sqrt{\frac{\mathbf{v}^T \mathbf{v}}{n - q}} \qquad (10.3\text{-}7)$$

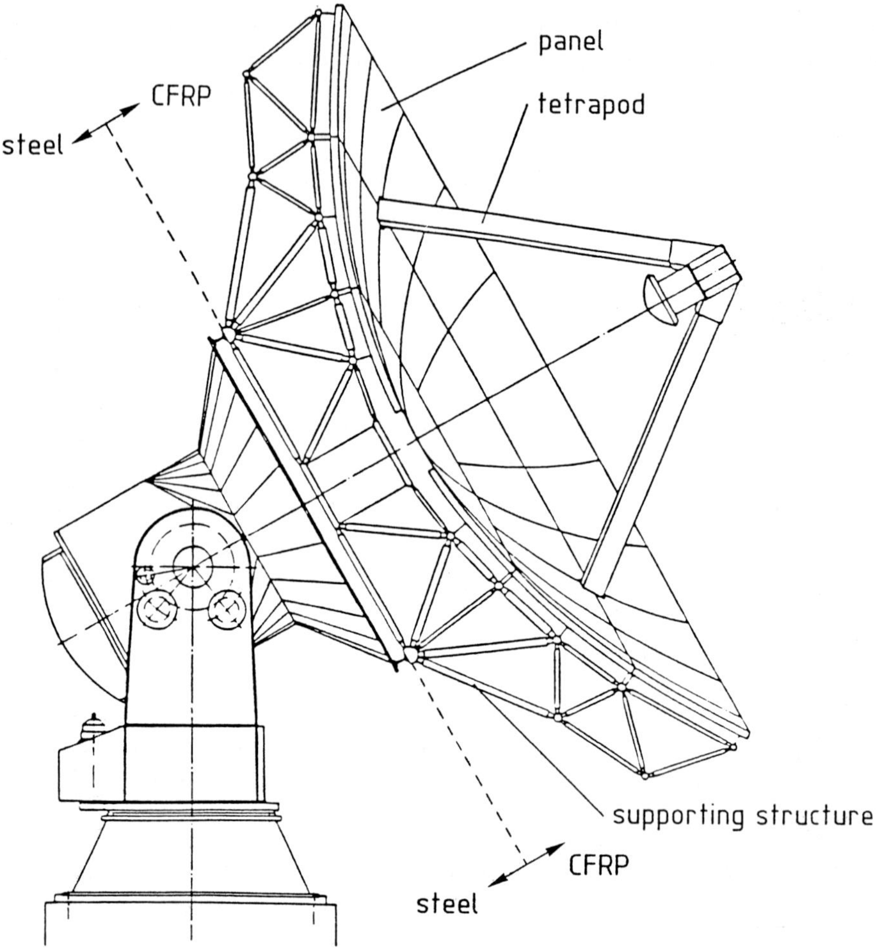

Fig. 10.3/2. Design of the 10-m-submillimeter radio telescope

with
- **v** vector of the remaining deviation from the best-fit surface,
- n number of the reflector surface points,
- q number of free parameters.

a) Telescope structure

The optimization process should start with establishing the best possible starting design. This demands the design engineer's creativity. Various concepts for the telescope structure have been studied. Fig. 10.3/2 shows the design which was judged to be the most suitable starting design as regards accuracy, panel shape and costs. The reflector surface (panels), the trusses, the central cylinder of the relector, and the tetrapod together with the head are made of CFRP.

The remaining part of the section will be confined to the subject of optimizing the telescope structure. Apart from an extremely high surface accuracy under various loads, minimal weight is required. The following objective functions are determined:

	rms-values due to
$f_1(\mathbf{x}) \hat{=} \sigma_H$	dead weight in horizontal position,
$f_2(\mathbf{x}) \hat{=} \sigma_Z$	dead weight in zenith position,
$f_3(\mathbf{x}) \hat{=} \sigma_{\Delta T_1}$	constant cooling of all antenna components by $\Delta T_i = 20$ K,
$f_4(\mathbf{x}) \hat{=} \sigma_{\Delta T_2}$	linear temperature distribution in the telescope from bottom to top in horizontal position with $\Delta T2 = 6$K,
$f_5(\mathbf{x}) \hat{=} \sigma_{\Delta T_3}$	linear temperature distribution in the reflector supporting structure from the reflecting side to the rear with $\Delta T3 = 4$ K,
$f_6(\mathbf{x}) \hat{=} \sigma_{\dot{T}}$	rapid change of the surrounding temperature,
$f_7(\mathbf{x}) \hat{=} \sigma_{wO}$	12 m/s wind at 0° elevation (horizontal position),
$f_8(\mathbf{x}) \hat{=} \sigma_{w50}$	12 m/s wind at 50° elevation,
$f_9(\mathbf{x}) \hat{=} \sigma_{w90}$	12 m/s wind at 90° elevation,
$f_{10}(\mathbf{x}) \hat{=} \sigma_F$	humidity influence,
	and
$f_{11}(\mathbf{x}) \hat{=} W$	dead weight.

For the telescope structure the following design variables were selected:

$x_1, \ldots, x_{60}$	stiffness of various strut types,
$x_{61}, \ldots, x_{63}$	wall thickness of central cylinder, tetrapod, and connecting elements,
$x_{64}, \ldots, x_{66}$	construction height of the reflector supporting structure,
$x_{67}, \ldots, x_{69}$	thermal expansion coefficients for the CFRP components,

and limited by manufacturing considerations:

$$x_{iu} < x_i < x_{iO} \qquad (i = 1,\ldots,69).$$

In order to determine objective functions, the deviation of the telescope established by the FEM-program ASKA is necessary. Fig. 10.3/3 shows

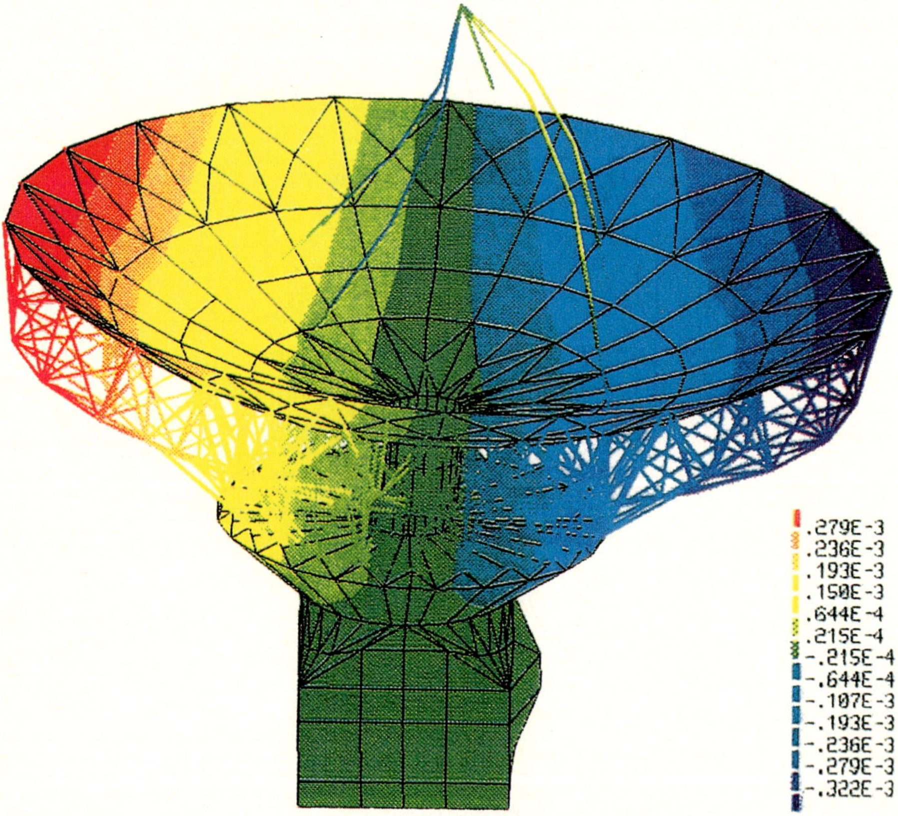

Fig. 10.3/3. Calculation model and deviation of the telescope structure under dead weight in horizontal position

the calculation model of the telescope and the deviation under dead weight in horizontal position. Deviation in vertical direction is highlighted in colour (numbers in m). An rms-program determining the best-fit surface for the deviating structure and calculating the rms-value according to equation (10.3-7) is coupled with the ASKA-program. Now, the optimization module performs iterative variations of the design variables until a termination criterion is reached.

Fig. 10.3/4 exemplifies the result of this optimization calculation for the objective function f_i and the dead weight in horizontal position. The contour deviations from the best-fit surface for the initial design with an rms of 14.9 μm and the optimal design with an rms of 5.2 μm is illustrated. Further results of the optimization calculation are presented in Table 10.3/1.

The contour deviation under dead weight can be influenced by adjusting the panels. An optimal adjustment position β_O is obtained by using the equation

$$\frac{1-2\cos\beta_O}{1-2\sin\beta_O} = \frac{\sigma_Z^2}{\sigma_H^2} . \tag{10.3-8}$$

The remaining contour deviation, depending on the elevation angle, is calculated as follows:

$$\sigma_\beta = \{[(\cos\beta - \cos\beta_O)\,\sigma_H]^2 + [(\sin\beta - \sin\beta_O)\,\sigma_Z\,]^2\}^{1/2} . \tag{10.3-9}$$

Fig. 10.3/5 shows this calculation for the optimized telescope structure. The total error of the telescope structure is composed of the quadratic addition of these loads which may occur simultaneously. They are:

$$\sigma_{tot}^2 = \sigma_{\beta=80}^2 + \sigma_{\Delta T_1}^2 + \sigma_F^2 + \sigma_{W50}^2 . \tag{10.3-10}$$

Based on the optimized values of Table 10.3/1 the telescope structure shows a deviation of 7.5 μm so that the desired specification of 10 μm is not exceeded.

b) Composite materials

When developing the SMT, it was important to fulfill the extreme demands made on the CFRP-materials such as stiffness, thermal expansion, form stability, and costs. It was realized that constant heating (or cooling) is a critical temperature load for composite materials. Thus, the CFRP-components had to be brought into line with each other as well as with the connecting elements made of INVAR (a special alloy with a thermal expansion coefficient of 1.3×10^{-6} 1/K). The position of the CFRP-reflector on the steel elevation part was designed in such a manner that the thermal expansion of CFRP could be decoupled from that of the steel. It was also realized that shape deviations of the CFRP-material due to humidity have to be considered. The objective functions of the CFRP-material read as follows:

- $f_1(\mathbf{x})$ minimum coefficient of thermal expansion,
- $f_2(\mathbf{x})$ maximum stiffness,
- $f_3(\mathbf{x})$ minimum shape deviations due to humidity,
- $f_4(\mathbf{x})$ minimum weight.

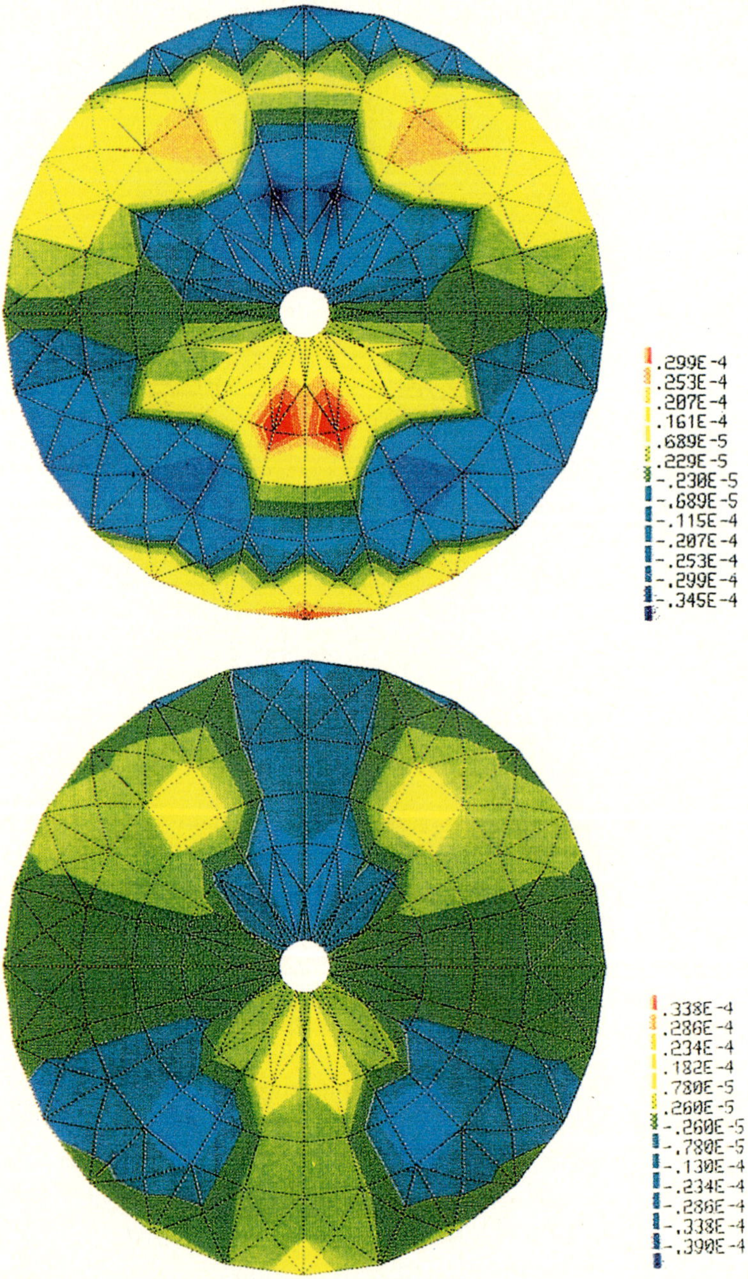

Fig. 10.3/4. Deviation from a best-fit surface under dead weight in horizontal position

a) Initial design with 14.9 μm rms

b) Optimal design with 5.2 μm rms

Objective functions	rms-value (μm)
Dead weight in horizontal position	5.2 (2.8)*
Dead weight in zenith position	3.1 (2.8)*
Constant cooling of all antenna components with ΔT_1 = 20 K	3.2
Linear temperature distribution in the telescope from bottom to top in horizontal position with ΔT_2 = 6 K	0.6
Linear temperature distribution in the reflector supporting structure from the reflecting surface to the rear with ΔT_3 = 4 K	0.6
Rapid change of the surrounding temperature	2.9
12 m/s wind at 0^0 elevation (horizontal p.)	2.0
12 m/s wind at 50^0 elevation	5.4
12 m/s wind at 90^0 elevation	4.0
Humidity influence	< 3.0 **

* under consideration of adjustment

** after conditioning of all CFRP-components

Table 10.3/1. rms-values of single loadings of the optimized 10-m-telescope sructure

However, the development of a mathematical model for the mechanics of fibre composite components requires a thorough knowledge of the material characteristics. For this purpose experiments on the designed components have to be carried out in order to verify the quality of the model. Computation models based on the works [9-11] proved to be efficient for the given problem.

A particular problem of calculating the fibre-composite components arises from the fact that particular parameters (e.g. fibre angle, layer thickness, fibre volume share etc.) are subject to certain variations. Slight changes of the input values of the calculation lead to changes of the desired characteristics of a CFRP-component. Fig. 10.3/6 shows a light-optical photo of a CFRP-laminate illustrating the irregularities of the material which depend on the construction method. Random parameters can be described by means of a probability calculation.

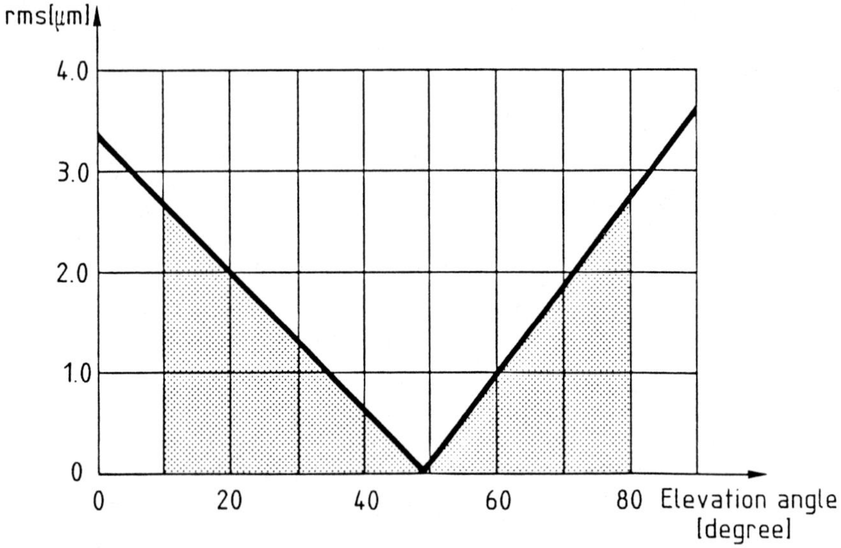

Fig. 10.3/5. Precision of the optimized SMT-reflector structure under dead weight with respect to the elevation angle including alignment

Certain assumptions regarding the probability model of the random variables make the calculation of the effect on the objective function possible [2].

Fig. 10.3/7 compares a stochastic calculation with the measured values of a CFRP-strut in an initial design. The E-modulus (parameter for stiffness) and the thermal expansion coefficient of the strut in the longitudinal direction of the tube and the corresponding probabilities are indicated so that information on the reliability of the required characteristics can be stated. Furthermore, the stochastic calculation shows that the probability variables influence the deviation of the objective functions, which allows the improvement of the quality of the mathematical model as well as the safety quality possible during the construction.

For the optimization of CFRP-material, the following design variables have been selected (Fig. 10.3/8):

$x_1, \ldots, x_{ns}$	layer thickness,
$x_{ns+1}, \ldots, x_{2ns}$	fibre angle,
x_{2ns+1}	fibre volume fraction,
x_{2ns+2}	fibre type,
x_{2ns+3}	matrix system.

M 100 : 1

— ≙ 50 µm

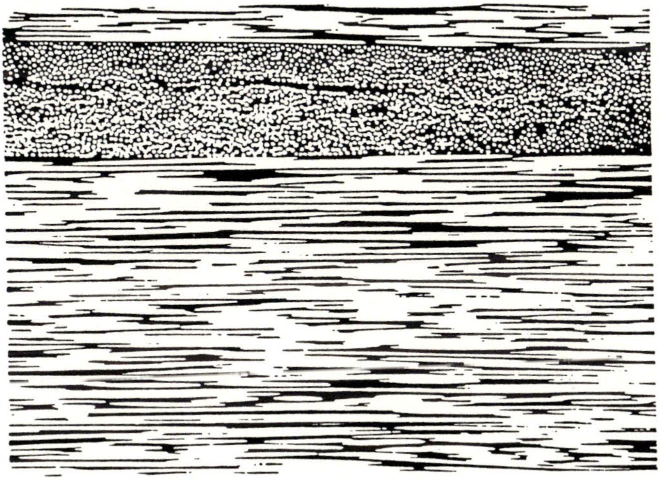

a)

M 500 : 1

— ≙ 10 µm

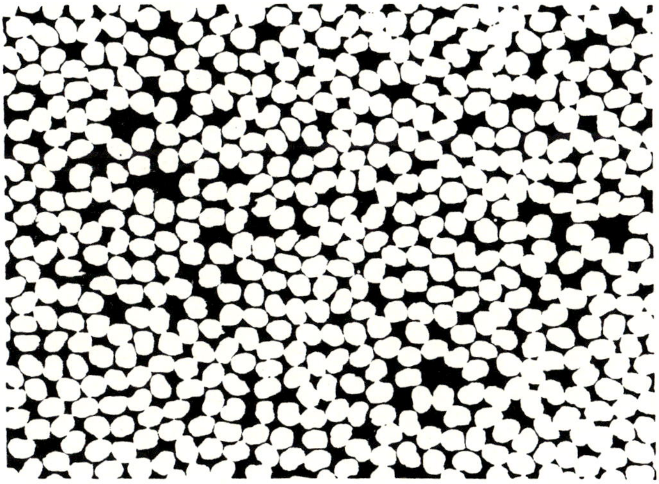

b)

Fig. 10.3/6. Light-optical photo of a CFRP-laminate at two different magnifications illustrating the micro-structure

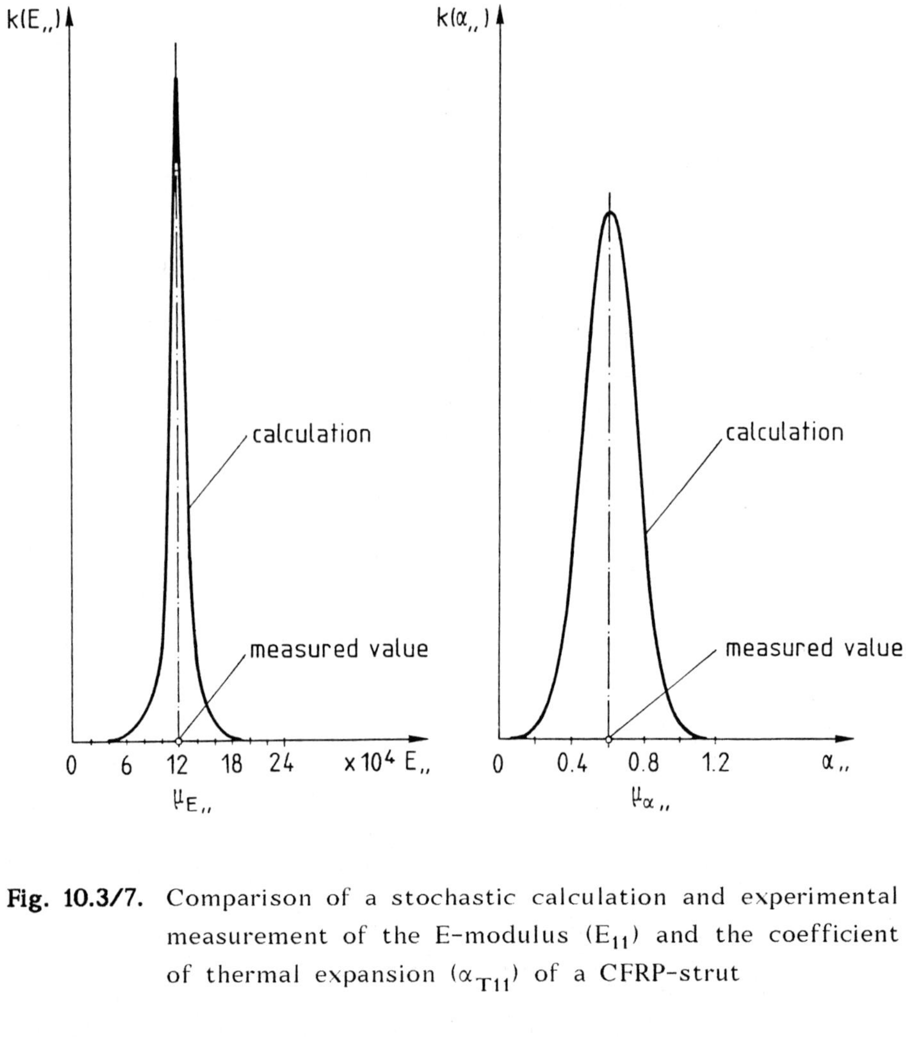

Fig. 10.3/7. Comparison of a stochastic calculation and experimental measurement of the E-modulus (E_{11}) and the coefficient of thermal expansion (α_{T11}) of a CFRP-strut

The results of the optimization of the CFC-struts are given in Fig. 10.3/9. The functional-efficient boundary can be established by different weighting of the single objective functions (f_1 and f_2) and the optimization calculation. The figure shows the theoretically optimal combination of stiffness and thermal expansion of a CFRP-strut for two different fibre types. The comparison with tested tubes from various suppliers which did not have the optimal laminate structure is very informative (indicated with 0 in Fig. 10.3./9). Here, the functional-efficient boundary is confirmed as the boundary curve of different designs.

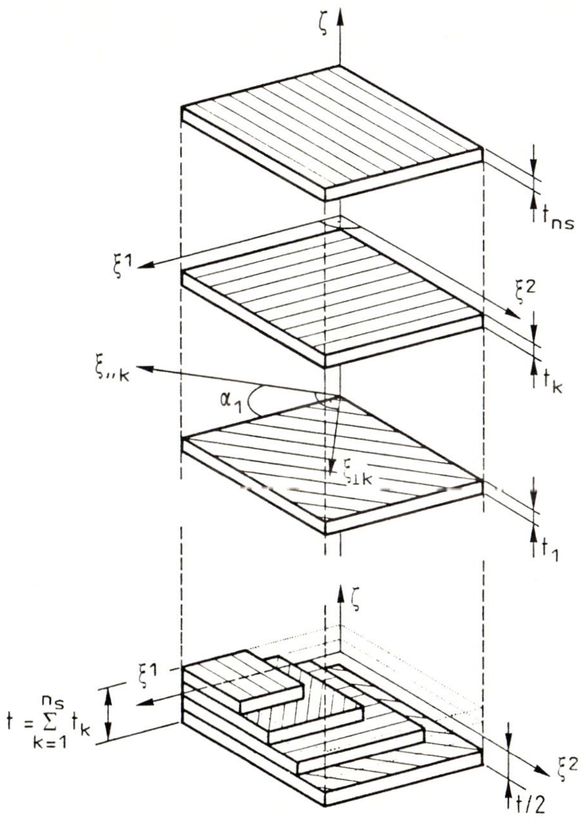

Fig. 10.3/8. Laminate composition and layer identification of a fibre composite material

Finally, theory and practice are compared with regard to an optimized CFRP-strut with connecting elements and nodal points made of INVAR. The values are given in Table 10.3/2. The measurement verification shows that a good correlation between the measured values and the predicted values can be achieved.

10.3.4 Conclusion and Future Prospects

When developing highly accurate radio telescopes, mathematical methods of structural optimization proved to be successful. In both, components design and dimensioning of the telescope structure, optimization procedures provide better results than the conventional trial and error

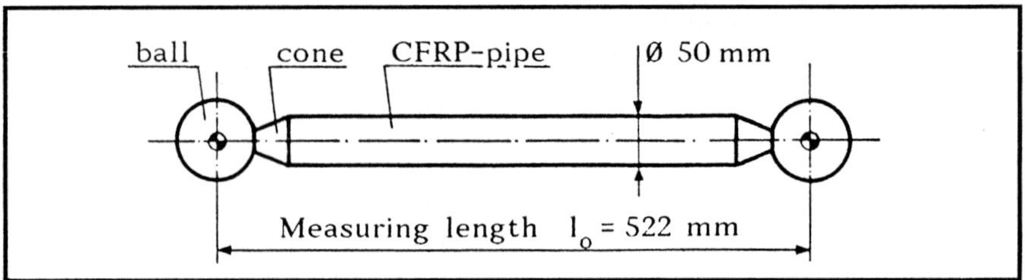

Objective Function	Dim.	Calculation	Measurement
Flexibility	μm/kN		
Total strut		10.75	10.76
- Ball		0.38	-
- Cone		1.44	-
- Slot		0.64	-
- CFRP-pipe		8.29	8.16
Thermal expansion coefficient	-10^{-6}[1/K]		
Total strut		0.62	0.65
- CFRP-pipe		0.33	0.31
Swell coefficient	10^{-6}[1/%]		
Total strut		121	-
- CFRP-pipe		158	~ 120

Table 10.3/2. Optimized CFRP-struts: comparison of calculated and measured results

methods. Thus, optimization is an efficient method for the design engineer - a fact which is confirmed by the results of telescope development.

Within the field of designing composite materials the humidity problem was covered by calculation. The accompanying experiments did not always provide results which could be interpreted and furthermore showed great variations with different matrix systems and design methods. Although a best-fit system was selected and an optimization was carried out, the humidity expansion of the SMT is still too large. Consequently, all CFRP-components were conditioned to avoid the humidity gradients of the component becoming too large. Here, the matrix system (CFRP-fibre does not absorb moisture) has to be improved in future as far as moisture absorption and swelling are concerned.

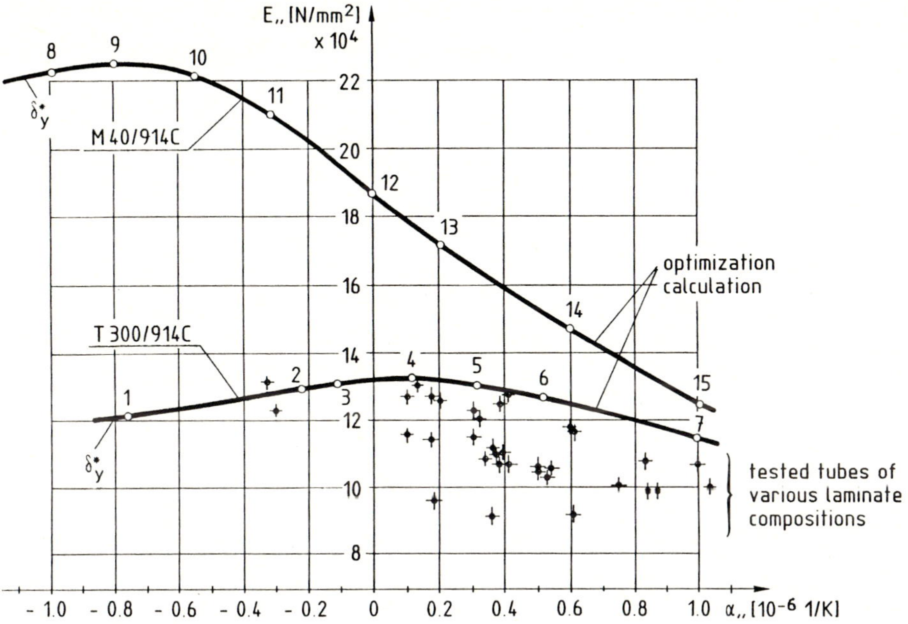

Fig. 10.3/9. Functional-efficient boundaries of CFRP-struts and comparison with measured sample tubes

Using stochastic methods, it was possible to calculate the initial parameters of a laminate quickly and to have them verified by measurements. This allowed us to predict the behaviour of the CFRP-components very precisely. In future it will be very important to minimize the calculated and measured deviations of the material characteristics for high-efficiency composite materials.

References

[1] Baars, J.W.M.; Hooghoudt, B.G.; Mezger, P.G.; De Jonge, J.J.: Das 30-m.-Radioteleskop für Millimeter-Astronomie. Sterne und Weltraum 7-8, 1986, 382-391

[2] Stenvers, K.H.: Stochastische Vektoroptimierung zur Auslegung von Tragwerkstrukturen. Dissertation Universität Siegen, 1985

[3] Bathe, K.J.; Peterson, F.E.; Wilson, E.L.: SAPIV, A Structural Analysis Program for Static and Dynamic Response of Linear Systems. University of California, Berkeley, 1974

[4] Saab-Scania; Ikoss: ASKA (Automatic System for Kinematic Analysis). Softwaresystem für statische, dynamische und thermische Strukturanalyse. Version 8.6. IKO Software Service GmbH, Stuttgart

[5] Haftka, R.T.; Starnes, J.H.: Applications of a Quadratic Extended Interior Penalty Function for Structural Optimization. AIAA Journal 14 (1976) 718-724

[6] Jacob, H.G.: Fortran-Program zur Ermittlung eines lokalen Optimums einer beschränkten Multivariablen Gütefunktion ohne Kenntnis ihrer Ableitungen. PDV-Bericht 36, 1975

[7] Dennis, J.E.; More, J.J.: Quasie-Newton Methods, Motivation and Theory. SIAM Review 19 (1977) 46-89

[8] Eschenauer, H.; Kneppe, G.; Stenvers, K.H.: Deterministic and Stochastic Multiobjective Optimization of Beam and Shell Structures. J. of Mechanisms, Transmissions and Automation in Design, ASME Vol. 108 (1986) 31-37

[9] Cuntze, R.: Grundlagen für die Berechnung von Rotationsschalen aus Faserverbundwerkstoff. Habilitationsschrift TU München 1978

[10] Niederstadt, G..; Nitsch, P.: Diffusion, Wärmedehnung und Quellung von CFK. DFVLR Braunschweig IB 131-84/06, 1984

[11] Gädke, M.; Geier, B.; Niederstadt, G.; Prinz, R.: Festigkeit und Gestaltung von Bauteilen aus Faserverbundwerkstoffen. Lehrgangsskript F7.04, 1986

10.4 Multicriteria Design of Ceramic Components

J. Koski, R. Silvennoinen

10.4.1 Introduction

Ceramic materials have recently emerged in various electrical and mechanical engineering applications where sufficient high-temperature capability, wear and corrosion resistance, certain optical or electrical properties are needed. Typical industrial products with ceramics are for example combustion engines, several bearing applications, cutting parts of tools, nozzles and microcircuit packages. Many of these components represent load supporting structures where previously more than one metal has been used. Although the strength of ceramics may be comparable to typical construction steels, their brittle behaviour even at high temperatures and the large variation in strength properties make the applications to load supporting structures quite complex. The number of ceramic materials is large, but a common feature for them is a relatively high price which is due to specialized manufacturing techniques [1]. It is very difficult to control the quality of a component during the manufacturing process, and thus flaws and other defects usually appear in the final product. By applying nondestructive flaw detection measurements and linear fracture mechanics it is possible to predict the life time of each individual component to some extent. However, it is necessary to evaluate the reliability of a structure during the design process by using the material data obtained from the manufacturers. A reliability evaluation based on the Weibull distribution has been generally accepted in ceramic design. This leads to a statistical approach where the designer may choose the probability of failure for each component separately, depending on its purpose of application. The reason which makes the multicriteria optimization so convenient for dealing with ceramics is that in practice it is extremely difficult to fix the allowable probability values.

Most evidently the application of ceramics offers totally new solution possibilities for many traditional constructions. For example the chance to leave out the cooling system in some engines or the possibility to improve the efficiency in gas turbines by raising the working temperature certainly adds interest in these alternative materials. The economic potential associated with ceramics seems large, but severe restrictions appear if high reliability for load supporting components is required. In particular, the shape optimization becomes an important part of the design process because structural reliability is closely associated with the geometric shape of the component. Much attention has been devoted to structural shape optimization, but applications to ceramics, as far as the authors know, have not been published. The purpose of this chapter is to present some basic ideas concerning the structural analysis and shape optimization of ceramic components. The theoretical background has been supplemented by a simple illustrative cantilever plate example as well as by a complicated real life application.

10.4.2 Structural Analysis and Reliability of Ceramic Components

Usually, the strength of a structure is evaluated by determining its stress distribution and by applying some suitable strength hypothesis, for example von Mises or Tresca yield condition for ductile materials, to compute the reference stress at the critical point. This value is compared to the allowable stress which is typical of each material and of the intended function and which can be found in design codes. Consequently, the strength of a component is generally determined by the stress state of one critical point only where the reduced stress has its largest value. Naturally, some other loading effects, such as fatigue, vibration, and the loss of stability for example, may damage the structure, but they are not considered in this context.

Constitutive linearity is characteristic of ceramics because practically no yielding, even at high temperatures, can be found during strength tests. Thus, the material behaviour is linearly elastic up to the fracture point where an unstable crack growth suddenly appears. A typical stress-strain line segment is shown in Fig. 10.4/1 where also the corresponding curve for a mild steel has been depicted for comparison. Fracture point A corresponds to mean fracture stress σ_c which may be high compared to yield stress σ_y of mild steel. Strength value σ_c

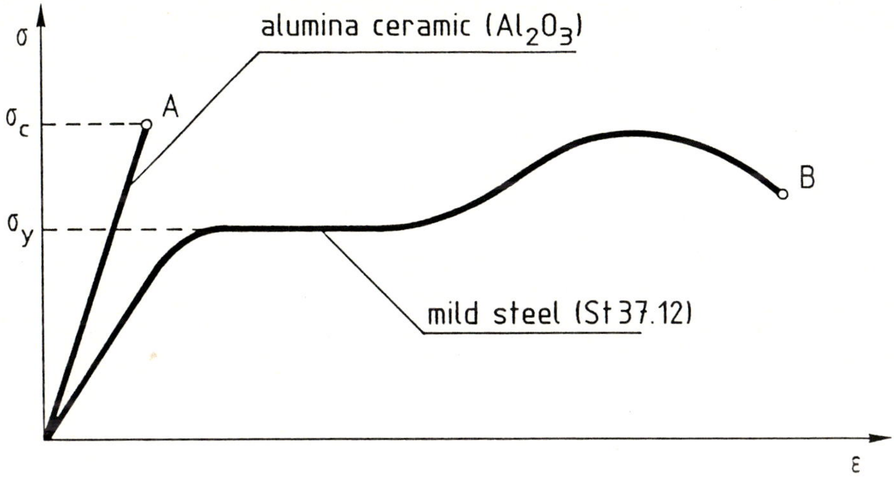

Fig. 10.4/1. Schematic representation of a stress-strain curve for ceramics and mild steel

depends on the material volume of the specimen. In spite of the relatively high mean fracture stress σ_c, problems arise in practice because the strength is extremely sensitive to small defects in the material which may be caused by the manufacturing process or by environmental effects during the everyday use. This leads to the application of a statistical approach in the structural analysis of ceramics which, on the other hand, can be performed by using a geometrical and constitutive linearity in most cases. This results in the numerical solution of the stiffness equation

$$\mathbf{K}\mathbf{u} = \mathbf{p} , \tag{10.4-1}$$

where **K** is the stiffness matrix, **u** is the nodal displacement vector and **p** is the nodal load vector. In the case of several loading conditions, as in the analysis of the piston crown discussed in the sequel, **u** and **p** are interpreted as matrices, each column corresponding to one loading condition.

Instead of considering the reference stress at one critical point only, it seems reasonable to determine the reliability of a ceramic component from the analysis results by using a statistical approach. Because of the large variation of the fracture stress obtained from strength tests it is very difficult to evaluate the real stength of each individual component, but instead the probability of failure can be assessed after computing the overall stress field. Among several possibilities a two parametric Weibull distribution has proved flexible and easy to use for this purpose [1-5]. The two material parameters used in this context are the mean fracture stress σ_c and the Weibull modulus m which are both obtained from the material data.

An interesting phenomenon observed in strength tests is the volume effect which appears such that the strength decreases when the size of a specimen is increased, unlike in metals where the yield or fracture stress is relatively independent of the material volume. The reason for this effect is that the probability of a fatal flaw in the component increases along with its material volume subject to tensile stresses. Usually, the effect of compressive stresses is neglected because they rather tend to close the flaw whereas crack opening tensile stresses are the main contributors to the fast fracture process of ceramics. Accordingly, the mean fracture stress σ_c is associated with certain reference volume V_c which usually comprises that part of a specimen where tensile stresses occur. The other material parameter, Weibull modulus m, can be interpreted as a measure for the narrowness of the strength distribution. The larger the value m, the smaller is the variation of the fracture stress of the material. Thus, it may be advisable in some applications to choose a material with a smaller σ_c if it has a larger Weibull modulus m because in this case the designer can better rely on the given strength value.

The probability of failure, denoted here by P_f, is the mathematical complement of the reliability P_r, i.e. $P_f = 1 - P_r$. These quantities can be computed for a ceramic component from its stress field by using the weakest link theory based on the Weibull distribution. Accordingly, it is assumed that the weakest crack-stress combination will cause the overall damage, but the location of this critical point is not known. In the case of a multiaxial stress state it is further assumed that the effect of each tensile principal stress at every material point is considered separately without any mutual interrelationship. The mathematical expression

$$P_f = 1 - e^{-\chi} \quad , \tag{10.4-2}$$

where the exponent is

$$\chi = \left(\frac{1}{m}!\right)^m \left(\frac{1}{\sigma_c}\right)^m \left(\frac{1}{V_c}\right) \iiint_V (\sigma_1^m + \sigma_2^m + \sigma_3^m)\, dV \, . \tag{10.4-3}$$

can be derived for the probability of failure of a ceramic component [1-5]. Here, σ_1, σ_2, and σ_3 are the principal stresses which are included in the volume integral at each point only if they are tensile (i.e. $\sigma_i < 0 \Rightarrow \sigma_i^m = 0$). Notations σ_c, V_c, and m are the material parameters obtained from the strength tests, as was already mentioned in the preceding text, whereas V denotes the volume of a component. The term

$$\left(\frac{1}{m}!\right) = \Gamma\left(\frac{1}{m} + 1\right) = \int_0^\infty t^{\frac{1}{m}} e^{-t}\, dt \,, \qquad (10.4\text{-}4)$$

which is the value of the gamma function at $\frac{1}{m} + 1$, can be easily computed, but problems may arise in determining the numerical value of the volume integral. From expression (10.4-3) it can be observed that those parts of the component with high tensile stresses are crucial from the failure point of view.

10.4.3 Multicriteria Shape Optimization Problem

Shape optimization has recently become an important tool in the finite element based structural design and applications to different mechanical, civil, aeronautical, and space engineering products can be found at the moment [6-11]. Usually, the purpose is to find such a boundary curve or surface for a structure as minimizes its material volume subject to some imposed constraints. Shape optimization can be performed independently, or can be combined with the commonly used cross-sectional optimization just by introducing the shape variables as additional design variables to the problem. Generally, constraints on stresses, displacements, natural frequencies, or some geometric properties are imposed, and a scalar weight minimization problem is obtained. One basic question, common to most engineering design tasks, still remains even if the scalar problem is solved numerically. It is the question whether the constraint limits could be chosen better. Multicriteria optimization is a flexible approach in this situation and in the sequel a general bicriterion problem for ceramics is formulated and its application to shape optimization is considered.

Obviously, the material volume or structural weight can be used as one important criterion which should be minimized also for ceramics. Another important physical quantity for the designer is the probability of failure of a component which presumably is strongly conflicting with the material volume. Instead of the stress constraints which are not sufficient for ceramics it seems reasonable to choose failure probability P_f given in (10.4-2) and (10.4-3) as another criterion in addition to the material volume V. Unlike stresses, it is not convenient to use P_f as a constraint function because its allowable value is difficult to fix in advance. This results in a bicriterion optimization problem

$$\min_{\mathbf{x} \in X} [V(\mathbf{x}),\ P_f(\mathbf{x})]^T ,$$

subject to (10.4-5)

$$x_i^l \le x_i \le x_i^u \quad , \quad i = 1,2,\ldots,n ,$$

which corresponds to the general multi-criteria problem formulated in (1-8) and represents a two-dimensional case with $m = 2$. Vector $\mathbf{x} = (x_1, x_2, \ldots, x_n)^T$ comprises the design variables x_i as components, and X represents the feasible set which is further cut by the side constraints shown separately in (10.4-5). Any nonlinear constraints for compressive stresses, displacements, or other quantities may be included in the feasible set X, but in the applications presented in this chapter only side constraints for the design variables have been imposed, i.e. $X = \mathbb{R}^n$. In principle, design variables x_i can be chosen freely, but in shape optimization applications they are usually represented by the chosen nodal coordinates or the locations of some control points which define the boundary.

That part of the boundary which is to be optimized can be approximated by some nodes of the elements at the boundary. The coordinates of these nodes can then be chosen as the design variables. However, much fewer design variables are needed if it is possible to determine the boundary by a small number of control points. This is the case when the boundary curves are described by B-splines. Then a great variety of boundary shapes can be found by using only a few control points. The nodes of the elements at the boundary are on the B-spline and can thus be moved by the same control points. Instead of nodal coordinates these are taken as the design variables for shape optimization.

A B-spline is defined by the $n + 1$ vertices (control points) $P_0, P_1, \ldots, P_n$ of a polygon. The order k of the B-spline can be chosen freely between 2 and $n + 1$, and then the curve is composed of different parametric curves of degree $k - 1$.

Thus, for example a B-spline of order 3 is constructed by different quadratic parametric curves. The domains of the parameter t for the parameter curves are subintervals [0, 1], [1, 2], ... of the interval $[0, t_{max}]$.

The B-spline curve is given by

$$P(t) = \sum_{i=0}^{n} P_i N_{i,k}(t) \quad , \qquad 0 \le t \le t_{max} \; , \qquad (10.4\text{-}6)$$

where the weighting functions $N_{i,k}$ can be found by the recursive relation:

$$N_{i,1}(t) = \begin{cases} 1, & \text{if } t_i \le t \le t_{i+1} \; , \\ 0 & \text{otherwise} \end{cases} \quad .$$

$$N_{i,k}(t) = \frac{(t-t_i)N_{i,k-1}}{t_{i+k-1} - t_i} + \frac{(t_{i+k} - t)N_{i+1,\,k-1}}{t_{i+k} - t_{i+1}} \qquad (k > 1) \quad . \qquad (10.4\text{-}7)$$

The integer values t_i, $t_0 \le t_1 \le \dots \le t_{n+k}$ form the so-called knot vector $[\, t_0 \; t_1 \; \dots \; t_{n+k} \,]$. Integers t_i are "parametric knots" and may have multiple values ($t_j = t_{j+1} = \dots$). It is necessary to specify the nodes of multiplicity k, both at the beginning and the end of the knot vector. When there are no intermediate multiple nodes $t_{max} = n - k + 2$. For example, a B-spline of the order 3 (k = 3) defined by four vertices (n = 3), with no intermediate multiplicities has the nodal vector [0 0 0 1 2 2 2] ($t_{max} = 3 - 3 + 2 = 2$). In the following two applications the shape of a structure has been described by the B-splines, and the control nodes have been chosen as the design variables.

10.4.4 Pareto-Optima of Ceramic Cantilever

As an introductory example of shape optimization, a ceramic cantilever, shown in Fig. 10.4/2, is considered. The thickness of the plate is assumed constant (h = 1 cm), and only the lower boundary of the cantilever is varied during the optimization. The purpose is to minimize the material volume and the probability of failure simultaneously, according to the bicriterion problem (10.4-5), by using the B-spline description for the lower boundary. A finite element mesh for this plane stress problem is presented in Fig. 10.4/2 a) where also the loading can be seen. The B-spline boundary curve construction is illustrated in Fig. 10.4/2 b). The finite element mesh consists of twelve eight-node isoparametric elements which are supposed to describe the structural behaviour of this beam-like cantilever relatively well.

The three control points P_0, P_1, and P_2 are used to form the B-spline of order k = 3 for the lower plate boundary. The x-coordinates of the control points, shown in Fig. 10.4/2 b), constitute the design variable vector

$$\mathbf{x} = [x_1, x_2, x_3]^T \tag{10.4-8}$$

for this problem. The parametric representation of the point P(t) = {x(t), y(t)} on the boundary curve is obtained from (10.4-6) as:

$$P(t) = \{(x_1 - 2x_2 + x_3)t^2 + (-2x_1 + 2x_2)t + x_1, \ 2Lt^2 + 4tL\} , \tag{10.4-9}$$

where parameter $0 \le t \le 1$ is chosen according to Fig. 10.4/2 b). Now, an analytic expression for the material volume of the plate can be derived as follows:

$$V(\mathbf{x}) = 12L^2 + \int_0^1 x(t)\, y'(t)\, dt = 12L^2 + \left(\frac{5}{3} x_1 + 2x_2 + \frac{7}{3} x_3\right)L \tag{10.4-10}$$

and from this also the gradient $\nabla_x V(\mathbf{x})$ may be seen immediately. Even in this simple case the probability of failure cannot be expressed analytically, but P_f and $\nabla_x P_f(\mathbf{x})$ must be computed numerically from the stress field at each iteration cycle. This has been carried out by integrating, first, the reliability of the whole structure as the product of all element reliabilities and finally by determining P_f as the complement of the overall reliability according to (10.4-2).

By imposing bounds for all the design variables one obtains the bicriterion cantilever problem

$$\min\ [V(\mathbf{x}), P_f(\mathbf{x})]^T , \tag{10.4-11}$$

subject to

$$x^l \le x_1 \le x^u ,$$
$$x^l \le x_2 \le x^u ,$$
$$x^l \le x_3 \le x^u .$$

The following design data for the problem are given:
F = 100kN, L = 25 cm, H = 30 cm, E = 34 000 kN/cm^2, ν = 0.22, σ_c = 300 MPa, V_c = 2.48 cm^3, m = 13 and the plate thickness is 1 cm. Upper and lower limits for the design variables are x^u = 100 cm and x^L = 5 cm. The expressions for the criteria can be found in (10.4-10) and (10.4-2).

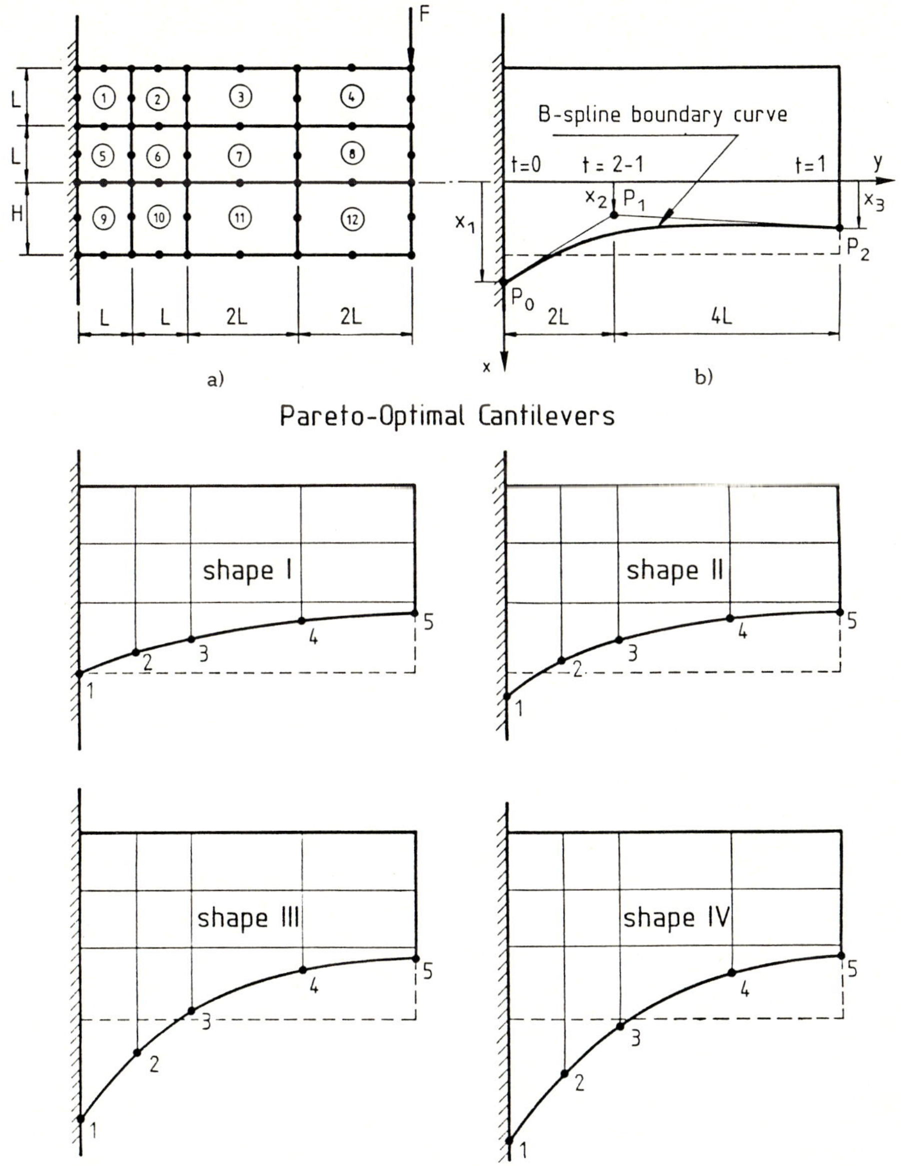

Fig. 10.4/2. Ceramic cantilever example:

a) structure, loading, and finite element mesh

b) description of shape curves by B-splines with the corresponding design variables x_1, x_2, x_3

Nodal coordinates [cm]							
Shape	1	2	3	4	5	V^*	P_f^*
I	32.15	22.55	16.04	8.34	5.00	9381	0.0025
II	40.11	26.52	17.62	7.99	5.00	9712	0.0005
III	75.33	47.27	29.13	10.05	5.00	11179	0.000095
IV	85.29	54.46	34.22	12.11	5.00	11767	0.000035

Table 10.4/1. Pareto optimal shapes and corresponding minimal solutions to bicriterion cantilever problem

The constraint method has been used to generate Pareto optima for this problem, and the numerical solution of each scalar subproblem has been performed by the SQP-algorithm of nonlinear programming [12]. The finite element program ANSYS has been combined with the optimization program to compute the stress field from which P_f and its gradient have been determined repeatedly during the optimization. About forty iteration cycles and half an hour CPU-time were needed to compute one Pareto optimum by the VAX-8800 computer when the initial shape given in Fig. 10.4/2 a) was used ($x_1^0 = x_2^0 = x_3^0 = 30$ cm).

Four Pareto-optimal shapes have been depicted in Fig. 10.4/2, and the corresponding nodal x-coordinates are given in cm-units in Table 10.4/1. The results resemble those obtained from a stress limited minimum weight problem because in this loading case only the elements 1 and 2, where high tensile stresses occur, determine the numerical value of the probability of failure.

10.4.5 Design of a Ceramic Piston Crown

High thermal efficiency and simple and reliable design are keys to success in the competition for a share of the marine, rail traction, and stationary engine market. The common means of improving the thermal efficiency of a combustion engine is to increase the temperature of the working medium. In a diesel engine the temperature increase can be

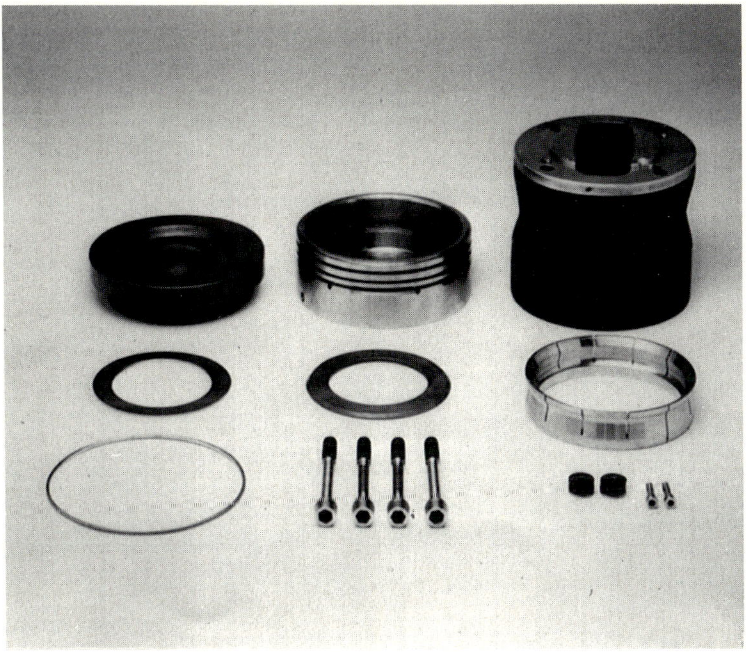

a)

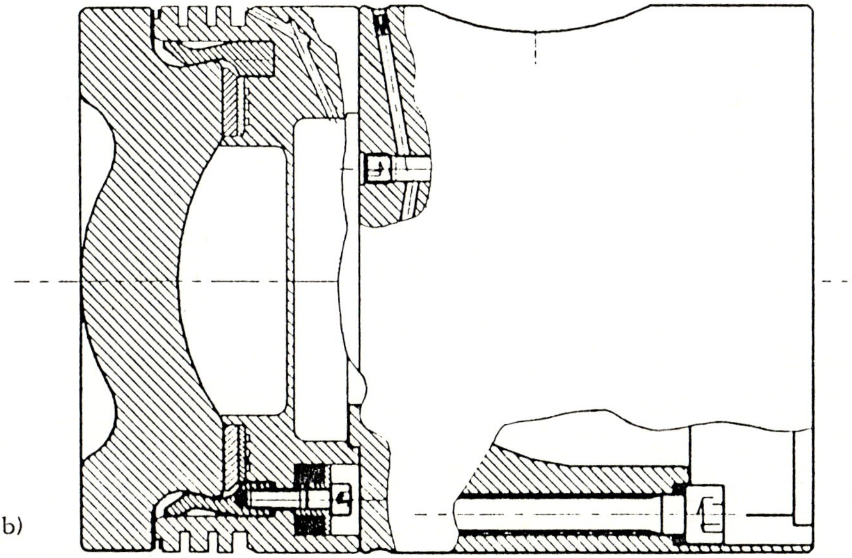

b)

Fig. 10.4/3. Prototype piston for the Wärtsilä Vasa 22 HF diesel engine:
a) cross-section of the piston where the topmost part is the ceramic crown
b) components of the prototype piston

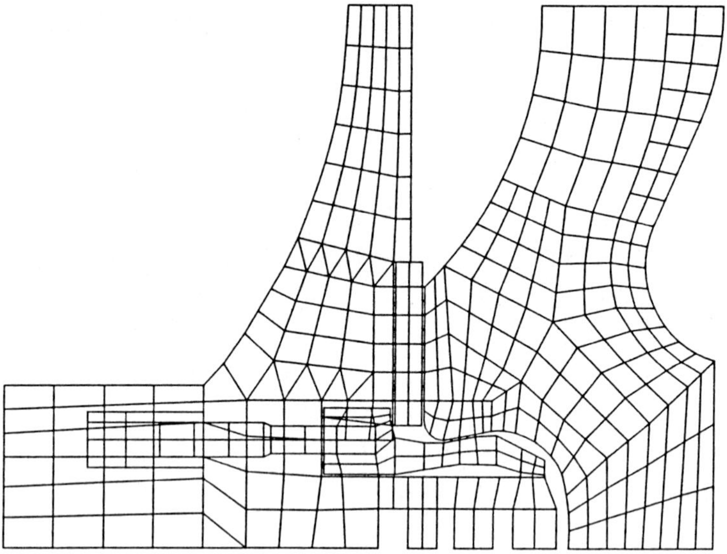

Fig. 10.4/4. Finite element model for the piston of diesel engine. Axisymmetric elements and ADINA-program were used to solve the stresses and deformations of the piston [13]

realized by using combustion chamber materials characterized by good thermal insulation and strength properties at elevated temperatures.

Ceramics are an alternative with great potential for this purpose. Due to the fact that the material price is high, it is normally not advisable to use monolithic ceramic engine components. This is of course especially important when designing such fairly big components as the piston of the medium-speed engine Wärtsilä Vasa 22HF. In this case a ceramic material is used only in the crown. From the two alternative materials zirconia (ZrO_2) and silicon nitride (Si_3N_4) the latter was chosen mainly because of the high density and thermal expansion coefficient values of zirconia. Silicon nitride has a considerable strength at high temperatures and in spite of its small thermal insulation capability, compared with zirconia, it allows the elevation of the temperature in the cylinder. In addition to the improved efficiency, another advantage is obtained simultaneously. It is the possibility of lowering the quality of the heavy fuel used in this engine brought about by favourable ignition conditions due to the higher surface temperature.

In addition, the cooling system of the piston becomes unnecessary and can be left out.

The structure of the new piston, which is under development at the moment, is shown in Fig. 10.4/3 where the ceramic crown, the intermediate steel part, and the aluminium skirt can be seen in addition to the smaller components. In the future this piston may be available in the diesel engine Vasa 22 HF or in some similar engine, the today's technical data of which is the following:

- cylinder bore 220 mm,
- piston stroke 240 mm,
- operating speed 900 - 1200 r/min ,
- maximum power 162.5 kW/cylinder (at speed 1000 r/min),
- maximum firing pressure 14 N/mm^2 (140 bar),
- mean effective pressure 2.12 - 1.91 N/mm^2,
- mean piston speed 7.2 - 9.6 m/s ,
- fuel consumption 200 g/kWh.

Both inline and V-types are available for this engine, the number of cylinders ranging from four to sixteen. Due to technical aging, the lifetime of the engine is estimated to be about twenty years or perhaps a little less. The heat transfer and structural analysis of the piston have been performed by using axisymmetrical elements, according to the mesh shown in Fig. 10.4/4, and finite element program ADINA. In this case, where the boundary conditions change during the deformation of the parts, a nonlinear analysis was applied. Several loading cases, consisting of different combinations of thermal load, gas pressure, inertia forces, and connection forces, caused by the flexible joining part, were studied in the analysis. The finite element model includes 353 isoparametric elements most of which are rings, with 1247 nodes. The most severe loading condition causing high tensile stresses in the crown was the combination of the thermal load and the connection forces. The details of the calculations and results of the analysis can be found in [13].

Shape optimization has only been performed for the ceramic piston crown leaving the other parts of the piston out. It is important to minimize the material volume of the crown because of two reasons. The first is the high price of the silicon nitride itself. The other reason is associated with the inertia forces which should be at the same relatively low level, as they are in the traditional cast iron pistons.

The other criterion to be minimized is the probability of failure of the crown which should be as small as possible in order to avoid expensive damages during the operation. Consequently, the general bicriterion problem (10.4-5) is applicable here as well, because a fixed constraint value cannot be imposed for either criterion.

The combustion conditions in the cylinder determine the shape of the upper boundary of the crown, whereas the shape of the lower boundary and the location of the support point, all shown schematically in Fig. 10.4/5a, can be varied during the optimization. B-spline representation for the description of the lower boundary has been used, and in addition to the x-coordinates of the control nodes x_1, x_2,...,x_n also an additional design variable x_{n+1} is needed to change the location of the support. Different numbers of control points and finite elements were used in the optimization process where only the most important axisymmetric loadings were considered. The finite element mesh shown in Fig. 10.4/4 is too expensive for this multicriteria aproach, and thus a more coarse finite element mesh, associated with a linear analysis (10.4-1), was used here. The existing design which weighs about five

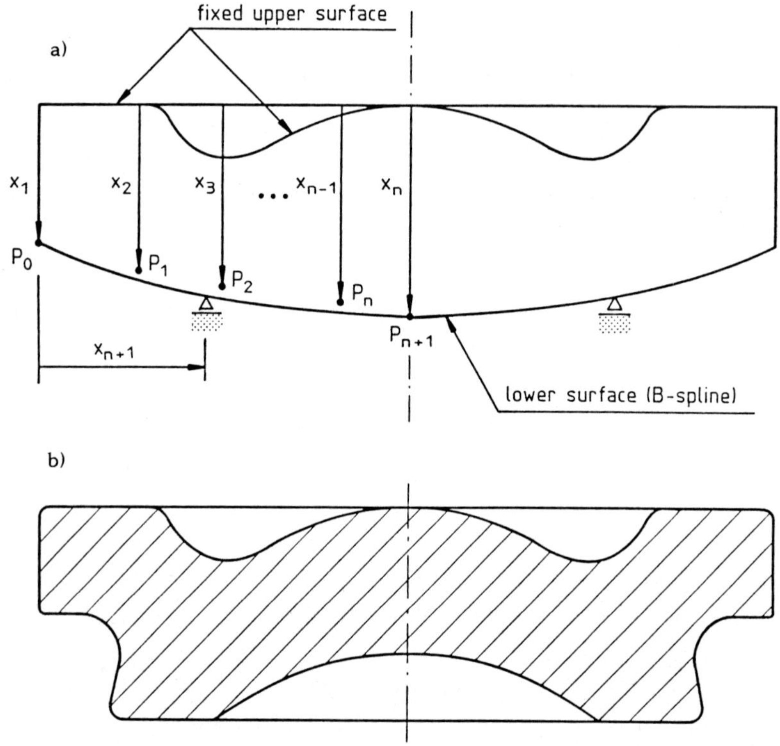

Fig. 10.4/5. Shape design of ceramic piston crown

kg is shown in Fig. 10.4/5 b) and is under the engine laboratory testing at the moment. The probability of failure for this crown has the value of $P_f = 0.19 \cdot 10^{-4}$ which means that under the assumed loading approximately one crown out of fifty thousand will fail.

10.4.6 Conclusion

Optimal shapes of ceramic components may differ considerably from those obtained for conventional materials. This is caused by the choice of a new criterion, the probability of failure of the component which is based on the use of the Weibull distribution in analysing the strength test data. Stress constraints are no longer sufficient for optimization, but the reliability should be considered in the design process.

The structural optimization problem for ceramics has been proposed in this chapter, and an introductory shape optimization example has been treated in detail to illustrate the presented theory. Also, a complex industrial application, which is the shape design problem of the ceramic piston crown, has been discussed to some extent. It has been emphasized in the text that the bicriterion problem, where the material volume and the probability of failure of the component are chosen as criteria, is useful for ceramics because these two quantities are inconvenient as constraints. The material volume is a natural criterion to be minimized, and the choice of the allowable value for the probability of failure is a sensitive area in the design process, especially if the damage may cause large economic or human losses.

The numerical computation of Pareto optima is usually expensive in shape optimization because the finite element analysis and the redesign step must be repeated several times to obtain one Pareto-optimal solution. The main difference in the computations, compared with optimization using other materials, is the determination of the value of the failure probability or its gradient with respect to the design variables. Efficient numerical integration and derivation techniques for this purpose might be expected to increase the accuracy as well as to speed up the computations. The constraint-oriented method (see (1-14)) used in the preceding applications is suitable for the generation of the Pareto optima for the bicriterion ceramic problem, and the trade-off information needed in the decision making process is easily available either graphically or from the Kuhn-Tucker multipliers

[14]. In optimizing the piston crown it sometimes appeared that also compressive stress should be considered in order to avoid inpractical shapes. In the case of a material with a high Weibull modulus m it may be reasonable to minimize the maximum tensile stress instead of the probability of failure, if this is easier.

At present, the development of ceramic materials is an intensive research field, and, without any doubt, cheaper and more reliable alternatives will appear on the market in the future. For example, the fracture toughness can be improved considerably by adding small fibres to the ceramic matrix. Along with the progress in the material development, ceramics may be expected to replace conventional materials in many new and unpredictable applications. As an answer to this challenge, both, the shape optimization and the multicriteria approach will together offer a modern and refined technique for the designer to achieve an economical and a reliable result.

References

[1] Creyke, W.E.C.; Sainsbury, I.E.J.: Morrell, R.: Design with Non-ductile Materials. London: Applied Science Publishers 1982

[2] Davidge, R.W.: Mechanical Behaviour of Ceramics. Cambridge University Press 1979

[3] Davidge, R.W.: Strength and Toughness in Ceramic Systems. Phil. Trans. R. Soc. London, A 310 (1983) 113-125

[4] Marshall, D.B.: Ritter, J.E.: Reliability of Advanced Structural Ceramics and Ceramic Matrix Composites - A Review. Ceramic Bulletin, 66(2) (1987) 309-317

[5] Stanley, P.; Chan, F.S.: A Probabilistic Treatment of Brittle Fracture under Nonmonotonically Increasing Stresses. International Journal of Fracture, 22 (1983) 187-202

[6] Braibant, V.; Fleury, C.: Shape Optimal Design Using B-Splines. Computer Methods in Applied Mechanics and Engineering, 44 (1984) 247-267

[7] Ding, Y.: Shape Optimization of Structures: A Literature Survey. Computers & Structures, 24(6) (1986) 985-1004

[8] Bäcklund, J.; Isby, R.: Shape Optimization of Holes in Composite Shear Panels. Structural Optimization. Proceedings of the IUTAM Symposium on Structural Optimization. Melbourne Australia, 3-9 February 1988. Dordrecht: Kluwer Academic Publishers 1988

[9] Esping, B.; Holm, D.: Structural Shape Optimization Using OASIS. Structural Optimization. Proceedings of the IUTAM Symposium on Structural Optimization. Melbourne Australia. 3-9 February, 1988. Dordrecht: Kluwer Academic Publishers (1988) 93-100

[10] Pedersen, P.: Design for Minimum Stress Concentration - Some Practical Aspects. Structural Optimization. Proceedings of the IUTAM Symposium on Structural Optimization. Melbourne Australia. 3-9 February, 1988. Dordrecht: Kluwer Academic Publishers (1988) 225-232

[11] Eschenauer, H.: Numerical and Experimental Investigations on Structural Optimization of Engineering Designs. A Research Project of the German Research Association (GRA) 1980-1984. University of Siegen, FR Germany. Siegen: Bonn + Fries Druckerei und Verlag 1986

[12] Schittkowski, K.: Software for Mathematical Programming. In: Computational Mathematical Programming. Berlin: Springer (1985) 383-451

[13] Silvonen, A.; Pisilä, S.: Reliability and its Determination for Medium Speed Diesel Engine Components. XVIIIth International Congress on Combustion Engines. CIMAC 1989, Tianjin, June 1989

[14] Koski, J.: Bicriterion Optimum Design Method for Elastic Trusses. Acta Polytechnica Scandinavica, Mechanical Engineering Series. No. 86, 68p. Helsinki, 1984

AUTHOR INDEX

Adali, S. 398; 417
Ambartsumyan, S.A. 397
Arora, J.S. 68
Atrek, E. 68

Baier, H. 244; 371
Balachandran, M. 115; 137
Barthelemy, J.-F.M. 301
Bartholomew, P. 68
Bayez 208
Benayoun, R. 71; 78
Benson, R.C. 302
Bendsøe, 29
Bernau, H. 72; 114
Bernoulli, D. 1; 323
Bernoulli, J. 1
Bert, C.W. 397
Box, G.E.P. 42
Braibant, V. 462
Brandt, A.M. 372
Bremicker, M. 5; 319
Brosowski, B. 113
Broyden 58

Carmichael, D.G. 371
Charnes, A. 32
Chebyshev, 12; 32; 171; 179
Claussen, T. 2
Cohon, J.L. 395
Cooper, W.W. 32
Currin, C. 208

Dantzig, G.B. 53
Davenpeck, M.L. 303
De Montgolfier, J. 78
Diaz, A. 72; 101
Ding, Y. 462
Draper, N.R. 217
Duckstein, L. 72
Dupre, R. 71
Dyer, J.S. 71; 251

Efroymson, M.A. 217
Eschenauer, H.A. 1; 35; 71; 230; 319; 371; 372; 397; 398
Esping, B. 69
Euler, L. 1; 152; 326

Fandel, G. 71; 74; 90; 100; 111
Fedorov, V.V. 217
Feinberg, A. 71
Fleury, C. 68
Fox, R.L. 32
Francis, P.H. 397
Frank, M. 83
Fuchs, W. 397; 398

Galerkin, B.G. 2
Gallagher, R.N. 69
Gauss, C.F. 21; 398
Geleji, A. 347
Geoffrion, A.M. 71; 83; 86; 97; 251;
Gerasimov, E.N. 371
Gero, J.S. 115; 137; 139; 376; 394
Gershon, M.E. 72
Goicoechea, A. 72
Griffith, R.E. 52
Guddat, J. 72

Habip, L.M. 397
Haftka, R.T. 397
Hamilton, W.R. 1; 309
Hansen, D.R. 72
Hati, S.K. 371
Haug, E.J. 68
Henrich, L. 339
Hesse 7; 58
Hill, R. 47; 304

Himmelblau, D.M. 282
Hoffman, O. 305
Hooke, R. 102
Huckert, K. 71
Hwang, C.L. 72

Isermann, H. 72

Jacob, H.G. 42
Jahn, J. 71; 80; 95; 111
Jeeves, T.A. 102
Jendo, S. 355; 372
Johnson, R.C. 302
Johnson, W. 304

Kneppe, G. 229; 372
Kobayashi, S. 304
Koski, J. 1; 151; 371; 447
Kowalik, I.. 114
Kronecker 215
Kuhn, H.W. 8; 14; 101

Lagrange, J.L. 1; 2; 58; 101; 230; 231
232; 243; 368
Laritchev, O. 78
Lebesgue 311
Lee, E.B. 308
Leibnitz, G.W. 1
Leitmann, G. 371
Lekhnitskii, S.G. 397
Malakooti, B. 113
Marks, W. 372
Markus, L. 308
Masud, A.S.M. 72
Mead, R.A. 161
Mellor, P.B. 304
Mises, von 304; 305
Mistree, F. 119
Mitchel, W.J. 127
Mitchell, T.J. 212
Montusiewicz, J. 282
Morris, A.J. 68; 372; 375
Morrison, H.L. 303
Müller-Slany, H.H. 193

Nagtegaal, R. 205; 217; 218
Nakayama, H. 72
Nelder; J.A. 161
Neuber, H. 322
Newton, Sir I. 1; 57; 152

Olhoff, N. 29
Osborne, M.R. 114
Osyczka, A. 1; 71; 151; 168; 282; 371

Papalambros, P.Y. 32
Pareto, V. 4; 5; 10; 13; 28; 116; 119;
137; 138; 151; 249; 266; 269; 274;
277; 289; 290; 297; 299; 311; 314
Pfeiffer, F. 184
Poisson, S.D. 327
Polanski, Z. 183
Post, P.U. 35; 397; 398
Potchman, Ju.M. 371
Powel, M.Y.D. 113
Prasad, B. 397

Radford, A.D. 119; 139
Ragsdell, K.M. 69
Rao, S.S. 280; 371
Rayleigh, Lord 2
Rechenberg 369
Reimers, U. 302
Richmond, O. 303
Ritz, W. 2
Rosenman, M.A. 376; 394
Roy, B. 71
Rozvany, G.I.N. 32
Ruze 23

Sachs, G. 305
Saint-Venant, B. 2; 306; 328
Sattler, H.J. 371
Save, M. 374
Sawaragy, Y. 92
Schäfer, E. 71
Schiffner, K. 339
Schittkowski, K. 463
Schmit, L.A. 229
Schoofs, A.J.G. 205
Silvennoinen, R. 371; 447
Skalazub, V.V. 371
Snedecor, G.W. 172
Sobieszanski-Sobieski, J. 68
Sortais, H.C. 304
Spyridonos, A. 23
Stadler, W. 2; 4; 303; 371
Stenvers, K.-H. 372; 429
Steuer, R.E. 71
Steward, R.A. 52
Sussman, B. 71
Szidarovsky, F. 72; 112

Tanino, T. 72
Taylor, J.E. 29; 52
Tergny, J. 78

Thiele, H. 351
Thierauf, G. 372
Thomson, S.G. 304
Tsai, S.W. 47
Tucker, A.W. 8; 14; 101

van Campen, D.H. 205

Wallenius, J. 71
Watkins, R.I 372; 375
Welch, W.J. 208; 211; 218
Wilde, D.J. 32
Wodtke, H.-W. 319
Wolfe, F. 83

Yoshimura, M. 261
Young, T. 319; 341

Zajac, J. 168
Zamorski, J. 282
Zeleny, M. 72
Zelikow, A.I. 347
Zienkiewicz, O.C. 67
Zionts, S. 71
Zoutendijk, G. 80

Collective names:
Broyden - Fletcher - Goldfarb - Shanno 58
Davidon - Fletcher - Powell 102
Kuhn - Tucker 8; 14; 101; 248; 255
Nelder - Mead 102; 161
Tsai - Hill 47
Tsai - Wu 47
von Mises - Hencky 305

SUBJECT INDEX

accumulator, hydraulic 175
accuracy
-, positioning 433
-, reflector surface 433
-, tracking 433
ADINA-program 458; 459
aeroelastic efficiency 232 236
aircraft, supersonic 232; 235
algorithm
-, "branch and bound" 211
-, Frank-Wolfe 83
-, gradient projection 59
-, (GRG) 55; 58; 59; 231
-, optimization 36; 39; 40; 43; 102; 211; 230; 232; 438
-, scalar optimization 73
-, SQP- 462
analysis 381
-, amplitude frequency 188
-, connection sensitivity 381
-, dynamic 193
-, exchange 212
-, modal 240
-, numerical sensitivity 48
-, postoptimality 380; 381; 388; 390
-, regression 217
-, sensitivity 36; 42; 43; 61; 230; 231; 321; 325
-, stability 380
-, structural 36; 43; 61; 230; 231; 321; 325
-, subsystem sensitivity 381
analysis of experiments, statistical 169; 171
analysis variables 37; 39; 45
angle(s)
-, coordinate 152
-, fibre 46; 440
ANSYS 456
antenna 20
-, cassegrain 256
-, efficiency of an ~ 23
-, parabolic 20; 22; 199
-, receiving 21
-, transmitting 20
antisymmetric angle-ply laminates 417
apparatus, measuring 279
approach
-, knowledge-based 301
-, Lagrangian 152
-, Newton-Euler 152
-, rule based 123
-, serial hybrid 42
-, statistical 447
-, statistical experimental 169
approximation
-, sequential quadratic 101
architecture, modularized 36
architecture of a knowledge-based system 118
area, cross-section 60; 196; 230; 239; 245; 248; 250; 254; 267; 292; 357; 365
ASKA (Automatic System for Kinematic Analysis) 430
astronomist, radio 21
astronomy, optical 22
automation of design 301

backpull 305; 306; 314
balancing 151
-, robot arm 165
band-width-minimizer 41
barrel core 319; 320
basic equations, augmented 403
basis, contravariant 400
beam, elastic-plastic 340
beam theory 323
bearings, hydrostatic 107; 282; 291
behaviour
-, brittle 441
-, convergence 409
-, dynamic 239
-, plastic 317
bend 341
bending equation, thermoelastic 418

bending problem 422
Bèzier curves 46
Bèzier surface 46
BFGS Update-Formula (**B**royden-**F**letcher-**G**oldfarb-**S**hanno) 58
binary tree 211
borehole rims 321
boundary , functional-efficient 90; 358; 361
bounds 47
-, design variable 323
-, lower 47; 52; 89
-, upper 47; 52; 89
-, variable 121
brachistochrone 1
B-spline 46; 452; 453; 460
buckling 230; 421
buckling equation , thermoelastic 418

cable , weight of 366; 368; 371
cable sag 365
cable system , single layer 366; 372
CAD (Computer Aided Design) 3; 280
CADE (Computer Aided Design of Experiment) 218
CAE (Computer Aided Engineering) 3; 232
CAE-integration 232
calculation
-, probability 439
-, stochastic 440
CAM (Computer Aided Manufacturing) 280
CAMOS (Computer Aided Multicriteria Optimization System) 101; 111; 160; 282
canonical form(s) 120; 121; 131; 133
cantilever, ceramic 453
cantilever truss 58; 60
cardan shaft
-, cast-iron 277; 319
cast metal technology 303
catenary 367
ceramics , alumina 277
CFC (Carbon Fibre Composite) material 413
CFRP (Carbon Fibre Reinforced Plastic) 436
CFRP-strut 442
chain rule 84
chatter, regenerative 264
Chebyshev metric 12; 17
civil engeneering structures 372
coefficients
-, damping 273
-, diffusion 405
-, hardening 341
-, thermal expansion 327; 406; 419; 430; 435
-, weighting 74; 76; 78; 79; 80; 92; 94; 100
compliance
-, minimum 356
-, static 263; 266; 267; 274
component(s), monolithic ceramic engine 458
composite(s)
-, laminated 372
-, multilayer 401
composite design(s) 42
composite fin 230; 232; 237
compromise decisions 139
compromise solutions 17; 18; 28; 63; 74; 97; 365
computation phase 71
computer experiments 208
computer graphic output 300
computing
-, inferential 117
-, parallel 36
computing time 40; 48; 50; 52; 63
concrete
-, epoxy resin 277
-, in-situ reinforced 387
conditions
-, compatibility 327
-, environmental 417
-, equilibrium 305; 327; 398; 403
-, Kuhn-Tucker 8; 14; 101; 248; 255; 268
-, loading 355; 356; 357; 371
-, necessary 7; 8; 362
-, sufficient 7; 8
conduction 240
conductivity
-, heat 239
-, thermal 324; 327
conflict, objective 9
consideration(s), physical 317
constraint (s) 6; 35; 37; 116; 121; 245; 251; 284; 286; 293; 323; 405
-, active 57
-, buckling 63
-, deformation 334
-, design 129

-, dynamic displacement 368
-, equality 345
-, explicit 378
-, geometrical 232
-, implicit 378
-, inequality 8; 14; 345
-, linear 134
-, nonlinear 134
-, redundant 117
-, side 360; 362
-, stress 323; 357; 360; 368; 461
constraint method 137
constructions, piping 26
continuous programming models 105
contraction
-, maximum 314; 315
-, total 307
control
-, knowledge-based 137; 142
-, optimal 19
-, optimization 133
-, Pareto-optimal 314; 316
control effectiveness, criterion of 10
controller
-, high level 140
-, knowledge-based 144
control mechanism, knowledge-based 133; 137; 144
control problem, multicriteria optimal 308
control theory 371
convection 240
convergence 53; 57
-, rate of 40; 52
convergence behaviour 409
convexity 7; 9
conveyer, rubber belt 25
conveyer belt drum 25
cooling system 459
coordinate, modal 194; 452
coordination 296
-, constraint 283; 285; 292; 296; 297
-, parametric 283; 285; 292; 297
correction rule 53
cost 437
-, aggregate 385
-, product manufacturing 280
-, unit 385
countershaft 188
counterweight (s) 151
counterweight balancing 151
crack growth, unstable 448
crankshaft 185
criterion space 10; 11; 16; 20; 75; 77; 90; 191
criterium, -a, -on 355; 357; 365
-, bonding break 47; 406
-, breaking-off 40; 78
-, cost 355
-, D- 211; 216
-, error 137
-, failure 47; 245
-, fibre break 47; 406
-, G- 211; 217
-, maximum critical load 356
-, maximum moment of inertia 357
-, maximum nastural frequency of free vibration 356
-, maximum reliability 356
-, maximum structural stiffness 356
-, max strain 47
-, max stress 47
-, minimum compliance 356
-, minimum cost 355; 371
-, minimum displacements 356; 371
-, minimum volume 355
-, minimum weight 355; 365; 371; 372
-, operational
-, optimality 2; 3; 14; 37; 59; 207; 212; 248
-, space 10; 11
-, V- 24; 217
-, weight 355
curvature 400; 402
curvature tensor
-, covariant 399
-, mixed 399
curves
-, Bezier 46
-, B-spline 46
-, functional-efficient 62
-, Pareto-optimal 62
cylinder
-, central 435
-, sandwich 254

data bases 41; 86; 117
data management 36; 41; 44
dead weight 435
decision, heuristic 289
decision making / maker (DM) 2; 12; 28; 29; 71; 277; 289
-, multicriteria 101

decision phase 71
decision process
decision support system 101
decision variables 115; 116; 120; 121; 287; 376
-, binary 134
-, continuous 116; 134
-, discrete 116; 134
-, identifying 130
decomposition 256; 259; 282; 301; 381
definition, recursive 134
deflection, rudder 233
deformation(s)
-, energy of elastic 356
-, large 403
-, residual 344
-, shear 400
deformation measurement 411
deformation process, plastic 317
density 321
-, bulk 358
-, fibre 405
-, matrix 405
derivative 49; 126; 213
-, covariant 399
-, partial 212
design
-, approximate 212
-, automation of 301
-, B_n- 171
-, B_4-experimental 179
-, ceramic 442
-, civil engineering 355
-, composite 42
-, die 303
-, D-optimal 211
-, experimental 169; 171; 177; 207
-, factorial 171
-, fully stressed 304
-, initial 37; 40
-, interactive 5
-, machine 188
-, machine tool spindle 282
-, minimum weight 232
-, Pareto-optimal 162
-, process 280
-, product 280
-, telescope 429
-, thermoelastic 417; 424
design constraint 129
design matrix 209
design model 37; 39; 42; 45
design object 129
design space 10; 129; 205
design variables 6; 35; 37; 39; 42; 45; 49; 121; 207; 245; 273; 291; 323; 346; 357; 360; 405
-, continuous 101; 103
-, discrete 101; 102; 103
-, global task 283
-, integer 101
-, mixed 101; 102
design variable bounds 323
deviation
-, mean quadratic 433
-, standard 178; 209; 215
DFP (Davidson-Fletcher-Powell Gradient Method) 102
diagram, weight-compliance 269; 270; 275
DIALOG (software package) 86; 111
diameter, aperture 23; 199
die 308
-, conical 301; 308
-, natural 313
-, perfect 303
diesel engine 465
differences
-, forward 48
-, systematic and statistical 22
differentiable 116; 134
differentiation
-, rules of 126
-, symbolic 126
diffusion process 403
direction of steepest ascent 84; 98
disk flywheel 191
displacement 245; 400
-, minimum 356
-, modal 241
-, perpendicular ~ to the middle surface 400
distortion
-, dynamic 245
-, static 245
distribution
-, aerodynamic pressure 234
-, hygrothermal 405
-, normal 178
-, optimal thickness 237
-, statistical 208
-, temperature 320
-, Weibull 447; 450
divergences 339
-, developable 339
-, nondevelopable 339

domain
-, feasible 8
-, infeasible 57
domain knowledge 118
drawing force, minimum 314
drive train 185; 190
-, automotive 184
DS (Direct Search Method) 102
ductility 319
dynamic characteristics 261
dynamic programming 376; 378; 380; 381
-, conventional 378
-, multicriteria 379
-, nonserial 378

EDT (Experimental Design Theory) 205
efficiency, thermal 456
efficient solution 364; 369; 370
eigenfrequency 194; 197; 245; 433
eigenmode 194; 197; 241; 366
eigenvalue (problem) 194; 241
element
-, isoparametric 454; 459
-, machine 261
element mesh 208
energy
-, complementary 2
-, maximum strain 322
-, potential 2; 266; 267; 272
-, strain 245; 266
-, total kinetic 152
-, total potential 152
energy of elastic deformation 356
engine, inference 125; 140
equation
-, adjoint 312; 314; 316
-, augmented basic 403
-, basic 398
-, diffusion 405
-, Euler 2; 326
-, governing 305
-, heat-flow 324
-, multiple regression 213
-, regression 207; 209
-, state 19; 312
-, strain-displacement 326; 400
-, system 245
equilibrium condition 305; 326; 398; 403
error
-, measurement 207
-, noncorrelating random 215
-, random 207; 208
-, residual 194
-, total 437
-, tracking 23
-, systematic 208
error criterion 137
error equation, Gaussian 23
error function 197
estimation 209
-, error 234
-, parameter 209
-, response 209
-, weighted least-squares 214
estimator 210
-, parameter 207; 210; 216
-, response 207; 210
Euclidean metric 12
excitation
-, high frequency 185
-, low frequency 185
-, motor 188
-, parametric 184
excursion 212
expansion 437
-, thermal 402
expansion coefficient, thermal 327; 402
experiments, simulation 178
expert system 41
extrusion 303

fabrication 245
factor
-, concentration 23
-, dynamic magnification 368
-, evaluative 266
-, intercept 23
-, stress concentration 322
-, weighting 84; 198; 249; 251; 268; 269
failure
-, bonding break (BBF) 406
-, fibre break (FBF) 406
-, machine 261
-, shear 406
failure load 406
fatigue 448
Feasible Directions method 59
feed motion 340; 344; 349; 352
FEM (Finite Element Method) 38; 43; 45; 59; 231; 245; 322
fibre angle 46; 403; 406; 440
Fibre Break Failure (FBF) 406

fibre criteria 47
fibre density 405
fibre orientation 417
fibres in the ceramic matrix 462
fibre volume fraction 405
FIZYL (analysis program) 329
flatness 339
flaw 447; 450
-, nondestructive 447
flexibility, modal 263; 274
flows, diagonal
focusing, highly accurate 20
force
-, connection 459
-, contact 346
-, drawing 309
-, inertia 459
-, minimum drawing 314
formulation
-, min-max 15; 29; 78; 189; 417; 423; 427
-, problem 115
fracture mechanics 477
frame 123; 128; 129;
frame formalism 122; 124
frequency 245
-, higher random 245
-, maximum natural 240; 356;
-, natural 239; 263; 265; 365
-, receptance 264
FT (Flexible Tolerance method) 102
function
-, affine-linear 9
-, continuous 378
-, displacement 399
-, distance 12; 29; 346; 361; 362; 364; 370
-, expensive 169
-, extended substitute objective 432
-, gamma 451
-, HELP 105
-, interpolating 207; 209; 252
-, irregular objective 378
-, Lagrangian 8; 14; 58; 152
-, main objective 14
-, modified ellipse 46
-, multiattribute 377
-, multicriteria objective 377
-, objective 6; 37; 59; 116; 121; 134; 245; 292; 345; 377; 378; 405
-, preference 10; 14; 28; 29; 37; 43; 50; 346
-, quadratic penalty 431
-, reduced objective 56
-, shape 39; 46
-, single-criterion objective 377
-, spline 432
-, switching 312
-, substitute-objective 10
-, unit step 404
-, utility 358; 360; 370
-, velocity 20
-, weighting 453
functional-efficient /-cy 5; 10
functions recognition, algebraic 134

gain
-, optimal antenna 21
-, ideal parabolic antenna 21
Gaussian error equation 23
Gaussian surface parameter 398
G-criterion 211; 217
gears , meshes of 184
generation process 138
global criterion method 104
gradient 42; 47; 116; 134
-, reduced 57
graph
-, linear 379
-, polygonal 379
GRG (Generalized Reduced Gradient) 55

Hamiltonian 309; 311
hardening 341
heat flow 324
heat flux 238; 239
Hessian-matrix 7; 58
heuristics 120; 379
honeycomb core 233; 257
humidity influence 435
hygrothermal 403
hypercube 52; 53
hyperplane 76;
hyperspace 16
hypersphere 433
hypersurface 52; 57
hypothesis , strength 448

I-beam, prestressed concrete 357
IBF (Inverse Barrier Function) 231
identification, layer 443

ill-conditioned 208
inferential computing 117
Infrared Space Observatory (ISO) 238;
insulation, thermal 10.4/12
interactive MO 90
interface
-, standard 36
-, user 140
INVAR 437
isoperimetric problem 1; 2
ISO-satellite 238
iterative method 105

journal bearing 295
-, end 293
-, front 293
-, hydrostatic 29; 293

kinematic 196
kinematic hardening 342
knowledge base 117; 118; 119; 133
knowledge engineering 115; 117

Lagrange multipliers 2; 8; 14
Lagrange polynomials 46
Lagrangian function 8; 14; 58; 101; 152
laminate 397; 401; 407
-, angle-ply 413
-, antisymmetric angle-ply 417; 418
-, carbon fibre 248
-, fibre 253
laminate plate theory 248
laminate stiffness 418
lathe
-, boring 262
-, turning 262
launcher 257
launching phase 239
layer
-, composite 253
-, martensite 321; 322
-, material 321
-, tangential stress-free 325
layer thickness 42; 46; 406; 417
layout
-, interactive MO- 107
-, optimal 23; 26
learning process 111
level 205
-, constraint 13
-, demand 12; 336
lifetime 239; 447; 459
linking
-, automatic 232
-, variable 232; 240
LISP 140
load
-, dynamic 238; 245
-, failure 406
-, maximum critical 356
-, static 245
-, thermal 245; 459
loading
-, aerodynamic 234
-, dynamic 356; 366
-, time-invariant 398
load vector, for aeroelastics 234
LPNLP (Lagrange Penalty Method for Nonlinear Problems) 42
LSAPL (Laminated Shell Analysis Program) 407

machine accuracy 262
machine, milling
machine productivity 262
machine tools 261; 279; 291
machine tool spindle system 261; 291; 279
mapping 10; 45
mass 245
-, centre of 194; 195
-, generalized 194
mass density 327
material(s)
-, advanced 429
-, anisotropic 400
-, Carbon Fibre Composite (CFC) 414
-, ceramic 447
-, combustion chamber 458
-, composite 46; 437
-, ductile 448
-, fibre composite 397; 443
-, glass fibre composite 239
-, hard 303
-, heterogeneous 400
-, ideally plastic 304
-, new 277
-, plastic 303
-, properties of 279
-, selection of structural 277
-, time-variant 403

material behaviour 404
-, linear viscoelastic 404
-, time-variant 412
material law 326; 400
mathematical programming method 37; 133
matrix
-, adjoint variable 49
-, aerodynamic influence coefficient 234
-, binary association 384
-, coordination 45; 46; 60
-, design 209
-, error-variance 216
-, Hessian 7; 58
-, mass 241
-, modal 194
-, of inertia terms 153
-, pseudo load 49; 50
-, sensitivity 48
-, stiffness 234; 241
-, system 49; 50
-, transfer 90
-, unit 60
-, variance-covariance 210; 216
-, weighted design 215
measurement
-, deformation 411
-, detection 447
-, experimental 442
-, strain 411
mechanics, fracture 447
mechanism
-, control 339
-, hydraulic spring balancing 178
-, inference 118
-, robot balancing 107
-, robot spring balancing 168; 169
members, structural 261
mesh, element 208
-, finite element 453; 460
method(s)
-, "ad hoc" 250
-, analytical 48; 59
-, appropriate 135
-, Chebyshev 171
-, combined 103
-, constrained 114; 137
-, cutting plane 52
-, decomposition 282
-, direct 50
-, direct search 276
-, discrete 59
-, distance function 12
-, dynamic programming 361
-, feasible directions 59; 271
-, finite difference 43; 45
-, finite element 38; 43; 45; 48; 59; 231; 245; 322
-, generalized reduced gradient 346
-, gradient 190
-, hybrid 55; 198
-, interactive 251
-, iterative 105
-, Lagrange multipliers 2; 368
-, least square 171; 209
-, Min-Max 104
-, Newton 57
-, non-preference 377
-, objective weighting 104; 189; 251; 268
-, penalty 59
-, preference 377
-, quasi Newton 59; 432
-, random search 104; 160
-, semi-analytical 50
-, SIMPLEX 52; 53; 161
-, STEP 71; 78; 93
-, state space 49; 50
-, stress-ratio 41; 42; 59
-, superior 111
-, transfer matrices 38; 43; 49; 89
-, "Trial and Error" 35
-, weighting 137; 377; 378; 380; 381
-, weighting least-squares 215
-, weighting min-max 104; 161; 289; 298
metric
-, Euclidian 12
-, Chebyshev 12; 17
minimization
-, cyclic 80
-, scalar weight 451
-, weight 320
minimum/a
-, constrained 8
-, global 7; 282
-, individual 75; 77; 91; 94;
-, local 7; 8; 165
-, unconstrained 7
min-max approach 408
min-max formulations 15; 29; 78; 423
-, with objective weighting 17
-, with selecting a demand-level vector 18

min-max method 104
-, weighting 104
min-max problem 79; 94
model
-, adapted 193; 194; 195; 197; 199
-, algebraic 127; 141
-, approximate analysis 182
-, canonical 140
-, design 36; 37; 39; 45
-, discrete 102
-, equilibrium 304
-, evaluation 36; 37; 46; 61
-, failure 304
-, kinetic 152
-, lumped mass 196
-, mathematical 141; 169
-, mechanical 185; 188; 340; 344
-, milling-machine 274; 275
-, optimization 36; 37; 88; 132; 230; 282; 320; 431
-, physical 193
-, regression 210; 213; 217
-, rigid-body 174
-, second order polynomial 170
-, simulation 262
-, stage-state 376
-, structural 37; 38; 57; 60; 267; 269
-, substitute 11; 194; 197
modelling
-, heuristic 193
-, optimization 29; 38; 39
-, structural 38
module, explanation 119
-, gradient 232
-, input 44
-, knowledge acquisition 119
-, output 44
-, user interface 119
modulus
-, elasticity 406
-, strain hardening 341
-, Weibull 450; 462
-, Young's 327; 341
MO-layout, interactive 107
moment of inertia, 357
MO-problem 7.2/5
multibody-system 185
multicriteria control problem 19; 308

NABLA-operator 80
Newton-method 57
NISE-method 137; 139; 142; 143; 144; 378
NLPQL (Nonlinear Program with Quadratic Line Search) 42
NM (Nelder-Mead Simplex Method) 102
noise level 187
nondifferentiable 116; 124
normal distribution 178
normalization 15

object, design 129
objective(s) 6; 35; 116; 321
-, conflicting 139
-, design 129
-, main 13
-, multiple 134
-, reference 84
-, secondary 13
-, single 134
objective function(s) 10; 37; 59; 116; 121; 134; 245; 292; 345; 377; 378
-, extended substitute 432
-, irregular 378
-, multicriteria 377
-, single criteria 397
objective weighting (method) 11; 15; 104; 189; 198; 251; 268; 370;
optimality criteria 2; 14; 37; 59; 207; 212; 248
optimization
-, frequency 238
-, geometric 42
-, interactive multicriteria 71; 178
-, Pareto 4; 119
-, scalar 10
-, shape 24; 42; 46; 50; 448; 451
-, simultaneous 62
optimization algorithm 36; 39; 40; 43; 102; 135; 211; 230; 231; 232; 438
optimization control 133
optimization formulation
-, knowledge-based 141
optimization model 36; 37; 88; 135; 136; 142; 230; 232, 282; 321; 431
optimization modelling 3; 29; 38
optimization model recognizer 140
optimization problem(s) 157; 174
-, bicriteria 366
-, large scale 230
-, multicriteria 159; 346
-, scalar 198
optimization procedure 36; 37
-, hierarchical 198

optimization recognition
-, knowledge-based 141
optimization strategy 29; 37; 230
-, Four-Stage Multicriteria 282; 283; 290
optimum/optimality
-, D- 211; 216
-, local 8
-, min-max 16; 300
-, Pareto 5; 10; 15; 28; 250; 266, 278
organization scheme, extended 73, 74; 101; 111
-, superior 73; 74; 90; 101; 111

PAFEC (Program for Automatic Finite Element Calculation) 43
-, analysis and optimization 430
parameters
-, adjustment 340
-, distinct 130
-, fixed 121; 130
-, random 439
-, shape 45
-, structural 37; 39; 45; 62
Pareto-optimal/ity 5; 10; 15, 28; 250; 266; 278
-, efficient 10
Pareto-optimal control 314
Pareto-optimal set 139
Pareto-optimal solution 71; 116; 266; 278; 289; 290; 311; 422
pattern matching 134; 140
pay-off table 75; 78
penalty function 75; 431
performance, product 279
plastic, carbon-fibre-reinforced 430
plasticity, von Mises-Hencky condition 305
plate(s) 403; 412
-, cantilever 448
-, fibre-reinforced 371
-, laminated 424
-, rectangular 398
-, viscoelastic 412
plate structure(s), fibre-reinforced 397
ply angles of a fibre composite lamina 42
point
-, candidate 210; 216
-, central 91
-, contact 271; 341
-, control 46; 452
-, cutting 262
-, intersection 19
-, reference 16
position, adjustment 437
postprocessor(s) 36
power loss 297
precision, high 279
predeformation 342
predicate logic, first order 125
preprocessor(s) 36
preference function, normalized 423
pressure, gas 459
principle
-, de Saint Venant 328
-, energy 1
-, integral 1
-, least action 1
-, maximum 312; 315; 316
-, minimum potential energy 1
probability calculation 439
probability model 440
probability of failure 447; 450; 452; 453; 460
problem, bending 422
-, bicriteria 451
-, buckling 421
-, conflicting 9
-, convex 8; 15
-, eigenvalue 194
-, isoperimetric 1; 2
-, large scale 243
-, large scale optimization 230
-, multicriteria control 19
-, multicriteria optimization 43; 159
-, nonconvex 12; 165; 251
-, saddle-type 422
-, scalar design 35
-, scalar optimization 10
-, scalar substitute 15; 61; 78; 346
-, substitute 10; 18
-, synthesis 188
procedure, line search 55
-, optimization 3
-, penalty 50
-, semi-analytical 62
-, sequential quadratic 57
-, transfer matrix 231
process(es), algebraic 117
-, decision 115; 289
-, forming 311; 339
-, generation 138
-, inferential 117
-, levelling 343
-, manufacturing 339

-, metal forming 339
-, prior 208
-, reduction 189
-, solution generation 133
-, substitution 132
production costs 333
program
-, ADINA 458; 459
-, conventional 117
-, laminated shell analysis 43
-, logic 135
-, RMS 436
-, sequential linear 59
-, sequential quadratic 59
programming
-, continuous nonlinear 103
-, discrete 102
-, dynamic 376; 378; 380; 381
-, goal 378
-, integer 103
-, mixed continuous-discrete 102; 103; 105; 109
-, mixed continuous-integer 105
-, quadratic 55
-, sequential quadratic 72
properties
-, damping 319
-, dynamic 197
-, inferential 207
-, material 324; 325
-, statistical 216
-, strength 458

quantity
-, cross-sectional 35
-, geometrical 35
-, material 45
-, state 89

radiation 240
-, cosmic infrared 238
radio-telescope 22; 356; 371
ratio
-, damping 263; 265; 272; 273; 275
-, fibre volume 440
-, modal flexibility 272
-, Poisson 327
-, trade-off 83; 84; 86
receptance frequency 263; 264; 275
reduction
-, model 185
-, rattling 188
reflector(s) 199
-, parabolic 21; 22
-, precision 247; 255
relation, stress-strain 341
reliability
-, maximum 356
-, overall 454
resistance
-, axial 306
-, total axial 306
-, wear 320
response
-, measured 210
-, multiple 213
-, structural 38
rigidity, static 261; 266
robot 178
-, industrial 169; 174; 279
robot arms 151; 152
robot balancing mechanism 107
robot manipulator 151
roller 319; 339
rule 138
-, correction 53
-, production 140

saddle-type problem 422
SAPOP (Structural Analysis Program and Optimization Procedure 331
satellite, X-ray 246; 254
satellite tanks 24
satellite technology 24
scalarization 12
search
-, random 160
-, sequential 160
search direction 56, 57
sensitivity analysis 388
-, connection 381
-, numerical 48
-, semi-analytical 51
-, subsystem 381
SEQLI (Sequential Linear Programming Method) 52; 331
SEQLI 2 (Sequential Linear Optimization Extended Version 42; 88
Sequential Linear Program 59; 231; 236
set
-, control constraint 20
-, feasible solution 368
-, Pareto-optimal 139; 143

-, natural structural 371
-, Pareto-optimal 456
shape design, optimal 25
shell 231
-, axisymmetric 245; 257
-, conical 88; 111; 258
-, spherical 398; 407
shell structure, fibre-reinforced 397
significance, level of 172
SIMPLEX-algorithm 95
SIMPLEX-method 52; 53; 275
sizing 42; 45
slewing crane 26
slope, cross section 400
slot 124
software package 72; 86; 111
software system 36; 40; 58
solar energy collector 23
solution
-, compromise 10; 18; 28; 63; 74; 97; 229; 360; 365
-, efficient 364; 369; 370
-, functional-efficient 4; 16; 97; 271; 331
-, non-Pareto-optimal 290
-, Pareto-optimal 71; 266; 278; 289; 290; 311; 422
-, Pareto-sub-optimal 380
space
-, criterion 10; 11; 16; 20; 75; 77; 90; 191
-, design 10; 89; 205
-, free 21
-, state 20
speed
-, divergence 230
-, flutter 230
spindle 291
-, machine tool 107; 108; 269
-, tool 111
spline
-, B- 452; 453; 460
-, cubic 340; 341
SQP (Sequential Quadratic Procedures) 57; 59; 72; 456
stability 274
-, form 437
-, loss of 448
state of strains, plane 320
static force loop 265; 267
step direction 81; 85; 95; 99
step length 81; 85; 99
STEP-method 71; 78; 93
stiffener
-, dimensions of 245
-, position of 245
stiffness 432
-, bearing 108
-, bending 402
-, coupling 402
-, laminate 418
-, shear 402
-, spring 274
-, strain 402
stochastic 207
strain 400
-, elastic 342
-, logorithmic principal 308
-, plastic 342
-, shearing 400
-, total 342
straining, proportional 307
strain measurements 411
strategy 10
-, "active set" 50
-, constraint linking 47
-, control 118
-, dual 59
-, optimization 29; 232; 431
stress
-, fracture 450
-, global principle 322
-, hypothesis of maximum principal 322
-, mean fracture 449; 450
-, normal 322
-, principal 306; 322
-, reference 322
-, shear 322
-, thermal 320
stress concentration 322
stress-ratio method 59; 231
stress-strain relation 306
structure
-, aircraft 229
-, cable 355
-, cable-suspended 356
-, civil engineering 372
-, composite 398
-, isostatic 153
-, laminate 245
-, lathe 291
-, load-bearing 397
-, machine-tool 273
-, sandwich 245

-, satellite 230; 239
-, spacecraft 229; 238; 244; 245
-, supporting 397
-, truss 199; 397
-, truss-type 254
strut 135
Submillimeter Radio Telescope (SMT) 429
subspace 75; 284
substitute problem 10; 18
-, scalar 15; 61; 78; 346
subsystem 381
SUMT (Sequential Unconstrained Minimization Technique) 59
support reactions, dynamic 165
surface
-, best-fit 434
-, Bèzier 46
-, B-spline 46
-, functional-efficient 190; 191
-, middle 398; 400
surface area 308
-, minimum 315
surface hardness 333
surface loads
-, normal 399
-, tangential 399
synthesis, optimal 184
system
-, aerodynamic 235
-, control 193
-, conventional computer 115
-, cooling 459
-, data-management 44
-, decision support 101
-, elasto-mechanical 193
-, equivalent 193
-, experimental 195
-, highly accurate 47
-, interactive design 165
-, knowledge-based 115; 117; 118; 140
-, large scale 36
-, lumped-mass 193; 195; 201
-, multi-body 193; 195
-, single-layer 364; 366; 372
-, substitute multi-body 193

task (s)
-, global 283; 284
-, leading 286
-, local 283; 284; 286
Taylor series 52; 58
technique (s)
-, clustering 378
-, multicriteria optimization 20
-, "offshore" 26
technology
-, cast metal 303
-, crane 26
-, satellite 24
telescope 238
-, large 429
-, millimeter-wave radio (MRT) 429
-, radio~ 356; 371
-, space 246
-, submillimeter 433
tensor
-, creep function 404
-, elasticity 402
-, elements of the inertia 197
-, force 399
-, hygrothermal stress 404
-, inertia 194
-, moment 399
-, permutation 399
-, thermal stress 404
-, transformation 403
thickness
-, laminate 400; 406
-, layer 42; 46; 406; 417; 439
-, plate 6.2/2
trade-off curves, optimal 427
trade-off factor 98
trade-off method 13; 29; 370
transfer matrix method 38; 43; 49; 89
transformation 37
-, constraint-oriented 13; 29; 61; 104; 236; 241; 412
transition zone 320; 324
trans-axle-gear 184
tree graph 289; 290
truss 355
-, isostatic 360
-, steel-staggered 387
tube-flange-structures 26

unit step function 404
utility function 358; 370

value
-, expected 209; 215

-, receptance 270; 273
-, response 209
-, specification 12
variable
-, analysis 37; 39; 45
-, basic 56; 57
-, binary decision 134
-, coordinate 284
-, decision 115; 120; 121; 134; 287; 376
-, dependent 56
-, design 6; 35; 37; 39; 42; 45; 49; 61; 121; 207; 245; 273; 291; 346; 357; 360; 405
-, discontinuous 378
-, discrete 378
-, independent 56
-, non basic 56; 57
-, parametric 175
-, parametric coordination 285
-, random 440
-, response 206; 207; 212
-, slack 16; 56
-, state 37; 46; 47; 61
-, transformation 39; 42; 61
variable fixing 39; 42; 45
variable linking 39; 42; 45
variance 178; 207; 209; 210; 215
V-criterion 211; 217
vector
-, acceleration 155
-, control 19
-, demand level 17; 19
-, displacement 234
-, elastic displacement 234
-, excitation 186
-, of constant parameters 169
-, of gravitational terms 153
-, state variable 46